Discrete-Time Linear Systems

Guoxiang Gu

Discrete-Time Linear Systems

Theory and Design with Applications

Springer

Guoxiang Gu
Department of Electrical and
Computer Engineering
Louisiana State University
Electrical Engineering Building
Baton Rouge, LA
USA

ISBN 978-1-4899-8907-9 ISBN 978-1-4614-2281-5 (eBook)
DOI 10.1007/978-1-4614-2281-5
Springer New York Dordrecht Heidelberg London

Softcover reprint of the hardcover 1st edition 2012

Printed on acid-free paper

Springer is part of Springer Science+Business Media (www.springer.com)

To my family and friends

Preface

Linear system theory and optimal estimation and control are taught at the graduate level in many universities. This book is written primarily for graduate students in electrical engineering who specialize in control. Yet, the mathematical theory applies not only to feedback control systems but also to communication and signal processing. Indeed, from two decades of my teaching in the advanced digital control class at LSU, I found that whenever I introduced applications to digital communications and signal processing such as equalization and precoding, more graduate students came to attend my class. Teaching of this new application stimulated more interests from students. Control students were eager to learn applications of control theory to communications that had more actions in the high-tech boom time, while students from communication and signal processing saw the new prosect of the mathematical system theory in their specialized area. These observations motivated me to seriously consider expansion of my teaching material to include new applications other than feedback control. However, it is also true that control and communication seldom interact in the last century, in spite of being two of the most successful areas in engineering system design. The two areas are more or less isolated from each other in terms of teaching and research.

Two major technological developments, namely, wireless communications and the Internet, in the last decade have changed the aforementioned isolatory phenomenon. Because of the existence of multipath and fading, the dynamic and random behaviors of the wireless channel cannot be ignored. And because of the multiuser nature for Internet, the communication channel is shared by more than one user. As such, the emergence of the wireless Internet has brought in design issues for dynamic multivariable communication systems. On the other hand, wireless communications and Internet have made remote and networked control systems possible, in which feedback controllers and physical processes are situated in two different locations and connected via wireless channels or Internet. Therefore, communication issues also need be addressed in control system design. It is felt strongly by this author that a unified approach is necessary to design of optimal MIMO dynamic systems in both control and communications. This text provides a platform for graduate students and researchers in these two different areas to study and work together. Such multidisciplinary interactions will greatly benefit each other and further advance the research frontier in engineering system design.

This text is focused on system and control theory with applications to design of feedback control systems and the signal processing aspect of the design issues. It is written for the first or second year graduate students, who are interested in general areas of control, or communications, and signal processing. The readers are assumed to have some basic knowledge on random processes and linear algebra, and have taken some basic undergraduate courses in discrete-time control, or digital communications and digital signal processing from electrical engineering. The appendices provide a quick review for the required mathematical background materials. Students are encouraged to read the related texts to strengthen their background knowledge.

This book consists of two parts. The first part presents linear system theory and theory of optimal estimation and optimal control for linear systems. State space is the main subject that provides not only the mathematical insight into the structure of linear systems but also the computational tool for obtaining the solution to optimal estimation and optimal control. The second part presents the design methodology for linear systems with feedback control and data communications as application areas. The design issues in modeling, system identification, channel estimation, symbol detection in data communications, and disturbance rejection in feedback control are addressed based on the theories from the first part.

It is observed by this author that design has been overlooked for the past two decades in engineering curricula at the graduate level. While courses on advanced research topics are important in training graduate students, the lack of design experience for graduate students may weaken the quality of our research programs. This text represents an effort to teach engineering system design at the graduate level, in hope to bring design into the graduate curriculum. The author welcomes any comments and suggestions regarding the materials in this text.

Baton Rouge, LA Guoxiang Gu

Acknowledgments

In writing this textbook, I received a lot of help from many people. The two who helped the most are John Baillieul and André Tits. John allowed me to step down in the middle of the second appointment from the job of Associate Editor for *SIAM Journal for Control and Optimization*, while André reduced my workload as an Associate Editor for *Automatica*. Both journals are premium control journals. I am really happy that I have worked with these two extraordinary people. I am also indebted to many other people, including Robert Bitmead, Xiren Cao, and Pramod Khargonkar.

I started writing this book more than 7 years ago. Many of my graduate students read its early drafts and helped me to revise and polish the book. Among them, I especially want to mention Jianqiang He, whose valuable inputs are crucial for improving the contents of the book. Other graduate students include Heshem Badr, Ehab Badran, Lijuan Li, and Zhongshan Wu, whose Ph.D. dissertations contribute to this textbook in various forms.

I would like to thank Jie Chen, Li Qiu, and Kemin Zhou for the long-lasting friendship over two decades. Their help and encouragement are extremely valuable to me. In addition, my family has put up with me with great patience. My children actually learned how to have weekends without me.

Finally, I would like to thank Steve Elliot, Merry Stuber, and Alison Waldron from Springer who helped to speed up the writing of this book.

Contents

Acronyms

ACS	Autocovariance sequence
AGM	Arithmetic-geometry mean
ARE	Algebraic Riccati equation
ARMA	Autoregressive moving average
AWGN	Additive, white, Gaussian noise
BER	Bit error rate
BIBO	Bounded input/bounded out
BPSK	Binary phase shift keying
BT	Bilinear transform
CDF	Cumulative distribution function
CDMA	Code division multiple access
CIR	Channel impulse response
CMOS	Complementary metal oxide semiconductor
CRLB	Cramér–Rao lower bound
CTFT	Continuous-time Fourier transform
DFE	Decision feedback equalization
DFT	Discrete Fourier transform
DMT	Digital multitone modulation
DRE	Difference Riccati equation
DS	Direct sequence
DTFT	Discrete-time Fourier transform
EIV	Error in variable
FBF	Feedback filter
FFF	Feedforward filter
FIM	Fisher information matrix
FIR	Finite impulse response
FLDFE	Finite length DFE
GCD	Greatest common divisor
GCLD	Greatest common left divisor
GCRD	Greatest common right divisor
IBT	Inverse balance truncation

ICI	Interchannel interference
IIR	Infinite impulse response
ISI	Intersymbol interference
KE	Kinetic energy
LAN	Local area network
LE	Linear equalization
LLR	Log-likelihood ratio
LQG	Linear quadratic Gaussian
LQR	Linear quadratic regulator
LS	Least squares
LTI	Linear time invariant
LTV	Linear time varying
MFD	Matrix fractional description
MIMO	Multi-input/multioutput
MLE	Maximum likelihood estimate
MMSE	Minimum mean-squared error
ODE	Ordinary differential equation
OFDM	Orthogonal frequency-division multiplexing
PDE	Partial differential equation
PDF	Probability density function
PE	Potential energy
PMF	Probability mass function
PSD	Power spectral density
QMF	Quadrature mirror filter
QoS	Quality of service
RF	Radio frequency
RLS	Recursive LS
RMSE	Root-mean-squared error
SISO	Single input/single output
SNR	Signal-to-noise ratio
SVD	Singular value decomposition
TLS	Total LS
TMA	Target motion analysis
WSS	Wide-sense stationary
ZF	Zero forcing

Chapter 1
Introduction

Digital technology has taken over many aspects of our society. From iPod to iPhone, and from eCommerce to eBook, such a digital take over has been one of the most fundamental changes in our time. Although digital take over began at the birth of digital computer, it has been accelerated in the past a few decades by the invention of personal computer (PC) and the development of Internet. Digital technology has conquered us all and changed not only our industry but also our daily life.

This textbook is aimed at covering both the theory and design for linear discrete-time systems built upon the digital technology with applications focusing on feedback control and wireless communications. The material of the book covers from mathematical models of discrete-time signals and systems to design of feedback control systems and wireless transceivers which are fairly extensive and are organized into eight different chapters. In order to help readers' mathematical background, three additional appendix chapters are prepared which make this text quite self-contained. While different opinions exist, it is believed that mathematics can and should be learned together with application examples for at least students in engineering.

Discrete-time/digital systems are originated from the continuous-time/analog world, the world we live in. It is the digital technology that transforms analog systems in continuous time into the digital ones in discrete time with the offer of unprecedented reliability and efficiency in large economic scale. We are compelled to get into the digital world without which we will not be able to compete in the global economy. For this reason, every college student in engineering needs to have some knowledge on linear discrete-time/digital systems. However, this textbook is not for every one. It is prepared for those readers who wish to have more solid theoretical understanding and more advanced design techniques than what they learned in their undergraduate studies. Even though this book can also be used as a text for senior undergraduate students specialized in control or communications, more suitable readers are first and second year graduate students. Because of the technological advances in the past half century, physical systems from controls and communications become increasingly multivariable and involve multi-input and multi-output (MIMO). MIMO systems are more difficult to analyze, and their design

G. Gu, *Discrete-Time Linear Systems: Theory and Design with Applications*,
DOI 10.1007/978-1-4614-2281-5_1,

poses considerable more challenges than single-input/single-output (SISO) systems. This textbook studies MIMO discrete-time systems, their mathematical properties, and various design issues. This introductive chapter presents some MIMO examples arising from controls and communications, and discretization of continuous-time systems.

1.1 Control Systems

Control is essential to modern technologies. Without it, there would be no running transportation vehicles in air or on ground or under water and no operating industrial machines, power plants, etc. In one word, no modern technology would function properly without control.

A physical system is a man-made device or machine. Control of the system requires understanding of its properties, its purpose, and its environment. In addition, sensors capable of measuring its state will have to be employed, and actuators capable of adjusting its state will have to be installed. These rough descriptions can all be made precise using mathematics. This section will present several physical processes. Moreover, discretization of continuous-time systems will be investigated in order that digital control can be applied.

1.1.1 MIMO Dynamic Systems

SISO control systems are abundant which can be traced back to ancient Greeks of 2,000 years ago who invented float regulators of water clocks. MIMO feedback control systems have a much shorter history which are the result of insatiable needs of our industry and ultimately our society. Consider an inverted pendulum mounted on a cart that is driven by a motor. See the schematic diagram in Fig. 1.1. The system shares similar dynamics to those of cranes in construction sites and rocket vehicles used in space exploration. Initially, the objective is confined to stabilization of the inverted pendulum by regulating the angle variable $\theta(t)$ to near zero. Later, people began to consider control of the position of the cart in addition to stabilization of the inverted pendulum which makes it a special MIMO system with one input/two output.

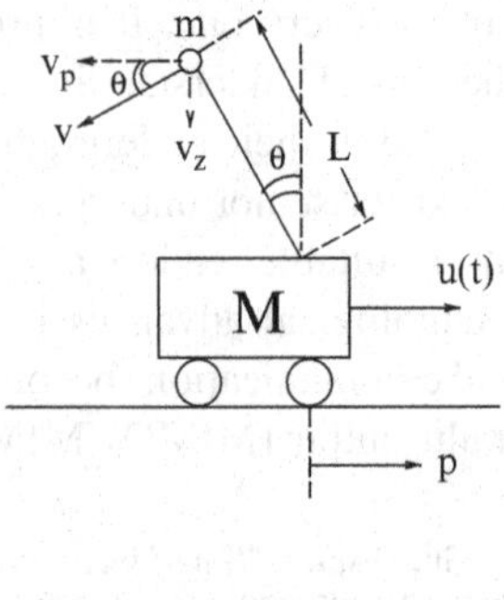

Fig. 1.1 Inverted pendulum

How can the inverted pendulum be controlled? Our past experience of education and practice in control indicates the necessity of the mathematical model that describes the pendulum-cart dynamics. A commonly adopted approach employs the Lagrange mechanics in modeling that is illustrated with a step by step procedure next.

Step 1 involves computation of the total kinetic energy (KE) of the system. Because of no vertical or rotational movement for the cart,

$$\mathrm{KE}_{\mathrm{cart}} = \frac{1}{2}M\dot{p}^2$$

with p the position of the cart. The KE for the pendulum is found to be

$$\mathrm{KE}_{\mathrm{pen}} = \frac{1}{2}m\left(v_p^2 + v_z^2\right).$$

In light of the schematic diagram in Fig. 1.1,

$$v_z = L\sin(\theta)\dot{\theta}, \quad v_p^2 = \dot{p}^2 + L^2\cos^2(\theta)\dot{\theta}^2,$$

due to both rotational and translational movement in p direction. Thus

$$\mathrm{KE}_{\mathrm{pen}} = \frac{1}{2}m\left(\dot{p}^2 + L^2\dot{\theta}^2 - 2\dot{p}L\cos(\theta)\dot{\theta}\right).$$

Since the total KE is the sum of $\mathrm{KE}_{\mathrm{cart}}$ and $\mathrm{KE}_{\mathrm{pen}}$, there holds

$$\mathrm{KE} = \frac{1}{2}\left\{M\dot{p}^2 + m\left(\dot{p}^2 + L^2\dot{\theta}^2 - 2\dot{p}L\cos(\theta)\dot{\theta}\right)\right\}.$$

Step 2 computes the total potential energy (PE). Clearly, only the pendulum admits the PE, and thus, $\mathrm{PE} = mgL\cos(\theta)$.

Step 3 forms $L_E = \mathrm{KE} - \mathrm{PE}$, the Lagrange of the system, as

$$L_E = \frac{1}{2}\left\{M\dot{p}^2 + m\left(\dot{p}^2 + L^2\dot{\theta}^2 - 2\dot{p}L\cos(\theta)\dot{\theta}\right)\right\} - mgL\cos(\theta),$$

and then computes the equations of motion in accordance with the following general principle:

$$\frac{\mathrm{d}}{\mathrm{d}t}\left[\frac{\partial L_E}{\partial \dot{p}}\right] - \left[\frac{\partial L_E}{\partial p}\right] = u, \tag{1.1}$$

$$\frac{\mathrm{d}}{\mathrm{d}t}\left[\frac{\partial L_E}{\partial \dot{\theta}}\right] - \left[\frac{\partial L_E}{\partial \theta}\right] = 0. \tag{1.2}$$

The right-hand side consists of "generalized forces" or external forces corresponding to each degree of freedom: θ and p. The only external force is $u(t)$ applied in direction of p. By direct calculations,

$$\begin{aligned}
\frac{\partial L_E}{\partial \theta} &= mL\sin(\theta)\dot{p}\dot{\theta} + mgL\sin(\theta),\\
\frac{\partial L_E}{\partial \dot{\theta}} &= mL^2\dot{\theta} - mL\cos(\theta)\dot{p},\\
\frac{\partial L_E}{\partial p} &= 0, \quad \frac{\partial L_E}{\partial \dot{p}} = (M+m)\dot{p} - mL\cos(\theta)\dot{\theta},\\
\frac{\mathrm{d}}{\mathrm{d}t}\left[\frac{\partial L_E}{\partial \dot{\theta}}\right] &= mL^2\ddot{\theta} - mL\cos(\theta)\ddot{p} + mL\sin(\theta)\dot{\theta}\dot{p},\\
\frac{\mathrm{d}}{\mathrm{d}t}\left[\frac{\partial L_E}{\partial \dot{p}}\right] &= (M+m)\ddot{p} - mL\cos(\theta)\ddot{\theta} + mL\sin(\theta)\dot{\theta}^2.
\end{aligned}$$

Substituting the above into (1.1) and (1.2) gives

$$\begin{aligned}
(M+m)\ddot{p} - mL\cos(\theta)\ddot{\theta} &= u - mL\sin(\theta)\dot{\theta}^2,\\
mL^2\ddot{\theta} - mL\cos(\theta)\ddot{p} &= mgL\sin(\theta),
\end{aligned}$$

which can be written into the matrix form:

$$\begin{bmatrix} -mL\cos(\theta) & M+m \\ mL^2 & -mL\cos(\theta) \end{bmatrix} \begin{bmatrix} \ddot{\theta} \\ \ddot{p} \end{bmatrix} = \begin{bmatrix} u - mL\sin(\theta)\dot{\theta}^2 \\ mgL\sin(\theta) \end{bmatrix}. \tag{1.3}$$

The matrix on left is called inertial matrix, and its determinant is given by

$$\Delta = (M+m)mL^2 - m^2L^2\cos^2(\theta) = (M + m\sin^2(\theta))mL^2. \tag{1.4}$$

Solving $\ddot{\theta}$ and $\ddot{p}$ in (1.3) yields

$$\ddot{\theta} = \frac{mL}{\Delta}\left(\cos(\theta)u - mL\sin(\theta)\cos(\theta)\dot{\theta}^2 + (M+m)g\sin(\theta)\right), \tag{1.5}$$

$$\ddot{p} = \frac{mL^2}{\Delta}\left(u - mL\sin(\theta)\dot{\theta}^2 + mg\sin(\theta)\cos(\theta)\right). \tag{1.6}$$

The three steps as described above are very general that applies not only to modeling of the inverted pendulum but also to modeling of other mechanical systems. The example of inverted pendulum alludes the ordinary different equations (ODEs) in describing the dynamic motions of physical systems. For the inverted pendulum, the ODEs are nonlinear functions of θ and p, but independent of time. For other mechanical systems, the ODE model may depend on time as well. More complex systems may have to be described by partial different equations (PDEs) which are beyond the scope of this text.

In design of MIMO control systems, state-space form of the mathematical model is more convenient. Denote vector by boldfaced letter and $\Longrightarrow$ for "imply." The state vector for the inverted pendulum can be defined as

$$\mathbf{x}(t) = \begin{bmatrix} x_1(t) \\ x_2(t) \\ x_3(t) \\ x_4(t) \end{bmatrix} = \begin{bmatrix} \theta(t) \\ \dot{\theta}(t) \\ p(t) \\ \dot{p}(t) \end{bmatrix} \Longrightarrow \dot{\mathbf{x}}(t) = \mathbf{f}[\mathbf{x}(t)] + \mathbf{g}[\mathbf{x}(t), u(t)]. \tag{1.7}$$

Let $f_i(\mathbf{x})$ and $g_i(\mathbf{x}, u)$ be the ith element of $\mathbf{f}(\cdot)$ and $\mathbf{g}(\cdot)$, respectively. It is easy to verify that $f_2(\mathbf{x})$ and $f_4(\mathbf{x})$ are given in (1.6) and (1.5), respectively,

$$f_1(\mathbf{x}) = x_2, \ f_3(\mathbf{x}) = x_4, \ g_1(\mathbf{x}, u) = g_3(\mathbf{x}, u) = 0,$$

and $g_2(\mathbf{x}, u) = mLu\cos(x_1)/\Delta, g_4(\mathbf{x}, u) = mL^2u/\Delta$ with Δ in (1.4).

In the general case, a MIMO system is described by state-space model of

$$\dot{\mathbf{x}} = \mathbf{f}(t, \mathbf{x}, \mathbf{u}), \ \mathbf{y} = \mathbf{h}(t, \mathbf{x}, \mathbf{u}), \tag{1.8}$$

where $\mathbf{u}$ and $\mathbf{y}$ are input and output vectors, respectively. For the inverted pendulum, input is scalar and output is a vector consisting of θ and p. The argument of t indicates the possible dependence of $\mathbf{f}(\cdot)$ and $\mathbf{h}(\cdot)$ on time.

Control of general nonlinear systems in form of (1.8) is extremely difficult. A tractable approach employs small signal analysis to obtain an approximate linear system prior to design of feedback controllers. This method computes the equilibrium points of $\mathbf{f}(t, \mathbf{x}_e, \mathbf{u}_e) = \mathbf{0}$ first and then chooses some pair of the roots $(\mathbf{x}_e, \mathbf{u}_e)$ as the desired operating point. Selection of $(\mathbf{x}_e, \mathbf{u}_e)$ depends on control objectives, system properties and other possible factors. A first-order approximation can be applied to linearize the system in (1.8). Define

$$A_t = \left.\frac{\partial \mathbf{f}}{\partial \mathbf{x}}\right|_{\mathbf{x}=\mathbf{x}_e, \mathbf{u}=\mathbf{u}_e}, \quad B_t = \left.\frac{\partial \mathbf{f}}{\partial \mathbf{u}}\right|_{\mathbf{x}=\mathbf{x}_e, \mathbf{u}=\mathbf{u}_e},$$

$$C_t = \left.\frac{\partial \mathbf{h}}{\partial \mathbf{x}}\right|_{\mathbf{x}=\mathbf{x}_e, \mathbf{u}=\mathbf{u}_e}, \quad D_t = \left.\frac{\partial \mathbf{h}}{\partial \mathbf{u}}\right|_{\mathbf{x}=\mathbf{x}_e, \mathbf{u}=\mathbf{u}_e}.$$

Let $\delta\mathbf{x} = \mathbf{x} - \mathbf{x}_e$ and $\delta\mathbf{u} = \mathbf{u} - \mathbf{u}_e$. Then the linear state-space system

$$\delta\dot{\mathbf{x}} = A_t\delta\mathbf{x} + B_t\delta\mathbf{u}, \ \mathbf{y} = C_t\delta\mathbf{x} + D_t\delta\mathbf{u} \tag{1.9}$$

represents a first-order approximation or linearization to the nonlinear system in (1.8). Such an approximation works well for small perturbations. The quadruplet (A_t, B_t, C_t, D_t) is called *realization* of the linearized system.

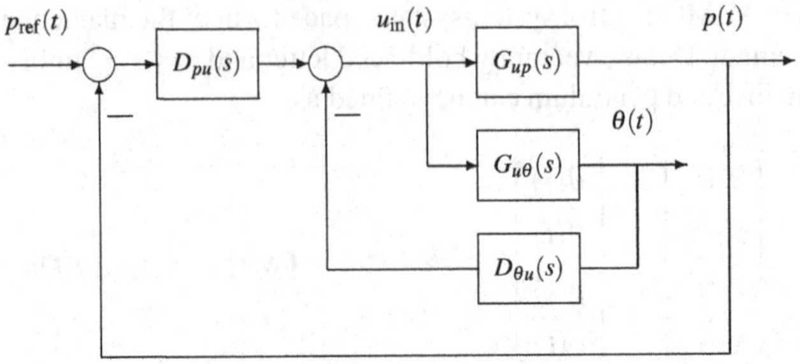

Fig. 1.2 Feedback control system for inverted pendulum

For inverted pendulum, the equilibrium point of $\mathbf{x}_{\text{e}} = \mathbf{0}$ and $u_{\text{e}} = 0$ is the meaningful one. Carrying out the linearization about the equilibrium point shows that the realization matrices are time invariant given by

$$A = \begin{bmatrix} 0 & 1\,0\,0 \\ \dfrac{(M+m)g}{ML} & 0\,0\,0 \\ 0 & 0\,0\,1 \\ \dfrac{mg}{M} & 0\,0\,0 \end{bmatrix}, \quad B = \begin{bmatrix} 0 \\ \dfrac{1}{ML} \\ 0 \\ \dfrac{1}{M} \end{bmatrix},$$

$$C = \begin{bmatrix} 1\,0\,0\,0 \\ 0\,0\,1\,0 \end{bmatrix}, \qquad D = \begin{bmatrix} 0 \\ 0 \end{bmatrix},$$

by output measurements $x_1 = \theta$ and $x_3 = p$. The inverted pendulum tends to fall down in the case of large signal sizes of $\mathbf{x}(t)$ which is captured by unstable eigenvalues of A. The linearized model works well for controlled pendulum in both Matlab simulations and in laboratory experiments.

Example 1.1. If the motor dynamics are included, the linearized state-space model has realization matrices given in Problem 1.2 of Exercises. The transfer functions from u_{in} to θ and to p can be obtained, respectively, as

$$G_{up}(s) = \frac{\frac{K_m K_g}{MRr}\left(s^2 - \frac{g}{L}\right)}{s\left(s^3 + \frac{K_m^2 K_g^2}{MRr^2}s^2 - \frac{(M+m)g}{ML}s - \frac{gK_m^2K_g^2}{MLRr^2}\right)},$$

$$G_{u\theta}(s) = \frac{\frac{K_m K_g}{MLRr}s}{s^3 + \frac{K_m^2 K_g^2}{MRr^2}s^2 - \frac{(M+m)g}{ML}s - \frac{gK_m^2K_g^2}{MLRr^2}}.$$

The block diagram of the pendulum control system is shown in Fig. 1.2. For the laboratory setup of the inverted pendulum system by the Quanzer Inc., the two transfer functions are given, respectively, by

$$G_{up}(s) = \frac{2.4805(s+5.4506)(s-5.4506)}{s(s+12.2382)(s-5.7441)(s+4.7535)}, \tag{1.10}$$

$$G_{u\theta}(s) = \frac{7.512s}{(s+12.2382)(s-5.7441)(s+4.7535)}. \tag{1.11}$$

If the following two simple dynamic compensators

$$D_{pu}(s) = K_{dy}\frac{s+z_y}{s+p_d}, \quad D_{\theta u}(s) = K_{d\theta}\frac{s+z_\theta}{s+p_d}, \tag{1.12}$$

are implemented in feedback block diagram of Fig. 1.2 with parameters specified in Problem 1.3 of Exercises, then the inverted pendulum is stabilized. This example exhibits the resilience of the linearization method. In fact, it is difficult to push down the inverted pendulum by hands in the lab experiment unless excessive forces are exerted to the pendulum.

While the linearization method works well for the control system of the inverted pendulum, it may not work well for other nonlinear control systems because of the large dynamic ranges of the equilibrium points. In this case, a set of linear feedback controllers can be designed for linearized systems at a number of equilibrium points which are then scheduled in operation. The actual controller is an interpolation of several controllers designed for equilibrium points close to the operating point. Such a method is often referred to as *gain schedule* which is widely used. The performance of the gain scheduled system is clearly hinged to that of each linear feedback control system. For this reason, linear systems play a fundamental role in feedback control design that is the focus of this text.

1.1.2 System Discretization

By definition, a discrete-time signal takes its values at only a set of discrete-time samples and is undefined elsewhere. This text considers signal samples taken at only equally spaced time instants. While continuous-time signals are abundant, discrete-time signals are often obtained from discretization of continuous-time ones. The ideal sampler is employed to take samples of the continuous-time signal $s_c(t)$ in accordance with

$$s(k) = s_c(kT_s), \ \ k = 0, \pm T_s, \pm 2T_s, \ldots, \tag{1.13}$$

where $T_s > 0$ is the sampling period. Discrete-time signals in practice involve quantization as well that is not addressed in this text.

For a given discrete-time signal $s(k) = s_c(kT_s)$, its frequency response is defined as the discrete-time Fourier transform (DTFT) given by

$$S\left(e^{j\omega}\right) = \mathscr{F}_d[s(k)] := \sum_{k=-\infty}^{\infty} s(k)e^{-jk\omega}. \tag{1.14}$$

A natural question is the relationship between the DTFT and continuous-time Fourier transform (CTFT) defined by

$$S_c(j\omega) = \mathscr{F}_c[s_c(t)] := \int_{-\infty}^{\infty} s_c(t) e^{-j\omega} \, d\omega. \tag{1.15}$$

The Dirac delta function, denoted by $\delta_D(t)$, turns out to be useful. This special function satisfies the following two properties:

$$\text{(i) } \delta_D(t) = 0 \ \forall\, t \neq 0, \quad \text{(ii) } \int_{-\infty}^{\infty} \delta_D(t) \, dt = 1. \tag{1.16}$$

It is a continuous-time signal but has some discrete-time flavor by noting that it takes zero value everywhere except at the origin. More importantly, it helps to connect the DTFT and CTFT as shown in the next result.

Lemma 1.1. *Denote $\omega_s = 2\pi/T_s$ as the sampling frequency. The* DTFT *as defined in (1.14) is related to the* CTFT *as defined in (1.15) according to*

$$S\left(e^{jT_s\omega}\right) = \frac{1}{T_s} \sum_{k=-\infty}^{\infty} S_c(j\omega - jk\omega_s). \tag{1.17}$$

Proof. Consider the periodic extension of the Dirac delta function

$$p_\delta(t) = \sum_{k=-\infty}^{\infty} \delta_D(t - kT_s). \tag{1.18}$$

Its Fourier series expansion is given by

$$p_\delta(t) = \frac{1}{T_s} \sum_{i=-\infty}^{\infty} e^{-ji\omega_s}. \tag{1.19}$$

See Problem 1.4 in Exercises. Let $s_p(t) = s_c(t) \times p_\delta(t)$. There holds

$$s_p(t) = \frac{1}{T_s} \sum_{k=-\infty}^{\infty} s_c(t) e^{-jk\omega_s t} \tag{1.20}$$

by (1.19). Applying the CTFT to $s_p(t)$ yields

$$\mathscr{F}_c[s_p(t)] = \frac{1}{T_s} \sum_{k=-\infty}^{\infty} \mathscr{F}_c\left[s_c(t) e^{-jk\omega_s t}\right] = \frac{1}{T_s} \sum_{k=-\infty}^{\infty} S_c(j\omega - jk\omega_s) \tag{1.21}$$

by Problem 1.5 in Exercises. On the other hand, there holds

$$s_p(t) = \sum_{k=-\infty}^{\infty} s_c(kT_s) \delta_D(t - kT_s) = \sum_{k=-\infty}^{\infty} s(k) \delta_D(t - kT_s). \tag{1.22}$$

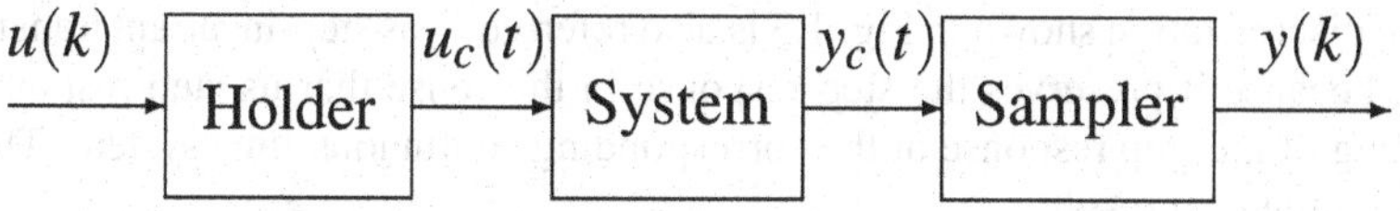

Fig. 1.3 Discretization with zero-order holder and ideal sampler

Hence applying the CTFT to $s_p(t)$ yields

$$\mathscr{F}_c[s_p(t)] = \sum_{k=-\infty}^{\infty} s(k)e^{-jkT_s\omega} = S\left(e^{jT_s\omega}\right). \tag{1.23}$$

The proof is complete. □

The well-known sampling theorem can be derived based on Lemma 1.1 by noting that $S(e^{jT_s\omega})$ consists of a train of shifted images of $S_c(j\omega)$. Hence, if $S_c(j\omega)$ admits low-pass characteristic, i.e., $|S_c(j\omega)|$ is symmetric about $\omega = 0$ and admits finite cut-off frequency $\omega_c < \omega_s/2$, then the train of shifted images of $S_c(j\omega)$ in (1.17) does not overlap. In practice, ω_c is taken as the first frequency beyond which the magnitude response $|S_c(j\omega)|$ falls within 3% of its peak.

Discretization of the continuous-time system takes more than the ideal sampler at the plant output. A zero-order holder is often employed at the plant input to convert the discrete-time signal into the continuous-time one. See the block diagram in Fig. 1.3. The zero-order holder is mathematically defined by

$$u_c(t) = u(k),\ kT_s \le t < (k+1)T_s, \tag{1.24}$$

where $k = 0, \pm1, \pm2, \ldots$. The above can also be written as

$$u_c(t) = \sum_{k=-\infty}^{\infty} u(k)\left[\mathbf{1}(t-kT_s) - \mathbf{1}(t-T_s-kT_s)\right], \tag{1.25}$$

where $\mathbf{1}(t)$ is the unit step function specified by

$$\mathbf{1}(t) = \begin{cases} 1, & t \ge 0, \\ 0, & t < 0. \end{cases} \tag{1.26}$$

Hence, the continuous-time signal at the output of the holder consists of piecewise constants and may have discontinuity at each sampling instant. The following result presents characterization of the zero-order holder. The proof is left as an exercise (Problem 1.6).

Lemma 1.2. *For the ideal holder defined in (1.24), there holds*

$$U_c(j\omega) = R(j\omega)U\left(e^{jT_s\omega}\right),\ R(j\omega) = \left(\frac{1-e^{-jT_s\omega}}{j\omega}\right). \tag{1.27}$$

The discretization shown in Fig. 1.3 is also referred to as step-invariant transform (SIT), because it preserves the step response in the sense that its step response is sampling of the step response of the corresponding continuous-time system. This is shown in the next result.

Theorem 1.2. *Let $G(s)$ be the transfer function of the continuous-time system and $G_d(z)$ be the transfer function for the corresponding discretized system in Fig. 1.3. Then there holds*

$$G_d\left(e^{j\omega}\right) = \sum_{k=-\infty}^{\infty} G(j\omega + jk\omega_s)R(j\omega + jk\omega_s). \tag{1.28}$$

Proof. In reference to Fig. 1.3 and in light of Lemma 1.2, there holds

$$Y_c(j\omega) = G(j\omega)U_c(j\omega) = G(j\omega)R(j\omega)U\left(e^{jT_s\omega}\right).$$

The discretized signal after the ideal sampler is $y(k) = y_c(kT_s)$, and thus

$$Y(e^{j\omega}) = \sum_{k=-\infty}^{\infty} Y_c(j\omega + jk\omega_s) = \sum_{k=-\infty}^{\infty} G(j\omega + jk\omega_s)R(j\omega + jk\omega_s)U\left(e^{jT_s\omega}\right)$$

by the periodicity of $U\left(e^{jT_s\omega}\right)$ with $\omega_s = 2\pi/T_s$ as the period and in light of Lemma 1.1. Hence, (1.28) holds that concludes the proof. □

Again, by the periodicity of $e^{jT_s\omega}$ with ω_s as the period, the step response of the discretized system in frequency domain is given by

$$Y_s\left(e^{jT_s\omega}\right) = \frac{G_d\left(e^{jT_s\omega}\right)}{1 - e^{-jT_s\omega}} = \sum_{k=-\infty}^{\infty} \frac{G(j\omega + jk\omega_s)}{j\omega + jk\omega_s}. \tag{1.29}$$

The right-hand side is the step response of the continuous-time system after the ideal sampling by Lemma 1.1. Hence, the discretization in Fig. 1.3 is indeed SIT and preserves the step response. Next, a numerical procedure is presented for computing the discretized system in Fig. 1.3 based on state-space models.

Consider the MIMO continuous-time state–space system described by

$$\dot{\mathbf{x}}(t) = A\mathbf{x}(t) + B\mathbf{u}(t), \quad \mathbf{y}(t) = C\mathbf{x}(t) + D\mathbf{u}(t).$$

It is known that for any $t_2 > t_1$, there holds

$$\mathbf{x}(t_2) = e^{A(t_2 - t_1)}\mathbf{x}(t_1) + \int_{t_1}^{t_2} e^{A(t_2 - t_1)}B\mathbf{u}(\tau)\,d\tau.$$

Taking $t_1 = kT_s$, and $t_2 = (k+1)T_s$ yields

$$\mathbf{x}[(k+1)T_s] = e^{Ah}\mathbf{x}(kT_s) + \int_{kT_s}^{(k+1)T_s} e^{A[(k+1)T_s - \tau]}B\mathbf{u}(\tau)\,d\tau.$$

If a zero-order holder is employed at the input, then $\mathbf{u}(t) = \mathbf{u}(kT_s)$ for $kT_s \le t < (k+1)T_s$, and thus

$$\begin{aligned}\mathbf{x}[(k+1)T_s] &= e^{AT_s}\mathbf{x}(kT_s) + \int_0^{T_s} e^{A\tau}\,d\tau B\mathbf{u}(kT_s)\\ &= A_d\mathbf{x}(kT_s) + B_d\mathbf{u}(kT_s).\end{aligned}$$

The ideal sampling at the output yields

$$\mathbf{y}(kT_s) = C\mathbf{x}(kT_s) + D\mathbf{u}(kT_s).$$

Hence, under SIT, a realization of the continuous-time system is mapped to a realization (A_d, B_d, C, D) for the discretized state-space system where

$$A_d = e^{AT_s},\ B_d = \int_0^{T_s} e^{A\tau}\,d\tau B. \tag{1.30}$$

If $|A| \neq 0$, then $B_d = A^{-1}(e^{AT_s} - I)B$. Although discretization was discussed for SISO systems earlier, the above shows that SIT is applicable to MIMO systems by employing the zero-order holder in each of the input channels and the ideal sampler in each of the output channels.

Example 1.3. Consider $G(s) = \frac{1}{s^3}$. It admits a realization (A,B,C,D) with $C = \begin{bmatrix}1 & 0 & 0\end{bmatrix}$, $D = 0$, and

$$A = \begin{bmatrix}0&1&0\\0&0&1\\0&0&0\end{bmatrix},\quad B = \begin{bmatrix}0\\0\\1\end{bmatrix}.$$

Since $A^3 = 0$, it is nilpotent. So for $T_s = h > 0$,

$$A_d = e^{Ah} = I + Ah + \frac{A^2h^2}{2} = \begin{bmatrix}1 & h & h^2/2\\0&1&h\\0&0&1\end{bmatrix},$$

$$B_d = \int_0^h \left(I + A\tau + \frac{A^2\tau^2}{2}\right) d\tau B = \begin{bmatrix}h^3/3!\\h^2/2!\\h\end{bmatrix}.$$

Via direct calculations, it can be verified that

$$G_d(z) = C(zI - A_d)^{-1}B_d = \frac{h^3(z^2+4z+1)}{3!(z-1)^3}. \tag{1.31}$$

Matlab command $[A_d, B_d] = \text{c2d}(A, B, h)$ can be used to obtain numerical values of (A_d, B_d) for a given $h > 0$.

Discretization under SIT can also be carried out directly for transfer function models that is given in the following result.

Corollary 1.1. *Assume that* $G(s)$ *is strictly proper. Then under the same hypothesis of Theorem 1.2, there holds*

$$G_d(z) = (1 - z^{-1})\text{Res}\left[\frac{e^{sT_s}z^{-1}}{1 - z^{-1}e^{sT_s}}\frac{G(s)}{s}\right], \tag{1.32}$$

where $\text{Res}[\cdot]$ *denotes operation of residues computed at poles of* $\frac{1}{s}G(s)$.

Proof. In light of (1.29), the proof amounts to showing

$$Y_s\left(e^{jT_s\omega}\right) = \sum_{k=-\infty}^{\infty} \frac{G(j\omega + jk\omega_s)}{j\omega + jk\omega_s} = \text{Res}\left[\frac{e^{sT_s}z^{-1}}{1 - z^{-1}e^{sT_s}}\frac{G(s)}{s}\right] \tag{1.33}$$

at $z = e^{jT_s\omega}$. By the inverse Laplace transform,

$$y_s(k) = \frac{1}{2\pi j}\int_{\sigma - j\infty}^{\sigma + j\infty} \frac{G(s)}{s} e^{skT_s}\, ds$$

for $k \geq 1$ by strict proper $G(s)$. Hence, for $\left|e^{sT_s}z^{-1}\right| < 1$, there holds

$$\begin{aligned} Y_s(z) &= \sum_{k=1}^{\infty} y_s(k)z^{-k} = \frac{1}{2\pi j}\int_{\sigma - j\infty}^{\sigma + j\infty} \frac{G(s)}{s} \sum_{k=1}^{\infty} e^{skT_s}z^{-k} ds \\ &= \frac{1}{2\pi j}\int_{\sigma - j\infty}^{\sigma + j\infty} \frac{e^{sT_s}z^{-1}}{1 - z^{-1}e^{sT_s}}\frac{G(s)}{s}\, ds. \end{aligned} \tag{1.34}$$

The relative degree of at least two for $\frac{1}{s}G(s)$ implies that

$$Y_s(z) = \frac{1}{2\pi j}\oint_{\Gamma_-} \frac{e^{sT_s}z^{-1}}{1 - z^{-1}e^{sT_s}}\frac{G(s)}{s}\, ds, \tag{1.35}$$

where Γ_- is the closed contour consisting of the vertical line $s = \sigma + j\omega$ for $\omega \in (-\infty, \infty)$ and the semicircle to the left of the vertical line. Hence, (1.33) holds that completes the proof. □

Example 1.4. Consider discretization of $G(s) = \frac{1}{s^3}$ using Corollary 1.1 with $T_s = h > 0$. Then $\frac{G(s)}{s} = \frac{1}{s^4}$, and thus

$$G_d(z) = \frac{z-1}{z}\text{Res}\left[\frac{e^{sh}}{z - e^{sh}}\frac{1}{s^4}\right] \tag{1.36}$$

where the residue is computed at $s = 0$ with multiplicity 4. It follows that

$$G_d(z) = \frac{z-1}{3!z}\frac{\mathrm{d}^3 f(s)}{\mathrm{d}s^3}\bigg|_{s=0}, \quad f(s) = \frac{e^{sh}}{z - e^{sh}}. \tag{1.37}$$

Direct calculation shows that

$$\frac{\mathrm{d}f(s)}{\mathrm{d}s} = \frac{hz^2}{(z-e^{sh})^2} - \frac{hz}{z-e^{sh}},$$

$$\frac{\mathrm{d}^2 f(s)}{\mathrm{d}s^2} = \frac{2h^2z^3}{(z-e^{sh})^3} - \frac{h^2e^{sh}z}{(z-e^{sh})^2},$$

$$\frac{\mathrm{d}^3 f(s)}{\mathrm{d}s^3}\bigg|_{s=0} = \frac{6h^3z^3}{(z-1)^4} - \frac{6h^3z^2}{(z-1)^3} + \frac{h^3z}{(z-1)^2}.$$

After substituting the above into (1.37) yields

$$G_d(z) = \frac{h^3(z^2+4z+1)}{3!(z-1)^3}$$

that is identical to the expression in (1.31).

A consequence of the SIT discretization is the nonminimum phase system under the high sampling frequency as shown next.

Corollary 1.2. *Consider SIT discretization for the continuous-time system represented by its transfer function $G(s)$ that has relative degree $\ell \geq 2$. Then as the sampling period $T_s \to 0$, $(\ell - 1)$ zeros of $G_d(z)$ tend to be unstable.*

Proof. Under the SIT, the $\mathscr{Z}$ transform of the discretized step response is given in (1.34). Instead of closing the contour to the left as in the proof of Corollary 1.1, consider closing contour to the right of $s = \sigma$, where σ is real and all poles of $\frac{1}{s}G(s)$ locate on left of $s = \sigma$. Recall that the relative degree $\ell \geq 2$. Hence, (1.35) is now replaced by

$$Y_s(z) = \frac{1}{2\pi j}\oint_{\Gamma_+} \frac{e^{sT_s}z^{-1}}{1 - z^{-1}e^{sT_s}}\frac{G(s)}{s}\,\mathrm{d}s, \tag{1.38}$$

where Γ_+ is the closed contour consisting of the vertical line $s = \sigma + j\omega$ for $\omega \in (-\infty, \infty)$ and the semicircle to the right of the vertical line. As a result, the residues associated with the above contour integral are those due to roots of $z = e^{sT_s}$, i.e.,

$$s_k = (\log(z) + j2k\pi)/T_s, \quad k = 0, \pm 1, \ldots.$$

Hence, in this case, by employing the L'Hospital's rule,

$$G_d(z) = \frac{1 - z^{-1}}{T_s}\sum_{k=-\infty}^{\infty}\frac{G(s_k)}{s_k}.$$

Note that if $T_s \approx 0$, then $s_k \approx \infty$. For the given $G(s)$, it has the form

$$G(s) = \frac{(s-z_1)\cdots(s-z_m)}{(s-p_1)\cdots(s-p_n)},$$

with relative degree $\ell = n - m \geq 2$. Let $G_1(s) = s^{-\ell}$ and

$$G_2(s) = \frac{s^\ell (s-z_1)\cdots(s-z_m)}{(s-p_1)\cdots(s-p_n)} \implies G_2(\infty) = 1.$$

Now take $|z-1| \geq \delta > 0$ with $\delta \approx 0$, and $|z| \geq 1$. Then

$$\begin{aligned} G_d(z) &= \frac{1-z^{-1}}{T_s} \sum_{k=-\infty}^{\infty} \frac{G_1(s_k)G_2(s_k)}{s_k} \\ &\approx \frac{1-z^{-1}}{T_s} \sum_{k=-\infty}^{\infty} \frac{G_1(s_k)}{s_k} = G_{1d}(z), \end{aligned}$$

where $G_2(s_k) \approx G_2(\infty) = 1$ for each k. Since $G_{1d}(z)$ is discretization of $s^{-\ell}$, it has the form

$$G_{1d}(z) = \frac{T_s^\ell z^{-1} \alpha_\ell(z)}{(1-z^{-1})^\ell}, \quad \alpha_\ell(z) = \sum_{i=0}^{\ell} \alpha_i z^{-i},$$

which has unstable roots for $\ell \geq 2$. The case with $\ell = 3$ is given in Example 1.3 in which $\alpha_\ell(z)$ indeed has unstable roots. □

Traditionally, SIT is associated with discretization of continuous-time systems based on which discrete-time feedback controllers are designed and implemented. However, continuous-time controllers are sometime designed first and discretized later prior to their implementation. Although SIT can be used, one may wish to preserve the frequency response, rather than step response, of the feedback controller. This gives rise to the bilinear transform (BT) method for discretization.

Let T_s be the sampling period. Consider approximation of integral $\frac{1}{s}$:

$$\begin{aligned} y(kT_s + T_s) &= y(kT_s) + \int_{kT_s}^{(k+1)T_s} u(\tau)\, d\tau \\ &= y(kT_s) + 0.5h[u(kT_s) + u(kT_s + T_s)]. \end{aligned}$$

Applying the $\mathscr{Z}$ transform to the above difference equation yields

$$\frac{Y(z)}{U(z)} = \frac{T_s}{2}\left(\frac{z+1}{z-1}\right) \implies s \approx \frac{2}{T_s}\left(\frac{z-1}{z+1}\right).$$

The above leads to the BT for discretization:

$$G_{\rm bt}(z) = G\left(\gamma\frac{z-1}{z+1}\right),\ \gamma = \frac{2}{T_{\rm s}}. \tag{1.39}$$

If $G(s)$ has realization (A,B,C,D) and γ is not an eigenvalue of A, then

$$\begin{aligned} A_{\rm bt} &= (\gamma I + A)(\gamma I - A)^{-1}, B_{\rm bt} = \sqrt{2\gamma}(\gamma I - A)^{-1}B, \\ C_{\rm bt} &= \sqrt{2\gamma}C(\gamma I - A)^{-1}, D_{\rm bt} = D + C(\gamma I - A)^{-1}B, \end{aligned} \tag{1.40}$$

are realization matrices for $G_{\rm bt}(z)$. The verification of the above expressions is left as an exercise (Problem 1.7).

Example 1.5. Consider first $G(s) = \frac{1}{s+1}$. Then the discretized transfer function based on BT is given by

$$\begin{aligned} G_{\rm bt}(z) &= \frac{1}{\gamma\left(\frac{z-1}{z+1}\right)+1} = \frac{z+1}{\gamma(z-1)+(z+1)} \\ &= \frac{z+1}{(\gamma+1)z-(\gamma-1)} = \frac{1}{\gamma+1}\frac{z+1}{z-\frac{\gamma-1}{\gamma+1}}. \end{aligned}$$

Substituting the relation $\gamma = \frac{2}{T_{\rm s}}$ yields

$$G_{\rm bt}(z) = \frac{T_{\rm s}}{2+T_{\rm s}}\frac{z+1}{z-\frac{2-T_{\rm s}}{2+T_{\rm s}}}. \tag{1.41}$$

If the SIT method is used to discretize $G(s)$, then by noting the poles of $\frac{1}{s}G(s)$ at $0,-1$, and in light of Corollary 1.1, there holds

$$\begin{aligned} G_d(z) &= (1-z^{-1})\,{\rm Res}\left[\frac{e^{sT_{\rm s}}z^{-1}}{1-z^{-1}e^{sT_{\rm s}}}\frac{G(s)}{s}\right]\Bigg|_{s=0,-1} \\ &= \frac{(1-z^{-1})e^{sT_{\rm s}}z^{-1}}{(s+1)(1-e^{sT_{\rm s}}z^{-1})}\Bigg|_{s=0} + \frac{(1-z^{-1})e^{sT_{\rm s}}z^{-1}}{s(1-e^{sT_{\rm s}}z^{-1})}\Bigg|_{s=-1} \\ &= z^{-1} - \frac{(1-z^{-1})e^{-T_{\rm s}}z^{-1}}{1-e^{-T_{\rm s}}z^{-1}} = \frac{1-e^{-T_{\rm s}}}{z-e^{-T_{\rm s}}}. \end{aligned}$$

Clearly, $G_{\rm bt}(z)$ and $G_d(z)$ are very different from each other.

While SIT-based discretization involves frequency distortion, it preserves step response. On the other hand, the BT-based discretization involves both frequency distortion and step response distortion. Specifically under the BT, the discretized frequency response is given by

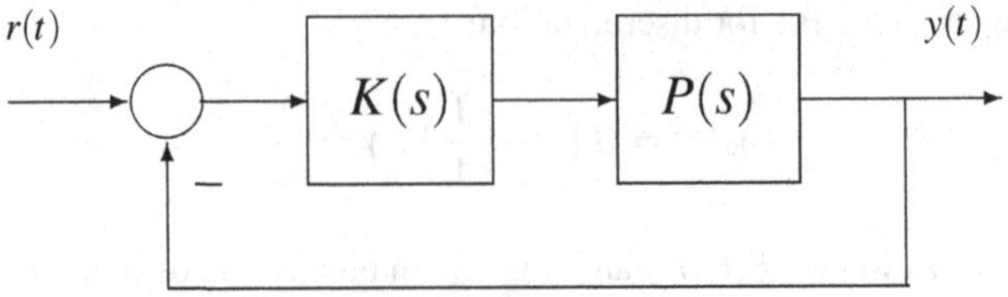

Fig. 1.4 Feedback control system in continuous time

$$G_{\text{bt}}\left(e^{j\omega}\right) = G\left(j\frac{2}{T_s}\tan\frac{\omega T_s}{2}\right).$$

The distortion is caused by the nonlinear map in frequency termed as *frequency warping*. It is interesting to note that if $\omega_c = \frac{2}{T_s}\tan\frac{\omega T_s}{2}$ were the continuous-time frequency, then there would be no error in frequency domain. The error in frequency response is given by

$$\left|G(j\omega) - G\left(j\frac{2}{T_s}\tan\frac{\omega T_s}{2}\right)\right|$$

which is zero at $\omega = 0$. The error at other frequencies is caused by frequency warping. The frequency warping can be avoided at $\omega = \omega_0 \neq 0$, if

$$s \mapsto \omega_0 \left(\tan\frac{\omega_0 T_s}{2}\right)^{-1} \frac{z-1}{z+1}$$

is used for BT-based discretization, in which case the frequency response at $\omega = \omega_0$ is preserved. This is the so-called frequency-prewarping BT.

Although the BT is different from the SIT in discretizing the continuous-time system, its implementation can be the same in the sense of the block diagram in Fig. 1.3 by using the zero-order holder and ideal sampler for the plant model. That is, the discrete-time controller is obtained via the BT based on the continuous-time controller and is implemented to control the discretized plant in Fig. 1.3. Since in this case the continuous-time controllers are often designed using the frequency domain technique, caution needs to be taken in frequency prewarping. For instance, one may wish to preserve the phase margin or gain margin of the continuous-time feedback system at certain critical frequency by using the frequency prewarping. However, because the frequency response of the discretized plant in Fig. 1.4 is altered (see Problem 1.8 in Exercises for the ideal case), the design of the continuous-time controller in frequency domain needs to take the frequency response error of the plant into consideration so that the phase margin or gain margin of the continuous-time feedback system at the chosen critical frequency can indeed be preserved.

Example 1.6. Consider the same transfer function $P(s) = \frac{1}{s+1}$ as in Example 1.5 for the plant model. An integral controller with $K(s) = \frac{\sqrt{2}}{s}$ in continuous time is designed that achieves the phase margin of 45° at the crossover frequency $\omega_0 = 1$ radian per second. The feedback control system in continuous time is shown in Fig. 1.4. The BT is used to discretize the continuous-time controller for its implementation in discrete time, while the SIT is employed to discretize the plant yielding

$$P_d(z) = \frac{1 - e^{-T_s}}{z - e^{-T_s}}, \; K_d(z) = \left(\tan\frac{T_s}{2}\right)\frac{z+1}{z-1},$$

where the frequency prewarping at $\omega_0 = 1$ is employed. If $T_s = 0.25$ is taken, i.e., the sampling frequency is 4 Hz, then the gain of $|P(j\omega_s/2)|$ is -22 dB at half the sampling frequency. Hence, possible aliasing due to sampling of the plant is suppressed significantly. However, $\left|K_d\left(e^{j\omega}\right)P_d\left(e^{j\omega}\right)\right| = 1$ takes place at $\omega_1 = 0.786$ rather than at $\omega_0 = 1$. If the gain of $K_d(z)$ is increased to assure the crossover frequency at 1 for the discretized feedback system, then the phase margin will be reduced to 37.54°. On the other hand, if 45° phase margin is required for the discretized feedback system, then the crossover frequency has to be smaller than 1, unless a more complicated controller is used.

Before ending this subsection, let us briefly discuss the pathological sampling as illustrated in the following example.

Example 1.7. Let T_s be the sampling period. Consider

$$G(s) = \frac{\omega_s s}{s^2 + \omega_s^2}, \; \omega_s = \frac{2\pi}{T_s}.$$

With step input $U(s) = 1/s$, the output of the plant is

$$Y(s) = G(s)U(s) = \frac{\omega_s}{s^2 + \omega_s^2} \implies y(t) = \sin(\omega_s t).$$

Hence, the sampling yields $y(kh) = 0$ for all integer-valued k. This is so-called pathological sampling. Note that $G(s)$ admits a realization:

$$G(s) = \left[\begin{array}{c|c} A & B \\ \hline C & D \end{array}\right] = \left[\begin{array}{cc|c} 0 & 1 & 0 \\ -\omega_s^2 & 0 & 1 \\ \hline 0 & \omega_s & 0 \end{array}\right],$$

and thus, A has eigenvalues at $\pm j\omega_s$. It can be verified that

$$A_d = e^{Ah} = \begin{bmatrix} \cos(2\pi) & \omega_s^{-1}\sin(2\pi) \\ -\omega_s\sin(2\pi) & \cos(2\pi) \end{bmatrix} = I,$$

$$B_d = \int_0^h e^{At}B\,\mathrm{d}t = A^{-1}[e^{Ah} - I] = 0.$$

Hence, sampling destroys the controllability.

Definition 1.1. The sampling frequency ω_s is pathological, if A or poles of the plant model has two eigenvalues with equal real parts and imaginary parts that differ an integral multiple of ω_s. Otherwise, the sampling frequency ω_s is called nonpathological.

The pathological can be easily avoided by taking the sampling frequency strictly greater than $\rho(A)$, the spectral radius of A. In fact, the sampling frequency smaller or equal to $\rho(A)$ causes the aliasing which should be avoided.

1.2 Communication Systems

Communication has enjoyed tremendous growth in the past century. The telephones invented by Bell have grown into communication networks and are now transformed into wireless handheld devices which can surf internet, play videos, and connect to computer networks. An important application content of this text focuses on design of wireless transceivers which are the building block and the physical layer of the wireless networks.

A digital communication system is linear and operates in discrete time. Its purpose lies in communication of digital data or symbols which have finite alphabet table. These data cannot be transmitted and received directly. They have to be modulated with continuous-time waveforms and converted to radio frequency (RF) signals prior to transmission and reception, leading to linear discrete-time systems for wireless data communications. A mathematical description is outlined next without providing the full detail.

1.2.1 Channel Models

Let $\tilde{s}(t)$ be the continuous-time waveforms carrying the data information. Its frequency response exhibits low-pass property normally which is often referred to as baseband signal. The RF signal to be transmitted has the form

$$s(t) = \text{Real}\left[e^{j\omega_c t}\tilde{s}(t)\right], \tag{1.42}$$

where ω_c is the carrier frequency and $\tilde{s}(t)$ is the complex envelope of the transmitted signal. Thus, the RF signal $s(t)$ admits bandpass property. Let $\tilde{s}(t) = \tilde{s}_I(t) + j\tilde{s}_Q(t)$ with $\tilde{s}_I(t)$ the real or in-phase part and $\tilde{s}_Q(t)$ the imaginary or quadrature part of the envelope signal. Simple calculation shows

$$s(t) = \cos(\omega_c t)\tilde{s}_I(t) - \sin(\omega_c t)\tilde{s}_Q(t).$$

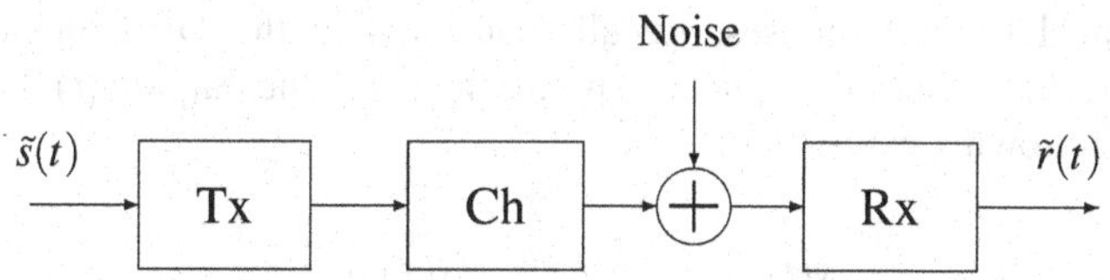

Fig. 1.5 Baseband signal model with Tx for transmitter and Rx for receiver antennas

Since $\cos(\omega_c t)$ and $\sin(\omega_c t)$ are orthogonal over the period $T_c = \frac{\omega_c}{2\pi}$, $\tilde{s}(t)$ can be reconstructed uniquely based on $s(t)$. Indeed, the in-phase component of $\tilde{s}(t)$ can be obtained by multiplying $s(t)$ with $\cos(\omega_c t)$ and by integration over $[0, T_c]$ in light of the fact that $\tilde{s}(t)$ is almost constant over $[0, T_c]$ due to the low-pass property of $\tilde{s}(t)$ and very high value of ω_c. Similarly, the quadrature component of $\tilde{s}(t)$ can be obtained by multiplying $s(t)$ with $-\sin(\omega_c t)$ and by the same integration. Consequently, $s(t)$ and $\tilde{s}(t)$ are equivalent in terms of communication.

Assume vertical polarization, plane waves, and N propagation paths for the RF signal transmitted over the space. The received signal in the kth path may experience magnitude distortion C_k, time delay d_k, and Doppler shift $\omega_{D,k}$ (that is positive if the motion is in the direction of the plane waves). Hence, the RF signal at the receiver in absence of the noise is specified by

$$r(t) = \sum_{k=1}^{N} \text{Real}\left[C_k e^{j(\omega_c + \omega_{D,k})(t-d_k)}\tilde{s}(t-d_k)\right]. \tag{1.43}$$

Let $\phi_k(t) = \omega_c d_k - \omega_{D,k}(t-d_k)$. Then $r(t) = \text{Real}\left[e^{j\omega_c t}\tilde{r}(t)\right]$ is in the same form of (1.42). The received complex envelope can find to be

$$\tilde{r}(t) = \sum_{k=1}^{N} C_k e^{-j\phi_k(t)}\tilde{s}(t-d_k). \tag{1.44}$$

The above baseband signal can be obtained by first multiplying $e^{-j\omega_c t}$ to the RF signal $r(t)$ in the receiver's antenna and then by filtering with a low-pass analog filter tailored to $\tilde{r}(t)$. The aforementioned discussions lead to the conclusion that there is no loss of generality to consider data communications in baseband with complex envelope signals as illustrated in Fig. 1.5 next.

The expression in (1.44) shows that the channel can be modeled by a linear time-varying (LTV) system with the complex impulse response:

$$g(t;\tau) = \sum_{k=1}^{N} C_k e^{-j\phi_k(t)}\delta_D(\tau - d_k). \tag{1.45}$$

See Problem 1.12 in Exercises. Recall that $\delta_D(\cdot)$ is the Dirac delta function satisfying (1.16). Under the impulse response in (1.45), the output $y(t)$ due to input $\tilde{s}(t)$ has the following expression:

$$y(t) = \int_{-\infty}^{\infty} g(t;t-\tau)\tilde{s}(\tau)\,\mathrm{d}\tau. \tag{1.46}$$

Hence, $g(t;\tau)$ is interpreted as the impulse response at time t due to the impulse input applied at time 0.

A channel represented by its impulse response in (1.45) suffers from both magnitude attenuation and phase distortion which are generally referred to as channel fading. Denote the duration of a modulated symbol by T_s. If $T_s \gg |d_i - d_k|$ for each pair of (i,k), then all frequency components of the transmitted signal experience the same random attenuation and phase shift due to multipath fading. Let μ_d be the mean of $\{d_k\}$. Then multipath delay spreads fluctuate about μ_d and

$$g(t;\tau) \approx \sum_{k=1}^{N} C_k e^{-j\phi_k(t)} \delta_D(\tau - \mu_d) = g(t)\delta_D(\tau - \mu_d). \tag{1.47}$$

In this case, the channel experiences flat fading and can be characterized by transmission of an unmodulated carrier. For this reason, $\tilde{s}(t) = 1$ can be assumed for (1.42), and $r(t)$ in (1.43) can be expressed as

$$r(t) = g_I(t)\cos(\omega_c t) - g_Q(t)\sin(\omega_c t), \tag{1.48}$$

where $g_I(t)$ and $g_Q(t)$ are the in-phase and quadrature components of the received signal, respectively, specified by

$$g_I(t) = \sum_{k=1}^{N} C_k \cos[\phi_k(t)], \; g_Q(t) = \sum_{k=1}^{N} C_k \sin[\phi_k(t)]. \tag{1.49}$$

The carrier frequency is often very high for data communications. Hence, small changes of the path delays $\{d_k\}$ will cause large variations in the phases $\{\phi_k(t)\}$ due to the terms $\omega_c d_k \gg 1$. At each time instant t, the random phases may result in constructive or destructive addition of the N multipath components. Consequently, $\phi_i(t)$ and $\phi_k(t)$ can be treated as uncorrelated random variables whenever $i \neq k$. It follows that for large N, $g_I(t)$ and $g_Q(t)$ are independent and approximately Gauss distributed with the same mean and variance in light of the central limit theorem. In the case of mean zero and variance σ^2, the magnitude $G = |g(t)|$ has a Rayleigh distribution with the probability density function (PDF) given by

$$p_G(x) = \frac{x}{\sigma^2}\exp\left\{-\frac{x^2}{2\sigma^2}\right\}, \; x \geq 0. \tag{1.50}$$

Denote $\mathrm{E}\{\cdot\}$ as the expectation. The average envelope power is given by

$$C_P = \mathrm{E}\{G^2\} = \mathrm{E}\{|g_I(t)|^2 + |g_Q(t)|^2\} = 2\sigma^2. \tag{1.51}$$

This type of fading with PDF in (1.50) is called Rayleigh fading. If the means of $g_I(t)$ or $g_Q(t)$ are nonzero and denoted by μ_I and μ_Q, respectively, then G is Ricean distributed or

$$p_G(x) = \frac{x}{\sigma^2}\exp\left\{-\frac{x^2+\mu^2}{2\sigma^2}\right\}\mathrm{I}_0\left(\frac{\mu x}{\sigma^2}\right) \tag{1.52}$$

for $x \geq 0$ where $\mu^2 = \mu_I^2 + \mu_Q^2$ and

$$\mathrm{I}_0(x) := \frac{1}{2\pi}\int_0^{2\pi} e^{x\cos(\theta)}\,\mathrm{d}\theta \tag{1.53}$$

is the zero-order modified Bessel function of the first kind. The type of fading as described in (1.52) is called Ricean fading.

Rayleigh and Ricean fading channels are common when line of sight (LoS) exists between the transmitter and receiver. On the other hand, complete obstructions often result in log-normal fading channels in which $\ln(G)$ is normal distributed. Another widely used channel is Nakagami fading in which the magnitude of the received envelope is described by the PDF

$$p_G(x) = \frac{2m^m x^{2m-1}}{\Gamma(m)C_P^m}\exp\left\{-\frac{mx^2}{C_P}\right\},\ m \geq \frac{1}{2}, \tag{1.54}$$

where $\Gamma(\cdot)$ is the Gamma function satisfying

$$\Gamma(x) = \int_0^{\infty} u^{x-1}e^{-u}\,\mathrm{d}u \tag{1.55}$$

that reduces to $n!$, if $x = n$ is an integer.

Nakagami fading is more versatile. In the case of $m = 1$, Nakagami distribution becomes Rayleigh distribution. For $m > 1$, it provides a close approximation to Ricean fading by taking

$$m = \frac{(K+1)^2}{2K+1} \iff K = \frac{\sqrt{m^2-m}}{m-\sqrt{m^2-m}}, \tag{1.56}$$

where $K = \frac{\mu^2}{2\sigma^2}$ is called Rice factor and $\iff$ stands for "equivalence." In the case of $m \to \infty$ or $K \to \infty$, the PDF approaches an impulse, and thus, the channel experiences no fading at all.

Recall that flat fading is associated with large T_s that is the duration of the modulated symbol. Since continuous-time waveforms are low-pass, the transmitted signal has narrowband if T_s is considerably greater than multipath delay spreads.

However, as T_s decreases, signal bandwidth increases. In the case of wide band signals, the frequency components in the transmitted signal experiences different phase shifts along the different paths. Under this condition, the channel introduces both magnitude and phase distortion into the message waveform. Such a channel exhibits *frequency-selective fading*.

When delay spreads are considerable greater than the symbol duration, the impulse response in (1.45) cannot be approximated as in (1.47). Instead, it is more conveniently represented by

$$g(t;\tau) = \sum_{k=1}^{N} g_k(t)\delta_D(\tau - d_k), \tag{1.57}$$

where $g_k(t) = C_k e^{-j[\omega_{D,k}d_k - \omega_c(t-d_k)]}$. Experimental tests show uncorrelated scattering (US) in the sense that $\{g_k(t)\}$ are uncorrelated for all time t. In addition, each path gain $g_k(t)$ is wide-sense stationary (WSS), i.e.,

$$R_{g_k}(t,\tau) = \mathrm{E}\{g_k(t)\bar{g}_k(\tau)\} = R_{g_k}(t-\tau), \tag{1.58}$$

with $\bar{\cdot}$ for complex conjugation. While wide-sense stationary and uncorrelated scattering (WSSUS) channels are very special, many radio channels satisfy the WSSUS assumption.

The wireless channels discussed thus far involve only single transmit/receive antenna. More and more mobile radio systems nowadays begin to employ antenna diversity by using multiple spatially separated antennas at both the transmitter and the receiver which result in MIMO channels. These antennas provide multiple faded replica of the same information bearing signal thereby increasing the channel capacity. The impulse response for the MIMO channel has the similar form as in (1.57), but the scalar path gain $g_k(t)$ is now replaced by the matrix function $G_k(t)$ of size $P \times M$, assuming M antennas at the transmitter and P antennas at the receiver, yielding the impulse response

$$G(t;\tau) = \sum_{k=1}^{N} G_k(t)\delta_D(\tau - d_k). \tag{1.59}$$

It is important to note that the impulse responses in (1.57) and (1.59) describe the channels in continuous time. Channel discretization will have to be carried out prior to processing the digital data which will be studied in the next subsection.

1.2.2 Channel Discretization

One of the fundamental issues in data communications is retrieval of the transmitted data at the receiver. Consider the schematic block diagram shown in Fig. 1.5.

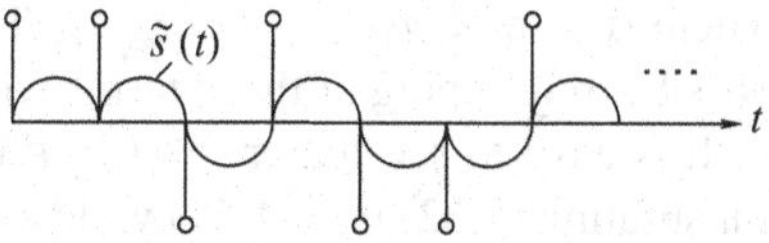

Fig. 1.6 A typical binary data signal and its associated waveform

At the transmitter site, the data are coded, modulated, and frequency shaped by the continuous-time waveform, and then sent through the channel. The transmitted signal experiences random distortion in the channel Ch due to various fading phenomena. At the receiver site, the observed noisy signal is processed, demodulated, and decoded by the receiver, in hope of recovering the original data.

The channel is a physical entity that cannot be altered by designers, even though no physical devices exist for channels in the case of wireless communications. In the case of wide band signals, the channel experiences frequency-selective fading. If channel distortions are function of time, then the channel experiences *time-selective* fading. Channel fading is one of the main impediments to data communications. In contrast to the channel, the transmitter and receiver can be altered by designers. For simplicity, it is assumed that both the transmitter and the receiver are linear. The nonlinear effects of modulation and codings are omitted, in order for us to focus on the basic issues in data communication. An objective of this text is design of wireless transceivers, i.e., the transmitters and receivers, to detect the transmitted data at the output of the receiver with the minimum achievable error probability.

The simplest data set consists of binary symbol of ± 1. Let $\{b(k)\}$ be coded and binary valued data. The transmitted signal shown in Fig. 1.5 is continuous and has the form of:

$$\tilde{s}(t) = \sum_{k=-\infty}^{\infty} b(k)\psi(t - kT_s), \tag{1.60}$$

where $\psi(\cdot)$ is the continuous-time waveform with support $[0,\ T_s]$ and $T_s > 0$ is the symbol period. Typical $\psi(\cdot)$ includes sinusoidal and rectangular functions (see Fig. 1.6).

Let $g(t;\tau)$ be the channel impulse response (CIR) in (1.45). The physical nature of the channel implies causality of $g(t;\tau)$, i.e., $g(t;\tau) = 0$ for $\tau < 0$. It follows that

$$\begin{aligned}\tilde{r}(t) &= \int_{-\infty}^{t} g(t;t-\tau)\tilde{s}(\tau)\,\mathrm{d}\tau + \eta(t) \\ &= \sum_{k=-\infty}^{\infty} b(k)\int_{-\infty}^{t} g(t;t-\tau)\psi(\tau - kT_s)\,\mathrm{d}\tau + \eta(t)\end{aligned} \tag{1.61}$$

is the signal at the receiver with $\eta(t)$ additive noises. For wireless communication systems with high data rate, T_s is rather small, and thus,

$$g(t;\tau) \approx \sum_{i=0}^{L_h-1} h_i(t)\delta_D(\tau - \tau_i), \tag{1.62}$$

where $0 \le \tau_0 < \tau_1 < \cdots < \tau_{L_h-1}$. The change of notation from (1.57) is due to possible regrouping of those terms with delay spreads close to each other.

It is easy to see that $\tau_i = (\ell_i + \varepsilon_i)T_s$ for some integer ℓ_i and $\varepsilon_i \in [0,\ 1)$. Upon substituting (1.62) into (1.61) yields:

$$\tilde{r}(t) = \sum_{k=-\infty}^{\infty} b(k) \sum_{i=0}^{L_h-1} h_i(t)\psi(t-(\ell_i+k)T_s - \varepsilon_i T_s) + \eta(t). \tag{1.63}$$

In order to recover the digital data $\{b(k)\}$, a common approach applies the matched filter to $\tilde{r}(t)$. Assume that $h_i(t) \approx h_i(n)$ over $[nT_s, (n+1)T_s)$, by an abuse of notation and the fact of small T_s due to high data rate. Thus

$$y(n) = \int_{nT_s}^{(n+1)T_s} \tilde{r}(t)\psi(t-nT_s)\,\mathrm{d}t = \sum_{k=-\infty}^{\infty} b(k) \sum_{i=0}^{L_h-1} h_i(n)J_i(n) + v(n) \tag{1.64}$$

with $J_i(n)$ the ith integral given by

$$J_i(n) = \int_{nT_s}^{(n+1)T_s} \psi(t-(\ell_i+k)T_s - \varepsilon_i T_s)\psi(t-nT_s)\,\mathrm{d}t,$$

and $v(n)$ the integral of $\psi(t-nT_s)\eta(t)$ over $[nT_s, (n+1)T_s]$.

Denote δ_k as the Kroneker delta function specified by

$$\delta_k = \begin{cases} 1, & \text{if } k = 0, \\ 0, & \text{otherwise.} \end{cases}$$

It can be verified that the integral $J_i(n)$ admits the expression

$$\begin{aligned} J_i(n) &= \delta_{n-k-\ell_i} \int_{\varepsilon_i T_s}^{T_s} \psi(\tau - \varepsilon_i T_s)\psi(\tau)\,\mathrm{d}\tau \\ &\quad + \delta_{n-k-\ell_i-1} \int_0^{\varepsilon_i T_s} \psi(\tau + (1-\varepsilon_i)T_s)\psi(\tau)\,\mathrm{d}\tau \\ &= \alpha_i(n)\delta_{n-k-\ell_i} + \beta_i(n)\delta_{n-k-\ell_i-1}. \end{aligned}$$

Substituting the above into (1.64) yields

$$\begin{aligned} y(n) &= \sum_{k=-\infty}^{\infty} b(k) \sum_{i=0}^{L_h-1} h_i(n)[\alpha_i(n)\delta_{n-k-\ell_i} + \beta_i(n)\delta_{n-k-\ell_i-1}] + v(n) \\ &= \sum_{i=0}^{L_h-1} h_i(n)[\alpha_i(n)b(n-\ell_i) + \beta_i(n)b(n-\ell_i-1)] + v(n). \end{aligned}$$

Let $L = 1 + \max\{\ell_i : 0 \le i < L\}$. Then

$$y(n) = \sum_{k=0}^{L} \phi(n;k)b(n-k) + v(n), \tag{1.65}$$

where $\phi(n;k)$ is the impulse response of the discretized channel at time index n and depends on $\{\alpha_i(n)\}$, $\{\beta_i(n)\}$, and $\{h_i(n)\}$. See Problem 1.16 in Exercises. In the MIMO case, the discretized CIR consists of matrices $\Phi(n;k)$. Both input and output are vector-valued, denoted by $\mathbf{b}(n)$, $\mathbf{v}(n)$, and $\mathbf{y}(n)$, respectively, leading to dynamic equation of

$$\mathbf{y}(n) = \sum_{k=0}^{L} \Phi(n;k)\mathbf{b}(n-k) + \mathbf{v}(n).$$

Generically, the wireless channel has the form of transversal filters. This fact is attributed to the finite duration of the CIR for wireless channels, contrast to the plant model in control systems.

1.3 Organization of the Book

This textbook is aimed at presentation of theory and design of linear discrete-time systems with applications to feedback control systems and wireless transceivers, although the primary audience is from the control area. This book probably represents the first attempt to present feedback control systems and wireless communication systems together, motivated by MIMO and linearity of the underlying dynamic systems. The author believes that readers from these two different areas will learn more by showing them new perspectives in linear systems. The book begins with warm-up examples and discretization in Chap. 1. Various signal and system models are introduced in Chap. 2. The core materials are Chaps. 3 and 5 in which the main results in linear system theory and optimal control/filtering are presented. These two are the most important chapters and are also the most theoretic part of this text. Chapter 4 is concerned with model reduction that is more oriented to feedback control systems. Compared with undergraduate control textbooks, this text clearly has more mathematical depth. The main reason lies in the MIMO nature of the modern control systems that are difficult to analyze and control. Chapters 3 and 5 provide mathematical notions and theory for us to understand MIMO linear systems.

While most textbooks on linear systems are focused on theory, this textbook makes an effort to present design, that is a tremendous challenge. Chapter 6 considers design of feedback control systems, while Chap. 7 is focused on design of wireless transceivers in data communications. Although the author wishes to present the state-of-art design techniques, these two chapters may not fulfill this wish. Design is not science. Various objectives arising from engineering practice are difficult to be modeled by a single performance index, and minimization of some mathematical functional may not lead to the desired feedback control system. However, the author hopes that Chap. 6 provides some guidelines for readers to learn design of MIMO feedback control systems, and Chap. 7 provides some mathematical tools for design of wireless transceivers. Chapter 8 is the final chapter that is aimed at modeling and identification of linear systems. It helps readers to know where the linear models come from and how to obtain them given input and output measurements of the MIMO systems. The textbook also includes three appendix chapters that are aimed at helping readers to prepare their mathematical background.

This book can be used as a textbook for the first and second year graduate students, although sophisticated senior students can also be the audience. It can be used either for one semester or for two semesters, dependent on the curriculum. If the book is used as a textbook for one semester, then the author suggests to focus on Chaps. 1–3, and 5. The instructors can use related material from other chapters as supplements. For instance, a class oriented to control may wish to use the control system design material from Chaps. 4 and 6, while a class oriented to DSP may wish to use the wireless transceiver design material from Chaps. 7 and 8. However, if a second semester is needed, then Chaps. 4, 6–8 can be taught together with suitable design projects. The author has taught several times at LSU as a one semester course, tested with different focuses, which result in this textbook. Many control students like the course that helps to diversify their background and to seek jobs in wireless communications and DSP. Indeed, many of them, including my own students, have taken other courses in digital and wireless communications afterward, and are currently working in high-tech companies such as Qualcomm, Broadcom, etc. Their control background actually helps them to do well in their new careers which is also one of the motivations for the author to write this textbook.

Each chapter of this book has a section of exercises. Problems in these exercise sections are carefully designed to help readers get familiar with the new concepts and knowledge presented in the main text. Some of the problems are part of the topics studied in different chapters. The author hopes that this will help readers study the material more actively. Some of the problems can be very hard. A solution manual will be worked out to aid instructors and readers in the future.

Notes and References

Discretization of continuous-time signals and systems are covered by many books. See for instance [23, 73, 90] and [64, 83–85]. The flight control example can be found in [80, 81]. The modeling of inverted pendulum employs the principle from Lagrange mechanics [17, 108]. For discretization and modeling of wireless communication channels, many papers and books are available. The books [19, 98, 107, 111] are recommended for further reading.

Exercises

1.1. For the inverted pendulum as described in (1.6) and (1.5), assume that the control force is generated by the actuator via

$$u(t) = \frac{K}{Rr}\left[u_{\text{in}}(t) - \frac{K}{r}\dot{p}(t)\right], \quad K = K_m K_g. \tag{1.66}$$

The above is a simple model for DC motor with $u_{\text{in}}(t)$ as the armature voltage. Show that the dynamic system is now described by

$$\ddot{\theta} = \frac{mL}{\Delta}\left(\frac{K\cos(\theta)}{Rr}u_{\text{in}} + (M+m)g\sin(\theta) - \frac{mL\sin(2\theta)}{2}\dot{\theta}^2 - \frac{K^2\cos(\theta)}{Rr^2}\dot{p}\right),$$

$$\ddot{p} = \frac{mL}{\Delta}\left(\frac{K}{Rr}u_{\text{in}}(t) + \frac{mg\sin(2\theta)}{2} - mL\sin(\theta)\dot{\theta} - \frac{K^2}{Rr^2}\dot{p}\right),$$

where $\Delta = \left(M + m\sin^2(\theta)\right)mL^2$.

1.2. For the nonlinear ODE in Problem 1.1, assume that θ and p are output measurements. Denote $\mathbf{x}(t)$ in (1.7) as the state vector. Show that its linearized system has realization matrices (A, B, C, D) given by

$$A = \begin{bmatrix} 0 & 1 & 0 & 0 \\ \frac{(M+m)g}{ML} & 0 & 0 & -\frac{K^2}{Rr^2} \\ 0 & 0 & 0 & 1 \\ \frac{mg}{M} & 0 & 0 & -\frac{K^2}{Rr^2} \end{bmatrix}, \quad B = \begin{bmatrix} 0 \\ \frac{K}{MLRr} \\ 0 \\ \frac{K}{MRr} \end{bmatrix},$$

$$C = \begin{bmatrix} 1 & 0 & 0 & 0 \\ 0 & 1 & 0 & 0 \end{bmatrix}, \qquad D = \begin{bmatrix} 0 \\ 0 \end{bmatrix}.$$

1.3. Use Matlab Simulink toolbox to simulate the controlled pendulum system in Example 1.1 by using parameters

$$K_{dy} = -524.4300, \quad z_y = 0.8200,$$
$$K_{d\theta} = 334.2439, \quad z_\theta = 6.2589,$$

and $p_d = 14.9672$.

1.4. For Dirac delta function defined in (1.16), show that

$$\int_{-\infty}^{\infty} f(t)\delta_D(t-\tau)\, dt = f(\tau)$$

for any function $f(t)$ that is continuous at $t = \tau$. In addition, show that the periodic Dirac delta function with period T satisfies

$$p_\delta(t) = \sum_{k=-\infty}^{\infty} \delta_D(t - kT) = \frac{1}{T} \sum_{i=-\infty}^{\infty} e^{ji\omega t},$$

where $\omega = 2\pi/T$.

1.5. Prove the CTFT for $s_p(t)$ in (1.21).

1.6. Prove Lemma 1.2 by first finding the impulse to the zero-order holder and then its CTFT.

1.7. Verify the expression in (1.40) for BT-based discretization.

1.8. Show that under the SIT discretization, the frequency response error between the continuous-time transfer function $G(s)$ and the discretized transfer function $G_d(z)$ is given by

$$\left|G(j\omega) - G_d\left(e^{j\omega}\right)\right| = \left|G(j\omega)\left[R(j\omega) - 1\right]\right|,$$

where $R(s) = \left(1 - e^{-sT_s}\right)/s$, provided that the cutoff frequency of $G(s)$ is smaller than half of the sampling frequency.

1.9. Suppose that A has eigenvalues at $0, 0, \pm j, 1 \pm 2j$. Show that the pathological sampling frequency ω_s is in the set of $\left\{\frac{4}{k} : k \text{ integers}\right\}$.

1.10. For the inverted pendulum plant model in (1.11) and (1.10), do the following:

(i) Discretize $G_{up}(s)$ and $G_{u\theta}(s)$ with sampling frequency $f_s = 10\,\text{Hz}$.
(ii) Discretize the above transfer functions with one sampling frequency $f_s < 10$ and the other $f_s > 10$.
(iii) Recommend one sampling frequency and detail your reason. In addition, make frequency domain analysis for the sampling frequency you recommend.

(*Note:* Matlab command "c2d" can be used to discretize the plant.)

1.11. Consider a flight control system with 3-input/3-output. This model is linearized from an aircraft motion in the vertical plane when small perturbations about a flight condition are considered. The linearized model is described by state-space realization given by

$$A = \begin{bmatrix} 0 & 0 & 1.132 & 0 & -1 \\ 0 & -0.0538 & -0.1712 & 0 & 0.0705 \\ 0 & 0 & 0 & 1 & 0 \\ 0 & 0.0485 & 0 & -0.8556 & -1.013 \\ 0 & -0.2909 & 0 & 1.0532 & -0.6859 \end{bmatrix},$$

$$B = \begin{bmatrix} 0 & 0 & 0 \\ -0.0012 & 1 & 0 \\ 0 & 0 & 0 \\ 0.4419 & 0 & -1.6646 \\ 0.1575 & 0 & -0.0732 \end{bmatrix}, \quad C = \begin{bmatrix} 1\,0\,0\,0\,0 \\ 0\,1\,0\,0\,0 \\ 0\,0\,1\,0\,0 \end{bmatrix}.$$

Repeat (i)–(iii) in the previous problem except that 10 Hz is replaced by 20 Hz, plus zero/pole plots.

1.12. For the time-varying impulse response in (1.45), show that the output response due to input $\tilde{s}(t)$ is given in (1.44).

1.13. For $r(t)$ in (1.48), assume that $r(t)$ is WSS, $\{\omega_{D,k}\}$ are identically distributed, and $\{\phi_k(t)\}$ are independent and uniformly distributed over $[-\pi, \pi]$. Show that

$$R_r(\tau) = \mathrm{E}[r(t)r(t+\tau)] = A(\tau)\cos(\omega_c\tau) - B(\tau)\sin(\omega_c\tau),$$

where expectation is taken for $V_k = \omega_c d_k$ which is uniformly distributed over $[-\pi,\ \pi]$ and for the Doppler frequency. The expressions of $A(\tau)$ and $B(\tau)$ are given by

$$A(\tau) = C_P\mathrm{E}[\cos(\omega_{D,k}\tau)],\ B(\tau) = C_P\mathrm{E}[\sin(\omega_{D,k}\tau)],$$

with expectation taken for only the Doppler frequency and C_P as in (1.51).

1.14. Suppose that X_1 and X_2 are independent Gauss with zero mean and common variance σ^2. Let

$$G = \sqrt{X_1^2 + X_2^2},\ P = \tan^{-1}(X_2/X_1). \tag{1.67}$$

Show that G is Rayleigh distributed as in (1.50) and P is uniformly distributed.

1.15. Consider the same X_1 and X_2 as in the previous problem except that $\mathrm{E}\{X_1\} = \mu_1$ and $\mathrm{E}\{X_2\} = \mu_2$. Define G and P as in (1.67), and $t = \tan^{-1}(\mu_2/\mu_1)$. Let $\mu = \sqrt{\mu_1^2 + \mu_2^2}$. Show that

$$p_{RP}(r,p) = \frac{r}{2\pi\sigma^2}\exp\left\{-\frac{r^2 + \mu^2 - 2r\mu\cos(p-t)}{2\sigma^2}\right\}$$

and then verify the Ricean PDF in (1.52) using

$$p_R(r) = \int_0^{2\pi} p_{RP}(r,p)\,\mathrm{d}p.$$

1.16. Find the expression of $\phi_i(n;k)$ in (1.65) in terms of $\{\alpha_i(n)\}$, $\{\beta_i(n)\}$, and $\{h_i(n)\}$, assuming $\ell_i = i$ for each i.

Chapter 2
Signals and Systems

It is a fact that signals and systems in feedback control are in continuous time and multivariable in nature. This is a contrast to data communications where the signals and systems are discrete time with single transmitter/receiver. But the wide use of digital computers in control systems and the emergence of wireless internet have diminished such differences between feedback control and data communications. Both are now in discrete time and both are MIMO systems. More importantly, they tend to use increasingly the same mathematical descriptions in modeling and share more and more mathematical tools in design. This text is aimed to provide the design theory and computational algorithms for MIMO dynamical systems in which either optimal disturbance rejection or optimal data detection is the main objective. This chapter introduces the background material for signals and systems.

Mathematical models are indispensable in design of both communication and control systems. To accommodate to data communications, signals are assumed to be discrete time and complex-valued. For MIMO systems, signals of interest are those having more than one component, each of which is random. Such signals are sequences of random vectors, assumed to be WSS, and have bounded power spectral densities (PSDs). Linear time-invariant (LTI) systems are capable of altering signals through modifying their PSDs in a transparent way. Commonly used dynamical models will be described and analyzed. LTV systems are also covered, albeit at a less degree. Although signals include noises in a larger sense, noise models will be discussed separately in this chapter, together with the bit error rate (BER) analysis in data communications.

2.1 Signals and Spectral Densities

For simplicity, the discrete-time variable t is measured in units of the sampling interval, and is thus integer-valued. That is, if $s(t)$ is a discrete-time signal obtained through sampling of the continuous-time signal $s_c(\cdot)$, then $s(t) = s_c(tT_s)$ for

G. Gu, *Discrete-Time Linear Systems: Theory and Design with Applications*,
DOI 10.1007/978-1-4614-2281-5_2,

$t = 0, \pm 1, \pm 2, \ldots$, with T_s the sampling period. In other words, a discrete-time signal can be viewed as a complex sequence $\{s(t)\}$ indexed by integer-valued time t.

2.1.1 Scalar Signals

Suppose that $\{s(t)\}$ is deterministic and has finite energy. Then

$$E_s := \sum_{t=-\infty}^{\infty} |s(t)|^2 < \infty, \tag{2.1}$$

where E_s is the energy of $\{s(t)\}$. In this case, there exists DTFT for $\{s(t)\}$, defined as

$$S\left(e^{j\omega}\right) := \sum_{t=-\infty}^{\infty} s(t)e^{-j\omega t}, \; j = \sqrt{-1}. \tag{2.2}$$

The corresponding inverse DTFT of $S\left(e^{j\omega}\right)$ is given by

$$s(t) := \frac{1}{2\pi} \int_{-\pi}^{\pi} S\left(e^{j\omega}\right) e^{j\omega t} \, d\omega. \tag{2.3}$$

The angular frequency $\omega = \omega_c T_s$ is normalized and measured in radians per sampling period where ω_c is the *physical frequency variable* measured in radians per second. Occasionally, $f = \omega/(2\pi)$ will be used, which has unit Hertz (Hz). If (2.1) holds, then by the well-known Parseval's theorem,

$$E_s = \sum_{t=-\infty}^{\infty} |s(t)|^2 = \frac{1}{2\pi} \int_{-\pi}^{\pi} \left|S(e^{j\omega})\right|^2 \, d\omega. \tag{2.4}$$

It follows that $\Phi_s(\omega) = \left|S(e^{j\omega})\right|^2$ represents the energy distribution of the sequence over frequency and is thus termed *energy spectral density* (ESD).

The ESD can be obtained through a different path. Define

$$\gamma_s(k) = \sum_{t=-\infty}^{\infty} s(t)\bar{s}(t-k), \; k = 0, \pm 1, \pm 2, \ldots. \tag{2.5}$$

The sequence $\{\gamma(k)\}$ resembles autocovariance sequence for random signals. For any energy-bounded signals $\{x(t)\}$ and $\{y(t)\}$, there holds

$$\left| \sum_{t=-\infty}^{\infty} x(t)\bar{y}(t) \right| \leq \sqrt{E_x E_y}, \tag{2.6}$$

which is the well-known Schwarz inequality. See Problem 2.2 in Exercises. Substituting $x(t) = s(t)$ and $y(t) = s(t-k)$ into (2.6) yields

$$E_s = \gamma_s(0) \geq \gamma_s(k) \text{ for } k = \pm 1, \pm 2, \ldots, \tag{2.7}$$

by noting $E_x = E_y = E_s$. Applying DTFT to $\{\gamma_s(k)\}$ yields

$$\begin{aligned}\sum_{k=-\infty}^{\infty} \gamma_s(k) e^{-j\omega k} &= \sum_{k=-\infty}^{\infty} \sum_{t=-\infty}^{\infty} \left[s(t) e^{-j\omega t}\right] \left[\bar{s}(t-k) e^{j\omega(t-k)}\right] \\ &= S\left(e^{j\omega}\right) S\left(e^{j\omega}\right)^* = \left|S(e^{j\omega})\right|^2 = \Phi_s(\omega)\end{aligned}$$

with * for conjugate transpose. Hence, the ESD is the DTFT of the sequence $\{\gamma_s(k)\}$.

In engineering practice, signals are often described by their probabilistic statements and are thus random sequences. Such a signal sequence consists of an ensemble of possible realizations, each of which has some associated probability to occur. However, even if the signal is taken to be deterministic, which is one realization from the whole ensemble, it may not have finite energy over the infinite time horizon. In particular, signals in data communications do not possess DTFTs in general. On the other hand, a random signal usually has a finite average power and thus admits PSD.

Denote $E\{\cdot\}$ as the expectation operator which averages over the ensemble of realizations. The discrete-time signal $\{s(t)\}$ is assumed to be a complex sequence of random variables, or *random process* with zero mean:

$$E\{s(t)\} = 0, \; t = 0, \pm 1, \pm 2, \ldots. \tag{2.8}$$

If, in addition, its *ACS* is given by

$$r_s(k) := E\{s(t)\bar{s}(t-k)\}, \; t = 0, \pm 1, \pm 2, \ldots, \tag{2.9}$$

which is independent of t, then $\{s(t)\}$ is called *WSS*. It is easy to see that $r_s(k) = \bar{r}_s(-k)$ and is left as an exercise to show that

$$r_s(0) \geq |r_s(k)| \text{ for } k = \pm 1, \pm 2, \ldots. \tag{2.10}$$

The PSD is defined as DTFT of ACS:

$$\Psi_s(\omega) := \sum_{k=-\infty}^{\infty} r_s(k) e^{-jk\omega}. \tag{2.11}$$

The inverse DTFT recovers $\{r_s(k)\}$ from the given $\Psi_s(\omega)$ via

$$r_s(k) = \frac{1}{2\pi} \int_{-\pi}^{\pi} \Psi_s(\omega) e^{j\omega k} \, d\omega. \tag{2.12}$$

The averaged power of $\{s(t)\}$ is thus

$$P_s := \mathrm{E}\left\{|s(t)|^2\right\} = r_s(0) = \frac{1}{2\pi}\int_{-\pi}^{\pi} \Psi_s(\omega)\,\mathrm{d}\omega, \tag{2.13}$$

which is also called mean-squared value of $s(t)$.

Example 2.1. Let ω_0 be real. Consider random signal

$$s(t) = A\cos(\omega_0 t + \Theta),\ 0 < \omega_0 < 2\pi, \tag{2.14}$$

where A and Θ are often employed to carry information bits in data communications. This example examines the case when A and Θ are real random variables, independent to each other, and uniformly distributed over $[0,\ 1]$ and $[0,\ 2\pi)$, respectively. The ensemble is a set of sinusoids with random amplitude and phase angle. Simple calculation shows

$$\begin{aligned}\mathrm{E}\{s(t)\} &= \mathrm{E}\{A\cos(\omega_0 t + \Theta)\}\\ &= \mathrm{E}\{A\cos(\omega_0 t)\cos(\Theta) - A\sin(\omega_0 t)\sin(\Theta)\}\\ &= \cos(\omega_0 t)\mathrm{E}\{A\}\mathrm{E}\{\cos(\Theta)\} - \sin(\omega_0 t)\mathrm{E}\{A\}\mathrm{E}\{\sin(\Theta)\} = 0,\end{aligned}$$

by independence, and $\mathrm{E}\{\cos(\Theta)\} = \mathrm{E}\{\sin(\Theta)\} = 0$. In addition,

$$\begin{aligned}\mathrm{E}\{s(t)\bar{s}(t-k)\} &= \mathrm{E}\left\{A^2\cos(\omega_0 t + \Theta)\cos(\omega_0(t-k) + \Theta)\right\}\\ &= \frac{1}{2}\mathrm{E}\left\{A^2\right\}\mathrm{E}\{\cos(\omega_0 k) + \cos(2\omega_0 t - \omega_0 k + 2\Theta)\}\\ &= \frac{1}{2}\mathrm{E}\left\{A^2\right\}\cos(\omega_0 k) = r_s(k).\end{aligned}$$

It follows that $\{s(t)\}$ is WSS. By the hypothesis on A, $\mathrm{E}\{A^2\} = 1/3$. Thus,

$$\Psi_s(\omega) = \frac{1}{6}\sum_{k=-\infty}^{\infty}\cos(\omega_0 k)\mathrm{e}^{-j\omega k} = \frac{1}{12}\left[\delta_D(\omega + \omega_0) + \delta_D(\omega - \omega_0)\right]$$

with $\delta_D(\cdot)$, the Dirac delta function, satisfying

$$\text{(i) } \delta_D(x) = 0 \ \text{ for } \ x \neq 0,\ \ \text{(ii) } \int_{\infty}^{\infty} \delta_D(x)\,\mathrm{d}x = 1. \tag{2.15}$$

The above indicates that there are two spectrum lines at $\pm\omega_0$, respectively. The analytical expression of $r_s(k)$ is useful in computing PSD of $\{s(t)\}$.

In practice, there is a difficulty in evaluating the PSD as defined in (2.11). Infinitely, many terms need be computed for ACS, which is not feasible. An approximate PSD is employed, consisting of finitely many signal samples:

$$\Psi_s^{(n)}(\omega) = \mathrm{E}\left\{\frac{1}{n}\left|\sum_{t=0}^{n-1} s(t)\mathrm{e}^{-j\omega t}\right|^2\right\}. \tag{2.16}$$

A natural question is whether or not $\Psi_s^{(n)}(\omega)$ converges to $\Psi_s(\omega)$ as $n \to \infty$. By straightforward calculation,

$$\begin{aligned}\Psi_s^{(n)}(\omega) &= \frac{1}{n}\sum_{t=0}^{n-1}\sum_{\tau=0}^{n-1} \mathrm{E}\{s(t)\bar{s}(\tau)\}\mathrm{e}^{-j\omega(t-\tau)} \\ &= \sum_{k=-n}^{n}\left(1-\frac{|k|}{n}\right) r_s(k)\mathrm{e}^{-jk\omega}.\end{aligned}$$

Since multiplication in time domain is the same as convolution in frequency-domain, the above yields

$$\Psi_s^{(n)}(\omega) = \frac{1}{2\pi}\int_{-\pi}^{\pi} F_n(\theta)\Psi_s(\omega-\theta)\,\mathrm{d}\theta, \tag{2.17}$$

where $F_n(\omega)$ is the nth order *Fejér's kernel* given by

$$F_n(\omega) := \sum_{k=-n}^{n}\left(1-\frac{|k|}{n}\right)\mathrm{e}^{-jk\omega} = \frac{1}{n}\left(\frac{\sin\frac{n}{2}\omega}{\sin\frac{1}{2}\omega}\right)^2. \tag{2.18}$$

Verification of the Fejér's kernel is left as an exercise (Problem 2.5). Before investigating the convergence issue for the approximate PSD in (2.17), it is illuminating to learn the useful properties of the Fejér's kernel.

Lemma 2.1. *Let $F_n(\omega)$ be defined as in (2.18). Then*

(i) $F_n(\omega) \geq 0 \ \forall \omega \in [0, 2\pi]$;

(ii) $\frac{1}{2\pi}\int_0^{2\pi} F_n(\omega)\,\mathrm{d}\omega = 1$ *for every* $n > 0$;

(iii) *For any closed interval* I *in* $(0, 2\pi)$, $\lim_{n\to\infty}\sup_{\omega\in \mathrm{I}}|F_n(\omega)| = 0$.

The proof of this lemma is again left as an exercise (Problem 2.5). The next theorem is the main result of this section.

Theorem 2.2. *Suppose that the random signal $\{s(t)\}$ has a finite averaged power. Then it admits the* PSD *as defined in (2.11). Let $\Psi_s(\omega)$ be continuous over $[0, 2\pi)$ and $\Psi_s(0) = \Psi_s(2\pi)$. Define $\Psi_s^{(n)}(\omega)$ as in (2.17). Then*

$$\lim_{n\to\infty} \Psi_s^{(n)}(\omega) = \lim_{n\to\infty} \mathrm{E}\left\{\frac{1}{n}\left|\sum_{t=0}^{n-1} s(t)\mathrm{e}^{-j\omega t}\right|^2\right\} = \Psi_s(\omega)$$

for all $\omega \in [0, 2\pi]$. In other words, $\Psi_s^{(n)}(\omega)$ converges uniformly to $\Psi(\omega)$.

Proof. By the expression of $\Phi_s^{(n)}(\omega)$ in (2.17) and (ii) of Lemma 2.1,

$$\Psi_s^{(n)}(\omega) - \Psi_s(\omega) = \frac{1}{2\pi}\int_{-\pi}^{\pi}[\Psi_s(\omega-\theta) - \Psi_s(\omega)]F_n(\theta)\,\mathrm{d}\theta.$$

Since $\Psi_s(\omega)$ is a continuous function of ω, there exists an $M > 0$ such that $|\Psi_s(\omega)| \leq M$ for all $\omega \in [-\pi, \pi]$. Take $\delta > 0$ and write

$$\Psi_s^{(n)}(\omega) - \Psi_s(\omega) = \frac{1}{2\pi}\int_{-\delta}^{\delta}[\Psi_s(\omega-\theta) - \Psi_s(\omega)]F_n(\theta)\,\mathrm{d}\theta + \frac{1}{2\pi}\int_{\delta\leq|\theta|\leq\pi}[\Psi_s(\omega-\theta) - \Psi_s(\omega)]F_n(\theta)\,\mathrm{d}\theta.$$

It follows from Lemma 2.1 that

$$\left|\Psi_s^{(n)}(\omega) - \Psi_s(\omega)\right| \leq \sup_{|\theta|\leq\delta}|\Psi_s(\omega-\theta) - \Psi_s(\omega)| + 2M \sup_{\delta\leq|\theta|\leq\pi} F_n(\theta). \tag{2.19}$$

According to property (3) of Lemma 2.1, and by the continuity of $\Psi_s(\omega)$, there exists an $N > 0$ such that for all $n \geq N$ and $\omega \in [0, 2\pi]$,

$$\left|\Psi_s^{(n)}(\omega) - \Psi_s(\omega)\right| \leq \varepsilon$$

for any given $\varepsilon > 0$. Therefore, $\Psi_s^{(n)}(\omega)$ converges uniformly to $\Psi_s(\omega)$. □

There is no loss of generality in using only the causal part of the signal for approximate PSD $\Psi_s^{(n)}(\omega)$ due to the WSS assumption for the random signal $\{s(t)\}$. In fact, (2.17) is also useful in the case when the PDF of $s(t)$ is unknown, in which case the PSD is often estimated using time averages instead of the ensemble average, by assuming ergodic process for $\{s(t)\}$. Since the measured signal data are always finitely many, they can be assumed to begin at time $t = 0$.

Theorem 2.2 reveals an important property of the PSD:

$$\Psi_s(\omega) \geq 0 \;\forall\, \omega. \tag{2.20}$$

That is, PSDs are positive real functions of frequency, even though $\Psi_s(\omega) = \Psi_s(-\omega)$ may not hold in the case of complex signals. If the random signals are real, then there holds $\Psi_s(\omega) = \Psi_s(-\omega) \geq 0$ for all ω.

2.1.2 Vector Signals

For MIMO communication channels, the data signals are vector-valued, denoted by boldfaced letters, at each sampling time t. Consider the vector signal $\{\mathbf{s}(t)\}$ with size $p > 1$. If $\{\mathbf{s}(t)\}$ is deterministic, then it is assumed that the energy of the vector signal is bounded. That is,

$$E_{\mathbf{s}} := \sum_{t=-\infty}^{\infty} \|\mathbf{s}(t)\|^2 < \infty, \tag{2.21}$$

where $\|\mathbf{s}(t)\| = \sqrt{\mathbf{s}(t)^*\mathbf{s}(t)}$ is the Euclidean norm of $\mathbf{s}(t)$. In this case, the DTFT of $\{\mathbf{s}(t)\}$ exists and has the same expression as (2.2):

$$\mathbf{S}\left(\mathrm{e}^{j\omega}\right) := \sum_{t=-\infty}^{\infty} \mathbf{s}(t)\mathrm{e}^{-j\omega t}. \tag{2.22}$$

Define the $p \times p$ matrices

$$\Gamma_{\mathbf{s}}(k) = \sum_{t=-\infty}^{\infty} \mathbf{s}(t)\mathbf{s}(t-k)^*, \; k = 0, \pm 1, \pm 2, \ldots. \tag{2.23}$$

Although each term in the summation has rank one, $\Gamma_{\mathbf{s}}(k)$ may have a rank greater than one and even be nonsingular. Moreover, $\{\Gamma_{\mathbf{s}}(k)\}$ is a bounded matrix sequence. There holds

$$E_{\mathbf{s}} = \mathrm{Tr}\{\Gamma_{\mathbf{s}}(0)\} \geq |\mathrm{Tr}\{\Gamma_{\mathbf{s}}(k)\}| \text{ for } k = \pm 1, \pm 2, \ldots. \tag{2.24}$$

Similar to the scalar case, the ESD can be defined as the DTFT of $\{\Gamma_{\mathbf{s}}(k)\}$:

$$\Phi_{\mathbf{s}}(\omega) = \sum_{k=-\infty}^{\infty} \Gamma_{\mathbf{s}}(k)\mathrm{e}^{-j\omega k} = \mathbf{S}\left(\mathrm{e}^{j\omega}\right)\mathbf{S}\left(\mathrm{e}^{j\omega}\right)^*. \tag{2.25}$$

The above can be obtained with a similar derivation as in the scalar case. It is interesting to observe that $\Phi_{\mathbf{s}}(\omega)$ always has rank one for all ω, even though $\Gamma_{\mathbf{s}}(k)$ may have a rank greater than one for each k. Consequently,

$$\mathrm{Tr}\{\Phi_{\mathbf{s}}(\omega)\} = \mathrm{Tr}\left\{\mathbf{S}\left(\mathrm{e}^{j\omega}\right)\mathbf{S}(\mathrm{e}^{j\omega})^*\right\} = \mathbf{S}\left(\mathrm{e}^{j\omega}\right)^*\mathbf{S}\left(\mathrm{e}^{j\omega}\right) = \left\|\mathbf{S}(\mathrm{e}^{j\omega})\right\|^2,$$

by properties of the trace. Parseval's theorem is extended to

$$E_{\mathbf{s}} = \sum_{t=-\infty}^{\infty} \|\mathbf{s}(t)\|^2 = \frac{1}{2\pi}\int_{-\pi}^{\pi} \left\|\mathbf{S}(\mathrm{e}^{j\omega})\right\|^2 \, \mathrm{d}\omega. \tag{2.26}$$

For the case of random vector signals, it is assumed that $\{\mathbf{s}(t)\}$ is WSS with mean $\mathrm{E}\{\mathbf{s}(t)\} = \mathbf{0}_p$ for all t. Then its ACS is given by

$$R_{\mathbf{s}}(k) := \mathrm{E}\{\mathbf{s}(t)\mathbf{s}(t-k)^*\}, \; k = 0, \pm 1, \pm 2, \ldots, \tag{2.27}$$

which is independent of t and has size $p \times p$. Similar to the deterministic case, $R_{\mathbf{s}}(k)$ can be nonsingular. It can be shown that (Problem 2.7 in Exercises)

$$\text{(i) } R_{\mathbf{s}}(k)^* = R_{\mathbf{s}}(-k), \;\; \text{(ii) } \mathrm{Tr}\{R_{\mathbf{s}}(0)\} \geq |\mathrm{Tr}\{R_{\mathbf{s}}(k)\}|. \tag{2.28}$$

Assume that $R_{\mathbf{s}}(0)$ exists and is bounded. Then the PSD of $\{\mathbf{s}(t)\}$ can be easily extended from (2.11) via the DTFT of ACS:

$$\Psi_{\mathbf{s}}(\omega) := \sum_{k=-\infty}^{\infty} R_{\mathbf{s}}(k) \mathrm{e}^{-jk\omega}. \tag{2.29}$$

Different from the ESD in the deterministic case, the PSD $\Psi_{\mathbf{s}}(\omega)$ is nonsingular generically. The inverse DTFT recovers $\{R_{\mathbf{s}}(k)\}$ via

$$R_{\mathbf{s}}(k) = \frac{1}{2\pi} \int_{-\pi}^{\pi} \Psi_{\mathbf{s}}(\omega) \mathrm{e}^{j\omega k} \, \mathrm{d}\omega. \tag{2.30}$$

The averaged power of $\{\mathbf{s}(t)\}$ is generalized as follows:

$$P_{\mathbf{s}} := \mathrm{E}\{\mathbf{s}(t)^*\mathbf{s}(t)\} = \mathrm{Tr}\{R_{\mathbf{s}}(0)\} = \mathrm{Tr}\left\{\frac{1}{2\pi} \int_{-\pi}^{\pi} \Psi_{\mathbf{s}}(\omega) \, \mathrm{d}\omega\right\}. \tag{2.31}$$

Example 2.3. Consider random vector signal

$$\mathbf{s}(t) = A\mathbf{x}(t), \;\; \mathbf{x}(t) = \begin{bmatrix} \cos(\omega_0 t + \Theta) \\ \sin(\omega_0 t + \Theta) \end{bmatrix}, \tag{2.32}$$

where A and Θ are real independent random variables uniformly distributed over $[0, \, 1]$ and $[0, \, 2\pi)$, respectively, as in Example 2.1. It is easy to show that $\mathrm{E}\{\mathbf{s}(t)\} = \mathbf{0}$ and

$$\mathrm{E}\{\mathbf{x}(t)\mathbf{x}(t-k)^*\} = \frac{1}{2} \begin{bmatrix} \cos(\omega_0 k) & -\sin(\omega_0 k) \\ \sin(\omega_0 k) & \cos(\omega_0 k) \end{bmatrix} =: R_{\mathbf{x}}(k).$$

See Problem 2.6 in Exercises. Thus, by independence and $\mathrm{E}\{A^2\} = 1/3$,

$$\mathrm{E}\{\mathbf{s}(t)\mathbf{s}(t-k)^*\} = \mathrm{E}\{A^2\}\mathrm{E}\{\mathbf{x}(t)\mathbf{x}(t-k)^*\} = \frac{1}{6} \begin{bmatrix} \cos(\omega_0 k) & -\sin(\omega_0 k) \\ \sin(\omega_0 k) & \cos(\omega_0 k) \end{bmatrix}.$$

It follows that $\{\mathbf{s}(t)\}$ is WSS with $R_{\mathbf{s}}(k) = \mathrm{E}\{\mathbf{s}(t)\mathbf{s}(t-k)^*\}$ as above, which is nonsingular for all k. Direct calculation yields

$$\Phi_s(\omega) = \frac{1}{6}\sum_{k=-\infty}^{\infty}\begin{bmatrix}\cos(\omega_0 k) & -\sin(\omega_0 k)\\ \sin(\omega_0 k) & \cos(\omega_0 k)\end{bmatrix}e^{-j\omega k}$$
$$= \frac{1}{12}\left(\delta_D(\omega+\omega_0)\begin{bmatrix}1 & -j\\ j & 1\end{bmatrix} + \delta_D(\omega-\omega_0)\begin{bmatrix}1 & j\\ -j & 1\end{bmatrix}\right).$$

Thus, each element of the PSD matrix $\Psi_s(\omega)$ contains two spectrum lines at $\pm\omega_0$, respectively, as in the scalar case. The power of the signal is given by $P_s = \mathrm{Tr}\{R_s(0)\} = 1/3$.

Approximate PSD can be employed if there are only finitely many terms of ACS available. The following

$$\Psi_s^{(n)}(\omega) = \mathrm{E}\left\{\frac{1}{n}\left(\sum_{t=0}^{n-1} s(t)e^{-j\omega t}\right)\left(\sum_{\tau=0}^{n-1} s(\tau)e^{-j\omega\tau}\right)^*\right\} \tag{2.33}$$

is generalized from (2.16). Straightforward calculation gives

$$\Psi_s^{(n)}(\omega) = \frac{1}{n}\sum_{t=0}^{n-1}\sum_{\tau=0}^{n-1}\mathrm{E}\{s(t)s(\tau)^*\}e^{-j\omega(t-\tau)}$$
$$= \sum_{k=-n}^{n}\left(1-\frac{|k|}{n}\right)R_s(k)e^{-jk\omega}$$
$$= \frac{1}{2\pi}\int_{-\pi}^{\pi} F_n(\theta)\Psi_s(\omega-\theta)\,d\theta,$$

where $F_n(\cdot)$ is the nth order Fejér's kernel as defined in (2.18). Because Fejér's kernel is a scalar function, the matrix-valued ACS and PSD do not pose any difficulty in extending Theorem 2.2 to the following.

Theorem 2.4. *Suppose that the random vector signal $\{s(t)\}$ has a finite averaged power. Then it admits the* PSD *as defined in (2.29). Let $\Psi_s(\omega)$ be continuous over $[0, 2\pi)$ and $\Psi_s(0) = \Psi_s(2\pi)$. Define $\Psi_s^{(n)}(\omega)$ as in (2.33). Then*

$$\lim_{n\to\infty}\Psi_s^{(n)}(\omega) = \Psi_s(\omega)\ \forall\,\omega\in[0, 2\pi].$$

Example 2.5. As an application example, consider estimation of the PSD based on a given set of N data samples $\{s(t)\}_{t=0}^{N-1}$ in the absence of the statistical information of the underlying signal. To employ $\Psi_s^{(n)}(\omega)$ in (2.33) as an approximation, it is assumed that $N = nm$ with n and m integers. The data set is partitioned into m disjoint subsets $\{s_i(t)\}_{t=0}^{n-1}$, where $i = 0, 1, \ldots, m-1$. A simple partition is

$$s_i(t) = s(in+t),\ 0\le i\le m-1,\ 0\le t\le n-1.$$

Let $W_n = e^{-j2\pi/n}$. For $i = 0, 1, \ldots, m-1$, compute

$$S_i(k) = \frac{1}{\sqrt{n}} \sum_{t=0}^{n-1} \mathbf{s}_i(t) W_n^{tk}, \; k = 0, 1, \ldots, n-1, \tag{2.34}$$

which is a modified discrete Fourier transform (DFT). That is, it computes n frequency response samples uniformly distributed over $[0,\ 2\pi]$ for $\{\mathbf{s}_i(t)\}_{t=0}^{n-1}$ modified by a factor of $\frac{1}{\sqrt{n}}$. The FFT (fast Fourier transform) algorithm can be used to implement the computation in (2.34). Now the ensemble average in (2.33) at $\omega = \omega_k = \frac{2k\pi}{n}$ is replaced by the time average as follows:

$$\Psi^{(n)}(\omega_k) \approx \frac{1}{m} \sum_{i=0}^{m-1} S_i(k) S_i(k)^*, \; k = 0, 1, \ldots, n-1. \tag{2.35}$$

If $\{\mathbf{s}(t)\}$ is an ergodic process, the right-hand side converges to $\Psi^{(n)}(\omega_k)$ as $m \to \infty$, which in turn converges to the true PSD $\Psi(\omega)$ uniformly as $n \to \infty$. Theorem 2.4 is the basis for such a spectral estimation technique.

2.2 Linear Systems

Systems can be viewed as operators which map input signals to output signals according to some mathematical mechanisms. A linear system is a linear map whose output is a linear function of the input. This text focuses on LTI systems that provide transparent relations between spectral densities of the input signals and output signals. In fact, LTI systems shape the spectral densities of the input signals through a simple multiplicative operation capable of producing entirely different spectral densities at the output. LTV systems will also be studied in this section, albeit at a less degree.

2.2.1 Transfer Functions and Matrices

An LTI scalar system can be represented by its transfer function which is the $\mathscr{Z}$ transform of its impulse response, as illustrated below (see Fig. 2.1).

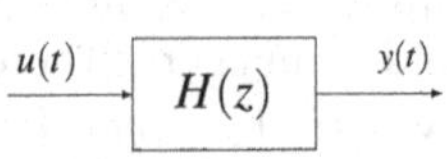

Fig. 2.1 An LTI system represented by its transfer function

That is, if $u(t) = \delta(t)$, which is the Kroneker delta function, then $y(t) = h(t)$ for $t = 0, \pm 1, \pm 2, \ldots$, with $\{h(t)\}$ the impulse response. The transfer function of the system is given by

$$H(z) := \sum_{t=-\infty}^{\infty} h(t) z^{-t}, \; z \in \mathbb{C}. \tag{2.36}$$

For any input $\{u(t)\}$, the output of the system is the *convolution* of the impulse response with the input:

$$y(t) = h(t) \star u(t) := \sum_{k=-\infty}^{\infty} h(t-k) u(k). \tag{2.37}$$

The system is said to be causal, if $h(t) = 0$ for $t < 0$, and strictly causal, if $h(t) = 0$ for $t \le 0$. Physical systems are causal in general, and often strictly causal. The following defines the notion of *stability*.

Definition 2.1. A system is said to be stable, if for every bounded input $\{u(t)\}$ (i.e., $|u(t)| \le M_u < \infty$ for all t, and some $M_u > 0$), the corresponding output $\{y(t)\}$ is bounded (i.e., $|y(t)| \le M_y < \infty$ for all t, and some $M_y > 0$).

The above stability is also termed BIBO (bounded-input/bounded-output) stability. The next result provides the stability criterion.

Theorem 2.6. *An* LTI *system with transfer function as in (2.36) is stable, if and only if*

$$\sum_{t=-\infty}^{\infty} |h(t)| < \infty. \tag{2.38}$$

Proof. For any bounded input $\{u(t)\}$ satisfying $|u(t)| \le M_u < \infty$ for all t, and some $M_u > 0$, the output satisfies

$$|y(t)| = \left| \sum_{k=-\infty}^{\infty} h(t-k) u(k) \right| \le \left(\sum_{k=-\infty}^{\infty} |h(k)| \right) M_u =: M_y < \infty$$

for $t = 0, \pm 1, \pm 2, \ldots$. Hence, (2.38) implies stability of the given LTI system. Conversely for the stable LTI system in (2.36), consider input $\{u(t)\}$ given by

$$u(k) = \begin{cases} \bar{h}(t_0 - k)/|h(t_0 - k)|, & h(t_0 - k) \ne 0, \\ 0, & h(t_0 - k) = 0, \end{cases}$$

with t_0 an integer. Then $|u(t)| \le 1$ for all t and

$$y(t_0) = \sum_{k=-\infty}^{\infty} h(t_0 - k) u(k) = \sum_{k=-\infty}^{\infty} |h(t_0 - k)| = \sum_{t=-\infty}^{\infty} |h(t)| < \infty$$

by the stability assumption. Thus, stability implies (2.38). □

Notice that if the LTI system is both stable and causal, then

$$H(z) = \sum_{t=0}^{\infty} h(t)z^{-t} \tag{2.39}$$

is analytic at $z \in \mathbb{C}$ such that $|z| > 1$ and continuous on the unit circle. In other words, the region of convergence (ROC) is $|z| \geq 1$. In the interest of this text, only a subset of causal and stable LTI systems will be studied. This is the set of causal and stable LTI systems that admit rational transfer functions, or have finitely many poles and zeros. The causality of such a system is equivalent to the properness of its transfer function. Moreover, there exists a positive number $r < 1$ such that its ROC is $|z| > r$. That is, it is also analytic on the unit circle. Various system models will be presented in the next subsection.

For MIMO LTI systems, both input and output are vector signals. Capital letters, for example, $\{H(t)\}$, are used to denote impulse responses. The $\mathscr{Z}$ transform of the impulse response $\{H(t)\}$ is a transfer function matrix, or simply called transfer matrix, denoted by boldfaced capital letter and defined by

$$\mathbf{H}(z) := \sum_{t=-\infty}^{\infty} H(t)z^{-t}. \tag{2.40}$$

If the input signal $\{\mathbf{u}(t)\}$ has dimension m and the output signal $\{\mathbf{y}(t)\}$ has dimension p, then $\mathbf{H}(z)$ has size $p \times m$ for each $z \in \mathbb{C}$. The input and the output are again governed by the convolution relation:

$$\mathbf{y}(t) = H(t) \star \mathbf{u}(t) := \sum_{k=-\infty}^{\infty} H(t-k)\mathbf{u}(k). \tag{2.41}$$

A vector signal $\{\mathbf{s}(t)\}$ is said to be bounded, if $\|\mathbf{s}(t)\| \leq M_s$ for all t and some bounded $M_s > 0$. The stability notion in Definition 2.1 can be easily generalized to MIMO systems.

Definition 2.2. A MIMO system is said to be stable, if for every bounded input $\{\mathbf{u}(t)\}$, the corresponding output $\{\mathbf{y}(t)\}$ is bounded.

Note that for vector equation $\mathbf{w} = A\mathbf{v}$ with A a fixed matrix, $\|\mathbf{w}\|$ is a function of $\mathbf{v}$. Recall that $\|\cdot\|$ is the Euclidean norm. There holds

$$\sup_{\|\mathbf{v}\|=1} \|\mathbf{w}\| = \sup_{\|\mathbf{v}\|=1} \|A\mathbf{v}\| = \overline{\sigma}(A) \tag{2.42}$$

with $\overline{\sigma}(\cdot)$ the maximum singular value (refer to Appendix A). The next result is extended from Theorem 2.6 by noting the equality (2.42) and by the fact that there exists $\mathbf{v}_0$ with $\|\mathbf{v}_0\| = 1$ such that $\|A\mathbf{v}_0\| = \overline{\sigma}(A)$.

Theorem 2.7. *The* LTI *system with transfer matrix as in (2.40) is stable, if and only if*

$$\sum_{t=-\infty}^{\infty} \overline{\sigma}(H(t)) < \infty. \tag{2.43}$$

The proof is left as an exercise (Problem 2.20). For deterministic input signals, output signals are deterministic as well. The convolution in time domain is translated into multiplication in $\mathscr{Z}$-domain or frequency domain:

$$\mathbf{Y}(z) = \mathbf{H}(z)\mathbf{U}(z), \ \mathbf{Y}\left(e^{j\omega}\right) = \mathbf{H}\left(e^{j\omega}\right)\mathbf{U}\left(e^{j\omega}\right). \tag{2.44}$$

Let $\Phi_{\mathbf{u}}(\omega) = \mathbf{U}\left(e^{j\omega}\right)\mathbf{U}\left(e^{j\omega}\right)^*$ be the ESD of the input. Then

$$\Phi_{\mathbf{y}}(\omega) = \mathbf{Y}\left(e^{j\omega}\right)\mathbf{Y}\left(e^{j\omega}\right)^* = \mathbf{H}\left(e^{j\omega}\right)\Phi_{\mathbf{u}}(\omega)\mathbf{H}\left(e^{j\omega}\right)^* \tag{2.45}$$

is the ESD of the output. As such, the frequency response of the system shapes the ESD of the input. It is appropriate to define the energy norm:

$$\|\mathbf{s}\|_{\mathscr{E}} := \sqrt{E_{\mathbf{s}}} = \sqrt{\sum_{t=-\infty}^{\infty} \|\mathbf{s}(t)\|^2}. \tag{2.46}$$

In light of (2.44), the energy norm of the output is given by

$$\|\mathbf{y}\|_{\mathscr{E}} = \sqrt{\operatorname{Tr}\left\{\frac{1}{2\pi}\int_{-\pi}^{\pi} \mathbf{H}(e^{j\omega})\Phi_{\mathbf{u}}(\omega)\mathbf{H}(e^{j\omega})^* \, d\omega\right\}}. \tag{2.47}$$

Generically, $\|\mathbf{y}\|_{\mathscr{E}} \neq \|\mathbf{u}\|_{\mathscr{E}}$, if $\mathbf{H}(z) \neq I$. Thus, the energy norm serves as an indicator on the frequency-shaping effect of the system frequency response.

For random signals, their DTFT may not exist and thus (2.44) may not hold, if the input is a random signal. Suppose that the input $\{\mathbf{u}(t)\}$ is a WSS random process with zero means and $\{R_{\mathbf{u}}(k)\}$ as the ACS. Then $\{\mathbf{y}(t)\}$ is a random process with zero mean due to $\mathrm{E}\{\mathbf{u}(t)\} = 0$ for all t and

$$\mathrm{E}\{\mathbf{y}(t)\} = \sum_{k=-\infty}^{\infty} H(t-k)\mathrm{E}\{\mathbf{u}(k)\} = 0 \ \forall \ t.$$

In fact, the output is also a WSS random process. Specifically,

$$\begin{aligned}
\mathrm{E}\{\mathbf{y}(t)\mathbf{y}(t-k)^*\} &= \sum_{\alpha=-\infty}^{\infty}\sum_{\beta=-\infty}^{\infty} H(t-\alpha)\mathrm{E}\{\mathbf{u}(\alpha)\mathbf{u}(\beta)^*\}H(t-k-\beta)^* \\
&= \sum_{\alpha=-\infty}^{\infty}\sum_{\beta=-\infty}^{\infty} H(t-\alpha)R_{\mathbf{u}}(\alpha-\beta)H(t-k-\beta)^*.
\end{aligned}$$

With variable substitution $\gamma = \alpha - \beta$, the above results in

$$\begin{aligned} \mathrm{E}\{\mathbf{y}(t)\mathbf{y}(t-k)^*\} &= \sum_{\beta=-\infty}^{\infty} \sum_{\gamma=-\infty}^{\infty} H(t-\beta-\gamma)R_{\mathbf{u}}(\gamma)H(t-\beta-k)^* \\ &= \sum_{\beta=-\infty}^{\infty} \tilde{R}_{\mathbf{y}}(t-\beta)H(t-\beta-k)^* \\ &= \sum_{\tau=-\infty}^{\infty} \tilde{R}_{\mathbf{y}}(\tau)H(\tau-k)^* = R_{\mathbf{y}}(k), \end{aligned}$$

which is independent of time t, where $\tilde{R}_{\mathbf{y}}(\tau) = H(\tau) \star R_{\mathbf{u}}(\tau)$ with $\tau = t - \beta$. Hence, it is concluded that the output of an LTI system is a WSS random process, provided that the input is. Let $\Psi_{\mathbf{u}}(\omega)$ be the PSD associated with input. Applying DTFT to the ACS of $\{\mathbf{y}(t)\}$ shows that the PSD of the output is given by (Problem 2.9 in Exercises)

$$\Psi_{\mathbf{y}}(\omega) = \mathbf{H}\left(e^{j\omega}\right)\Psi_{\mathbf{u}}(\omega)\mathbf{H}\left(e^{j\omega}\right)^*. \tag{2.48}$$

The resemblance of (2.48) to (2.45) is obvious, implying that LTI systems are capable of shaping the PSD of the input signal through multiplicative operations. The frequency response of the underlying system determines how much frequency shaping the system can exert to the input PSD. A useful measure is the power norm:

$$\|\mathbf{s}\|_{\mathscr{P}} = \sqrt{P_{\mathbf{s}}} = \sqrt{\mathrm{E}\{\|\mathbf{s}(t)\|^2\}} = \sqrt{\mathrm{Tr}\{R_{\mathbf{s}}(0)\}}. \tag{2.49}$$

Thus, the power norm of the output is

$$\|\mathbf{y}\|_{\mathscr{P}} = \sqrt{\mathrm{Tr}\left\{\frac{1}{2\pi}\int_{-\pi}^{\pi} \mathbf{H}(e^{j\omega})\Psi_{\mathbf{u}}(e^{j\omega})\mathbf{H}(e^{j\omega})^*\, d\omega\right\}}, \tag{2.50}$$

which indicates the shaping effect of the system frequency response to the input PSD. More investigation will be carried out in later sections. If the input is white noise with zero mean and identity covariance, i.e., $\Phi_{\mathbf{u}}(e^{j\omega}) \equiv I$, then (2.49) provides one way to compute the system norm defined by

$$\|\mathbf{H}\|_2 := \sqrt{\mathrm{Tr}\left\{\frac{1}{2\pi}\int_{-\pi}^{\pi} \mathbf{H}(e^{j\omega})\mathbf{H}(e^{j\omega})^*\, d\omega\right\}} = \sqrt{\mathrm{Tr}\left\{\sum_{t=-\infty}^{\infty} H(t)H(t)^*\right\}} \tag{2.51}$$

in light of the Parseval's theorem. Such a system norm is sometime called Frobenius norm of the system. A more general system norm is

$$\|\mathbf{H}\|_{\mathrm{p}} := \left[\frac{1}{2\pi}\int_{-\pi}^{\pi}\left(\sqrt{\mathrm{Tr}\{\mathbf{H}(e^{j\omega})\mathbf{H}(e^{j\omega})^*\}}\right)^p d\omega\right]^{1/p}$$

for $1 \le p < \infty$, which reduces to $\|\mathbf{H}\|_2$ for $p = 2$.

Example 2.8. Consider the scalar system with transfer function

$$H(z) = K\left(1 - 2\cos(\omega_h)z^{-1} + z^{-2}\right) = K\left(z - e^{j\omega_h}\right)\left(z - e^{-j\omega_h}\right),$$

where K is a real constant gain. Simple calculations show that

$$\left|H(e^{j\omega})\right| = 2|K|\sqrt{\left|\sin\left(\frac{\omega+\omega_h}{2}\right)\sin\left(\frac{\omega-\omega_h}{2}\right)\right|}$$

and $\|H\|_2 = 2|K|\sqrt{1+\cos^2(\omega_h)/2}$. If the input to the system is $u(t) = s(t)$ with $s(t)$ as given in Example 2.1, then the input PSD is

$$\Psi_u(\omega) = \frac{1}{12}\left[\delta_D(\omega+\omega_0) + \delta_D(\omega-\omega_0)\right].$$

In the scalar case, (2.48) reduces to $\Psi_y(\omega) = \left|H\left(e^{j\omega}\right)\right|^2 \Psi_u(\omega)$, and thus,

$$\begin{aligned}\Psi_y(\omega) &= \frac{K^2}{3}\left|\sin\left(\frac{\omega+\omega_h}{2}\right)\sin\left(\frac{\omega-\omega_h}{2}\right)\right|\left[\delta_D(\omega+\omega_0)+\delta_D(\omega-\omega_0)\right]\\ &= \frac{K^2}{3}\left|\sin\left(\frac{\omega_0+\omega_h}{2}\right)\sin\left(\frac{\omega_0-\omega_h}{2}\right)\right|\left[\delta_D(\omega+\omega_0)+\delta_D(\omega-\omega_0)\right]\end{aligned}$$

is the output PSD. It follows that the amplitude of the two spectrum lines of the input PSD is shaped by the frequency response $H\left(e^{j\omega}\right)$ at frequency ω_0. Indeed, if $\omega_h = \pm\omega_0$, then $\Psi_y(\omega) \equiv 0$ and there are no spectrum lines for the output PSD. On the other hand, if $\omega_h \neq \pm\omega_0$, the maximum amplitude of the two spectrum lines at the output is given by (with either $\omega_h = 0$, or $\omega_h = \pi$)

$$\frac{K^2}{3}\max\left\{\cos^2(\omega_0/2), \sin^2(\omega_0/2)\right\} \geq \frac{K^2}{6},$$

which can be large if the gain K is large.

2.2.2 *System Models*

The systems under consideration are causal and stable LTI systems, which have finitely many poles. Such systems form a *dense set* in the class of all causal and stable systems having continuous frequency responses. In other words, any causal and stable LTI system which admits continuous frequency response can be approximated arbitrarily well by a causal and stable LTI system which has finitely many poles, provided that the number of poles is adequately large in light of Weierstrass Theorem from calculus. Such systems are also called *finite-dimensional* due to their finitely many poles and, more importantly, that they can be implemented or realized with finitely many arithmetic and delay operations. This subsection will provide a brief review of commonly used mathematical models for finite-dimensional LTI systems.

FIR or MA Models

For MIMO systems with m input and p output, the FIR model, also called transversal filter, refers to the transfer matrices of the form

$$\mathbf{H}(z) = \sum_{k=0}^{\ell} H(k) z^{-k}, \tag{2.52}$$

where $H(k)$ is a matrix of size $p \times m$ and is the impulse response at time $t = k$. In obtaining the impulse response of the system, the m impulse inputs need be applied one by one. The corresponding m output signals of size p can then be packed together column-wise to form $\{H(t)\}_{t=0}^{\ell}$. Since the impulse response dies out in finitely many samples, it acquires the name FIR (finite impulse response).

Consider the system with FIR model in (2.52). Let $\{\mathbf{u}(t)\}$ and $\{\mathbf{y}(t)\}$ be the associated input and output signals, respectively. Then

$$\mathbf{y}(t) = \sum_{k=0}^{\ell} H(k)\mathbf{u}(t-k) = \sum_{k=t-\ell}^{t} H(t-k)\mathbf{u}(k). \tag{2.53}$$

That is, the output is the (weighted) moving average (MA) of the input. Hence, the input/output description in (2.53) for the FIR model is also called the MA model. FIR or MA models are the simplest, yet extremely important, for wireless communication systems. The wireless channels are characterized by multipath, of which gains of each path can be regarded as the impulse responses of the channels and are often complex valued.

IIR or ARMA Models

The IIR model for SISO systems has the fractional form

$$H(z) = \frac{N(z)}{M(z)} = \frac{\nu_0 + \nu_1 z^{-1} + \cdots + \nu_{n_\nu} z^{-n_\nu}}{1 - \mu_1 z^{-1} - \cdots - \mu_{n_\mu} z^{-n_\mu}}, \tag{2.54}$$

where $\mu_k \neq 0$ for at least one integer $k > 0$ and $M(z) \neq 0$ for all z outside and on the unit circle. It follows that the system is stable and has a causal and infinite impulse response (IIR). Let $\{u(t)\}$ and $\{y(t)\}$ be the associated input and output signals, respectively. Then

$$y(t) = \sum_{k=1}^{n_\mu} \mu_k y(t-k) + \sum_{k=0}^{n_\nu} \nu_k u(t-k). \tag{2.55}$$

That is, the output $y(t)$ consists of two parts: the autoregressive (AR) part in the first summation and the MA part in the second summation. If $\nu_k = 0$ for $k = 1, 2, \ldots, n_\nu$, then the ARMA model reduces to the AR model, in which case the system admits an all-pole model. Hence, the ARMA model includes the AR model as a special case. In light of (2.55), the computational complexity in computing the output $y(t)$ is dependent on the degrees of the numerator and denominator polynomials in the ARMA model (2.54). There is an incentive to minimize n_ν and n_μ, which can be carried out through cancelation of the common factors or common roots of $M(z)$ and $N(z)$. If $M(z)$ and $N(z)$ do not share common roots, then $\{M(z), N(z)\}$ is called *relative coprime* or simply *coprime*. In this case, the roots of $N(z)$ are called *zeros*, the roots of $M(z)$ are called *poles*, and $n = \max\{n_\nu, n_\mu\}$ is called the *degree* of the system.

For MIMO systems with m input and p output, the transfer matrices for the IIR model can be extended to the left fractional form

$$\mathbf{H}(z) = \mathbf{M}(z)^{-1}\mathbf{N}(z) = \left(M_0 - \sum_{k=1}^{n_\mu} M_k z^{-k}\right)^{-1} \left(\sum_{k=0}^{n_\nu} N_k z^{-k}\right), \tag{2.56}$$

where M_k is a $p \times p$ matrix and N_k is a $p \times m$ matrix for each integer k. Again, $M_k \neq \mathbf{0}_{p\times p}$ for at least one integer $k > 0$, assuming M_0 is nonsingular. If M_0 is an identity, then the following describes the MIMO ARMA model:

$$\mathbf{y}(t) = \sum_{k=1}^{n_\mu} M_k \mathbf{y}(t-k) + \sum_{k=0}^{n_\nu} N_k \mathbf{u}(t-k), \tag{2.57}$$

where $\{\mathbf{u}(t)\}$ and $\{\mathbf{y}(t)\}$ are the input and output signals, respectively. For MIMO systems, there exists right fractional form

$$\mathbf{H}(z) = \tilde{\mathbf{N}}(z)\tilde{\mathbf{M}}(z)^{-1} = \left(\sum_{k=0}^{\tilde{n}_\nu} \tilde{N}_k z^{-k}\right) \left(\tilde{M}_0 - \sum_{k=1}^{\tilde{n}_\mu} \tilde{M}_k z^{-k}\right)^{-1}, \tag{2.58}$$

which can be entirely different from the one in (2.56).

Several notions need be introduced for MIMO systems. For the left fraction in (2.56), $\{\mathbf{M}(z), \mathbf{N}(z)\}$ is called left coprime, if

$$\operatorname{rank}\left\{\left[\mathbf{M}(z)\ \mathbf{N}(z)\right]\right\} = p \quad \forall\, z \in \mathbb{C}. \tag{2.59}$$

For the right fraction in (2.58), $\{\tilde{\mathbf{N}}(z), \tilde{\mathbf{M}}(z)\}$ is called right coprime, if

$$\text{rank}\left\{\begin{bmatrix}\tilde{\mathbf{M}}(z)\\ \tilde{\mathbf{N}}(z)\end{bmatrix}\right\} = m \quad \forall z \in \mathbb{C}. \tag{2.60}$$

In practice, it is unnecessary to test the rank conditions in (2.59) and (2.60) at all $z \in \mathbb{C}$. For the left fraction, one needs test (2.59) only at those z which are roots of $\det(\mathbf{M}(z)) = 0$, and in the case of the right fraction, one needs test (2.60) only at those z which are roots of $\det\left(\tilde{\mathbf{M}}(z)\right) = 0$.

Suppose that $\{\mathbf{M}(z), \mathbf{N}(z)\}$ and $\{\tilde{\mathbf{N}}(z), \tilde{\mathbf{M}}(z)\}$ are left and right coprimes, respectively. A complex number p_0 is called pole of $\mathbf{H}(z)$, if:

$$\lim_{z \to p_0} \text{rank}\{\mathbf{M}(z)\} < p \iff \lim_{z \to p_0} \text{rank}\left\{\tilde{\mathbf{M}}(z)\right\} < m. \tag{2.61}$$

That is, some elements of $\mathbf{H}(z)$ become unbounded as $z \to p_0$. A complex number z_0 is called zero of $\mathbf{H}(z)$, if with $\rho = \min\{p, m\}$,

$$\lim_{z \to z_0} \text{rank}\{\mathbf{N}(z)\} < \rho \iff \lim_{z \to z_0} \text{rank}\left\{\tilde{\mathbf{N}}(z)\right\} < \rho. \tag{2.62}$$

That is, $\text{rank}\{\mathbf{H}(z)\} < \rho = \min\{p, m\}$ as $z \to z_0$. The system is stable, if $\mathbf{H}(z)$ has all its poles strictly inside the unit circle. The system is called *minimum phase*, if $\mathbf{H}(z)$ has no zeros outside the unit circle, and called *strict minimum phase*, if $\mathbf{H}(z)$ has no zeros on and outside the unit circle.

Example 2.9. For systems with single input ($m = 1$), their right coprime fractions can be easily obtained, which amounts to computing the greatest common divisor (GCD). Specifically, consider the case of $p = 2$ with

$$\begin{aligned}
\mathbf{H}(z) = \tilde{\mathbf{N}}(z)\tilde{M}(z)^{-1} &= \begin{bmatrix} a_1 + b_1 z^{-1} + c_1 z^{-2} \\ a_2 + b_2 z^{-1} + c_2 z^{-2}\end{bmatrix} \left(1 - \alpha z^{-1} - \beta z^{-2}\right)^{-1} \\
&= \begin{bmatrix} a_1\left(1 - s_{1,1} z^{-1}\right)\left(1 - s_{1,2} z^{-1}\right) \\ a_2\left(1 - s_{2,1} z^{-1}\right)\left(1 - s_{2,2} z^{-1}\right)\end{bmatrix} \left[\left(1 - p_1 z^{-1}\right)\left(1 - p_2 z^{-1}\right)\right]^{-1}.
\end{aligned}$$

The system has a zero at z_0, if and only if $z_0 \neq p_1$, $z_0 \neq p_2$, and $z_0 = s_{1,i} = s_{2,k}$ for some i and k. The right coprimeness condition in (2.60) is equivalent to whether or not the two numerator and one denominator polynomials have common roots. Thus, $\{\tilde{\mathbf{N}}(z), \tilde{M}(z)\}$ is not right coprime, if and only if

$$\begin{aligned}
\mathbf{H}(z) &= \begin{bmatrix} a_1\left(1 - s_1 z^{-1}\right) \\ a_2\left(1 - s_2 z^{-1}\right)\end{bmatrix} \frac{\left(1 - s z^{-1}\right)}{\left(1 - s z^{-1}\right)\left(1 - p z^{-1}\right)} \\
&= \begin{bmatrix} a_1\left(1 - s_1 z^{-1}\right) \\ a_2\left(1 - s_2 z^{-1}\right)\end{bmatrix} \frac{1}{1 - p z^{-1}},
\end{aligned}$$

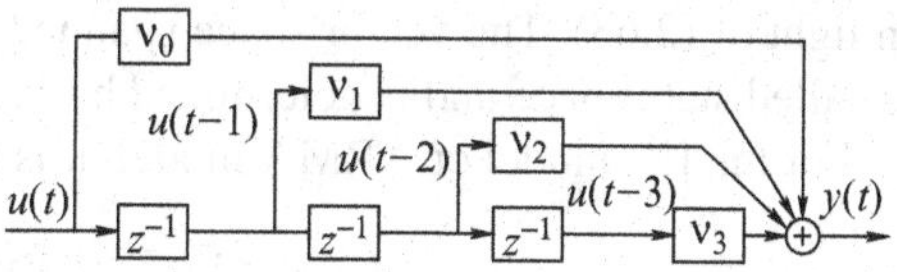

Fig. 2.2 Block diagram for FIR models of degree $n = 3$

for some s. If $\left(1 - sz^{-1}\right)$ is the GCD, then the last expression of the above equation provides a right coprime fraction, which has a zero if and only if $s_1 = s_2$. This procedure can be easily generalized to other single input systems. As a result, the procedure for obtaining the right coprime fractions for single-input systems is quite similar to that for SISO systems, which can be extended to compute the left coprime fractions for single output ($p = 1$) systems.

It is possible for $\mathbf{H}(z)$ to have common poles and zeros while its fractions are coprime. Generically, it is difficult to obtain coprime fractions for MIMO systems, and coprime fractions are not possible, if $M_0 = I_p$ and $\tilde{M}_0 = I_m$ are required. Consequently, it is considerably more difficult to minimize the computational complexity associated with the right-hand side of (2.57) than the case of SISO systems. For this and other reasons, state-space models are more preferred for MIMO systems to be discussed next.

State-Space Models

State-space models describe dynamic systems with *state variables*. Let FIR models of degree 3 be realized as in the following block diagram.

Then the input/output relation in Fig. 2.2 satisfies (2.55) for $n = n_v = 3$ and $\mu_k = 0 \; \forall \; k \geq 1$. A common practice in the state-space description is to take the output of each delay device as the state variable. Thus, for the SISO MA model, one may define $n = n_v$ state variables $\{x_k(t)\}_{k=1}^{n}$ via

$$\mathbf{x}(t) = \begin{bmatrix} x_1(t) \\ x_2(t) \\ \vdots \\ x_n(t) \end{bmatrix} = \begin{bmatrix} u(t-1) \\ u(t-2) \\ \vdots \\ u(t-n) \end{bmatrix} \implies \mathbf{x}(t+1) = \begin{bmatrix} u(t) \\ x_1(t) \\ \vdots \\ x_{n-1}(t) \end{bmatrix}. \tag{2.63}$$

Let $d = v_0$. Then the state-space equations

$$\mathbf{x}(t+1) = A\mathbf{x}(t) + \mathbf{b}u(t), \; y(t) = \mathbf{c}\mathbf{x}(t) + du(t) \tag{2.64}$$

hold, where $(A, \mathbf{b}, \mathbf{c}, d)$ is called a *realization* of the system, given by

$$A = \begin{bmatrix} \mathbf{0}_{n-1}^{*} & 0 \\ I_{n-1} & \mathbf{0}_{n-1} \end{bmatrix}, \; \mathbf{b} = \begin{bmatrix} 1 \\ \mathbf{0}_{n-1} \end{bmatrix}, \; \mathbf{c} = \begin{bmatrix} v_1 & \cdots & v_n \end{bmatrix}, \tag{2.65}$$

in light of (2.63). The vector space spanned by state vectors $\mathbf{x}(t)$ at different time t is called state space and is determined by the pair $(A, \mathbf{b})$.

For the IIR model or ARMA model, it is assumed that

$$H(z) = d + \frac{\tilde{v}_1 z^{-1} + \tilde{v}_2 z^{-2} + \cdots + \tilde{v}_n z^{-n}}{1 - \mu_1 z^{-1} - \cdots - \mu_n z^{-n}}, \tag{2.66}$$

where $n = \max\{n_v, n_\mu\}$. The conversion from (2.54) to (2.66) is always possible by zero-padding either the AR coefficients or MA coefficients. In this case, $H(z)$ admits a realization $(A, \mathbf{b}, \mathbf{c}, d)$ with

$$A = \begin{bmatrix} \mathbf{v}_{n-1} & \mu_n \\ I_{n-1} & \mathbf{0}_{n-1} \end{bmatrix}, \ \mathbf{b} = \begin{bmatrix} 1 \\ \mathbf{0}_{n-1} \end{bmatrix}, \ \mathbf{c} = \begin{bmatrix} \tilde{v}_1 & \cdots & \tilde{v}_n \end{bmatrix}, \tag{2.67}$$

where $\mathbf{v}_{n-1} = \begin{bmatrix} \mu_1 & \cdots & \mu_{n-1} \end{bmatrix}$. The above is termed *canonical controller form* or simply *controller form*. To verify that $(A, \mathbf{b}, \mathbf{c}, d)$ is indeed a realization for $H(z)$ in (2.66), denote $\{x_k(t)\}$ as the corresponding state variables. Then for $1 \le k < n$,

$$x_{k+1}(t+1) = x_k(t) \implies x_k(t) = x_1(t-k+1).$$

Hence, for $d = 0$, the expressions in (2.67) and (2.64) yield

$$x_1(t+1) = \sum_{k=1}^{n} \mu_k x_k(t) + u(t) = \sum_{k=1}^{n} \mu_k x_1(t-k+1) + u(t),$$

$$y(t) = \sum_{k=1}^{n} \tilde{v}_k x_k(t) = \sum_{k=1}^{n} \tilde{v}_k x_1(t-k+1).$$

Applying $\mathscr{Z}$ transform to the above with zero initial conditions yields

$$X_1(z) = \frac{z^{-1} U(z)}{1 - \mu_1 z^{-1} - \cdots - \mu_n z^{-n}},$$

$$Y(z) = \left(\tilde{v}_1 + \tilde{v}_2 z^{-1} + \cdots + \tilde{v}_n z^{-n+1}\right) X_1(z),$$

which verifies that the transfer function from $u(t)$ to $y(t)$ is indeed $H(z)$.

For MIMO FIR systems, a simple realization (A, B, C, D) is given by

$$\begin{aligned} A &= \begin{bmatrix} \mathbf{0}_{m\times(n-m)} & \mathbf{0}_{m\times m} \\ I_{(n-m)m} & \mathbf{0}_{(n-m)\times m} \end{bmatrix}, \quad B = \begin{bmatrix} I_m \\ \mathbf{0}_{(n-m)\times m} \end{bmatrix}, \\ C &= \begin{bmatrix} H_1 & H_2 & \cdots & H_\ell \end{bmatrix}, \qquad D = H_0, \ n = m\ell, \end{aligned} \tag{2.68}$$

which is generalized from (2.65) and termed block controller form. Extension of the above realization to include the IIR MIMO system in (2.56) is left as an exercise (Problem 2.23). Its state-space system is described by

$$\mathbf{x}(t+1) = \mathbf{A}\mathbf{x}(t) + \mathbf{B}\mathbf{u}(t), \ \mathbf{y}(t) = \mathbf{C}\mathbf{x}(t) + \mathbf{D}\mathbf{u}(t), \tag{2.69}$$

where A, B, C and D have appropriate dimensions. Applying $\mathscr{Z}$ transform to (2.69) with zero initial condition $\mathbf{x}(0) = \mathbf{0}_n$ yields the transfer matrix

$$\mathbf{H}(z) = D + C(zI_n - A)^{-1}B. \tag{2.70}$$

Its impulse response $\{H(t)\}$ is given by

$$H(0) = D,\ H(t) = CA^{t-1}B,\ t \geq 1. \tag{2.71}$$

State-space realizations are not unique. For the state-space equation (2.69), let the linear transform be $\mathbf{x}_T(t) = T\mathbf{x}(t)$ with T square and nonsingular. Then $\mathbf{x}(t) = T^{-1}\mathbf{x}_T(t)$, which upon substituted into (2.69), yields

$$\mathbf{x}_T(t+1) = TAT^{-1}\mathbf{x}_T(t) + TBu(t),\ y(t) = CT^{-1}\mathbf{x}_T(t) + Du(t). \tag{2.72}$$

Hence, a different realization $\left(TAT^{-1}, TB, CT^{-1}, D\right)$ is obtained for the same system. The transform in (2.72) is called *similarity transform*. Since T is an arbitrary nonsingular matrix, a system can have infinitely many different realizations. Moreover, realizations with different state dimensions may exist. Minimal realizations are preferred due to the obvious reason of complexity. The dimension of the state vector $\mathbf{x}(t)$ is called *order* of the state-space system. If the order n is minimum among all possible realizations for the same system, then (A, B, C, D) is called a *minimal* realization.

Let p_0 be a pole of $\mathbf{H}(z)$. Then it is an eigenvalue of A. The converse may not be true in general unless the realization is minimal. Let z_0 be a zero of $\mathbf{H}(z)$. Then

$$\text{rank}\left\{\begin{bmatrix} A - z_0 I_n & B \\ C & D \end{bmatrix}\right\} < n + \min\{p, m\}. \tag{2.73}$$

Again, the converse holds for only minimal realizations in general. The state-space system (2.69) is said to be internally stable, if all eigenvalues of A are strictly inside the unit circle. A formal definition for stability will be delayed to the next chapter. It is worth pointing out that the stability notion for state-space systems is stronger than the stability notion for ARMA models or transfer functions and matrices. The two coincide with each other when the state-space system has a minimal realization.

Example 2.10. The following transfer function

$$H(z) = \frac{-3z^{-1} + 6z^{-2}}{1 - 2z^{-1}} = \frac{-3z + 6}{z^2 - 2z} \tag{2.74}$$

admits a realization $(A, \mathbf{b}, \mathbf{c}, d)$ with

$$A = \begin{bmatrix} 2 & 0 \\ 1 & 0 \end{bmatrix},\ \mathbf{b} = \begin{bmatrix} 1 \\ 0 \end{bmatrix},\ \mathbf{c} = \begin{bmatrix} -3 & 6 \end{bmatrix},\ d = 0.$$

The state-space system is unstable as A has two eigenvalues with one at 2 and the other at 0. In absence of the input, the recursive computation yields

$$\mathbf{x}(t+1) = \begin{bmatrix} x_1(t+1) \\ x_2(t+1) \end{bmatrix} = A\mathbf{x}(t) = \begin{bmatrix} 2 \\ 1 \end{bmatrix} x_1(t) = \begin{bmatrix} 2^{t+1} \\ 2^t \end{bmatrix} x_1(0)$$

with $x_1(0)$ the first component of $\mathbf{x}(0)$. Hence, if $x_1(0) \neq 0$, each element of $\mathbf{x}(t)$ diverges as $t \to \infty$. On the other hand,

$$y(t) = \mathbf{c}\mathbf{x}(t) = \begin{bmatrix} -3 & 6 \end{bmatrix} \mathbf{x}(t) = -3 \times 2^t + 6 \times 2^{t-1} = 0.$$

So the unstable mode 2^t does not show up at the output. Alternatively, $H(z)$ in (2.74) admits a different realization with

$$A = \begin{bmatrix} 2 & 1 \\ 0 & 0 \end{bmatrix}, \ \mathbf{b} = \begin{bmatrix} -3 \\ 6 \end{bmatrix}, \ \mathbf{c} = \begin{bmatrix} 1 & 0 \end{bmatrix}, \ d = 0.$$

Again, A has eigenvalues at 2 and 0. Moreover, $x_1(t) = 2^t x_1(0) - 3u(t)$ and $y(t) = x_1(t)$ based on the recursive state-space equation. In this case, the unstable mode 2^t does show up at the output, but cannot be removed from both $x_1(t)$ and $x_2(t)$, i.e., stabilized by any bounded control input $\{u(t)\}$.

It is important to note $H(z) = -3z^{-1}$ after canceling the common factor $(z-2)$. Thus, the system is BIBO stable. The unstable eigenvalue at 2 is not a pole of $H(z)$. In fact, a minimal realization of $H(z)$ is $(A, \mathbf{b}, \mathbf{c}, d) = (0, 1, -3, 0)$, which is stable, coinciding with the stability of $H(z)$. This example illustrates a serious issue in realizations: It is possible for a system to be internally unstable while being externally or BIBO stable. Such realizations are harmful in the sense that unstable modes of the state-space system are either not detectable via the measured output or not stabilizable via the control input, which will be investigated thoroughly in the next chapter.

To summarize, the LTI models can be basically classified into two categories. The first one includes FIR or MA and IIR or ARMA models, which emphasizes input/output descriptions for dynamic systems. Its advantages lie in the simplicity and clear notions of poles, zeros, and stability. Such models are well studied for SISO systems. However, the coprime fractions for MIMO systems such as ARMA or IIR models are not easy to obtain, especially if $M_0 = I_p$ or $\tilde{M}_0 = I_m$ is required. The second category is the state-space models, which provide internal descriptions for dynamic systems in terms of state vectors. The dynamic behavior of the system is completely specified by the state variables and the input. Although more parameters are used, minimal realizations are always possible. Thus, poles, zeros, and stability can be characterized as well. More importantly, state-space models reveal internal structural information of the underlying systems and introduce new concepts and results for system design, which are especially suitable to MIMO systems. Hence, this text will focus on state-space models.

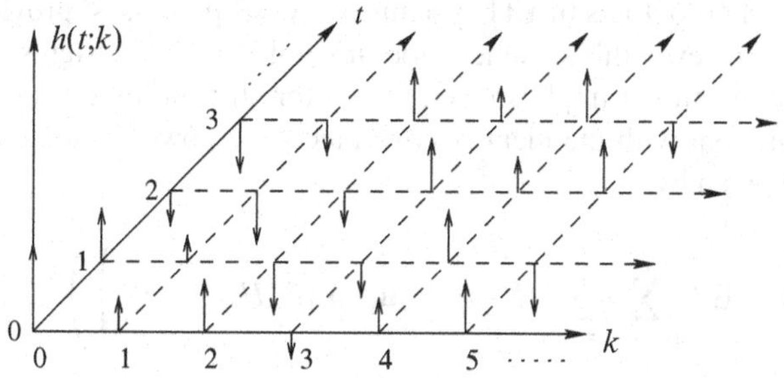

Fig. 2.3 Impulse response for SISO LTV systems

2.2.3 *Time-Varying Systems*

A LTV system can be viewed as a family of LTI systems parameterized by the time index t. As such, its impulse response is denoted by $\{h(t;k)\}$. Basically, $\{h(t_0;k)\}$ is an impulse response of the system at time t_0 with the impulse input applied at $k = 0$. An illustrative plot is shown in Fig. 2.3. At each integer-valued time t, $h(t;k)$ is shown horizontally from left to right. For MIMO systems, impulse responses are denoted by $\{H(t;k)\}$. Let $\{\mathbf{u}(t)\}$ and $\{\mathbf{y}(t)\}$ be the input and output, respectively. Then

$$\mathbf{y}(t) = H(t;k) \star \mathbf{u}(t) = \sum_{k=-\infty}^{\infty} H(t;t-k)\mathbf{u}(k) = \sum_{k=-\infty}^{\infty} H(t,k)\mathbf{u}(k). \tag{2.75}$$

In the control literature, $H(t,k) = H(t;t-k)$ is the standard notation. If the impulse responses are all the same at different time index t, then (2.75) becomes the same as in (2.41) for LTI systems (see Fig. 2.3).

For LTV systems, transfer functions or transfer matrices do not exist. As a consequence, notions of poles and zeros are lost, and frequency-domain analysis is inapplicable, which are negative. On the positive side, the BIBO stability condition can be derived in a similar way to that for LTI systems, as shown in the following result.

Theorem 2.11. *The LTV system with impulse response $\{H(t;k)\}$ is BIBO stable, if and only if*

$$\sum_{k=-\infty}^{\infty} \overline{\sigma}(H(t;k)) < \infty, \quad \forall\, t. \tag{2.76}$$

Basically, the stability condition in Theorem 2.11 treats the LTV impulse response as a "frozen time" LTI system indexed by time t. Thus, its proof is similar to that for LTI systems and is left as an exercise (Problem 2.26). While the stability condition for LTV systems is simple and resembles that for LTI systems, difficulties exist to apply it in practice due to the lack of analytic form of $\{H(t;k)\}$ and verification of (2.76) for each t. It needs to be pointed out that for causal LTV systems, $H(t;k) = \mathbf{0}$ for $k < 0$, which is assumed in the rest of this subsection.

As shown earlier, outputs of LTI systems are WSS processes, provided that the inputs are also. However, this statement does not hold for LTV systems, even though $\mathrm{E}\{\mathbf{u}(t)\} = \mathbf{0}$ for all time t implies $\mathrm{E}\{\mathbf{y}(t)\} = \mathbf{0}$ for all t, in light of (2.75). Indeed, for white noise input with the identity covariance, the power of the output is time dependent and given by

$$\begin{aligned} P_{\mathbf{y}}(t) &= \mathrm{Tr}\left(\mathrm{E}\left\{\sum_{k=-\infty}^{t}\sum_{i=-\infty}^{t} H(t;t-k)\mathbf{u}(k)\mathbf{u}(i)^* H(t;t-i)^*\right\}\right) \\ &= \mathrm{Tr}\left\{\sum_{k=-\infty}^{t} H(t;t-k)H(t;t-k)^*\right\} = \mathrm{Tr}\left\{\sum_{k=0}^{\infty} H(t;k)H(t;k)^*\right\}. \end{aligned} \tag{2.77}$$

Basically, $P_{\mathbf{y}}(t)$ quantifies the energy of the impulse response at time t. The above suggests that the system norm in (2.51) for LTI systems be generalized to LTV systems as

$$\|\mathbf{H}_t\|_2 = \sqrt{\mathrm{Tr}\left\{\sum_{k=0}^{\infty} H(t;k)H(t;k)^*\right\}} \tag{2.78}$$

which is time dependent.

Even though LTV systems are considerably more difficult to analyze, MA, ARMA, and state-space models are still effective for the class of systems emphasized in this text. Specifically, the ARMA model in (2.57) for MIMO systems can be adapted to

$$\mathbf{y}(t) = -\sum_{k=1}^{n_\mu} M_k(t)\mathbf{y}(t-k) + \sum_{k=0}^{n_\nu} N_k(t)\mathbf{u}(t-k), \tag{2.79}$$

where the AR and MA coefficient matrices are function of time t. If $M_k(t) = 0$ for $1 \le k \le n_\mu$ and all time t, then the above is collapsed to the MA model

$$\mathbf{y}(t) = \sum_{k=0}^{n_\nu} N_k(t)\mathbf{u}(t-k)$$

and $\{N_k(t)\}$ can be viewed as an impulse response of the LTV system at time t. That is, $\{H(t,k) = N_k(t)\}$ is parameterized by time index t. It is noted that MA models, time varying or not, are always stable.

For state-space descriptions, state-space models are adapted to

$$\mathbf{x}(t+1) = A_t\mathbf{x}(t) + B_t\mathbf{u}(t), \ \mathbf{y}(t) = C_t\mathbf{x}(t) + D_t\mathbf{u}(t), \tag{2.80}$$

where (A_t, B_t, C_t, D_t) can be viewed as a realization for the underlying MIMO system at time t. For LTV MA models, a realization with time-invariant A can be used. But for general LTV systems, a time-varying A_t needs to be assumed. Similarity transform can also be applied to obtain a new realization $\left(TA_tT^{-1}, TB_t, C_tT^{-1}, D_t\right)$

for the same system, where T is square and nonsingular. If a time-varying nonsingular matrix T_t is used as transform, then similarity $\left(T_{t+1}A_tT_t^{-1}, T_{t+1}B_t, C_tT_t^{-1}, D_t\right)$ is a new realization.

Different from LTI systems, a clear relation is lacking between the impulse response $\{H(t;k)\}$ and the realization (A_t, B_t, C_t, D_t). Hence, "frozen time" analysis as in Theorem 2.11 cannot be used to study stability for LTV state-space systems. In fact, the stability notion for LTI state-space models is generalized to the following.

Definition 2.3. The state-space system (2.80) is said to be exponentially stable, if there exist some α and β with $\alpha > 0$ and $0 < \beta < 1$ such that

$$\rho\left(A_{t+N}A_{t+N-1}\cdots A_{t+1}A_t\right) \leq \alpha\beta^N$$

for all time t and $N > 0$, where $\rho(\cdot)$ denotes the spectral radius.

In general, stability for each A_t, i.e., $\rho(A_t) < 1$ for each t, does not ensure stability of the state-space system (refer to Problem 2.27 in Exercises). It is worth to pointing out that if the state-space system is exponentially stable, then the state response to zero input with initial condition $\mathbf{x}(t_0) \neq \mathbf{0}_n$ is given by

$$\mathbf{x}(T) = \left(A_{t_0+T-1}A_{t_0+T-2}\cdots A_{t_0+1}A_{t_0}\right)\mathbf{x}(t_0) \to \mathbf{0}_n$$

for any $\mathbf{x}(t_0) \neq \mathbf{0}_n$, as $T \to \infty$. It is noted that in the case $A_t = A$ for all t, exponential stability reduces to the known condition that all eigenvalues of A are strictly inside the unit circle.

2.3 Noise Processes and BER Analysis

One of the impediments to data detection is the contamination of random noises at the receiver site. In most situations, observation noises can be assumed to be additive, white, and Gaussian noise (AWGN). Consider the signal model as illustrated below (see Fig. 2.4).

Let $\{\mathbf{v}(t)\}$ be AWGN. Then for each time index t, $\mathbf{v}(t)$ is a Gaussian random vector, i.e., $\mathbf{v}(t)$ is normal distributed. It is assumed that $\mathrm{E}[\mathbf{v}(t)] = \mathbf{0}$ for all t. The white assumption implies that the autocovariance matrix is given by

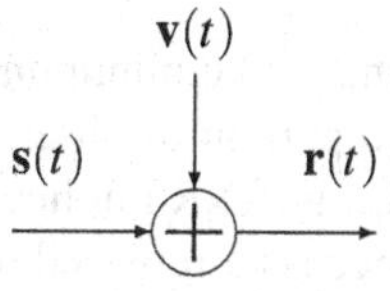

Fig. 2.4 Observed signal with contaminated noise

Fig. 2.5 PDF for received signal

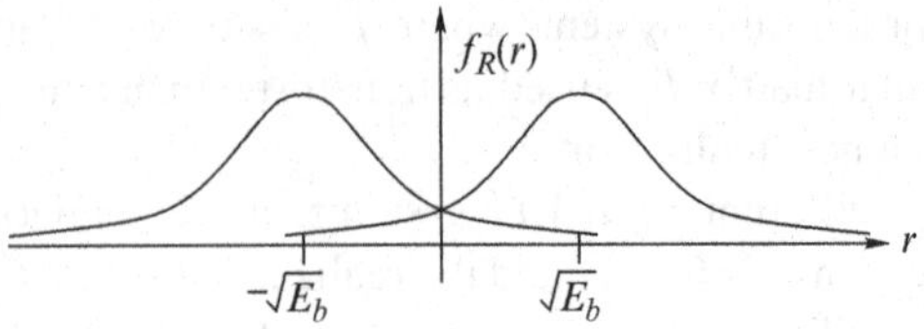

$$\mathrm{E}\{\mathbf{v}(t)\mathbf{v}(t-k)^*\} = R_\mathbf{v}(t)\delta(k) = \begin{cases} R_\mathbf{v}(t), & \text{if } k=0, \\ 0, & \text{if } k \neq 0. \end{cases} \tag{2.81}$$

If $R_\mathbf{v}(t) \equiv R_\mathbf{v}$ is a constant nonnegative matrix, then the AWGN $\{\mathbf{v}(t)\}$ is WSS. Otherwise, the AWGN is nonstationary.

For the simple case of scalar signals and noises, $s(t) = \pm\sqrt{E_\mathrm{b}}$ and $v(t)$ is a Gaussian random variable with zero mean and variance σ_v^2. That is, $s(t)$ carries only one bit of information which is either $+1$ or -1, and E_b is the bit energy of $s(t)$. The data detection problem aims to detect the sign of $s(t)$ based on the observed signal $r(t)$ at each time index t. Clearly, $r(t)$ is also a Gaussian random variable and has PDF

$$f_R(r) = \frac{1}{\sqrt{2\pi}\sigma_v}\exp\left\{-\frac{(r-s)^2}{2\sigma_v^2}\right\},$$

where the time index t is skipped due to the stationarity of $s(t)$ and $v(t)$. The figure below shows the PDFs of the received signal $r(t)$ for both $s(t) = \sqrt{E_\mathrm{b}}$ and $s(t) = -\sqrt{E_\mathrm{b}}$. Note that there is a symmetry about $r(t) = 0$ (see Fig. 2.5).

For the case of equal probable $s(t)$, i.e., $s(t)$ takes equal number of positive and negative values, a moment of reflection indicates that the optimal detection rule is

$$\check{s}(t) = \begin{cases} +1, & \text{if } r(t) > 0, \\ -1, & \text{if } r(t) < 0. \end{cases} \tag{2.82}$$

Indeed, by symmetry, the probability of the BER is given by

$$\begin{aligned} \varepsilon_\mathrm{b} &= \int_0^\infty \frac{1}{\sqrt{2\pi}\sigma_v}\exp\left\{-\frac{(r+\sqrt{E_\mathrm{b}})^2}{2\sigma_v^2}\right\}dr \\ &= \int_{\sqrt{E_\mathrm{b}/\sigma_v^2}}^\infty \frac{1}{\sqrt{2\pi}}\exp\left\{-\frac{r^2}{2}\right\}dr =: Q\left(\sqrt{E_\mathrm{b}/\sigma_v^2}\right) \end{aligned} \tag{2.83}$$

that is the minimum. The quantity E_b/σ_v^2 is called signal-to-noise ratio (SNR). It is important to observe that the BER performance is determined solely by the SNR. Large SNR implies small BER and vice versa. If $s(t)$ is taken to be random, then E_b needs be replaced by bit power $P_\mathrm{b} = \mathrm{E}\{|s(t)|^2\}$.

The case when $s(t)$ carries more than one bit information is not pursued in this text due to two reasons. First, any data can be represented by binary codes. There is no loss of generality in investigating the case of binary data. Second, generalization from binary data to the case of multiple bits does not involve new concepts and knowledge for data detection. Focusing on the binary case will help illuminate the basic issues and the essential difficulties and understand the approaches to optimal data detection.

For vector signals of size m, the noise $\mathbf{v}(t)$ is again assumed to be AWGN with mean zero and covariance $\Sigma_{\mathbf{v}}$. Suppose that $\Sigma_{\mathbf{v}}$ is nonsingular. Then the observed signal $\mathbf{r}(t)$ admits Gaussian distribution with PDF

$$f_R(\mathbf{r}) = \frac{1}{\sqrt{(2\pi)^m \det(\Sigma_{\mathbf{v}})}} \exp\left\{-\frac{1}{2}(\mathbf{r}-\mathbf{s})^* \Sigma_{\mathbf{v}}^{-1}(\mathbf{r}-\mathbf{s})\right\}. \tag{2.84}$$

Let $s_i(t)$ and $r_i(t)$ be the ith component of $\mathbf{s}(t)$ and $\mathbf{r}(t)$, respectively. For the equal probable case, the detection rule (2.82) can be adapted to

$$\check{s}_i(t) = \begin{cases} +1, & \text{if} \quad r_i(t) > 0, \\ -1, & \text{if} \quad r_i(t) < 0, \end{cases} \quad 1 \le i \le m. \tag{2.85}$$

Unfortunately, the above detection rule is not optimal anymore. The reason lies in the correlation of the noise components. For instance, the detected symbol, if correct, may help to detect other symbols. This problem will be studied in Chap. 7. Assume temporarily that $\Sigma_{\mathbf{v}}$ is diagonal. With $P_{\mathrm{s}}(i) = \mathrm{E}\left\{|s_i(t)|^2\right\}$ and $\sigma_{\mathbf{v}}^2(i) = \mathrm{E}\left\{|v_i(t)|^2\right\}$ (the ith diagonal element of $\Sigma_{\mathbf{v}}$), the corresponding BER is given by

$$\varepsilon_{\mathrm{b}}(i) = Q\left(P_{\mathrm{s}}(i)/\sigma_{\mathbf{v}}^2(i)\right), \; i = 1, 2, \ldots, m, \tag{2.86}$$

under the detection rule (2.85) where $Q(\cdot)$-function is defined as in (2.83). The average BER for detection of $\mathbf{s}(t)$ can be calculated according to

$$\overline{\varepsilon}_{\mathrm{b}} = \frac{1}{m}\sum_{i=1}^{m} Q\left(P_{\mathrm{s}}(i)/\sigma_{\mathbf{v}}^2(i)\right). \tag{2.87}$$

Gauss noise is a legitimate assumption in data communications, but the white assumption may not be, due to frequency-selective fading and the presence of the receiver. A common hypothesis is that the noise $\mathbf{v}(t)$, if colored, is generated by an LTI filter driven by a Gaussian white noise process $\mathbf{w}(t)$ of zero mean and identity covariance. That is, the PSD of $\mathbf{v}(t)$ is given by

$$\Psi_{\mathbf{v}}(\omega) = \mathbf{G}(e^{j\omega})\Psi_{\mathbf{w}}(\omega)\mathbf{G}(e^{j\omega})^*, \; \Psi_{\mathbf{w}}(\omega) \equiv I,$$

where $\mathbf{G}(z)$ can be assumed to be stable and minimum phase without loss of generality. It should be clear that the transmitted signal at the receiver site is also distorted, giving rise to the following signal model for data detection in Fig. 2.6.

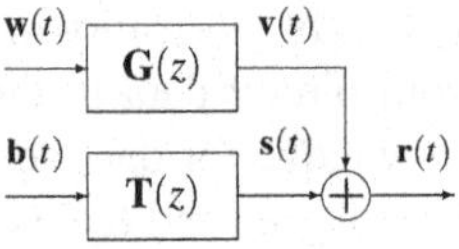

Fig. 2.6 Baseband signal model for data detection

This signal model is quite general in which $\mathbf{b}(t)$ is the original binary data signal at the transmitter, and $\mathbf{w}(t)$ is the AWGN with zero vector mean and identity covariance. The transfer matrices $\mathbf{T}(z)$ and $\mathbf{G}(z)$ are both causal, stable, and rational. Assume that $\mathbf{b}(t)$ and $\mathbf{s}(t)$ have the same size $m > 1$. Then

$$\mathbf{r}(t) = \mathbf{b}(t) + [T(t) - \delta(t)I_m] \star \mathbf{b}(t) + G(t) \star \mathbf{w}(t), \tag{2.88}$$

where $\{T(t)\}$ and $\{G(t)\}$ are impulse responses of $\mathbf{T}(z)$ and $\mathbf{G}(z)$, respectively. Even though $\mathbf{v}(t) = G(t) \star \mathbf{w}(t)$ can be treated as Gaussian distributed, the second term on the right-hand side of (2.88) does not have a normal distribution, in general.

Denote $D(t) = T(t) - \delta(t)I_m$ and $\mathbf{d}(t) = D(t) \star \mathbf{b}(t)$. Let $D_{i,\ell}(t)$ denote the (i,ℓ)th element of $D(t)$ and $d_i(t)$ the ith element of $\mathbf{d}(t)$. Then

$$d_i(t) = \sum_{k=-\infty}^{t} \sum_{\ell=1}^{m} D_{i,\ell}(k) b_\ell(t-k), \ 1 \le i \le m, \tag{2.89}$$

where $b_\ell(t)$ is the ℓth element of $\mathbf{b}(t)$, assumed to be equal probable and independent with respect to both ℓ and t. As such, one may conjecture that $\{d_i(t)\}$ is Gaussian distributed for each i by the *central limit theorem*. Unfortunately, it is not. The main reason is stability and rationality of $\mathbf{T}(z)$, two good properties as entailed for data communications, which imply the existence of $M > 0$ such that $|d_i(t)| \le M < \infty$ for all i and t by the boundedness of the input $\mathbf{b}(t)$. It follows that the support of PDF for $d_i(t)$ in (2.89) is finite precluding it from having normal distribution.

Although $\{\mathbf{d}(t)\}$ does not have a Gaussian distribution, it is close to being normal distributed, provided that impulse response $\{D(t)\}$ or equivalently $\{T(t)\}$ does not die out too quickly. Otherwise, m, the size of the data vector, needs to be adequately large. The next example illustrates this fact.

Example 2.12. Let Z be a random variable generated via

$$Z = \sum_{k=0}^{n-1} \rho^k Y_k, \ \rho = 0.8, \ n = 50,$$

where $\{Y_k\}$ is an i.i.d. sequence with an equal probability of 0.5 at ± 1. Clearly, Z has a zero mean and a variance

$$\mathrm{E}\left\{|Z|^2\right\} = \sum_{k=0}^{n-1} \rho^{2k} \mathrm{E}\left\{|Y_k|^2\right\} = \sum_{k=0}^{n-1} \rho^{2k} \le \frac{1}{1-\rho^2} = \frac{1}{0.36}.$$

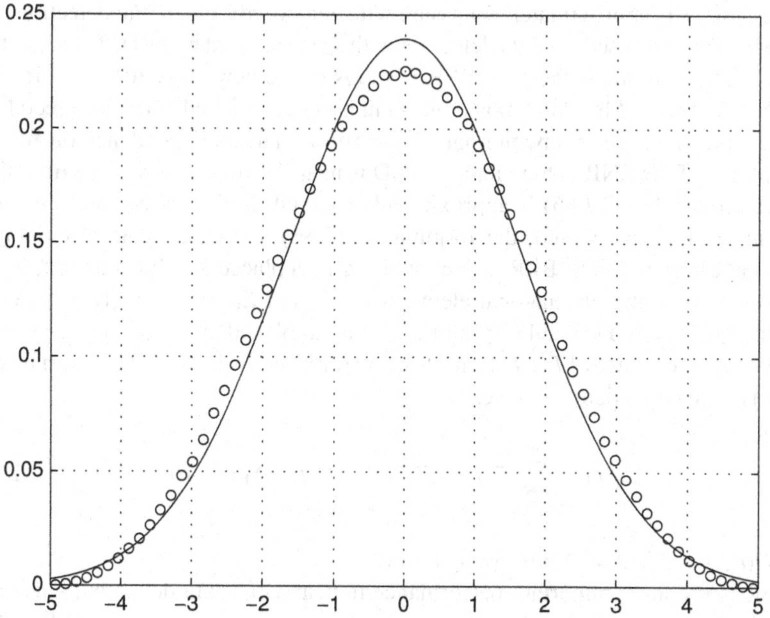

Fig. 2.7 Approximate PDF compared with normal distribution

One million samples of Z are obtained with Matlab, which produce an approximate PDF in Fig. 2.7, plotted with "o," based on the periodogram method. It can be observed that the curve is close to the Gaussian PDF with zero mean and variance $\frac{1}{0.36}$, plotted in solid line.

The aforementioned discussions are summarized next.

Proposition 2.1. *Suppose that* $\mathbf{G}(z)$ *and* $\mathbf{T}(z)$ *as in Fig. 2.6 are causal, stable, and rational. Let* $\{\mathbf{w}(t)\}$ *be AWGN of mean zero and covariance identity and* $\{\mathbf{b}(t)\}$ *be equal probable and independent. Denote* $\mathbf{D}(z) = \mathbf{T}(z) - I_m$ *and* $\Sigma_{\mathbf{b}} = \mathrm{E}\{\mathbf{b}(t)\mathbf{b}(t)^*\}$. *If the impulse response of* $\mathbf{D}(z)$ *does not die out too quickly or the size of the data vector is adequately large, then the observed signal* $\mathbf{r}(t)$ *in Fig. 2.6 consists of two parts: the transmitted data signal* $\mathbf{b}(t)$ *and a fictitious additive noise* $\{\mathbf{n}(t)\}$ *which has an approximate normal distribution with mean vector zero and covariance*

$$\Sigma_{\mathbf{n}} = \frac{1}{2\pi}\int_{-\pi}^{\pi}\left[\mathbf{G}\left(e^{j\omega}\right)\mathbf{G}\left(e^{j\omega}\right)^* + \mathbf{D}\left(e^{j\omega}\right)\Sigma_{\mathbf{b}}\mathbf{D}\left(e^{j\omega}\right)^*\right]\,d\omega. \tag{2.90}$$

In light of (2.88), the fictitious additive noise is given by

$$\mathbf{n}(t) = D(t)\star\mathbf{b}(t) + G(t)\star\mathbf{w}(t)$$

with $D(t) = T(t) - \delta(t)I_m$. Its covariance matrix $\Sigma_{\mathbf{n}}$ can be computed according to (2.90) by the fact that $\{\mathbf{b}(t)\}$ and $\{\mathbf{w}(t)\}$ are independent of each other. Even though $\{\mathbf{n}(t)\}$ is approximately Gaussian with zero vector mean, the detection rule in (2.85) cannot be used for data detection with $\hat{s}_i(t)$ replaced by $\hat{b}_i(t)$. There are two reasons. The first one is the poor SNR in terms of the new noise $\mathbf{n}(t)$, considering that $\mathbf{D}(z) = \mathbf{T}(z) - I$ has large power norm in absence of equalization or precoding. The second one is the nonwhite nature for $\{\mathbf{n}(t)\}$ and its dependence on $\mathbf{b}(t)$ in general. But if the SNR is high and the PSD is near flat for the new noise $\mathbf{n}(t)$, then the detection rule in (2.85) is approximately optimal. In this case, $\{\mathbf{n}(t)\}$ is close to being normal distributed, the formulas in (2.86) and (2.87) can be employed to estimate approximate BER values with $P_{\mathbf{s}}(i)$ replaced by $P_{\mathbf{b}}(i)$ and $\sigma_{\mathbf{v}}(i)$ by $\sigma_{\mathbf{n}}(i)$, which are the ith diagonal elements of $\Sigma_{\mathbf{b}}$ and $\Sigma_{\mathbf{n}}$, respectively. It is worth pointing out that, if $\mathbf{T}(z)$ and $\mathbf{G}(z)$ in Fig. 2.6 are replaced by time-varying systems with impulse responses $\{T(t,k)\}$ and $\{G(t,k)\}$, respectively, then the covariance in (2.90) is time dependent and given by:

$$\Sigma_{\mathbf{n}}(t) = \sum_{k=0}^{\infty} \left[G(t;k)G(t;k)^* + D(t;k)\Sigma_{\mathbf{b}}D(t;k)^*\right] \tag{2.91}$$

with $D(t,k) = T(t,k) - \delta(k)I_m$ in light of (2.77).

BER is the most important performance indicator for data detection, but it can be difficult to minimize in design of optimal receivers. It can also be difficult to compute, if the detection error does not have normal distribution. As shown earlier, the BER is hinged to the error variance, if the signal power is kept constant. For this reason, a closely related performance indicator, root-mean-squared error (RMSE), is often employed for data detection, which does not require the knowledge of distribution of the noise, provided that the PDF of the noise is symmetric about the origin. For the case in Proposition 2.1, the RMSE is simply $\varepsilon_{\mathrm{p}} = \sqrt{\mathrm{Tr}\{\Sigma_{\mathbf{n}}\}}$, i.e.,

$$\varepsilon_{\mathrm{p}} = \sqrt{\mathrm{Tr}\left\{\frac{1}{2\pi}\int_{-\pi}^{\pi}\left[\mathbf{G}\left(e^{j\omega}\right)\mathbf{G}\left(e^{j\omega}\right)^* + \mathbf{D}\left(e^{j\omega}\right)\Sigma_{\mathbf{b}}\mathbf{D}\left(e^{j\omega}\right)^*\right]\,d\omega\right\}}. \tag{2.92}$$

A receiver design algorithm that achieves the minimum RMSE is called minimum mean-squared-error (MMSE) algorithm. Although the RMSE performance is different from the BER performance, they are closely related. In the case of Gaussian processes, they are equivalent to each other in the sense that the detection rule remains the same. Data detection is often carried out after equalization or precoding which will be studied in Chap. 7.

Notes and References

Many books provide excellent coverage of signals and systems in the case of discrete-time. A sample of such textbooks are [7, 8, 69, 90, 100]. The BER analysis is based on textbooks for digital communications such as [92, 116].

Exercises

2.1. Prove (2.4) by assuming that $\{s(t)\}$ is deterministic and has finite energy.

2.2. Prove the Schwarz inequality (2.6).

2.3. Compute ESD for the discrete-time signal

$$s(t) = \mathrm{e}^{-\alpha|t|}\cos(\omega_0 t + \pi/2),\ \alpha > 0,\ \omega_0 \neq 0.$$

Compute energy of $\{s(t)\}$ in both time domain and frequency domain.

2.4. Let $\{r_s(k)\}$ be ACS of $\{s(t)\}$ as defined in (2.9). Show that for each integer k, $|r_s(k)| \leq r_s(0)$.

2.5. Verify the expression of the nth order Fejér's kernel in (2.18) and prove Lemma 2.1. (*Hint:* Note that for deterministic $\{s(t)\}$,

$$\Psi^{(n)}(\omega) = \frac{1}{n}\left|\sum_{t=0}^{n-1} s(t)\mathrm{e}^{-j\omega t}\right|^2$$

which is the same as $F_n(\omega)$ for $s(t) \equiv 1$).

2.6. Suppose that Θ is a uniformly distributed random variable over $[0,\ 2\pi]$. Show that

(a) $\mathrm{E}\{\cos^2(\Theta)\} = \mathrm{E}\{\sin^2(\Theta)\} = \frac{1}{2}$,
(b) $\mathrm{E}\{\cos(\Theta)\} = \mathrm{E}\{\sin(\Theta)\} = \mathrm{E}\{\cos(\Theta)\sin(\Theta)\} = 0$, and
(c) for $\mathbf{x}(t) = \left[\cos(\omega_0 t + \Theta)\ \sin(\omega_0 t + \Theta)\right]'$,

$$R_{\mathbf{x}}(k) = \mathrm{E}\{\mathbf{x}(t)\mathbf{x}(t-k)^*\} = \frac{1}{2}\begin{bmatrix} \cos(\omega_0 k) & -\sin(\omega_0 k) \\ \sin(\omega_0 k) & \cos(\omega_0 k) \end{bmatrix}.$$

2.7. Let $\{R_{\mathbf{s}}(k)\}$ be ACS of the WSS vector process $\{\mathbf{s}(t)\}$ as defined in (2.27). Show that for each integer k,

$$\text{(i) } R_{\mathbf{s}}(k)^* = R_{\mathbf{s}}(-k), \quad \text{(ii) } \mathrm{Tr}\{R_{\mathbf{s}}(0)\} \geq |\mathrm{Tr}\{R_{\mathbf{s}}(k)\}|.$$

(*Hint:* Note that $\mathrm{Tr}\{R_{\mathbf{s}}(0)\} = \mathrm{E}\{\mathbf{s}(t)^*\mathbf{s}(t)\} = \mathrm{E}\{\mathbf{s}(t-k)^*\mathbf{s}(t-k)\}$, and $|\mathrm{Tr}\{R_{\mathbf{s}}(k)\}| = |\mathrm{E}\{\mathbf{s}(t-k)^*\mathbf{s}(t)\}| = |\mathrm{E}\{\mathbf{s}(t)^*\mathbf{s}(t-k)\}|$, as well as

$$\mathrm{E}\left\{\begin{bmatrix} \mathbf{s}(t)^* \\ \mathbf{s}(t-k)^* \end{bmatrix}\begin{bmatrix} \mathbf{s}(t) & \mathbf{s}(t-k) \end{bmatrix}\right\} \geq \mathbf{0}$$

for each integer k).

2.8. For Example 2.3, set $\omega_0 = 0.25\pi$. Use Simulink to generate a set of $N = 2^{14}$ data samples for $\{\mathbf{s}(t)\}$ in (2.32):

1. Follow the estimation scheme outlined in Example 2.5 with $n = 2^8$ and $m = 2^6$ to estimate the PSD with comparison to the true PSD.
2. Consider the use of $\omega_0 = 0.25\pi + \pi/n$ with different values of (n,m), but with $N = nm = 2^{14}$ and the same data samples set. Compare the estimation results with that in (i).

The quantity π/n is called resolution in spectrum estimation which is the possible maximum error for the location of the spectrum lines.

2.9. Consider the system in Fig. 2.1. Let

$$r_{yu}(k) = \mathrm{E}\{y(t)u(t-k)^*\}$$

be the cross covariance sequence and $\Psi_{yu}(\omega)$ be the DTFT of $\{r_{yu}(k)\}$. Let $\Psi_u(\omega)$ and $\Psi_y(\omega)$ be the DTFT of ACS $\{r_u(k)\}$ and $\{r_y(k)\}$, respectively. Show that

$$\Psi_{yu}(\omega) = H\left(\mathrm{e}^{j\omega}\right)\Psi_u(\omega), \; \Psi_y(\omega) = \Psi_{yu}(\omega)H\left(\mathrm{e}^{j\omega}\right)^*.$$

Generalize the above to MIMO systems.

2.10. Let $\{y(t)\}_{t=0}^{n-1}$ and $\{u(t)\}_{t=0}^{n-1}$ be input and output measurement data. Approximate $r_{yu}(k)$ as in the previous problem by

$$\hat{r}_{yu}(k) = \begin{cases} \frac{1}{n}\sum_{t=k}^{n-1} y(t)u(t-k)^*, & k \geq 0, \\ \frac{1}{n}\sum_{t=0}^{n+k-1} y(t)u(t-k)^*, & k \leq 0. \end{cases}$$

Show that the DTFT of $\{\hat{r}_{yu}(k)\}$ is given by

$$\hat{\Psi}_{yu}(\omega) = \frac{1}{n}\left(\sum_{t=0}^{n-1} y(t)\mathrm{e}^{-jt\omega}\right)\left(\sum_{k=0}^{n-1} u(k)\mathrm{e}^{-jk\omega}\right)^*.$$

2.11. For the system in Fig. 2.1, propose an algorithm to estimate the system frequency response $\left|H(\mathrm{e}^{j\omega})\right|$ and $\angle H(\mathrm{e}^{j\omega})$. (*Hint:* Use the results in Problems 2.8, 2.9, and 2.10).

2.12. For the block diagram in Fig. 2.8, the block with L is an interpolator or upper sampler where $L > 1$ is an integer. The output of the interpolator is governed by

$$s(t) = \begin{cases} u(k), & \text{if } t = Lk, \\ 0, & \text{if } t \neq Lk. \end{cases}$$

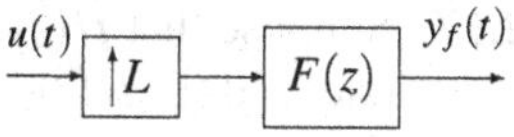

Fig. 2.8 Interpolator followed by filter

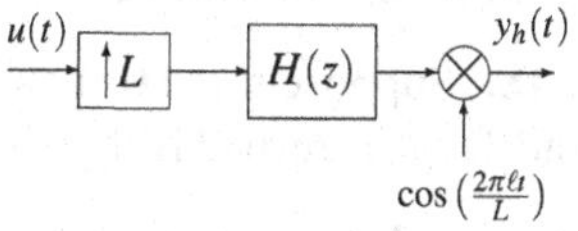

Fig. 2.9 Filtering followed by modulation

Show that (i) the system in Fig. 2.8 is linear but not time invariant, and (ii) $\Psi_s(\omega) = \Psi_u(L\omega)$ and $\Psi_{y_f}(\omega) = \left|F(e^{j\omega})\right|^2 \Psi_u(L\omega)$ where $\Psi(\cdot)$ is PSD.

2.13. Compute impulse responses for BIBO stable systems which admit the following transfer functions:

$$H_1(z) = \frac{2.5}{z^2 + 1.5z - 1}, \quad H_2(z) = \frac{z+1}{z^2 + 3.5z - 2}.$$

2.14. Let $\mathbf{H}(z)$ be a causal transfer function matrix of size $p \times p$. Let $\mathbf{D}(z) = \text{diag}\left(z^{-d_1}, z^{-d_2}, \ldots, z^{-d_p}\right)$ with $d_i \geq 0$ integers. Show that $\tilde{\mathbf{H}}(z) = \mathbf{D}(z)^{-1}\mathbf{H}(z)\mathbf{D}(z)$ may not be causal.

2.15. Consider the two systems in Figs. 2.8 and 2.9. Suppose that the impulse responses of $F(z)$ and $H(z)$ are $\{f(t)\}_{t=0}^n$ and $\{h(t)\}_{t=0}^n$, respectively, where $n > 1$, and ($\ell > 0$ is any integer)

$$f(t) = h(t)\cos\left(\frac{2\pi\ell t}{L}\right). \tag{2.93}$$

(i) Show that for $L > 1$, the two system block diagrams are equivalent, or $y_f(t) = y_h(t)$ for all time t. That is, filtering followed by cosine modulation has the same effect as filtering with cosine-modulated impulse response.
(ii) Show that, if we remove the interpolator, the two signal block diagrams are not equivalent or $y_f(t) \neq y_h(t)$ for at least some time t.

2.16. Consider again the systems in Figs. 2.8 and 2.9 where the impulse responses of $H(z)$ and $F(z)$ are $\{h(t)\}_{t=0}^n$ and $\{f(t)\}_{t=0}^n$, respectively, satisfying (2.93) and

$$\left|H\left(e^{j\omega}\right)\right| \approx \begin{cases} 1, & |\omega| \leq \pi/5, \\ 0, & \text{elsewhere.} \end{cases}$$

For $L = 5$, $\ell = 1$, and the input PSD

$$\Psi_u(\omega) = |\omega| \quad \text{for} \quad |\omega| \leq \pi,$$

give rough sketches for the output PSDs for $\{y_f(t)\}$ and $\{y_h(t)\}$.

2.17. Suppose that $H(z)$ in (2.39) is stable. Show that (i) it is analytic outside the unit circle, and (ii) $H(z)$ is continuous on the unit circle, i.e., $H(e^{j\omega})$ is a continuous function of ω. (*Hint:* A transfer function $H(z)$ is analytic at $z = z_0$, if it admits the (continuous) derivative at $z = z_0$).

2.18. Suppose that $\mathbf{H}(z)$ and $\mathbf{G}(z)$ are BIBO stable with impulse response $\{H(t)\}$ and $\{G(t)\}$, respectively. Show that

$$\frac{1}{2\pi}\int_{-\pi}^{\pi} \mathbf{H}\left(e^{j\omega}\right)\mathbf{G}\left(e^{j\omega}\right)^* \,d\omega = \sum_{t=-\infty}^{\infty} H(t)G(t)^*$$

and conclude (i) the Parseval's relation (2.51), and (ii) if $\mathbf{H}(z)$ is causal and $\mathbf{G}(z)$ is anticausal, then there holds the orthogonality relation

$$\frac{1}{2\pi}\int_{-\pi}^{\pi} \mathbf{H}\left(e^{j\omega}\right)\mathbf{G}\left(e^{j\omega}\right)^* \,d\omega = \mathbf{0}.$$

2.19. Suppose that $\mathbf{H}(z)$ is BIBO stable. Show that $\mathbf{H}(z) = \mathbf{H}_A(z) + \mathbf{H}_C(z)$ with $\mathbf{H}_A(z)$ anticausal, $\mathbf{H}_C(z)$ causal, and

$$\|\mathbf{H}\|_2^2 = \|\mathbf{H}_A\|_2^2 + \|\mathbf{H}_C\|_2^2.$$

2.20. Prove Theorem 2.7.

2.21. (i) If there exists a square polynomial matrix $\mathbf{R}(z)$ such that

$$\mathbf{M}(z) = \mathbf{R}(z)\mathbf{M}_c(z), \quad \mathbf{N}(z) = \mathbf{R}(z)\mathbf{N}_c(z),$$

where $\det(\mathbf{R}(z)) = 0$ for some $z \in \mathbb{C}$, show that $\{\mathbf{M}(z), \mathbf{N}(z)\}$ is not left coprime. If $\mathbf{R}(z)$ is the GCD, show that $\{\mathbf{M}_c(z), \mathbf{N}_c(z)\}$ is left coprime.

(ii) If there exists a square polynomial matrix $\tilde{\mathbf{R}}(z)$ such that

$$\tilde{\mathbf{M}}(z) = \tilde{\mathbf{M}}_c(z)\tilde{\mathbf{R}}(z), \quad \tilde{\mathbf{N}}(z) = \tilde{\mathbf{N}}_c(z)\tilde{\mathbf{R}}(z),$$

where $\det(\tilde{\mathbf{R}}(z)) = 0$ for some $z \in \mathbb{C}$, show that $\{\tilde{\mathbf{N}}(z), \tilde{\mathbf{M}}(z)\}$ is not right coprime. If $\tilde{\mathbf{R}}(z)$ is the GCD, show that $\{\tilde{\mathbf{N}}_c(z), \tilde{\mathbf{M}}_c(z)\}$ is right coprime.

2.22. Find the relation between the two IIR models in (2.54) and (2.66), and draw a similar block diagram to the one in Fig. 2.2 for the state-space realization in (2.67) with $n = 3$.

2.23. Extend the realization in (2.68) (canonical controller form) to cover the MIMO IIR model (2.58) by assuming that $\tilde{M}_0 = I_m$, $\tilde{N}_0 = \mathbf{0}$, and $\ell = \max\{\tilde{n}_\nu, \tilde{n}_\mu\}$.

2.24. Consider a 2×2 MIMO IIR model

$$\mathbf{H}(z) = \frac{z^{-1}}{3+2.5z^{-1}+0.5z^{-2}} \begin{bmatrix} 3(z^{-1}+2) & 6(1+0.5z^{-1}) \\ 3z^{-1} & 3(1+0.5z^{-1}) \end{bmatrix}.$$

(i) Find an ARMA description in the form of (2.57) and the corresponding IIR in the form of left fraction.
(ii) Show that a right fraction is given by

$$\mathbf{H}(z) = \begin{bmatrix} z^{-2}+2z^{-1} & 2z^{-1} \\ z^{-2} & z^{-1} \end{bmatrix} \begin{bmatrix} 1+\frac{2.5}{3}z^{-1}+\frac{1}{6}z^{-2} & 0 \\ 0 & 1+\frac{1}{3}z^{-1} \end{bmatrix}^{-1}$$

which is coprime.
(iii) Compute poles and zeros of $\mathbf{H}(z)$.
(iv) Show that with

$$A = \begin{bmatrix} -\frac{2.5}{3} & -\frac{1}{6} & 0 \\ 1 & 0 & 0 \\ 0 & 0 & -\frac{1}{3} \end{bmatrix}, \quad B = \begin{bmatrix} 1 & 0 \\ 0 & 0 \\ 0 & 1 \end{bmatrix}, \quad C = \begin{bmatrix} 2 & 1 & 2 \\ 1 & 0 & 1 \end{bmatrix},$$

and $D = \mathbf{0}_{2\times 2}$, (A,B,C,D) is a minimal realization.

2.25. (i) Find a right coprime fraction for

$$\mathbf{H}_1(z) = \begin{bmatrix} 1-z^{-2} \\ 1-3z^{-1}+2z^{-2} \end{bmatrix} \frac{2}{1-1.8z^{-1}+0.8z^{-2}}.$$

(ii) Find a left coprime fraction for

$$\mathbf{H}_2(z) = \frac{2}{1+0.4z^{-1}-0.6z^{-2}} \begin{bmatrix} 2+3z^{-1}+z^{-2} & 1-z^{-2} \end{bmatrix}.$$

(iii) Find minimal realizations for $\mathbf{H}_1(z)$ and $\mathbf{H}_2(z)$. (*Hint:* Use canonical controller form.)

2.26. (i) For an LTV system with impulse response $\{h(t;\tau)\}$, show that it is BIBO stable, if and only if

$$\sum_{\tau=-\infty}^{\infty} |h(t;\tau)| < \infty \quad \forall\, t.$$

(ii) Prove the similar result in (2.76) for MIMO systems.

2.27. Let $|a| > 1$ and $|b| < 1$. Consider state-space model

$$\mathbf{x}(t+1) = A_t\mathbf{x}(t), \quad A_t = \begin{cases} \begin{bmatrix} 0 & a \\ ba^{-1} & 0 \end{bmatrix}, & \text{if } t \text{ is even,} \\ \\ \begin{bmatrix} 0 & ba^{-1} \\ a & 0 \end{bmatrix}, & \text{if } t \text{ is odd.} \end{cases}$$

(a) Compute eigenvalues of A_t and verify that $\rho(A_t) < 1 \; \forall \, t$. (b) Show that

$$\mathbf{x}(t) = \begin{cases} \begin{bmatrix} (b/a)^{2k} & 0 \\ 0 & a^{2k} \end{bmatrix} \mathbf{x}(0), & \text{if } t = 2k, \\ \\ \begin{bmatrix} 0 & a^{2k+1} \\ (b/a)^{2k+1} & 0 \end{bmatrix} \mathbf{x}(0), & \text{if } t = 2k+1. \end{cases}$$

That is, $\|\mathbf{x}(t)\| \to \infty$ as $t \to \infty$, if the second element of $\mathbf{x}(0)$ is nonzero.

2.28. Consider the signal model as in Fig. 2.4, where the noise is AWGN with zero mean and variance σ_v^2. Suppose that the binary data source $\{s(t)\}$ is not equal probable and has probability distribution

$$P_S[s(t) = +1] = p > 0, \quad P_S[s(t) = -1] = 1 - p > 0,$$

and thus, $\mathrm{E}\{|s(t)|^2\} = 1$. Modify the detection rule in (2.85) as

$$\hat{s}(t) = \begin{cases} +1, & \text{if } r(t) > \rho, \\ -1, & \text{if } r(t) < \rho, \end{cases}$$

with ρ a threshold. Then the BER is a function of ρ. Show that

$$\rho = \rho_{\text{opt}} = \frac{\sigma_v^2}{2} \log_e \left(\frac{1-p}{p} \right)$$

minimizes the BER. It is noted that for equal probable case, $\rho_{\text{opt}} = 0$ which coincides with the detection rule (2.85). (*Hint:* Show that

$$\varepsilon_b = (1-p) P_{R|S}[r(t) > \rho \, | \, s(t) = -1] + p P_{R|S}[r(t) < \rho \, | \, s(t) = +1]$$

is a function of ρ. Find its expression and then compute its minimum).

Chapter 3
Linear System Theory

For SISO systems, input/output descriptions, such as FIR/IIR or MA/ARMA models, have been effective in modeling, analysis, and design of LTI systems. Coprime fractions and impulse responses are easy to obtain for given transfer functions. Poles and zeros determine not only stability but also performance of the system completely. Many analytical and empirical methods have been developed and are powerful design tools for LTI systems. Nonetheless, such design tools are not as effective for MIMO systems. For instance, two MIMO systems may have the same poles and zeros but behave entirely differently as shown in the following example.

Example 3.1. Consider two transfer matrices of size 2×2:

$$\mathbf{H}_1(z) = \begin{bmatrix} 2\frac{z-1}{z+r} & 0 \\ 0 & 2\frac{z+1}{z-r} \end{bmatrix}, \quad \text{and} \quad \mathbf{H}_2(z) = \begin{bmatrix} \frac{z+1}{z+r} & 0 \\ 0 & \frac{z-1}{z-r} \end{bmatrix}.$$

These two transfer matrices have identical poles $\{\pm r\}$ and zeros $\{\pm 1\}$. Due to the decoupling between the two inputs and the two outputs, each system basically represents two SISO systems. For $r \approx 1$ and $|r| < 1$, both transfer matrices are stable, and at $z = e^{j\omega}$,

$$\left|2\frac{z-1}{z+r}\right| = \frac{2|\sin(\frac{\omega}{2})|}{\sqrt{1+r^2+2r\cos(\omega)}}, \quad \left|\frac{z+1}{z+r}\right| \approx 1,$$

$$\left|2\frac{z+1}{z-r}\right| = \frac{2|\cos(\frac{\omega}{2})|}{\sqrt{1+r^2-2r\sin(\omega)}}, \quad \left|\frac{z-1}{z-r}\right| \approx 1.$$

It follows that the two diagonal transfer functions of $\mathbf{H}_1(z)$ are approximately high-pass and low-pass filters, respectively, while the two diagonal transfer functions of $\mathbf{H}_2(z)$ behave like allpass filters.

G. Gu, *Discrete-Time Linear Systems: Theory and Design with Applications*,
DOI 10.1007/978-1-4614-2281-5_3, © Springer Science+Business Media, LLC 2012

The above example is notwithstanding. Traditional approaches based on input and output descriptions have their limitations, in addition to the difficulty of computing coprime fractions as discussed in Sect. 2.2.2. What it lacks is the structural information of the system, which are unimportant to SISO systems but crucial to MIMO systems. Such structural information is more suitably described by internal descriptive models of state space.

Linear system theory is developed for state-space models and centered right at the intrinsic structural properties of linear systems in relation to their inputs and outputs. It examines basic issues such as realization, observation, and control, and studies system structures from inputs to state vectors and from state vectors to outputs via rigorous mathematical analysis. However, it is the conceptual notions of reachability, observability, and Lyapunov stability that are at the heart of the linear system theory and are the central theme of this chapter.

3.1 Realizations

Given a transfer function or transfer matrix, how to realize it with a digital circuit composed of delays, multipliers, and adders? More importantly, how to search for a realization which deploys the minimum number of delay devices? These realization issues are investigated in this section.

Example 3.2. Consider a realization of the transfer function

$$H(z) = \frac{\nu_1 z^{-1} + \nu_2 z^{-2} + \nu_3 z^{-3}}{1 - \mu_1 z^{-1} - \mu_2 z^{-2} - \mu_3 z^{-3}}, \tag{3.1}$$

which is shown in the following block diagram:

It can be verified that the transfer function for Fig. 3.1 is indeed the same as $H(z)$ in (3.1). Moreover

$$\begin{aligned} x_1(t+1) &= \mu_1 x_1(t) + x_2(t) + \nu_1 u(t), \\ x_2(t+1) &= \mu_2 x_1(t) + x_3(t) + \nu_2 u(t), \\ x_3(t+1) &= \mu_3 x_1(t) + \nu_3 u(t). \end{aligned}$$

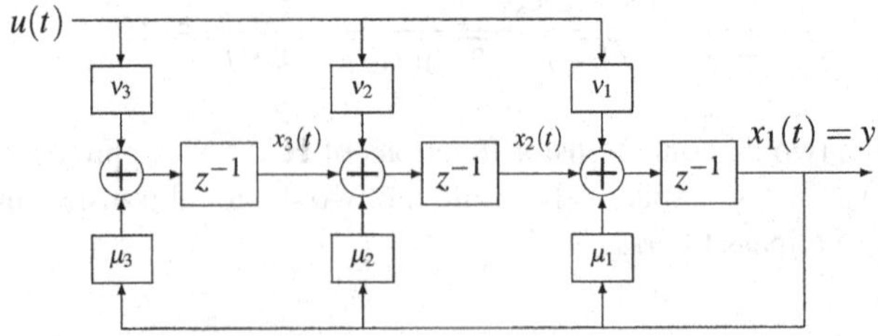

Fig. 3.1 Block diagram for implementation of $H(z)$

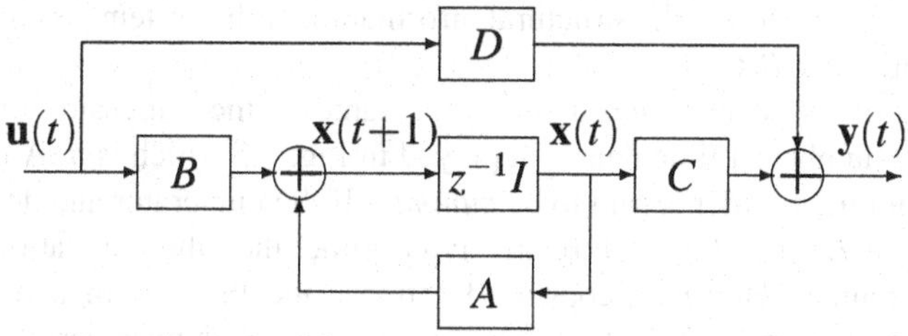

Fig. 3.2 Block diagram for state-space systems

Hence, with $\mathbf{x}(t) = \begin{bmatrix} x_1(t) & x_2(t) & x_3(t) \end{bmatrix}'$ where $'$ stands for transpose, the dynamic system in Fig. 3.1 satisfies the state-space equation

$$\mathbf{x}(t+1) = A\mathbf{x}(t) + \mathbf{b}u(t), \quad y(t) = \mathbf{c}\mathbf{x}(t) + du(t),$$

where $(A, \mathbf{b}, \mathbf{c}, d)$ is a realization of $H(z)$, given by $d = 0$ and

$$A = \begin{bmatrix} \mu_1 & 1 & 0 \\ \mu_2 & 0 & 1 \\ \mu_3 & 0 & 0 \end{bmatrix}, \quad \mathbf{b} = \begin{bmatrix} \nu_1 \\ \nu_2 \\ \nu_3 \end{bmatrix}, \quad \mathbf{c} = \begin{bmatrix} 1 \\ 0 \\ 0 \end{bmatrix}'. \tag{3.2}$$

The above can be extended to MIMO systems of size $p \times m$. Let

$$\mathbf{H}(z) = \left(I_p - M_1 z^{-1} - M_2 z^{-2} - M_3 z^{-3}\right)^{-1} \left(N_1 z^{-1} + N_2 z^{-2} + N_3 z^{-3}\right).$$

Then $\mathbf{H}(z)$ admits a realization (A, B, C, D) with $D = \mathbf{0}_{p \times m}$, and

$$A = \begin{bmatrix} M_1 & I_p & \mathbf{0}_{p\times p} \\ M_2 & \mathbf{0}_{p\times p} & I_p \\ M_3 & \mathbf{0}_{p\times p} & \mathbf{0}_{p\times p} \end{bmatrix}, \quad B = \begin{bmatrix} N_1 \\ N_2 \\ N_3 \end{bmatrix}, \quad C = \begin{bmatrix} I_p \\ 0 \\ 0 \end{bmatrix}'. \tag{3.3}$$

It can also be verified that $\mathbf{H}(z) = D + C(zI_n - A)^{-1}B$ with $n = 3p$.

In general, a rational transfer matrix admits a realization (A, B, C, D) and can be described by the state-space model

$$\mathbf{x}(t+1) = A\mathbf{x}(t) + B\mathbf{u}(t), \quad \mathbf{y}(t) = C\mathbf{x}(t) + D\mathbf{u}(t).$$

The following block diagram illustrates its implementation.

As seen in Fig. 3.2, the matrix D represents a direct transmission from the input to the output. The matrix B maps the input to the state-space spanned by state vectors, the matrix C maps the space space to the output, and the matrix A is a map from the state-space to itself. Hence, the structural information of the system is fully captured by its realization (A,B,C,D).

The order of the state-space model is the same as the dimension of the state vector, or the number of time delays deployed in Fig. 3.2 which is n. A realization having the minimum order is said to be *minimal*. If the numerator and denominator polynomials of $H(z)$ in (3.1) are relatively coprime, then the realization as given in (3.2) is minimal. However, coprime fractions, and thus minimal realizations, are difficult to obtain for MIMO systems in general. Redundancies are likely to exist for the corresponding realizations. It turns out that such redundancies are associated with either (C,A) in terms of the observation of the state vector at the output or (A,B) in terms reaching the desired state vector from the input which can be removed via numerically efficient algorithms. The notions of observability and reachability play the pivotal role and help to deepen the understanding of the structural properties of linear systems as studied in this section.

3.1.1 Observability

Consider the unforced state-space system described by

$$\mathbf{x}(t+1) = A\mathbf{x}(t), \quad \mathbf{y}(t) = C\mathbf{x}(t), \quad \mathbf{x}(0) = \mathbf{x}_0. \tag{3.4}$$

An important problem for the unforced system (3.4) is the reconstruction of the initial state vector $\mathbf{x}_0$ based on the output measurements for $t \geq 0$, which gives rise of the notion of *observability* as defined next.

Definition 3.1. The system (3.4), or simply (C,A), is observable, if given true output data $\{\mathbf{y}(t)\}_{t=0}^{\ell-1}$, the initial state $\mathbf{x}_0$ can be reconstructed uniquely for some $\ell > 0$.

The following result is well known in the linear system theory.

Theorem 3.3. *The pair (C,A) is observable, if and only if*

$$\text{rank}\{\mathscr{O}_n(C,A)\} = n, \quad \mathscr{O}_n(C,A) = \begin{bmatrix} C \\ CA \\ \vdots \\ CA^{n-1} \end{bmatrix},$$

where n is the dimension of the state vector $\mathbf{x}(t)$.

Proof. For simplicity, $\mathscr{O}_n(C,A)$ is denoted by $\mathscr{O}_n$. Direct computation gives

$$\mathbf{y}(0) = C\mathbf{x}_0, \quad \mathbf{y}(1) = CA\mathbf{x}_0, \quad \ldots, \quad \mathbf{y}(n-1) = CA^{n-1}\mathbf{x}_0.$$

Let vec(M) be the column vector by stacking columns of M sequentially. Then

$$\mathscr{Y}_n = \text{vec}\left(\begin{bmatrix} \mathbf{y}(0) & \mathbf{y}(1) & \cdots & \mathbf{y}(n-1) \end{bmatrix}\right) = \mathscr{O}_n x_0. \tag{3.5}$$

Let the dimension of the output be p. Then $\mathscr{O}_n$ has size $pn \times n$. The full rank condition for $\mathscr{O}_n$ and the hypothesis on $\{\mathbf{y}(t)\}$ imply that there is a unique solution $\mathbf{x}_0$ to (3.5) from which the observability follows. Conversely, assume that (C,A) is observable. Then $\mathscr{Y}_\ell = \mathscr{O}_\ell \mathbf{x}_0$ has a unique solution $\mathbf{x}_0$ for some $\ell > 0$, implying that $\mathscr{O}_\ell$ has the full column rank. In light of the Cayley Hamilton Theorem (refer to Appendix A), A^k is a linear combination of $\{A^i\}_{i=0}^{n-1}$ for any $k \geq n$, i.e.,

$$A^k = \sum_{i=0}^{n-1} \alpha_i A^i, \quad k \geq n,$$

where $\{\alpha_i\}$ are not identically zero. Thus, each row of CA^k is in the row space spanned by row vectors of CA^i for $0 \leq i < n$. Consequently,

$$\text{rank}\{\mathscr{O}_\ell\} = \text{rank}\{\mathscr{O}_n\} = n$$

for all $\ell \geq n$, which completes the proof. □

The matrix $\mathscr{O}_\ell$ is called the *observability matrix* of size ℓ. It is noted that there is no loss of generality in studying the observability for unforced system (3.4). Specifically for the general state-space system

$$\mathbf{x}(t+1) = A\mathbf{x}(t) + B\mathbf{u}(t), \quad \mathbf{y}(t) = C\mathbf{x}(t) + D\mathbf{u}(t) \tag{3.6}$$

with $\mathbf{x}(0) = \mathbf{x}_0$, the observability can be modified to reconstructability of the initial state $\mathbf{x}_0$, based on the true input/output data $\{\mathbf{u}(t)\}_{t=0}^{\ell-1}$ and $\{\mathbf{y}(t)\}_{t=0}^{\ell-1}$ for some $\ell > 0$. In this case, (3.5) needs to be replaced by

$$\mathscr{Y}_\ell = \mathscr{O}_\ell \mathbf{x}_0 + \mathscr{T}_\ell \mathscr{U}_\ell, \tag{3.7}$$

where $\ell > 0$ is a given integer and

$$\mathscr{T}_\ell = \begin{bmatrix} D & \mathbf{0} & \cdots & \mathbf{0} \\ CA^0B & D & \ddots & \vdots \\ \vdots & \ddots & \ddots & \mathbf{0} \\ CA^{\ell-2}B & \cdots & CA^0B & D \end{bmatrix}, \quad \mathscr{U}_\ell = \begin{bmatrix} \mathbf{u}(0) \\ \mathbf{u}(1) \\ \vdots \\ \mathbf{u}(\ell-1) \end{bmatrix}. \tag{3.8}$$

The matrix $\mathscr{T}_\ell$ is lower block triangular, called *Toeplitz matrix*, and is uniquely specified by its first block column. Hence, the uniqueness of the solution $\mathbf{x}_0$ is tied to the full rank condition of $\mathscr{O}_\ell$ for all $\ell \geq n$, which is equivalent to the full rank condition of $\mathscr{O}_n$ by again the Cayley Hamilton Theorem. The existence of the solution to (3.7) or (3.5) should be clear, because the measurement noises are assumed to be zero, and $\{\mathbf{u}(t)\}_{t=0}^{\ell-1}$ and $\{\mathbf{y}(t)\}_{t=0}^{\ell-1}$ are the true input and output data, respectively.

Example 3.4. A simple state-space realization for the SISO IIR model

$$H(z) = d + \frac{\nu_1 z^{-1} + \nu_2 z^{-2} + \cdots + \nu_n z^{-n}}{1 - \mu_1 z^{-1} - \mu_2 z^{-2} - \cdots - \mu_n z^{-n}} \tag{3.9}$$

is $(A_o, \mathbf{b}_o, \mathbf{c}_o, d_o)$ generalized from (3.2) and given by

$$A_o = \begin{bmatrix} \mathbf{v}_{n-1} & I_{n-1} \\ \mu_n & \mathbf{0}_{n-1}^* \end{bmatrix}, \quad \mathbf{v}_{n-1} = \begin{bmatrix} \mu_1 \\ \vdots \\ \mu_{n-1} \end{bmatrix}, \quad \mathbf{b}_o = \begin{bmatrix} \nu_1 \\ \vdots \\ \nu_n \end{bmatrix},$$

$$\mathbf{c}_o = \begin{bmatrix} 1 & 0 & \cdots & 0 \end{bmatrix}, \quad d_o = d.$$

A direct computation shows that the observability matrix is

$$\mathscr{O}_n = \begin{bmatrix} 1 & 0 & \cdots & 0 \\ * & 1 & \ddots & \vdots \\ \vdots & \ddots & \ddots & 0 \\ * & \cdots & * & 1 \end{bmatrix}$$

and nonsingular where the elements marked with $*$ are not relevant to the observability. Thus, the state-space system with realization $(A_o, \mathbf{b}_o, \mathbf{c}_o, d_o)$ is always observable even if the numerator and denominator polynomials are not relative coprime. For this reason, realization $(A_o, \mathbf{b}_o, \mathbf{c}_o, d_o)$ is called canonical *observer form*, or simply observer form. For the $p \times m$ transfer matrix

$$\mathbf{H}(z) = D + \left(I_p - \sum_{k=1}^{\ell} M_k z^{-k} \right)^{-1} \left(\sum_{k=1}^{\ell} N_k z^{-k} \right), \tag{3.10}$$

a realization (A_o, B_o, C_o, D_o) in block observer form can be obtained with

$$A_o = \begin{bmatrix} \mathbf{V}_{\ell-1} & I_{(\ell-1)p} \\ M_\ell & \mathbf{0}_{\ell \times (\ell-1)p} \end{bmatrix}, \quad \mathbf{V}_{\ell-1} = \begin{bmatrix} M_1 \\ \vdots \\ M_{\ell-1} \end{bmatrix}, \quad B_o = \begin{bmatrix} N_1 \\ \vdots \\ N_\ell \end{bmatrix},$$

$$C_o = \begin{bmatrix} I_p & \mathbf{0}_{p \times (\ell-1)p} \end{bmatrix}, \quad D_o = D.$$

The above is a generalization of the realization in (3.3). It can be verified that the observability matrix $\mathcal{O}_n$ is again lower triangular with 1 on the diagonal and $n = p\ell$. Hence, the state-space system with realization (A_o, B_o, C_o, D_o) is always observable as well.

Example 3.4 shows that any rational transfer function or matrix admits a state-space realization. In fact, observable realizations can always be obtained even if the fractions are not coprime. On the other hand, for a given transfer matrix $\mathbf{H}(z)$, not all its realizations are observable. If the given realization is unobservable, is it possible to obtain a different realization for the same transfer matrix which is observable? This question is answered by *Kalman decomposition* as given next.

Theorem 3.5. *Let (A,B,C,D) be a realization of $\mathbf{H}(z)$ with size $n \times n$ for A and $p \times m$ for $\mathbf{H}(z)$. Suppose that (C,A) is unobservable. Then there exists a similarity transform T such that*

$$\text{(i) } \tilde{A} = TAT^{-1} = \begin{bmatrix} \tilde{A}_o & \mathbf{0}_{r\times(n-r)} \\ \tilde{A}_{o\bar{o}} & \tilde{A}_{\bar{o}} \end{bmatrix}, \quad \tilde{B} = TB = \begin{bmatrix} \tilde{B}_o \\ \tilde{B}_{\bar{o}} \end{bmatrix},$$

$$\text{(ii) } \tilde{C} = CT^{-1} = \begin{bmatrix} \tilde{C}_o & \mathbf{0}_{p\times(n-r)} \end{bmatrix}, \quad \tilde{D} = D,$$

where $(\tilde{C}_o, \tilde{A}_o)$ is observable with $r \times r$ the dimension of $\tilde{A}_o$. Moreover

$$\mathbf{H}(z) = D + C(zI_n - A)^{-1}B = \tilde{D} + \tilde{C}_o\left(zI_r - \tilde{A}_o\right)^{-1}\tilde{B}_o. \tag{3.11}$$

Proof. Since (C,A) is unobservable, the observability matrix $\mathcal{O}_n$ has rank $r < n$. By singular value decomposition (SVD), $\mathcal{O}_n = US_nV^*$ where U and V are unitary matrices and S_n of size $pn \times n$ has nonzero singular values $\{\sigma_k\}_{k=1}^r$ on its diagonal with the rest elements zero. Hence, $\mathcal{O}_nV = \begin{bmatrix} \tilde{\mathcal{O}}_{n_r} & \mathbf{0} \end{bmatrix}$, and $\tilde{\mathcal{O}}_{n_r}$ has r columns with r the rank of $\mathcal{O}_n$. Let $T = V^*$. Then with $\tilde{A} = TAT^{-1}$ and $\tilde{C} = CT^{-1}$,

$$\mathcal{O}_nV = \mathcal{O}_nT^{-1} = \begin{bmatrix} CT^{-1} \\ CAT^{-1} \\ \vdots \\ CA^{n-1}T^{-1} \end{bmatrix} = \begin{bmatrix} \tilde{C} \\ \tilde{C}\tilde{A} \\ \vdots \\ \tilde{C}\tilde{A}^{n-1} \end{bmatrix} = \begin{bmatrix} \tilde{\mathcal{O}}_{n_r} & \mathbf{0} \end{bmatrix}.$$

Therefore, the expression of $\tilde{C}$ in (ii) follows. Partition $\tilde{A}$ into a 2×2 block matrix compatibly with $\{\tilde{A}_{i,k}\}$, the block at the (i,k)th position for $i,k = 1,2$, and $\tilde{A}_{1,1}$ of size $r \times r$. Then by induction, there holds

$$\mathcal{O}_nT^{-1} = \begin{bmatrix} \tilde{\mathcal{O}}_{n_r} & \tilde{\mathcal{O}}_{n_r}\tilde{A}_{12} \end{bmatrix} = \begin{bmatrix} \tilde{\mathcal{O}}_{n_r} & \mathbf{0} \end{bmatrix},$$

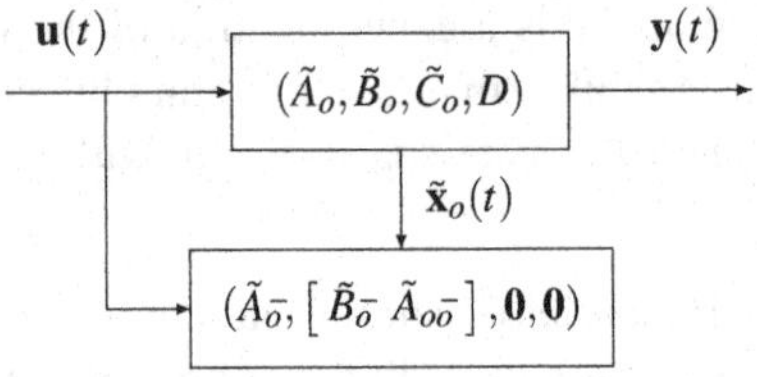

Fig. 3.3 Observability Kalman decomposition

where $\tilde{\mathscr{O}}_{n_r}$ is the observability matrix of $(\tilde{C}_o, \tilde{A}_{1,1})$ of size $np \times r$. Since $\tilde{\mathscr{O}}_{n_r}$ has full column rank, $\tilde{A}_{12} = \mathbf{0}_{r\times(n-r)}$. The expression of $\tilde{A}$ in (i) follows by taking $\tilde{A}_o = \tilde{A}_{1,1}$, $\tilde{A}_{o\bar{o}} = \tilde{A}_{2,1}$, and $\tilde{A}_{\bar{o}} = \tilde{A}_{2,2}$. The full column rank condition for $\tilde{\mathscr{O}}_{n_r}$ implies the observability of $(\tilde{C}_o, \tilde{A}_o)$. The expression for the transfer matrix in (3.11) can then be verified via straightforward computations. □

Let $\mathbf{x}(t)$ be the state vector associated with realization (A,B,C,D). Then Theorem 3.5 illustrates that the linear transform

$$\tilde{\mathbf{x}}(t) = T\mathbf{x}(t) = \begin{bmatrix} \tilde{\mathbf{x}}_o(t)^* & \tilde{\mathbf{x}}_{\bar{o}}(t)^* \end{bmatrix}^*$$

decomposes the state vector into two parts: One is $\tilde{\mathbf{x}}_o(t)$ with dimension r which is observable, and the other is $\tilde{\mathbf{x}}_{\bar{o}}(t)$ with dimension $(n-r)$ which is unobservable. Such a decomposition via similarity transforms is called Kalman decomposition as illustrated in Fig. 3.3. As it is seen, the subsystem $(\tilde{A}_o, \tilde{B}_o, \tilde{C}_o, D)$ is observable from the output $\mathbf{y}(t)$, while the subsystem with inputs $\mathbf{u}(t)$ and $\tilde{\mathbf{x}}_o(t)$ and with realization $(\tilde{A}_{\bar{o}}, [\tilde{B}_{\bar{o}}\ \tilde{A}_{o\bar{o}}], \mathbf{0}, \mathbf{0})$ is unobservable. It is now clear why the expression for the transfer matrix in (3.11) is true. Kalman decomposition offers a procedure to eliminate the unobservable states, or unobservable subsystem of the state-space model.

It is noted that testing of observability via the rank condition of $\mathscr{O}_n$ suffers a numerical problem, if n is large. The following PBH test is more convenient.

Theorem 3.6. (PBH test) *The state-space system (3.6) or simply (C,A) is unobservable, if and only if there exists a vector $\mathbf{v} \neq \mathbf{0}$ such that*

$$A\mathbf{v} = \lambda\mathbf{v}, \quad C\mathbf{v} = \mathbf{0}. \tag{3.12}$$

Proof. If there exists $\mathbf{v} \neq \mathbf{0}$ such that (3.12) holds, then for any $k > 0$,

$$CA^k\mathbf{v} = \lambda CA^{k-1}\mathbf{v} = \lambda^2 CA^{k-2}\mathbf{v} = \cdots = \lambda^k C\mathbf{v} = \mathbf{0}.$$

As a consequence, $\mathscr{O}_n\mathbf{v} = \mathbf{0}$ and (C,A) is unobservable. Conversely, if (C,A) is unobservable, then Kalman decomposition in Theorem 3.5 can be applied to obtain a new realization $(\tilde{A}, \tilde{B}, \tilde{C}, \tilde{D})$ in (i) and (ii) with $(\tilde{C}_o, \tilde{A}_o)$ observable. Set $\tilde{\mathbf{v}} = \begin{bmatrix} \mathbf{0}^* & \tilde{\mathbf{v}}_{\bar{o}}^* \end{bmatrix}^*$ with $\tilde{\mathbf{v}}_{\bar{o}}$ an eigenvector of $\tilde{A}_{\bar{o}}$. Then

$$\tilde{C}\tilde{\mathbf{v}} = 0, \quad \tilde{A}\tilde{\mathbf{v}} = \begin{bmatrix} \tilde{A}_{\mathrm{o}} & 0 \\ \tilde{A}_{\mathrm{o}\bar{\mathrm{o}}} & \tilde{A}_{\bar{\mathrm{o}}} \end{bmatrix} \begin{bmatrix} \mathbf{0} \\ \tilde{\mathbf{v}}_{\bar{\mathrm{o}}} \end{bmatrix} = \lambda \begin{bmatrix} 0 \\ \tilde{\mathbf{v}}_{\bar{\mathrm{o}}} \end{bmatrix} = \lambda \tilde{\mathbf{v}}.$$

Thus, (3.12) holds by taking $\mathbf{v} = T^{-1}\tilde{\mathbf{v}}$. The theorem is true. □

The PBH test converts the observability test into an equivalent eigenvalue and eigenvector problem for which effective algorithms exist. Furthermore, it offers a powerful tool for theoretical analysis because A and C appear linearly in the PBH test. If (3.12) holds for some $\mathbf{v} \neq \mathbf{0}$, then λ is often loosely called an unobservable eigenvalue of (C,A). Assume that all eigenvalues of A are distinct. The subspace spanned by those eigenvectors satisfying (3.12) is termed unobservable subspace. If matrix A has multiple eigenvalues, then the unobservable subspace may contain generalized eigenvectors.

3.1.2 *Reachability*

The notion of reachability is dual to that of observability. However, its physical meanings are rather different.

Definition 3.2. The pair (A,B), or the state-space system

$$\mathbf{x}(t+1) = A\mathbf{x}(t) + B\mathbf{u}(t), \quad \mathbf{x}(0) = \mathbf{x}_0, \tag{3.13}$$

is reachable, if given any desired state target $\mathbf{x}_T$, there exists a bounded control input $\{\mathbf{u}(t)\}_{t=0}^{\ell-1}$ such that $\mathbf{x}(\ell) = \mathbf{x}_T$ for some $\ell > 0$.

The following result holds for which the proof is left as an exercise.

Theorem 3.7. *The pair (A,B) is reachable, if and only if*

$$\operatorname{rank}\{\mathscr{R}_n(A,B)\} = n, \quad \mathscr{R}_n(A,B) = \begin{bmatrix} B & AB & \cdots & A^{n-1}B \end{bmatrix},$$

where n is the dimension of the state vector $\mathbf{x}(t)$.

For simplicity, $\mathscr{R}_n(A,B)$ is denoted by $\mathscr{R}_n$, which is called *reachability matrix* of size n. If the input $\mathbf{u}(t)$ has size m, then $\mathscr{R}_n$ has dimension $n \times nm$. The result in Theorem 3.7 shows that (A,B) is reachable, if and only if (B',A') or (B^*,A^*) is observable. Hence, notions of reachability and observability are dual to each other.

Example 3.8. Consider the SISO transfer function $H(z)$ in (3.9). Let

$$A_{\mathrm{c}} = \begin{bmatrix} \tilde{\mathbf{v}}_{n-1} & \mu_n \\ I_{n-1} & \mathbf{0}_{n-1} \end{bmatrix}, \quad \mathbf{b}_{\mathrm{c}} = \begin{bmatrix} 1 \\ \mathbf{0}_{n-1} \end{bmatrix}, \quad d_{\mathrm{c}} = d$$

$$\mathbf{c}_{\mathrm{c}} = \begin{bmatrix} v_1 & \cdots & v_n \end{bmatrix}, \quad \tilde{\mathbf{v}}_{n-1} = \begin{bmatrix} \mu_1 & \cdots & \mu_{n-1} \end{bmatrix}.$$

Then $(A_c, \mathbf{b}_c, \mathbf{c}_c, d_c)$ is dual to $(A_o, \mathbf{b}_o, \mathbf{c}_o, d_o)$ in the sense that

$$A_c = A_o', \quad \mathbf{b}_c = \mathbf{c}_o', \quad \mathbf{c}_c = \mathbf{b}_o', \quad d_c = d_o.$$

It is recognized that $(A_c, \mathbf{b}_c, \mathbf{c}_c, d_c)$ is the canonical controller form or simply controller form. Reachability matrix $\mathscr{R}_n$ is upper triangular with 1 on the diagonal. It follows that the state-space system with realization $(A_c, \mathbf{b}_c, \mathbf{c}_c, d_c)$ is always reachable. For the $p \times m$ transfer matrix

$$\mathbf{H}(z) = D + \left(\sum_{k=1}^{\ell} \tilde{N}_k z^{-k}\right)\left(I_m - \sum_{k=1}^{\ell} \tilde{M} z^{-k}\right)^{-1}, \tag{3.14}$$

a realization (A_c, B_c, C_c, D_c) in block controller form can be obtained with

$$A_c = \begin{bmatrix} \tilde{\mathbf{V}}_{\ell-1} & \tilde{M}_\ell \\ I_{(\ell-1)m} & \mathbf{0}_{(\ell-1)m\times m} \end{bmatrix}, \quad B_c = \begin{bmatrix} I_m \\ \mathbf{0}_{(\ell-1)m\times m} \end{bmatrix},$$
$$\tilde{\mathbf{V}}_{\ell-1} = \begin{bmatrix} \tilde{M}_1 & \cdots & \tilde{M}_{\ell-1} \end{bmatrix}, \quad C_c = \begin{bmatrix} \tilde{N}_1 & \cdots & \tilde{N}_\ell \end{bmatrix},$$

and $D_c = D$. It can be verified that the reachability matrix $\mathscr{R}_n$ is again upper triangular with 1 on the diagonal and $n = m\ell$. Hence, the state-space system with realization (A_c, B_c, C_c, D_c) is always reachable as well.

For a given transfer matrix $\mathbf{H}(z)$, not all its realizations are reachable. As in the previous subsection, Kalman decomposition can be applied to unreachable realizations to eliminate the unreachable subsystem. The following is dual to Theorem 3.5 for which the proof is again left as an exercise.

Theorem 3.9. *Let (A, B, C, D) be a realization of $\mathbf{H}(z)$ with size $n \times n$ for A, and $p \times m$ for $\mathbf{H}(z)$. Suppose that (A, B) is unreachable. Then there exists a similarity transform T such that*

$$\text{(i) } \tilde{A} = TAT^{-1} = \begin{bmatrix} \tilde{A}_c & \tilde{A}_{c\bar{c}} \\ \mathbf{0}_{(n-r)\times r} & \tilde{A}_{\bar{c}} \end{bmatrix}, \quad \tilde{B} = TB = \begin{bmatrix} \tilde{B}_c \\ \mathbf{0}_{(n-r)\times m} \end{bmatrix},$$
$$\text{(ii) } \tilde{C} = CT^{-1} = \begin{bmatrix} \tilde{C}_c & \tilde{C}_{\bar{c}} \end{bmatrix}, \qquad \tilde{D} = D,$$

where $(\tilde{A}_c, \tilde{B}_c)$ is reachable with $r \times r$ the dimension of $\tilde{A}_c$. Moreover

$$\mathbf{H}(z) = D + C(zI_n - A)^{-1}B = \tilde{D} + \tilde{C}_c\left(zI_r - \tilde{A}_c\right)^{-1}\tilde{B}_c.$$

Let $\mathbf{x}(t)$ be the state vector associated with realization (A, B, C, D). Similar to Sect. 3.1.1, Theorem 3.9 illustrates that the linear transform

$$\tilde{\mathbf{x}}(t) = T\mathbf{x}(t) = \begin{bmatrix} \tilde{\mathbf{x}}_c(t)^* & \tilde{\mathbf{x}}_{\bar{c}}(t)^* \end{bmatrix}^*$$

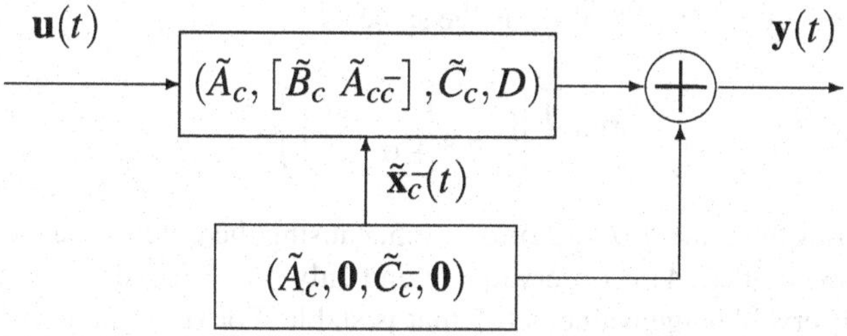

Fig. 3.4 Reachability Kalman decomposition

decomposes the state vector into two parts. The reachable one is $\tilde{\mathbf{x}}_c(t)$ with dimension r. The unreachable one is $\tilde{\mathbf{x}}_{\bar{c}}(t)$ with dimension $(n-r)$. Such a decomposition is shown in Fig. 3.4.

As it is seen, the subsystem $(\tilde{A}_c, \tilde{B}_c, \tilde{C}_c, D)$ is reachable from the input $\mathbf{u}(t)$, while the subsystem with realization $(\tilde{A}_{\bar{c}}, \mathbf{0}, \tilde{C}_{\bar{c}}, \mathbf{0})$ is unreachable. Kalman decomposition offers a procedure to eliminate the unreachable states or unreachable subsystem of the state-space model. The next result is dual to Theorem 3.6, and thus, the proof is omitted.

Theorem 3.10. (PBH test) *The state-space system (3.13), or simply* (A,B), *is unreachable, if and only if there exists a row vector* $\mathbf{q} \neq \mathbf{0}$ *such that*

$$\mathbf{q}A = \lambda \mathbf{q}, \quad \mathbf{q}B = \mathbf{0}. \tag{3.15}$$

Example 3.11. Let α be a real parameter and

$$\mathbf{H}(z) = \begin{bmatrix} -z^{-1} + \alpha z^{-2} \\ 2z^{-1} - 2\alpha z^{-2} \end{bmatrix} \frac{1}{1 - 2.5z^{-1} + z^{-2}}.$$

The poles of $\mathbf{H}(z)$ are roots of $1 - 2.5z^{-1} + z^{-2} = 0$ which are at 0.5 and 2. A realization in the canonical controller form for $\mathbf{H}(z)$ is given by

$$A = \begin{bmatrix} 2.5 & -1 \\ 1 & 0 \end{bmatrix}, \quad B = \begin{bmatrix} 1 \\ 0 \end{bmatrix}, \quad C = \begin{bmatrix} C_1 \\ C_2 \end{bmatrix} = \begin{bmatrix} -1 & \alpha \\ 2 & -2\alpha \end{bmatrix},$$

and $D = \mathbf{0}$. Clearly, (A,B) is reachable. To determine the observability of (C,A), its associated observability matrix is computed as follows:

$$\mathscr{O}_n = \begin{bmatrix} C \\ CA \end{bmatrix} = \begin{bmatrix} -1 & \alpha \\ 2 & -2\alpha \\ -2.5 + \alpha & 1 \\ 5 - 2\alpha & -2 \end{bmatrix}.$$

Because $C_2 = -2C_1$, the rank of $\mathscr{O}_n$ is the same as

$$\text{rank}\left\{\begin{bmatrix} -1 & \alpha \\ -2.5+\alpha & 1 \end{bmatrix}\right\}.$$

The above loses the rank at $\alpha = 2$ or $\alpha = 0.5$, causing the pole/zero cancellation in $\mathbf{H}(z)$. It follows that (C,A) is observable, if and only if $\alpha \neq 2$ and $\alpha \neq 0.5$. For $\alpha = 0.5$, the unobservable eigenvalue is 0.5 that is stable. For $\alpha = 2$, the unobservable eigenvalue is 2 that is unstable. Since unstable and unobservable eigenvalues can cause severe problems in system design, Kalman decomposition is employed to eliminate the unobservable eigenvalue at 2 via SVD $\mathscr{O}_n = USV^*$ and taking the similarity transform $T = V^*$. Hence, it yields a new state-space realization for $\mathbf{H}(z)$ at $\alpha = 2$:

$$TAT^{-1} = \begin{bmatrix} 0.5 & 0 \\ -2.0 & 2 \end{bmatrix}, \quad TB = \begin{bmatrix} 0.45 \\ -0.89 \end{bmatrix}, \quad CT^{-1} = \begin{bmatrix} -2.24 & 0 \\ 4.47 & -0 \end{bmatrix},$$

which has the same form as in Theorem 3.5. Consequently,

$$A_o = 0.5, \quad B_o = 0.45, \quad C_o = \begin{bmatrix} -2.24 \\ 4.47 \end{bmatrix}, \quad D = \begin{bmatrix} 0 \\ 0 \end{bmatrix}$$

constitute a minimal realization for $\mathbf{H}(z)$ in the case $\alpha = 2$, in light of the fact that $\mathbf{H}(z)$ still has another pole at 0.5.

3.1.3 Minimal Realization

As discussed at the beginning of the section, a minimal realization refers to the realization which has the minimal order among all realizations for the same transfer function or matrix. For SISO systems, coprime fractions are easy to obtain. One may simply cancel out the common poles and zeros to arrive at coprime fractions. As such, minimal realizations can also be obtained rather easily by employing either the observer form (refer to Example 3.4) or the controller form (refer to Sect. 2.2.2) for the coprime fractions. Clearly, such realizations employ the minimum number of time delays, and are thus minimal. However, coprime fractions are much more difficult to obtain for MIMO systems. As a result, minimal realizations are difficult to obtain directly from transfer matrices. The following observation is crucial.

Theorem 3.12. *A realization (A,B,C,D) is not minimal, if and only if either (A,B) is unreachable, or (C,A) is unobservable, or both.*

Proof. If (A,B) is unreachable, or (C,A) is unobservable, or both, Kalman decomposition can be applied to find a different state-space realization with a smaller order for the same transfer matrix. This implies that (A,B,C,D) is not a minimal realization. Conversely assume that A has dimension $n \times n$ and (A,B,C,D) is not a minimal realization. Then a different realization $\left(\tilde{A},\tilde{B},\tilde{C},\tilde{D}\right)$ exists for the same transfer matrix, say, $\mathbf{H}(z)$, with order $\tilde{n} < n$. Because physical systems are causal, there holds, for z in the region of convergence,

$$\mathbf{H}(z) = D + \sum_{i=1}^{\infty} CA^{i-1}Bz^{-i} = D + \sum_{i=1}^{\infty} \tilde{C}\tilde{A}^{i-1}\tilde{B}z^{-i}.$$

It is noted that the impulse response $H(t) = CA^{t-1}B = \tilde{C}\tilde{A}^{t-1}\tilde{B}$ for $t > 0$ is independent of realizations. Hence, its Hankel matrix of size $n \times n$ satisfies

$$\mathscr{H}_n = \begin{bmatrix} CB & CAB & \cdots & CA^{n-1}B \\ CAB & CA^2B & \cdots & CA^nB \\ \vdots & \cdots & \cdots & \vdots \\ CA^{n-1}B & \cdots & \cdots & CA^{2(n-1)}B \end{bmatrix} \tag{3.16}$$

$$= \begin{bmatrix} \tilde{C}\tilde{B} & \tilde{C}\tilde{A}\tilde{B} & \cdots & \tilde{C}\tilde{A}^{n-1}\tilde{B} \\ \tilde{C}\tilde{A}\tilde{B} & \tilde{C}\tilde{A}^2\tilde{B} & \cdots & \tilde{C}\tilde{A}^n\tilde{B} \\ \vdots & \cdots & \cdots & \vdots \\ \tilde{C}\tilde{A}^{n-1}\tilde{B} & \cdots & \cdots & \tilde{C}\tilde{A}^{2(n-1)}\tilde{B} \end{bmatrix}.$$

Then it can be verified that there holds factorization

$$\mathscr{H}_n = \mathscr{O}_n(C,A)\mathscr{R}_n(A,B) = \mathscr{O}_n\left(\tilde{C},\tilde{A}\right)\mathscr{R}_n\left(\tilde{A},\tilde{B}\right).$$

The rank of $\mathscr{H}_n$ is thus no more than $\tilde{n}$, implying that either $\mathscr{O}_n(C,A)$, $\mathscr{R}_n(A,B)$, or both have rank smaller than n. As a consequence, there is a redundancy in its corresponding state vector, which is either unreachable, or unobservable, or both. The proof is now completed. □

Theorem 3.12 indicates that a realization (A,B,C,D) is minimal, if and only if it is both reachable and observable. The next example serves to illustrate minimal realizations in relation to reachability and observability.

Example 3.13. (Gilbert realization) This example considers the case when all poles of $\mathbf{H}(z)$ are distinct. Thus, there holds partial fraction

$$\mathbf{H}(z) = \Theta_0 + \sum_{i=1}^{\ell} \frac{\Theta_i}{z - p_i},$$

where Θ_i has the same size as $\mathbf{H}(z)$ and $\{p_i\}$ are all distinct. If $\ell = 1$, then the order of minimal realizations is the same as n_1, the rank of Θ_1. In this case, there holds decomposition $\Theta_1 = C_1 B_1$ where C_1 and B_1 have n_1 columns and n_1 rows, respectively. Setting $A_1 = p_1 I_{n_1}$ yields a minimal realization $(A_1, B_1, C_1, \Theta_0)$ of order n_1. Since both C_1 and B_1 have rank n_1,

$$\text{rank}\{\mathscr{O}_{n_1}\} \geq n_1, \quad \text{rank}\{\mathscr{R}_{n_1}\} \geq n_1.$$

It follows that the realization $(A_1, B_1, C_1, \Theta_0)$ is both reachable and observable. For $\ell > 1$, decomposition $\Theta_i = C_i B_i$ holds for $1 \leq i \leq \ell$, where C_i and B_i have n_i columns and n_i rows, respectively, and n_i is the rank of Θ_i. Let

$$A = \text{diag}(A_1, \ldots, A_\ell), \quad B = \begin{bmatrix} B_1^* & \cdots & B_\ell^* \end{bmatrix}^*, \quad C = \begin{bmatrix} C_1 & \cdots & C_\ell \end{bmatrix},$$

and $D = \Theta_0$ with $A_i = p_i I_{n_i}$ for $1 \leq i \leq \ell$. Because (A_i, B_i, C_i) is a minimal realization for each i, and $\{p_i\}$ are all distinct, (A, B, C, D) is a minimal realization for $\mathbf{H}(z)$. Thus, (A, B) is reachable, and (C, A) is observable. Such realizations are called Gilbert realization.

Suppose that (A, B, C, D) is both unreachable and unobservable. Then a similarity transform can be applied to eliminate the unreachable and the unobservable subsystems to obtain a minimal realization as shown next.

Theorem 3.14 (Kalman canonical decomposition). *Let (A, B, C, D) be a realization of $\mathbf{H}(z)$ with order n. If it is unreachable and unobservable, then there exists a similarity transform T such that*

$$\tilde{A} = TAT^{-1} = \begin{bmatrix} \tilde{A}_{co} & \mathbf{0} & \tilde{A}_{13} & \mathbf{0} \\ \tilde{A}_{21} & \tilde{A}_{c\bar{o}} & \tilde{A}_{23} & \tilde{A}_{24} \\ \mathbf{0} & \mathbf{0} & \tilde{A}_{\bar{c}o} & \mathbf{0} \\ \mathbf{0} & \mathbf{0} & \tilde{A}_{43} & \tilde{A}_{\bar{c}\bar{o}} \end{bmatrix}, \quad \tilde{B} = TB = \begin{bmatrix} \tilde{B}_{co} \\ \tilde{B}_{c\bar{o}} \\ \mathbf{0} \\ \mathbf{0} \end{bmatrix},$$

$$\tilde{C} = CT^{-1} = \begin{bmatrix} \tilde{C}_{co} & \mathbf{0} & \tilde{C}_{\bar{c}o} & \mathbf{0} \end{bmatrix}, \quad \tilde{D} = D,$$

where the subsystem $\left(\tilde{A}_{co}, \tilde{B}_{co}, \tilde{C}_{co}, \tilde{D}\right)$ is both reachable and observable and $\mathbf{H}(z) = \tilde{D} + \tilde{C}_{co}\left(zI_n - \tilde{A}_{co}\right)^{-1}\tilde{B}_{co}$.

The proof can proceed in two steps. The first step employs the similarity transform T_1 to convert the system into two subsystems with one reachable and the other unreachable. The second step uses the similarity transform T_2 to convert each of the subsystems into two parts with one observable and the other unobservable. The similarity transform $T = T_2 T_1$ can then accomplish the Kalman decomposition in Theorem 3.14. The details are omitted.

Let $\mathbf{x}(t)$ be the state vector associated with (A, B, C, D). The similarity transform T in Theorem 3.14 yields

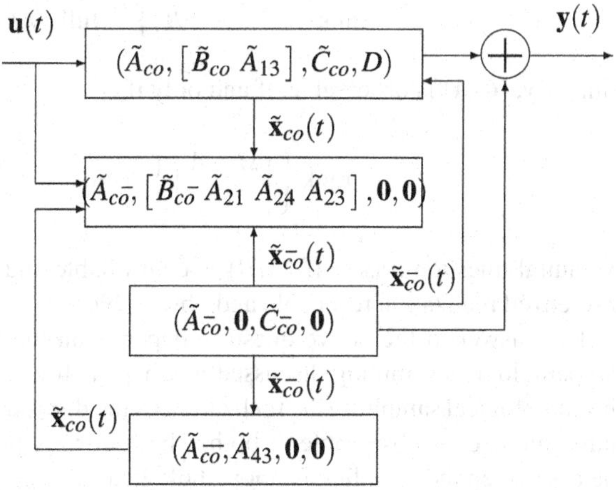

Fig. 3.5 Kalman canonical decomposition

$$\tilde{\mathbf{x}}(t) = T\mathbf{x}(t) = \begin{bmatrix} \tilde{\mathbf{x}}_{\text{co}}^*(t) & \tilde{\mathbf{x}}_{c\bar{o}}^*(t) & \tilde{\mathbf{x}}_{\bar{c}\text{o}}^*(t) & \tilde{\mathbf{x}}_{\bar{c}\bar{o}}^*(t) \end{bmatrix}^*$$

with compatible partitions to those of $\tilde{A}$, $\tilde{B}$, and $\tilde{C}$. The above results in the complete Kalman decomposition in Fig. 3.5. The first subsystem with state vector $\tilde{\mathbf{x}}_{\text{co}}$ is both reachable and observable. The other three subsystems with state vectors $\tilde{\mathbf{x}}_{c\bar{o}}(t), \tilde{\mathbf{x}}_{\bar{c}\text{o}}(t)$, and $\tilde{\mathbf{x}}_{\bar{c}\bar{o}}(t)$ are unreachable, unobservable, and both unreachable and unobservable, respectively. It is emphasized that only $\tilde{\mathbf{x}}_{\text{co}}$ or $(\tilde{A}_{\text{co}}, \tilde{B}_{\text{co}}, \tilde{C}_{\text{co}}, \tilde{D})$ has contributions to the input/output behavior of the system, while the other three do not due to their unreachability, unobservability, or both.

The complete Kalman decomposition in Theorem 3.14, illustrated in Fig. 3.5, is important. It offers a numerically efficient procedure to obtain minimal realizations, given any initial realization regardless of its reachability and observability, and is considerably simpler than coprime fractions for transfer matrices. Moreover, Kalman decomposition demonstrates the pivotal role of the conceptual notions of reachability and observability, which are also crucial to future developments of this text in later sections and chapters.

3.1.4 Sampling Effects

This subsection considers the effect of sampling on controllability and observability of the continuous-time state-space systems. Consider the following system

$$\dot{\mathbf{x}}(t_{\text{c}}) = A\mathbf{x}(t_{\text{c}}) + B\mathbf{u}(t_{\text{c}}), \quad \mathbf{y}(t_{\text{c}}) = C\mathbf{x}(t_{\text{c}}) + D\mathbf{u}(t_{\text{c}})$$

that is in continuous time where t_{c} is real-valued. Its realization matrices are $\{A, B, C, D\}$. It is well known that (A, B) is controllable, if and only if

$$\text{rank}\left\{\begin{bmatrix} sI-A & B \end{bmatrix}\right\} = \text{full} \quad \forall\, s \in \mathbb{C}.$$

Similarly, (C,A) is observable, if and only if

$$\text{rank}\left\{\begin{bmatrix} sI-A \\ C \end{bmatrix}\right\} = \text{full} \quad \forall\, s \in \mathbb{C}.$$

A natural question arises: If (A,B) is controllable and (C,A) is observable, is the discretized realization reachable and observable?

The answer to the above question depends on the sampling frequency. Recall the pathological sampling discussed in Chap. 1. It is not difficult to see that under the pathological sampling, the realization of the discretized state-space may become unreachable or unobservable or both. The following provides a positive answer to the case when the sampling is not pathological.

Theorem 3.15. *Let (A_d, B_d) be discretization of (A,B) as in (1.30) under the nonpathological sampling period T_s. Then*

(i) *(A,B) being controllable implies that (A_d, B_d) is reachable;*
(ii) *(C,A) being observable implies that (C, A_d) is observable.*

Proof. Only the proof for (ii) is given, since (i) is dual to (ii). Let λ be an eigenvalue of A. The proof amounts to show that

$$\text{rank}\begin{bmatrix} A-\lambda I \\ C \end{bmatrix} = \text{full} \implies \text{rank}\begin{bmatrix} A_d - e^{\lambda T_s} I \\ C \end{bmatrix} = \text{full}.$$

Denote $\Delta = A_d - e^{\lambda T_s} I = e^{AT_s} - e^{\lambda T_s} I$. Then

$$\begin{aligned} \Delta &= \sum_{k=1}^{\infty}\left(\frac{T_s^k A^k}{k!} - \frac{T_s^k \lambda^k}{k!} I\right) \\ &= \sum_{k=1}^{\infty}\frac{T_s^k}{k!}\left(A^k - \lambda^k I\right) = g(A)(A-\lambda I), \end{aligned}$$

where by induction it can be shown that

$$g(A) = \sum_{k=1}^{\infty}\frac{T_s^k}{k!}\left(A^{k-1} + A^{k-2}\lambda + \cdots + \lambda^{k-1} I\right) = \sum_{k=1}^{\infty}\frac{T_s^k}{k!}\left(\sum_{i=1}^{k} A^{k-i}\lambda^{i-1}\right).$$

By nonpathological sampling, $g(\lambda) \neq 0$ for any λ eigenvalue of A where

$$g(s) = \frac{e^{sT_s} - e^{\lambda T_s}}{s-\lambda} \neq 0 \ \text{ if } \ s = \lambda.$$

Moreover, the set of zeros of $g(s)$, denoted by $\mathscr{S}_g$, is given by

$$\begin{aligned}
\mathscr{S}_g &= \left\{s:\ e^{sT_s} = e^{T_s\lambda},\ s \neq \lambda\right\} \\
&= \{s:\ T_s s = T_s\lambda + j2k\pi,\ k = \pm 1, \pm 2, \ldots\} \\
&= \{s:\ s = \lambda + jk\omega_s,\ k = \pm 1, \pm 2, \ldots\}
\end{aligned}$$

that is disjoint from any eigenvalue of A, by again the nonpathological sampling. As a result, $g(A)$ is nonsingular. Now it is straightforward to show that

$$\begin{aligned}
\text{rank}\left\{\begin{bmatrix} e^{AT_s} - e^{\lambda T_s} I \\ C \end{bmatrix}\right\} &= \text{rank}\left\{\begin{bmatrix} g(A)(A - \lambda I) \\ C \end{bmatrix}\right\} \\
&= \text{rank}\left\{\begin{bmatrix} g(A) & 0 \\ 0 & I \end{bmatrix}\begin{bmatrix} A - \lambda I \\ C \end{bmatrix}\right\}
\end{aligned}$$

that is full by nonsingularity of $g(A)$ and observability of (C, A). □

Example 3.16. For the flight control system from Problem 1.11 in Exercises of Chap. 1, its discretized realization can be easily obtained via Matlab. By the expressions of (A, B, C) in Problem 1.11, (A, B) is controllable and (C, A) is observable. Moreover, A has eigenvalues 0, $-0.7801 \pm j1.0296$ and $-0.0176 + j0.1826$. Hence, the sampling period of $T_s = 0.025$ is nonpathological. Matlab command $[A_d, B_d] = \text{c2d}(A, B, T_s)$ yields the disctretized realization matrices given by

$$A_d = \begin{bmatrix} 1 & 0.0001 & 0.0283 & 0 & -0.0248 \\ 0 & 0.9986 & -0.0043 & 0 & 0.0017 \\ 0 & 0 & 1 & 0.0247 & -0.0003 \\ 0 & 0.0013 & 0 & 0.9785 & -0.0248 \\ 0 & -0.0072 & 0 & 0.0258 & 0.9827 \end{bmatrix},$$

$$B_d = \begin{bmatrix} 0 & 0 & 0 \\ 0 & 0.0250 & 0 \\ 0.0001 & 0 & -0.0005 \\ 0.0109 & 0 & -0.0411 \\ 0.0040 & -0.0001 & -0.0024 \end{bmatrix}.$$

It can be easily verified that (A_d, B_d) is reachable and (C, A_d) is observable, which follow from Theorem 3.15.

The test of reachability and observability can be carried out with the PBH test that is easy to implement in the Matlab environment. The next example considers the inverted pendulum system.

Example 3.17. The linearized model for inverted pendulum from Example 1.1 is represented by transfer functions $G_{up}(s)$ and $G_{u\theta}(s)$ for the position of the cart, and angle of the pendulum, respectively. For the laboratory setup of the inverted pendulum system by the Quanzer Inc., the two transfer functions are given by

$$G_{up}(s) = \frac{2.4805(s+5.4506)(s-5.4506)}{s(s+12.2382)(s-5.7441)(s+4.7535)},$$

$$G_{u\theta}(s) = \frac{7.512s}{(s+12.2382)(s-5.7441)(s+4.7535)}.$$

Since all poles are real, pathological sampling is not involved for the inverted pendulum system. If the c2d command is used in Matlab for each of the above transfer functions with sampling period $T_s = 0.25$, the following discretized models

$$G_{up}^{(d)}(z) = \frac{0.03833z^3 - 0.143z^2 - 0.02844z + 0.01603}{z^4 - 5.556z^3 + 6.048z^2 - 1.553z + 0.06009},$$

$$G_{u\theta}^{(d)}(z) = \frac{0.1408z^2 - 0.08757z - 0.05321}{z^3 - 4.556z^2 + 1.493z - 0.06009},$$

are resulted in. The poles of the two transfer functions are the same except that $G_{up}^{(d)}(z)$ has an additional pole at 1. In addition, numerator polynomials of $G_{up}^{(d)}(z)$ and $G_{up}^{(d)}(z)$ are coprime, which follow from Theorem 3.15 and the nonpathological sampling. A minimal realization can be obtained for

$$G_d(z) = \begin{bmatrix} G_{up}^{(d)}(z) \\ \frac{(z-1)G_{u\theta}^{(d)}(z)}{z-1} \end{bmatrix}$$

by employing the canonical controller form, leading to the realization matrices:

$$A = \begin{bmatrix} 5.556 & -6.048 & 1.553 & -0.0601 \\ 1 & 0 & 0 & 0 \\ 0 & 1 & 0 & 0 \\ 0 & 0 & 1 & 0 \end{bmatrix}, \quad B = \begin{bmatrix} 1 \\ 0 \\ 0 \\ 0 \end{bmatrix},$$

$$C = \begin{bmatrix} 0.0383 & -0.143 & -0.0284 & 0.0160 \\ 0.1408 & -0.2283 & 0.0344 & 0.0532 \end{bmatrix}, \quad D = \begin{bmatrix} 0 \\ 0 \end{bmatrix}.$$

Clearly (C,A) is observable.

3.2 Stability

Stability is a primary concern in dynamic system design. This section investigates stability for state-space systems focusing on Lyapunov stability. In addition, stabilization will be investigated for state feedback, state estimation, as well as observer-based feedback control systems.

3.2.1 Lyapunov Criterion

Stability was briefly discussed in Chap. 2 for state-space systems. Such a stability notion is referred to as *asymptotic stability* or simply internal stability, and is different from the BIBO stability or simply external stability. The following defines formally stability for state-space systems.

Definition 3.3. Let (A,B,C,D) be a realization. The corresponding state-space system is asymptotically stable, or stable in the sense of Lyapunov, if the solution $\mathbf{x}(t)$ to

$$\mathbf{x}(t+1) = A\mathbf{x}(t), \quad \mathbf{x}(0) = \mathbf{x}_0 \tag{3.17}$$

satisfies $\|\mathbf{x}(t)\| \to 0$ as $t \to \infty$ for each nonzero initial condition $\mathbf{x}_0$.

The solution to (3.17) is clearly given by $\mathbf{x}(t) = A^t\mathbf{x}(0)$ for $t > 0$. Since eigenvalues of A^t are $\{\lambda_k^t\}$ with $\{\lambda_k\}$ eigenvalues of A, the internal stability is equivalent to $\rho(A) < 1$ with $\rho(\cdot)$ the spectral radius. The following result is obvious in light of the definition and thus the proof is omitted.

Proposition 3.1. *The state-space system with realization (A,B,C,D) is asymptotically stable, if and only if $\rho(A) < 1$.*

Let (A,B,C,D) be a realization of $\mathbf{H}(z)$. Then its asymptotic stability implies its BIBO stability in light of the fact that poles of $\mathbf{H}(z)$ are eigenvalues of A which are all stable. However, the converse may not be true, as discussed in Sect. 2.2.2, unless (A,B,C,D) is a minimal realization, in which case eigenvalues of A are poles of $\mathbf{H}(z)$ as well. As the asymptotic stability is related to the matrix A only, it will be referred to as stability of A for simplicity.

It needs to be pointed out that the aforementioned stability is different from *marginal stability* which allows nonrepeated eigenvalues on the unit circle. In the case of marginal stability, solution $\mathbf{x}(t)$ to (3.17) remains bounded as $t \to \infty$, albeit not approaching zero in general.

The following stability result is called Lyapunov criterion.

Theorem 3.18. *A matrix A is a stability matrix, if and only if for each given matrix $Z = Z^* > 0$ (i.e., hermitian positive definite) there exists a unique matrix $P = P^* > 0$ such that*

$$P = A^*PA + Z. \tag{3.18}$$

Proof. If A is a stability matrix, then $\rho(A) < 1$, and

$$P = \sum_{k=0}^{\infty} (A^*)^k Z A^k$$

exists and satisfies (3.18). The hypothesis on Z implies that P is positive definite hermitian. To show that P is unique, suppose that P_o also satisfies

$$P_o = A^* P_o A + Z.$$

Taking the difference between the above equation and (3.18) yields

$$\begin{aligned}(P - P_o) &= A^*(P - P_o)A = (A^*)^2 (P - P_o) A^2 = \cdots \\ &= \lim_{k\to\infty} (A^*)^k (P - P_o) A^k = \mathbf{0}\end{aligned}$$

by stability of A. It follows that $P_o = P$ which is unique. Note that a weaker condition exists for the uniqueness of the solution P. See Problem 3.19 in Exercises. Conversely, if (3.18) has a unique solution P which is positive definite hermitian, then by the state-space equation (3.17), the energy function (called Lyapunov function) $V[\mathbf{x}(t)] = \mathbf{x}(t)^* P \mathbf{x}(t)$ satisfies the difference equation

$$\begin{aligned}\Delta V[\mathbf{x}(t)] &= \mathbf{x}(t+1)^* P \mathbf{x}(t+1) - \mathbf{x}(t)^* P \mathbf{x}(t) \\ &= \mathbf{x}(t)^* (A^* P A - P) \mathbf{x}(t) = -\mathbf{x}(t)^* Z \mathbf{x}(t) < 0\end{aligned}$$

whenever $\mathbf{x}(t) \neq \mathbf{0}$. Therefore, $V[\mathbf{x}(t)]$ is monotonically decreasing, and thus, $V[\mathbf{x}(t)] \to 0$ and $\mathbf{x}(t) \to \mathbf{0}$ as $t \to \infty$ by the hypothesis on P, concluding stability of A. □

Equation (3.18) is called Lyapunov equation. In the Lyapunov criterion, positive definite Z is required which is unnecessary and restrictive. The following stability criterion is more useful.

Theorem 3.19. *Suppose that (C, A) is observable. Then A is a stability matrix, if and only if there exists a unique matrix $P = P^* > 0$ such that*

$$P = A^* P A + C^* C. \tag{3.19}$$

Proof. If A is stable, then

$$P = \sum_{k=0}^{\infty} (A^*)^k C^* C A^k = \lim_{\ell\to\infty} \mathscr{O}_\ell^* \mathscr{O}_\ell \tag{3.20}$$

is the solution to (3.19) with $\mathscr{O}_\ell$ the observability matrix of size ℓ. The hypothesis on the observability implies that $\mathscr{O}_\ell$ has rank n, the order of the corresponding state-space system, for any $\ell \geq n$. Thus, P is positive definite hermitian. The uniqueness of P can be shown with the same argument as in the proof of Theorem 3.19. Conversely, assume that (3.19) holds with P positive definite. Let λ be an eigenvalue of A with eigenvector $\mathbf{v}$. Then multiplying both sides of (3.19) with $\mathbf{v}$ from right and $\mathbf{v}^*$ from left yields

$$\left(1-|\lambda|^2\right)\mathbf{v}^*P\mathbf{v} = \|C\mathbf{v}\|^2. \tag{3.21}$$

The PBH test implies that $\|C\mathbf{v}\|^2 > 0$, which in turn implies that $|\lambda| < 1$ by the fact that P is positive definite. Since λ is an arbitrary eigenvalue of A, $\rho(A) < 1$ is true. That is, A is a stability matrix. □

A dual result to Theorem 3.19 is the following for which the proof is omitted.

Theorem 3.20. *Suppose that (A,B) is reachable. Then A is a stability matrix, if and only if there exists a unique positive definite hermitian matrix Q such that*

$$Q = AQA^* + BB^*. \tag{3.22}$$

The matrix P in (3.19) is called the observability gramian, and Q in (3.22) is called the reachability gramian. In light of Theorems 3.19 and 3.20, a realization (A,B,C,D) is minimal, if and only if their associated observability and reachability gramians are nonsingular.

The results in Theorems 3.19 and 3.20 show that stability testing requires that either (C,A) be observable or (A,B) be reachable. Although Kalman decomposition can be used to obtain observable and reachable realizations, such stability criteria are inconvenient in their usage. For this reason, it is beneficial to examine the PBH test again. Recall that (C,A) is unobservable, if and only if there exists $\mathbf{v} \neq \mathbf{0}$ such that

$$A\mathbf{v} = \lambda\mathbf{v}, \quad C\mathbf{v} = \mathbf{0}. \tag{3.23}$$

Consider the case when λ is a simple eigenvalue of A. The solution $\mathbf{x}(t)$ to (3.17) contains the term λ^t (called *mode*), which does not show up at the output $\mathbf{y}(t)$. In other words, the mode corresponding to eigenvalue λ is unobservable. For the case when λ is a repeated eigenvalue of A with multiplicity $m > 1$, there are m modes corresponding to eigenvalue λ, which are in the form of $P_k(t)\lambda^t$ shown up in the state vector $\mathbf{x}(t)$ with $P_k(t)$ a kth degree polynomial of t and $k = 0, 1, \ldots, m-1$. At least one of the modes $\{P_k(\lambda)\lambda^t\}_{k=0}^{m-1}$ is unobservable at the output due to (3.23).

Example 3.21. Consider the unforced state-space equation (3.17) with

$$A = \begin{bmatrix} \alpha & 1 \\ 0 & \alpha \end{bmatrix}, \quad C = \begin{bmatrix} 0 & 1 \end{bmatrix}.$$

Then $\lambda = \alpha$ and $\mathbf{v} = \begin{bmatrix} 1 & 0 \end{bmatrix}^*$ satisfy (3.23), implying that the pair (C,A) is unobservable. Simple calculations show that

$$\mathbf{x}(t+1) = A^t \mathbf{x}_0 = \begin{bmatrix} \alpha^t & t\alpha^{t-1} \\ 0 & \alpha^t \end{bmatrix} \begin{bmatrix} x_{01} \\ x_{02} \end{bmatrix} = \begin{bmatrix} (a+bt) \\ c \end{bmatrix} \alpha^t,$$

where $a = x_{01}$, $b = x_{02}/\alpha$, and $c = x_{02}$. There are two modes in the state response at time $t: c\alpha^{t-1}$ and $(a+bt)\alpha^{t-1}$. However, $y(t) = c\alpha^{t-1}$. Thus, the mode $(a+bt)\alpha^{t-1}$ does not show up at the output which is unobservable.

In general, there are n modes in an nth order state-space system. It is the collective behaviors of the n modes that determine the dynamic behavior of the state-space system. Some of the modes are observable while others may not be. For stable modes (the corresponding eigenvalues are strictly inside the unit circle), their unobservability does not hinder stability testing based on the output response. But unstable modes have to be dealt with more carefully. The following introduces the notion for observability of unstable modes.

Definition 3.4. The pair (C,A) is called *detectable*, if for any unstable eigenvalue of A, denoted by λ, there does not exist a vector $\mathbf{v} \neq \mathbf{0}$ such that (3.23) holds.

The stability criterion in Theorem 3.19 is now generalized to unobservable realizations.

Theorem 3.22. *Suppose that (C,A) is detectable. Then A is a stability matrix, if and only if there exists a unique positive semidefinite matrix P, denoted as $P \geq \mathbf{0}$, such that $P = A^*PA + C^*C$, i.e., (3.19) holds.*

Proof. If A is stable, then P has the same expression as in (3.20). Though (C,A) may not be observable, P is positive semidefinite in general and is unique by the same argument as before. Conversely, assume that a positive semidefinite P exists and satisfies (3.19). Then with the same procedure as in the proof of Theorem 3.19, (3.21) can be obtained. Assume that $|\lambda| \geq 1$. Since any mode corresponding to eigenvalue λ with $|\lambda| \geq 1$ is observable, (3.21) implies that $\|C\mathbf{v}\|^2 > 0$ leading to $|\lambda| < 1$ by $P \geq 0$, which is a contradiction to the hypothesis $|\lambda| \geq 1$. Hence, stability of A follows. □

A dual notion to the detectability is the following.

Definition 3.5. The pair (A,B) is called *stabilizable*, if for any unstable eigenvalue of A, denoted by λ, there does not exist a nonzero row vector $\mathbf{q}$ such that

$$\mathbf{q}A = \lambda\mathbf{q}, \quad \mathbf{q}B = \mathbf{0}^*.$$

Similarly, the stability criterion in Theorem 3.20 is now generalized to unreachable realizations.

Theorem 3.23. *Suppose that (A,B) is stabilizable. Then A is a stability matrix, if and only if there exists a unique positive semi-definite hermitian matrix Q such that $Q = AQA^* + BB^*$.*

Theorems 3.22 and 3.23 point out that the positivity of the observability and the reachability gramians does not necessarily imply stability of A unless the unobservable and unreachable modes are stable, respectively. On the other hand, if the observability or reachability gramian has strictly negative eigenvalues, then A cannot be a stability matrix, which can be inferred from Theorem 3.22 or Theorem 3.23, respectively.

The notions of detectability and stabilizability will be explored further in the next subsection and help deepen the understanding of modal observability and reachability. This subsection is concluded with the following example.

Example 3.24. Let β be a real parameter and

$$A = \begin{bmatrix} \beta & 1 & 0 \\ 0 & \beta & 0 \\ 0 & 1 & \beta \end{bmatrix}, \quad B = \begin{bmatrix} 1 & 0 \\ 0 & 0 \\ 0 & 1 \end{bmatrix}, \quad C = \begin{bmatrix} 0 & 1 & 0 \\ 0 & 0 & 1 \end{bmatrix}$$

be realization matrices for some given state-space system. The eigenvalues of A are β with multiplicity 3. The observability gramian of (C,A) is

$$P = \begin{bmatrix} 0 & \mathbf{0} \\ \mathbf{0} & P_o \end{bmatrix}, \quad P_o = \begin{bmatrix} \frac{1+(1+\beta-\beta^2)^2}{(1-\beta^2)^3} & \frac{1+\beta-\beta^2}{(1-\beta^2)^2} \\ \frac{1+\beta-\beta^2}{(1-\beta^2)^2} & \frac{1}{(1-\beta^2)} \end{bmatrix}.$$

Since P is singular, (C,A) is unobservable (refer to Problem 3.20). The question is stability of A. It is noted that $\det(P_o) = 1/(1-\beta^2)^4 > 0$. Thus, $P \geq 0$, if and only if the two nonzero diagonal elements of P are positive, which is in turn equivalent to that $|\beta| < 1$ consistent with the fact that β is the only eigenvalue of A. For $|\beta| > 1$, $P \leq \mathbf{0}$, and thus, (C,A) is not detectable, implying that A is unstable. For $|\beta| = 1$, elements of P are unbounded, implying that A has eigenvalues either on the unit circle or in mirror pattern with respect to the unit circle (refer to Problem 3.19). It can also be verified that the reachability gramian is

$$Q = \begin{bmatrix} 1 & 0 & 0 \\ 0 & 0 & 0 \\ 0 & 0 & 1 \end{bmatrix} \frac{1}{1-\beta^2}.$$

So, (A,B) is not reachable by $\det(Q) = 0$. Moreover, $Q \geq \mathbf{0}$, if and only if $|\beta| < 1$ which is identical to that for P. In this case, A is a stability matrix, although such a conclusion cannot be obtained from Q or P alone. For $|\beta| > 1$, $Q \leq \mathbf{0}$, and thus, A is not a stability matrix, consistent with the discussion earlier.

3.2.2 Linear Stabilization

Stabilization is an integral part of the linear system theory. For instance, if A is an instability matrix in the state-space model

$$\mathbf{x}(t+1) = A\mathbf{x}(t) + B\mathbf{u}(t), \quad \mathbf{x}(0) = \mathbf{x}_0, \tag{3.24}$$

one may wish to employ state-feedback control $\mathbf{u}(t) = F\mathbf{x}(t)$ and hope that the resultant feedback control system

$$\mathbf{x}(t+1) = (A+BF)\mathbf{x}(t), \quad \mathbf{x}(0) = \mathbf{x}_0, \tag{3.25}$$

is asymptotically stable. If such a state-feedback gain F exists, then it is called *stabilizing*. Clearly, stabilizing state-feedback gains do not always exist. Indeed, when (A,B) is not stabilizable, then at least one unstable mode is unreachable, which remains a mode for the closed-loop system in (3.25) in light of the PBH test. A pertinent question is then whether the stabilizability of (A,B) ensures the existence of stabilizing state-feedback control.

A dual case to the above is estimation of the state vector $\mathbf{x}(t)$, based on the observation of the input $\mathbf{u}(t)$ and the output

$$\mathbf{y}(t) = C\mathbf{x}(t) + D\mathbf{u}(t). \tag{3.26}$$

One may wish to consider the state estimator or observer of the form

$$\hat{\mathbf{x}}(t+1) = A\hat{\mathbf{x}}(t) + B\mathbf{u}(t) + L[\hat{\mathbf{y}}(t) - \mathbf{y}(t)], \quad \hat{\mathbf{y}}(t) = C\hat{\mathbf{x}}(t) + D\mathbf{u}(t), \tag{3.27}$$

where $\hat{\mathbf{x}}(0) = \hat{\mathbf{x}}_0$. Note that $\hat{\mathbf{x}}(t) = \mathbf{x}(t)$ is an equilibrium to (3.24) and (3.27), providing some justifications for the use of such state estimators. The matrix L is called state estimation gain. Taking difference between (3.24) and (3.27) with $\mathbf{x}_e(t) = \mathbf{x}(t) - \hat{\mathbf{x}}(t)$ yields

$$\mathbf{x}_e(t+1) = (A+LC)\mathbf{x}_e(t), \quad \mathbf{x}_e(0) = \mathbf{x}_{e0}, \tag{3.28}$$

which is dual to (3.25). Whether or not the estimation error $\|\mathbf{x}_e(t)\|$ tends to zero as $t \to \infty$ amounts to stability of $(A+LC)$. Again, L is said to be stabilizing, if $(A+LC)$ is a stability matrix. Clearly, detectability of (C,A) is a necessary condition for the existence of stabilizing L.

Due to the linearity of feedback gains F and L, stabilization as posed in (3.25), and (3.28) is referred to as linear stabilization. It is related to the more general problem of eigenvalues assignment, which investigates relocating eigenvalues of A via state-feedback gain F and state estimation gain L. That is, eigenvalues assignment considers shifting all eigenvalues of A, not only the unstable ones, to the desired locations. It can be expected that reachability of (A,B) and observability of (C,A) play the decisive role in eigenvalues assignment as shown in the following result.

Theorem 3.25. *Let (A,B,C,D) be a realization of $\mathbf{G}(z)$.* (i) *Eigenvalues of $(A+BF)$ can be arbitrarily assigned by some F, if and only if (A,B) is reachable.* (ii) *Eigenvalues of $(A+LC)$ can be arbitrarily assigned by some L, if and only if (C,A) is observable.*

Proof. Because (i) and (ii) are dual to each other, only the proof for (ii) is presented. If (C,A) is unobservable, then there exists an eigenvalue λ and an eigenvector $\mathbf{v}$ such that

$$A\mathbf{v} = \lambda\mathbf{v}, \quad C\mathbf{v} = \mathbf{0}_p$$

assuming dimension p for the output. Thus, for any state estimation gain L,

$$(A+LC)\mathbf{v} = A\mathbf{v} = \lambda\mathbf{v}.$$

That is, the eigenvalue λ remains an eigenvalue for $(A+LC)$. Conversely, assume that (C,A) is observable. An induction process will be employed to show that eigenvalues of $(A+LC)$ can be arbitrarily assigned with respect to the dimension of the output. So for $p=1$, a similarity transform exists which transforms (C,A) into the observer form (refer to Problem 3.18 in Exercises):

$$A = A_o = \begin{bmatrix} -a_1 & 1 & 0 & \cdots & 0 \\ -a_2 & 0 & \ddots & \ddots & \vdots \\ \vdots & \vdots & \ddots & \ddots & 0 \\ \vdots & \vdots & \ddots & \ddots & 1 \\ -a_n & 0 & \cdots & \cdots & 0 \end{bmatrix}, \quad C = \mathbf{c}_o = \begin{bmatrix} 1 & 0 & \cdots & 0 \end{bmatrix}.$$

Hence, the characteristic polynomial is

$$\mathbf{a}(z) = \det(zI_n - A) = z^n + a_1 z^{n-1} + \cdots + a_n.$$

The eigenvalues of A are identical to the roots of $\mathbf{a}(z) = 0$. Let

$$L = \begin{bmatrix} \ell_1 \\ \ell_2 \\ \vdots \\ \vdots \\ \ell_n \end{bmatrix} \implies A+LC = \begin{bmatrix} \ell_1 - a_1 & 1 & 0 & \cdots & 0 \\ \ell_2 - a_2 & 0 & 1 & \ddots & \vdots \\ \vdots & \vdots & \ddots & \ddots & 0 \\ \vdots & \vdots & \ddots & \ddots & 1 \\ \ell_n - a_n & 0 & \cdots & \cdots & 0 \end{bmatrix}.$$

As a result, the eigenvalues of $(A+LC)$ are roots of

$$\det(zI_n - A - LC) = z^n + (a_1 - \ell_1)z^{n-1} + \cdots + (a_n - \ell_n),$$

which can be arbitrarily assigned by choosing $\{\ell_i\}$ correctly. For $p = k > 1$, assume that observable (C,A) implies that eigenvalues of $(A+LC)$ can be arbitrarily assigned. To complete the induction process, consider the case $p = k+1$. Denote the submatrix of C with the first k rows by C_1. If (C_1,A) is observable, then the proof can be concluded by taking $L = \begin{bmatrix} L_1 & \mathbf{0}_n \end{bmatrix}$, for, in this case, $(A+LC) = (A+L_1C_1)$ whose eigenvalues can be arbitrarily assigned by the hypothesis for $p = k$. If (C_1,A) is unobservable, then Kalman decomposition can be applied to (C_1,A), leading to

$$CT^{-1} = \begin{bmatrix} \tilde{C}_{11} & \mathbf{0} \\ \tilde{C}_{21} & \tilde{C}_{22} \end{bmatrix}, \quad TAT^{-1} = \begin{bmatrix} \tilde{A}_{11} & \mathbf{0} \\ \tilde{A}_{21} & \tilde{A}_{22} \end{bmatrix},$$

where $C_1T^{-1} = \begin{bmatrix} \tilde{C}_{11} & \mathbf{0} \end{bmatrix}$, and $(\tilde{C}_{11},\tilde{A}_{11})$ is observable. Let the dimension of $\tilde{A}_{11}$ be $r \times r$ with $r < n$. Then $\tilde{L}_{11}$ exists such that the r eigenvalues of $(\tilde{A}_{11} + \tilde{L}_{11}\tilde{C}_{11})$ can be arbitrarily assigned, by again the hypothesis for $p = k$. Note that $\begin{bmatrix} \tilde{C}_{21} & \tilde{C}_{22} \end{bmatrix}$ is a row vector. For (C,A) is observable, $(\tilde{C}_{22},\tilde{A}_{22})$ is observable as well. The procedure for $p = 1$ can be employed to compute $\tilde{L}_{22}$ such that the remaining $(n-r)$ eigenvalues of $\tilde{A}$ or eigenvalues of $(\tilde{A}_{22} + \tilde{L}_{22}\tilde{C}_{22})$ can be assigned arbitrarily. Taking

$$L = T^{-1}\tilde{L} = T^{-1}\begin{bmatrix} \tilde{L}_{11} & \mathbf{0} \\ \mathbf{0} & \tilde{L}_{22} \end{bmatrix}$$

yields the characteristic polynomial for the closed-loop system:

$$\begin{aligned} \tilde{\mathbf{a}}(z) &= \det[zI_n - (A+LC)] = \det\left[zI_n - (\tilde{A} + \tilde{L}\tilde{C})\right] \\ &= \det\left(\begin{bmatrix} zI_r - (\tilde{A}_{11} + \tilde{L}_{11}\tilde{C}_{11}) & \mathbf{0} \\ -(\tilde{A}_{21} + \tilde{L}_{22}\tilde{C}_{21}) & zI_{n-r} - (\tilde{A}_{22} + \tilde{L}_{22}\tilde{C}_{22}) \end{bmatrix}\right) \\ &= \det\left[zI_r - (\tilde{A}_{11} + \tilde{L}_{11}\tilde{C}_{11})\right] \det\left[zI_{n-r} - (\tilde{A}_{22} + \tilde{L}_{22}\tilde{C}_{22})\right], \end{aligned}$$

concluding the fact that the eigenvalues of $(A+LC)$ can be arbitrarily assigned for some matrix L, provided that (C,A) is observable. □

It is noted that the procedure employing similarity transforms and Kalman decompositions may not be efficient in relocating eigenvalues of A. However, it is not the intention of this text to provide an efficient algorithm for eigenvalues assignment. Design of optimal state feedback and optimal state estimation gains will be studied in the next chapter. Rather, Theorem 3.25 provides the insight to modal reachability and observability in relation to shifting the corresponding eigenvalues. The following result follows from Theorem 3.25, and thus, the proof is omitted.

Theorem 3.26. *Let* (A,B,C,D) *be a realization of* $\mathbf{G}(z)$. (i) *There exists a stabilizing state-feedback gain F, if and only if* (A,B) *is stabilizable.* (ii) *There exists a stabilizing state estimation gain L, if and only if* (C,A) *is detectable.*

Theorem 3.26 is practically more useful in design of state-feedback and state estimation gains, because the minimality of the realization is removed.

Example 3.27. Suppose that (A,B,C,D) is a realization of $\mathbf{G}(z)$ with

$$A = \begin{bmatrix} -0.6 & 0 & 0 \\ 0 & -0.6 & 2 \\ 0 & 0 & 2.2 \end{bmatrix}, \quad B = \begin{bmatrix} 1 & 0 \\ 0 & 1 \\ 2 & 1 \end{bmatrix},$$

$$C = \begin{bmatrix} 1 & -0.5 & 1 \\ 2 & -1 & 2 \end{bmatrix}, \quad D = \begin{bmatrix} 0 & 0 \\ 0 & 0 \end{bmatrix}.$$

There are two different eigenvalues $\{-0.6, 2.2\}$. To determine reachability of (A,B), the PBH test shows that

$$\operatorname{rank}\left\{\begin{bmatrix} \lambda I_3 - A & B \end{bmatrix}\right\} = \operatorname{rank}\left\{\begin{bmatrix} \lambda+0.6 & 0 & 0 & 1 & 0 \\ 0 & \lambda+0.6 & -2 & 0 & 1 \\ 0 & 0 & \lambda-2.2 & 2 & 1 \end{bmatrix}\right\}$$

is 3 at both eigenvalues $\lambda = -0.6$ and $\lambda = 2.2$. Hence, (A,B) is reachable. For observability, the PBH test gives

$$\operatorname{rank}\left\{\begin{bmatrix} \lambda I_3 - A \\ C \end{bmatrix}\right\} = \operatorname{rank}\left\{\begin{bmatrix} \lambda+0.6 & 0 & 0 \\ 0 & \lambda+0.6 & -2 \\ 0 & 0 & \lambda-2.2 \\ 1 & -0.5 & 1 \\ 2 & -1 & 2 \end{bmatrix}\right\}$$

$$= \begin{cases} 2, & \text{at } \lambda = -0.6, \\ 3, & \text{at } \lambda = 2.2. \end{cases}$$

It is thus concluded that (C,A) is detectable but unobservable. It can be verified that both gains

$$F = \begin{bmatrix} 0 & 0 & 0 \\ 0 & 0 & -2 \end{bmatrix}, \quad L = \begin{bmatrix} 0 & 0 \\ -2 & 0 \\ -2 & 0 \end{bmatrix}$$

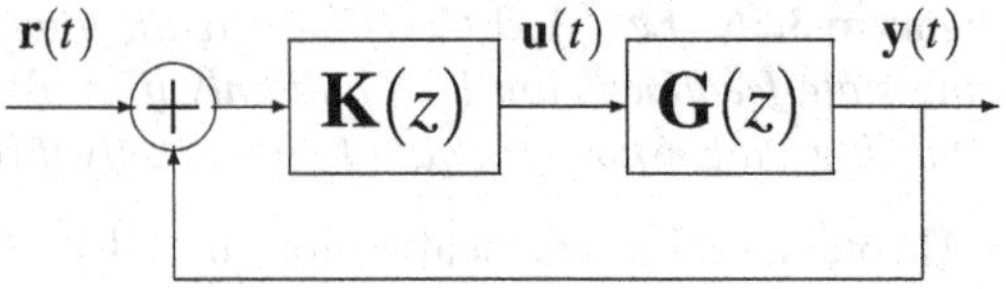

Fig. 3.6 Feedback control system

are stabilizing, yielding

$$A+BF = \begin{bmatrix} -0.6 & 0 & 0 \\ 0 & -0.6 & 0 \\ 0 & 0 & 0.2 \end{bmatrix}, \quad A+LC = \begin{bmatrix} -0.6 & 0.0 & 0.0 \\ -2.0 & 0.4 & 0.0 \\ -2.0 & 1.0 & 0.2 \end{bmatrix},$$

which are indeed stability matrices.

3.2.3 *All Stabilizing Controllers*

State feedback and state estimation are fundamental to MIMO feedback control system design. Indeed, all stabilizing controllers can be parameterized once the stabilizing gains for state feedback and state estimation are available. Consider the feedback control system as depicted in Fig. 3.6 where the physical plant represented by $\mathbf{G}(z)$ has m input/p output and the feedback controller represented by $\mathbf{K}(z)$ has p-input/m-output.

Note that the negative sign of the feedback path is absorbed in $\mathbf{K}(z)$, shown in its expression later. The feedback system in Fig. 3.6 is called *well posed*, if

$$\lim_{z\to\infty} \det\left[I_m - \mathbf{K}(z)\mathbf{G}(z)\right] \neq 0. \tag{3.29}$$

The well-posed condition ensures that all signals in the feedback system shown in Fig. 3.6 are causal (refer to Problem 3.27 in Exercises).

Let (A,B,C,D) be a realization of $\mathbf{G}(z)$ with A possibly unstable. Then the first priority in control system design is stabilization in the sense of Lyapunov or internal stability. For convenience, the reference input is taken to be $\mathbf{r}(t) = \mathbf{0}$. Design of stabilizing feedback controllers can proceed in two steps. In the first step, a state-feedback control law $\mathbf{u}(t) = F\mathbf{x}(t)$ is designed such that it stabilizes $\mathbf{x}(t+1) = (A+BF)\mathbf{x}(t)$. This is always possible provided that (A,B) is stabilizable. Since the measured output is $\mathbf{y}(t)$ not $\mathbf{x}(t)$, an estimator as in (3.27) is designed in the second step from which the control law $\mathbf{u}(t) = F\hat{\mathbf{x}}(t)$ is applied with $\hat{\mathbf{x}}(t)$ the estimated state vector. It turns out that such a design method works, if, in addition, (C,A) is detectable for which L exists such that $(A+LC)$ is a stability matrix. Indeed, by $\mathbf{u}(t) = F\hat{\mathbf{x}}(t)$ and (3.27),

$$\begin{aligned} \hat{\mathbf{x}}(t+1) &= A\hat{\mathbf{x}}(t) + B\mathbf{u}(t) + L\left[\hat{\mathbf{y}}(t) - \mathbf{y}(t)\right] \\ &= (A+BF+LC+LDF)\hat{\mathbf{x}}(t) - L\mathbf{y}(t). \end{aligned} \tag{3.30}$$

Thus, $\mathbf{K}(z) = -F(zI - A - BF - LC - LDF)^{-1}L$ is the feedback controller, which is often referred to as *observer-based* controller.

Lemma 3.1. *Let* $\mathbf{G}(z) = D + C(zI - A)^{-1}B$ *and* $\mathbf{K}(z)$ *be in (3.30) for the feedback system in Fig. 3.6. The closed-loop system in internally stable, if and only if* $(A+BF)$ *and* $(A+LC)$ *are stability matrices.*

Proof. Since $\mathbf{u}(t) = F\hat{\mathbf{x}}(t)$ and $y(t) = C\mathbf{x}(t) + D\mathbf{u}(t) = C\mathbf{x}(t) + DF\hat{\mathbf{x}}(t)$,

$$\begin{bmatrix} \mathbf{x}(t+1) \\ \hat{\mathbf{x}}(t+1) \end{bmatrix} = \begin{bmatrix} A & BF \\ -LC & A+LC+BF \end{bmatrix} \begin{bmatrix} \mathbf{x}(t) \\ \hat{\mathbf{x}}(t) \end{bmatrix}. \tag{3.31}$$

Internal stability of the feedback system is equivalent to asymptotic stability of the above state-space system. Define similarity transform via

$$\begin{bmatrix} \mathbf{x}_s(t) \\ \hat{\mathbf{x}}_s(t) \end{bmatrix} = T \begin{bmatrix} \mathbf{x}(t) \\ \hat{\mathbf{x}}(t) \end{bmatrix}, \quad T = \begin{bmatrix} I & -I \\ \mathbf{0} & I \end{bmatrix}. \tag{3.32}$$

The above similarity transform yields the following equivalent "A" matrix

$$\begin{bmatrix} I & -I \\ \mathbf{0} & I \end{bmatrix} \begin{bmatrix} A & BF \\ -LC & A+LC+BF \end{bmatrix} \begin{bmatrix} I & I \\ \mathbf{0} & I \end{bmatrix} = \begin{bmatrix} A+LC & \mathbf{0} \\ -LC & A+BF \end{bmatrix}.$$

for the closed-loop system. Hence, the feedback system in Fig. 3.6 is internally stable, if and only if $(A+BF)$ and $(A+LC)$ are stability matrices. □

At this moment, it is worth taking another look at the coprime fraction issue for MIMO systems. Instead of coprime fractions over polynomials of z^{-1}, coprime factorizations over proper and stable transfer matrices will be pursued. That is, factorizations

$$\mathbf{G}(z) = \mathbf{M}(z)^{-1}\mathbf{N}(z) = \tilde{\mathbf{N}}(z)\tilde{\mathbf{M}}(z)^{-1} \tag{3.33}$$

are searched for with $\{\mathbf{M}(z), \mathbf{N}(z)\}$ left coprime and $\{\tilde{\mathbf{N}}(z), \tilde{\mathbf{M}}(z)\}$ right coprime for all $|z| \geq 1$, and $\mathbf{M}(z), \mathbf{N}(z), \tilde{\mathbf{N}}(z), \tilde{\mathbf{M}}(z)$ are all proper and stable transfer matrices. By an abuse of notation denote

$$\mathbf{G}(z) = D + C(zI - A)^{-1}B = \left[\begin{array}{c|c} A & B \\ \hline C & D \end{array}\right]. \tag{3.34}$$

The 2×2 block matrix on the right-hand side of (3.34) represents the transfer matrix $\mathbf{G}(z)$ and helps related computations in terms of its realization. Suppose that $\mathbf{G}(z)$ has size $p \times m$. Then

$$D = \begin{bmatrix} D_{1,1} & \cdots & D_{1,m} \\ \vdots & \ddots & \vdots \\ D_{p,1} & \cdots & D_{p,m} \end{bmatrix}, \quad C = \begin{bmatrix} C_1 \\ \vdots \\ C_p \end{bmatrix}, \quad \begin{bmatrix} B_1 & \cdots & B_m \end{bmatrix}.$$

Thus, the (i,k)th element of $\mathbf{G}(z)$ is given by

$$G_{i,k}(z) = D_{i,k} + C_i(zI - A)^{-1}B_k = \left[\begin{array}{c|c} A & B_k \\ \hline C_i & D_{i,k} \end{array}\right]$$

for $1 \le i \le p$ and $1 \le k \le m$. Note that it is considerably more difficult to compute $G_{i,k}(z)$ based on the coprime fraction model which shows the power of the state-space method. More importantly, coprime factorization can be easily obtained with observer-based controller in (3.30). In fact, the existence of left coprime factorization is hinged to the detectability of (C,A) and right coprime factorization to the stabilizability of (A,B). The following result is true.

Theorem 3.28. *Let $\mathbf{G}(z)$ be a $p \times m$ transfer matrix as in (3.34) with a stabilizable and detectable realization (A,B,C,D). Suppose that F and L are stabilizing, i.e., $(A+BF)$ and $(A+LC)$ are stability matrices. Define*

$$\begin{bmatrix} \mathbf{V}(z) & \mathbf{U}(z) \\ -\mathbf{N}(z) & \mathbf{M}(z) \end{bmatrix} = \left[\begin{array}{c|cc} A+LC & -(B+LD) & L \\ \hline F & I_m & \mathbf{0} \\ C & -D & I_p \end{array}\right], \tag{3.35}$$

$$\begin{bmatrix} \tilde{\mathbf{M}}(z) & -\tilde{\mathbf{U}}(z) \\ \tilde{\mathbf{N}}(z) & \tilde{\mathbf{V}}(z) \end{bmatrix} = \left[\begin{array}{c|cc} A+BF & B & -L \\ \hline F & I_m & \mathbf{0} \\ C+DF & D & I_p \end{array}\right]. \tag{3.36}$$

Then $\mathbf{G}(z) = \mathbf{M}(z)^{-1}\mathbf{N}(z) = \tilde{\mathbf{N}}(z)\tilde{\mathbf{M}}(z)^{-1}$. Furthermore, there holds the double Bezout identity

$$\begin{bmatrix} \mathbf{V}(z) & \mathbf{U}(z) \\ -\mathbf{N}(z) & \mathbf{M}(z) \end{bmatrix} \begin{bmatrix} \tilde{\mathbf{M}}(z) & -\tilde{\mathbf{U}}(z) \\ \tilde{\mathbf{N}}(z) & \tilde{\mathbf{V}}(z) \end{bmatrix} = I_{m+p}. \tag{3.37}$$

Proof. Clearly, the coprime factors are all internally stable, by stability of $(A+BF)$ and $(A+LC)$. To show that $\mathbf{G}(z) = \mathbf{M}(z)^{-1}\mathbf{N}(z)$, it is noted that

$$\mathbf{M}(z)^{-1} = \left[\begin{array}{c|c} A+LC & L \\ \hline C & I_p \end{array}\right]^{-1} = \left[\begin{array}{c|c} A & L \\ \hline -C & I_p \end{array}\right].$$

See Problem 3.22 in Exercises. It follows that

$$\begin{aligned}
\mathbf{M}(z)^{-1}\mathbf{N}(z) &= \left[\begin{array}{c|c} A & L \\ \hline -C & I_p \end{array}\right] \left[\begin{array}{c|c} A+LC & (B+LD) \\ \hline C & D \end{array}\right] \\
&= \left[\begin{array}{cc|c} A+LC & \mathbf{0} & B+LD \\ LC & A & LD \\ \hline C & -C & D \end{array}\right] \quad \left\{ T = \begin{bmatrix} I_n & -I_n \\ \mathbf{0} & I_n \end{bmatrix} \right\} \\
&= \left[\begin{array}{cc|c} A & \mathbf{0} & B \\ LC & A+LC & LD \\ \hline C & \mathbf{0} & D \end{array}\right] = \left[\begin{array}{c|c} A & B \\ \hline C & D \end{array}\right] = \mathbf{G}(z),
\end{aligned}$$

where n is the order of (A,B,C,D). The formula for cascade realizations in Problem 3.21 is employed to obtain the realization for $\mathbf{M}(z)^{-1}\mathbf{N}(z)$, and the similarity transform T is used to arrive at the observability Kalman decomposition, which then leads to the realization of $\mathbf{G}(z)$. The proof for the right coprime factorization is similar and thus omitted. The proof for the double Bezout identity (3.37) is left as an exercise (Problem 3.22). □

There is an obvious difference between coprime factorizations and coprime fractions. The former requires relative coprimeness for $|z| \geq 1$, while the latter requires relative coprimeness for all $z \in \mathbb{C}$. It is claimed that coprime factorizations require only stabilizability and detectability of the realization as shown in Theorem 3.28, but coprime fractions require reachability and observability of the realization. Indeed, suppose that (A,B,C,D) is a minimal realization. Then (refer to (2.73) in Sect. 2.2.2 and Problem 3.26 in Exercises)

$$\text{rank}\left\{\begin{bmatrix} zI_n - (A+LC) & (B+LD) & L \\ -C & D & I_p \end{bmatrix}\right\} = n+p \;\; \forall\, z \in \mathbb{C}, \tag{3.38}$$

that is equivalent to $\{\mathbf{M}(z), \mathbf{N}(z)\}$ being left coprime $\forall\, z \in \mathbb{C}$. Similarly,

$$\text{rank}\left\{\begin{bmatrix} zI_n - (A+BF) & B \\ -F & I_m \\ -(C+DF) & D \end{bmatrix}\right\} = n+m \;\; \forall\, z \in \mathbb{C} \tag{3.39}$$

is equivalent to $\{\tilde{\mathbf{M}}(z), \tilde{\mathbf{N}}(z)\}$ being right coprime $\forall z \in \mathbb{C}$. It follows that polynomial matrices $\det(zI_n - A - LC)\mathbf{M}(z)$ and $\det(zI_n - A - LC)\mathbf{N}(z)$ are left coprime fractions of $\mathbf{G}(z)$ and, dually, $\det(zI_n - A - BF)\tilde{\mathbf{N}}(z)$ and $\det(zI_n - A - BF)\tilde{\mathbf{M}}(z)$ are right coprime fractions of $\mathbf{G}(z)$. In fact, the above coprime fractions are all polynomials of z^{-1} and have the least degrees (see Appendix for the definition of degree of polynomial matrices), by the minimality assumptions. Therefore, state-space realization theory also provides a numerical procedure in computing coprime fractions for transfer matrices.

Theorem 3.28 shows that the observer-based controller $\mathbf{K}(z)$ in (3.30) admits coprime factorizations

$$\mathbf{K}(z) = -\mathbf{V}(z)^{-1}\mathbf{U}(z) = -\tilde{\mathbf{U}}(z)\tilde{\mathbf{V}}(z)^{-1} \tag{3.40}$$

with $\{\mathbf{V}(z), \mathbf{U}(z)\}$ and $\{\tilde{\mathbf{V}}(z), \tilde{\mathbf{U}}(z)\}$ given in (3.35) and (3.36), respectively. The proof is left as an exercise (Problem 3.23). More importantly, the observed-based controller helps to parameterize all the stabilizing controllers as shown in the next result.

Theorem 3.29. *Under the same hypotheses of* Theorem *3.28 and that the feedback system in Fig. 3.6 is well posed, all stabilizing controllers are given by*

$$\begin{aligned} \mathbf{K}(z) &= -[\mathbf{V}(z) - \mathbf{Q}(z)\mathbf{N}(z)]^{-1}[\mathbf{U}(z) + \mathbf{Q}(z)\mathbf{M}(z)] \\ &= -\left[\tilde{\mathbf{U}}(z) + \tilde{\mathbf{M}}(z)\mathbf{Q}(z)\right]\left[\tilde{\mathbf{V}}(z) - \tilde{\mathbf{N}}(z)\mathbf{Q}(z)\right]^{-1} \end{aligned} \tag{3.41}$$

for some $\mathbf{Q}(z)$ *that is a proper and stable transfer matrix.*

Proof. In light of the double Bezout identity (3.37), there hold

$$[\mathbf{V}(z) - \mathbf{Q}(z)\mathbf{N}(z)]\,[I_m - \mathbf{K}(z)\mathbf{G}(z)]\,\tilde{\mathbf{M}}(z) = I_m,$$

$$\mathbf{M}(z)\,[I_m - \mathbf{G}(z)\mathbf{K}(z)]\left[\tilde{\mathbf{V}}(z) - \tilde{\mathbf{N}}(z)\mathbf{Q}(z)\right] = I_p.$$

Therefore, the controller $\mathbf{K}(z)$ in (3.41) is indeed stabilizing. Conversely, any stabilizing controller $\mathbf{K}_0(z)$ admits coprime factorizations

$$\mathbf{K}_0(z) = -\mathbf{V}_0(z)^{-1}\mathbf{U}_0(z) = -\tilde{\mathbf{U}}_0(z)\tilde{\mathbf{V}}_0(z)^{-1}. \tag{3.42}$$

One may always find a stabilizable and detectable realization for $\mathbf{K}_0(z)$ and then calculate its left and right coprime factors by applying Theorem 3.28 to $\mathbf{K}_0(z)$. Furthermore, coprime factors of $\mathbf{K}_0(z)$ can always be sought to satisfy

$$\mathbf{R}(z) = \mathbf{V}_0(z)\tilde{\mathbf{M}}(z) + \mathbf{U}_0(z)\tilde{\mathbf{N}}(z) = I_m,$$

$$\tilde{\mathbf{R}}(z) = \mathbf{M}(z)\tilde{\mathbf{V}}_0(z) + \mathbf{N}(z)\tilde{\mathbf{U}}_0(z) = I_p.$$

If not, the well-posed assumption implies that both $\mathbf{R}(z)^{-1}$ and $\tilde{\mathbf{R}}(z)^{-1}$ are proper and stable. Thus, $\mathbf{R}(z)^{-1}$ can be subsumed into $\mathbf{V}_0(z)$ and $\mathbf{U}_0(z)$, and $\tilde{\mathbf{R}}(z)^{-1}$ can be subsumed into $\tilde{\mathbf{V}}_0(z)$ and $\tilde{\mathbf{U}}_0(z)$. Notice that

$$\begin{bmatrix} \mathbf{V}_0 & \mathbf{U}_0 \end{bmatrix} \begin{bmatrix} \tilde{\mathbf{M}} & -\tilde{\mathbf{U}} \\ \tilde{\mathbf{N}} & \tilde{\mathbf{V}} \end{bmatrix} = \begin{bmatrix} I_m & \mathbf{U}_0\tilde{\mathbf{V}} - \mathbf{V}_0\tilde{\mathbf{U}} \end{bmatrix},$$

$$\left(\begin{bmatrix} \mathbf{V} & \mathbf{U} \end{bmatrix} + \mathbf{Q}\begin{bmatrix} -\mathbf{N} & \mathbf{M} \end{bmatrix}\right) \begin{bmatrix} \tilde{\mathbf{M}} & -\tilde{\mathbf{U}} \\ \tilde{\mathbf{N}} & \tilde{\mathbf{V}} \end{bmatrix} = \begin{bmatrix} I_m & \mathbf{Q} \end{bmatrix}.$$

by the double Bezout identity (3.37). Taking $\mathbf{Q}(z) = \mathbf{U}_0(z)\tilde{\mathbf{V}}(z) - \mathbf{V}_0(z)\tilde{\mathbf{U}}(z)$ leads to $\mathbf{V}_0(z) = \mathbf{V}(z) - \mathbf{Q}(z)\mathbf{N}(z)$ and $\mathbf{U}_0(z) = \mathbf{U}(z) + \mathbf{Q}(z)\mathbf{M}(z)$. Dually, the double Bezout identity (3.37) implies

$$\begin{bmatrix} \mathbf{V}(z) & \mathbf{U}(z) \\ -\mathbf{N}(z) & \mathbf{M}(z) \end{bmatrix} \begin{bmatrix} -\tilde{\mathbf{U}}_0(z) \\ \tilde{\mathbf{V}}_0(z) \end{bmatrix} = \begin{bmatrix} \mathbf{U}(z)\tilde{\mathbf{V}}_0(z) - \mathbf{V}(z)\tilde{\mathbf{U}}_0(z) \\ I_p \end{bmatrix}$$

$$\text{or} \quad \begin{bmatrix} \mathbf{V}(z) & \mathbf{U}(z) \\ -\mathbf{N}(z) & \mathbf{M}(z) \end{bmatrix} \begin{bmatrix} -\tilde{\mathbf{U}}(z) - \tilde{\mathbf{M}}(z)\tilde{\mathbf{Q}}(z) \\ \tilde{\mathbf{V}}(z) - \tilde{\mathbf{N}}(z)\tilde{\mathbf{Q}}(z) \end{bmatrix} = \begin{bmatrix} -\tilde{\mathbf{Q}}(z) \\ I_p \end{bmatrix},$$

where $\tilde{\mathbf{Q}}(z) = \mathbf{V}(z)\tilde{\mathbf{U}}_0(z) - \mathbf{U}(z)\tilde{\mathbf{V}}_0(z) = \mathbf{U}_0(z)\tilde{\mathbf{V}}(z) - \mathbf{V}_0(z)\tilde{\mathbf{U}}(z) = \mathbf{Q}(z)$ (Problem 3.31 in Exercises). Hence

$$\tilde{\mathbf{V}}_0(z) = \tilde{\mathbf{V}}(z) - \tilde{\mathbf{N}}(z)\mathbf{Q}(z), \quad \tilde{\mathbf{U}}_0(z) = \tilde{\mathbf{U}}(z) + \tilde{\mathbf{M}}(z)\mathbf{Q}(z).$$

Clearly, $\mathbf{Q}(z)$ is proper and stable. Therefore, the stabilizing controller (3.42) has the form in (3.41). □

The results of this subsection highlight the fact that the state space is also a computational tool for analysis and design of feedback control systems. The representation of the transfer matrix $\mathbf{G}(z)$ as in (3.34) facilitates the computations for coprime factorizations, coprime fractions, and parameterizations of all stabilizing controllers. Hence, design of optimal feedback control systems is made possible by searching stable $\mathbf{Q}(z)$ or $\tilde{\mathbf{Q}}(z)$ which are free parameters to be designed.

3.3 Time-Varying Systems

For a LTV system, its impulse response $\{h(t;k)\}$ has two arguments: One is the time index t and the other is the index of the impulse response at time t. As such, an LTV system is a family of (frozen time) LTI systems, indexed by time t. For MIMO LTV systems, impulse responses are denoted by $H(t;k)$. Often it is convenient to use the following form of the impulse response:

$$H(t,k) = H(t;t-k) \quad \forall\, t,k. \tag{3.43}$$

If $H(t,k) = H(t-k)$ for all t,k, then it represents an LTI system. By the causality, the input/output relation is governed by

$$\mathbf{y}(t) = \sum_{k=-\infty}^{t} H(t;t-k)\mathbf{u}(k) = \sum_{k=-\infty}^{t} H(t,k)\mathbf{u}(k). \tag{3.44}$$

LTV systems are more difficult to study than LTI systems mainly due to their time-varying nature. Their impulse responses are two-dimensional arrays as shown in Fig. 2.3. Commonly used methods include frozen time analysis for slowly time-varying systems and short-time Fourier analysis for nonstationary signals. This section investigates state-space representations for LTV systems rather than the external description in (3.44). It will be shown that those familiar notions in the previous sections can be generalized to LTV state-space systems. The emphasis will again be placed on observability, reachability, and stability. Nevertheless, many results in the linear system theory will have to be confined to local time intervals.

3.3.1 Realizations

A MIMO LTV state-space system is described by:

$$\mathbf{x}(t+1) = A_t\mathbf{x}(t) + B_t\mathbf{u}(t), \quad \mathbf{y}(t) = C_t\mathbf{x}(t) + D_t\mathbf{u}(t) \tag{3.45}$$

in which the realization matrices are time dependent. Moreover, there does not exists a simple relation between the realization and its impulse response.

Example 3.30. Consider the third-order ARMA model:

$$y(t) = \sum_{k=1}^{3} \mu_k(t)y(t-k) + \sum_{k=1}^{3} \nu_k(t)u(t-k). \tag{3.46}$$

If the frozen time method is used, then the transfer function at each frozen time t is

$$H_t(z) = \frac{\nu_1(t)z^{-1} + \nu_2(t)z^{-2} + \nu_3(t)z^{-3}}{1 - \mu_1(t)z^{-1} - \mu_2(t)z^{-2} - \mu_3(t)z^{-3}}. \tag{3.47}$$

The impulse response $\{h(t;k)\}$ is basically the inverse $\mathscr{Z}$ transform of $H_t(z)$ with t an integer parameter. Define $x_1(t) = y(t)$ and let

$$\begin{aligned}
x_3(t+1) &= \mu_3(t+3)x_1(t) + \nu_3(t+3)u(t) \\
&= \mu_3(t+3)y(t) + \nu_3(t+3)u(t), \\
x_2(t+1) &= \mu_2(t+2)x_1(t) + x_3(t) + \nu_2(t+2)u(t) \\
&= \mu_2(t+2)y(t) + \mu_3(t+2)y(t-1) \\
&\quad + \nu_3(t+2)u(t-1) + \nu_2(t+2)u(t), \\
x_1(t+1) &= \mu_1(t+1)x_1(t) + x_2(t) + \nu_1(t+1)u(t) \\
&= \mu_1(t+1)y(t) + \mu_2(t+1)y(t-1) + \mu_3(t+1)y(t-2) \\
&\quad + \nu_1(t+1)u(t) + \nu_2(t+1)u(t-1) + \nu_3(t+1)u(t-2).
\end{aligned}$$

Thus, $x_1(t+1) = y(t+1)$ has the same expression as in (3.46) except that t is replaced by $(t+1)$. So setting the state vector at time t as

$$\mathbf{x}(t) = \begin{bmatrix} x_1(t) & x_2(t) & x_3(t) \end{bmatrix}'$$

yields a time-varying state-space model in the same form as in (3.45) with

$$A_t = \begin{bmatrix} \mu_1(t+1) & 1 & 0 \\ \mu_2(t+2) & 0 & 1 \\ \mu_3(t+3) & 0 & 0 \end{bmatrix}, \quad B_t = \mathbf{b}_t = \begin{bmatrix} \nu_1(t+1) \\ \nu_2(t+2) \\ \nu_3(t+3) \end{bmatrix},$$
$$C_t = \mathbf{c}_t = \begin{bmatrix} 1 & 0 & 0 \end{bmatrix}, \quad D_t = d_t = 0.$$

It is seen that the pair (A_t, B_t) depends on the frozen time impulse responses at $(t+1), (t+2)$, and $(t+3)$. The transfer function for the frozen time state-space realization is $D_t + C_t(zI - A_t)^{-1}B_t$, which is different from the frozen time transfer function $H_t(z)$ as in (3.47).

It can be expected that the relation between the impulse response $H(t,k)$ and its state-space realization (A_t, B_t, C_t, D_t) is a complex one. For instance, it is difficult to obtain a state-space realization based on the impulse response directly. Nevertheless, ARMA models for LTV systems offer a procedure to obtain simple state-space realizations similarly to those for LTI systems. The details are skipped in order to keep the subsection concise. Interested readers are referred to other texts on linear system theory.

For the state-space system in (3.45) with order n, it is convenient to define the state transition matrix

$$\Phi_A(t,k) = A_{t-1}A_{t-2}\cdots A_k, \quad t > k. \tag{3.48}$$

It is a convention that $\Phi_A(t,k) = I_n$ whenever $t = k$ with n the order of the state-space system. For convenience, $\Phi(t,k) = \Phi_A(t,k)$ is taken. The solution $\mathbf{x}(t)$ to (3.45) at $t = t_f$ with the initial time $t_0 < t_f$ is given by

$$\mathbf{x}(t_f) = \Phi(t_f, t_0)\mathbf{x}(t_0) + \sum_{k=t_0}^{t_f-1} \Phi(t_f, k+1)B_k\mathbf{u}(k). \tag{3.49}$$

The time-varying nature of the system indicates that notions of reachability and observability are time dependent. Denote the input and output over the time interval $[t_0, t_f)$ by column vectors

$$\mathscr{Y}_{t_0,t_f-1} = \begin{bmatrix} \mathbf{y}(t_0) \\ \mathbf{y}(t_0+1) \\ \vdots \\ \mathbf{y}(t_f-1) \end{bmatrix}, \quad \tilde{\mathscr{U}}_{t_0,t_f-1} = \begin{bmatrix} \mathbf{u}(t_f-1) \\ \vdots \\ \mathbf{u}(t_0+1) \\ \mathbf{u}(t_0) \end{bmatrix}. \tag{3.50}$$

Then $\mathbf{x}(t_f)$ as in (3.49) has the expression

$$\mathbf{x}(t_f) = \Phi(t_f,t_0)\mathbf{x}(t_0) + \mathscr{R}_{t_0,t_f}(A_t,B_t)\tilde{\mathscr{U}}_{t_0,t_f-1}, \tag{3.51}$$

where $\mathscr{R}_{t_0,t_f}(A_t,B_t)$, denoted by $\mathscr{R}_{t_0,t_f}$ for simplicity, is the reachability matrix over the time interval $[t_0, t_f)$ given by

$$\mathscr{R}_{t_0,t_f} := \begin{bmatrix} B_{t_f-1} & \Phi(t_f,t_f-1)B_{t_f-2} & \cdots & \Phi(t_f,t_0+1)B_{t_0} \end{bmatrix}. \tag{3.52}$$

It is noted that if the realization is time invariant, then $\Phi(t_f,t_f-k) = A^k$, $B_t = B$, and thus, $\mathscr{R}_{t_0,t_f}(A_t,B_t)$ reduces to $\mathscr{R}_\ell(A,B)$ as in the previous sections with $\ell = t_f - t_0$. The observability matrix over $[t_0, t_f)$ is defined dually as

$$\mathscr{O}_{t_0,t_f} := \mathscr{O}_{t_0,t_f}(C_t,A_t) = \begin{bmatrix} C_{t_0} \\ C_{t_0+1}\Phi(t_0+1,t_0) \\ \vdots \\ C_{t_f-1}\Phi(t_f-1,t_0) \end{bmatrix}. \tag{3.53}$$

There holds equality

$$\mathscr{Y}_{t_0,t_f-1} = \mathscr{O}_{t_0,t_f}\mathbf{x}(t_0) + \mathscr{T}_{t_0,t_f}\mathscr{U}_{t_0,t_f-1}, \tag{3.54}$$

where $\mathscr{U}_{t_0,t_f-1}$ is defined in the same way as $\mathscr{Y}_{t_0,t_f-1}$ with $\mathbf{y}(t)$ replaced by $\mathbf{u}(t)$, and $\mathscr{T}_{t_0,t_f}$ is a lower block triangular matrix of size $\ell p \times \ell m$ with $\ell = t_f - t_0$, and p and m the dimensions of the output and the input, respectively. Denote $\mathscr{T}(i,k)$ as the (i,k)th block of $\mathscr{T}_{t_0,t_f}$ with size $p \times m$. Then

$$\mathscr{T}(i,k) = \begin{cases} \mathbf{0}_{p\times m}, & i < k, \\ D_{t_0+k-1}, & i = k, \\ C_{t_0+i-1}\Phi(t_0+i-1,t_0+k)B_{t_0+k-1}, & i > k. \end{cases}$$

Note that (3.51) reduces to (3.7) for the case $t_0 = 0$ and $t_f = \ell$, if the realization (A_t,B_t,C_t,D_t) is independent of time t.

Different from LTI systems, reachability and observability are defined locally in terms of time t.

Definition 3.6. Consider the time-varying state-space equations in (3.45). Let $t_f > t_0$. (i) The pair (A_t, B_t) is said to be reachable over $[t_0, t_f)$, if for any given state target $\mathbf{x}_T$, there exist bounded control inputs $\{\mathbf{u}(t)\}_{t=t_0}^{t_f-1}$ such that $\mathbf{x}(t_f) = \mathbf{x}_T$. (ii) The pair (C_t, A_t) is said to be observable over $[t_0, t_f)$, if given true input and output data $\{\mathbf{u}(t)\}_{t=t_0}^{t_f-1}$ and $\{\mathbf{y}(t)\}_{t=t_0}^{t_f-1}$, the initial condition $\mathbf{x}(t_0)$ can be reconstructed uniquely.

The following result regards reachability and observability for LTV state-space systems.

Theorem 3.31. *The pair (A_t, B_t) is reachable over $[t_0, t_f)$, if and only if the reachability matrix $\mathscr{R}_{t_0,t_f}$ as defined in (3.52) has the full row rank. The pair (C_t, A_t) is observable over $[t_0, t_f)$, if and only if the observability matrix $\mathscr{O}_{t_0,t_f}$ as defined in (3.53) has the full column rank.*

Proof. Suppose that the reachability matrix $\mathscr{R}_{t_0,t_f}$ has the full row rank. Then $\mathscr{R}_{t_0,t_f}\mathscr{R}_{t_0,t_f}^*$ is a nonsingular matrix of size $n \times n$. For any target state vector $\mathbf{x}_T$, there exists control input $\tilde{\mathscr{U}}_{t_0,t_f-1}$ such that (3.51) holds for $\mathbf{x}(t_f) = \mathbf{x}_T$. That is, the pair (A_t, B_t) is reachable over the time interval $[t_0, t_f)$. One such control input is given by

$$\tilde{\mathscr{U}}_{t_0,t_f-1} = \mathscr{R}_{t_0,t_f}^* \left(\mathscr{R}_{t_0,t_f}\mathscr{R}_{t_0,t_f}^*\right)^{-1} \left[\mathbf{x}_T - \Phi(t_0,t_f)\mathbf{x}(t_0)\right].$$

Conversely, if the pair (A_t, B_t) is reachable over $[t_0, t_f)$, then there exists control input sequence $\{\mathbf{u}(t)\}_{t=t_0}^{t_f-1}$ such that (3.51) holds for any target $\mathbf{x}(t_f) = \mathbf{x}_T$. Consequently, $\mathscr{R}_{t_0,t_f}$ has the full row rank. The proof for the observability over the time interval $[t_0, t_f)$ is similar and is skipped. □

Similar to time-invariant systems, Lyapunov equations are introduced:

$$Q_{t,t_f} = A_t Q_{t+1,t_f} A_t^* + B_t B_t^*, \quad Q_{t_f,t_f} = \mathbf{0}, \tag{3.55}$$

$$P_{t_0,t+1} = A_t^* P_{t_0,t} A_t + C_t^* C_t, \quad P_{t_0,t_0} = \mathbf{0}, \tag{3.56}$$

where $t_0 \le t < t_f$. These are difference Lyapunov equations with opposite time recursions: One is forward and the other backward. The solutions at the two boundaries are given by the reachability and observability gramians:

$$Q_{t_0,t_f} = \sum_{k=t_0}^{t_f-1} \Phi(t_f,k+1) B_k B_k^* \Phi(t_f,k+1)^*, \tag{3.57}$$

$$P_{t_0,t_f} = \sum_{k=t_0}^{t_f-1} \Phi(k,t_0)^* C_k^* C_k \Phi(k,t_0), \tag{3.58}$$

over the time interval $[t_0, t_f)$, respectively. The next result is true.

Theorem 3.32. *The pair (A_t, B_t) is reachable over $[t_0, t_f)$, if and only if the reachability gramian Q_{t_0,t_f} in (3.57) is positive definite. The pair (C_t, A_t) is observable over $[t_0, t_f)$, if and only if the observability gramian P_{t_0,t_f} in (3.58) is positive definite.*

Proof. The results in (i) and (ii) can be easily established by noting that

$$Q_{t_0,t_f} = \mathscr{R}_{t_0,t_f}\mathscr{R}^*_{t_0,t_f}, \quad P_{t_0,t_f} = \mathscr{O}^*_{t_0,t_f}\mathscr{O}_{t_0,t_f},$$

and by applying the results in Theorem 3.31. □

As discussed earlier, there is no simple relation between the realization and the impulse response for the same time-varying system. However, there exists an algebraic relation between the two as stated next.

Lemma 3.2. *Let $\{H(t,k)\}$ be the impulse response. Then (A_t, B_t, C_t, D_t) is its corresponding realization, if and only if*

$$H(t,k) = D_t\delta(t-k) + C_t\Phi(t,k+1)B_k[1-\delta(t-k)], \quad \forall\, t \geq k. \tag{3.59}$$

Proof. Recall the definition of the transition matrix $\Phi(t,k)$. If (3.59) holds, then in light of the convolution relation in (3.44), (A_t, B_t, C_t, D_t) is a realization for the given input/output description of the time-varying system via its impulse response $\{H(t,k)\}$. Conversely, given a state-space realization (A_t, B_t, C_t, D_t), its state response has the expression in (3.49). Taking $t_0 \to -\infty$ with $\mathbf{x}(t_0) = \mathbf{0}$ yields

$$\mathbf{x}(t_f) = \sum_{k=-\infty}^{t_f-1} \Phi(t_f, k+1)B_k\mathbf{u}(k).$$

Hence, the output of the underlying state-space system at time t is given by

$$\mathbf{y}(t) = D_t\mathbf{u}(t) + \sum_{k=-\infty}^{t-1} C_t\Phi(t,k+1)B_k\mathbf{u}(k) = \sum_{k=-\infty}^{t} H(t,k)\mathbf{u}(k).$$

It follows that $H(t,k)$ in (3.59) is the corresponding impulse response. □

Even though the algebraic relation in (3.59) is not of much help in finding state-space realizations based on impulse responses, it does help to characterize the minimal realizations for time-varying systems. For this purpose, it is beneficial to consider time-dependent similarity transforms $\{T_t\}$ satisfying $\det(T_t) \neq 0$ for all t. Consider time-varying state-space model in (3.45). The similarity transform $\tilde{\mathbf{x}}(t) = T_t\mathbf{x}(t)$ yields a new realization

$$\left(\tilde{A}_t, \tilde{B}_t, \tilde{C}_t, \tilde{D}_t\right) = \left(T_{t+1}A_tT_t^{-1}, T_{t+1}B_t, C_tT_t^{-1}, D_t\right). \tag{3.60}$$

It is left as an exercise (Problem 3.33) to show that

$$T_{t_f}\mathscr{R}_{t_0,t_f} = T_{t_f}\mathscr{R}_{t_0,t_f}(A_t,B_t) = \mathscr{R}_{t_0,t_f}\left(\tilde{A}_t,\tilde{B}_t\right) =: \tilde{\mathscr{R}}_{t_0,t_f}, \tag{3.61}$$

$$\mathscr{O}_{t_0,t_f}T_{t_0}^{-1} = \mathscr{O}_{t_0,t_f}(C_t,A_t)T_{t_0}^{-1} = \mathscr{O}_{t_0,t_f}\left(\tilde{C}_t,\tilde{A}_t\right) =: \tilde{\mathscr{O}}_{t_0,t_f}. \tag{3.62}$$

The next result is useful.

Lemma 3.3. *Let G and F be two matrices with dimensions $p \times n$ and $n \times q$, respectively, where $q < n$. Suppose that $r_g + q \leq n$ with r_g the rank of G. Then $GF = \mathbf{0}_{p\times q}$, if and only if there exists a nonsingular matrix T such that*

$$GT^{-1} = \begin{bmatrix} G_o & \mathbf{0}_{p\times q} \end{bmatrix}, \quad TF = \begin{bmatrix} \mathbf{0}_{(n-q)\times q} \\ F_o \end{bmatrix}. \tag{3.63}$$

Proof. It is easy to see that (3.63) implies $GF = \mathbf{0}_{p\times q}$. The difficult part is the converse. So suppose that $GF = \mathbf{0}_{p\times q}$. The proof needs show the existence of T such that (3.63) holds. As r_g is the rank of G, $r_g \leq p$. Let r_f be the rank of F. Then $r_f \leq q$. In light of QR factorizations,

$$G = R_g V^*, \quad F = UR_f, \quad V^*V = I_{r_g}, \quad U^*U = I_{r_f}.$$

Such U and V are called orthogonal matrices. Since R_g and R_f have full ranks, the assumption $GF = \mathbf{0}_{p\times q}$ yields

$$V^*U = \mathbf{0}_{r_g\times r_f}, \quad r_f + r_g \leq q + r_g \leq n.$$

It follows that orthogonal $U_\perp$ and $V_\perp$ exist such that

$$\tilde{U} = \begin{bmatrix} U & U_\perp \end{bmatrix}, \quad \tilde{V} = \begin{bmatrix} V & V_\perp \end{bmatrix}$$

are unitary matrices of size $n \times n$. The hypothesis $r_g + q \leq n$ implies that $\rho = n - r_g \geq q \geq r_f$. Setting $T_1 = \tilde{V}^*$ gives

$$GT_1^{-1} = G\tilde{V} = R_g V^* \begin{bmatrix} V & V_\perp \end{bmatrix} = \begin{bmatrix} R_g & \mathbf{0}_{p\times(n-r_g)} \end{bmatrix}, \tag{3.64}$$

$$T_1 F = \tilde{V}^* U R_f = \begin{bmatrix} V^* \\ V_\perp^* \end{bmatrix} U R_f = \begin{bmatrix} \mathbf{0}_{r_g\times q} \\ V_\perp^* U R_f \end{bmatrix} \tag{3.65}$$

by $V^*V_\perp = \mathbf{0}$ and $V^*U = \mathbf{0}$. Thus, if $\rho = n - r_g = q$, then (3.63) holds true by taking $T = T_1$. If $\rho > q$, then there exists factorization $V_\perp^* U R_f = Q\tilde{R}_f$ where Q is an orthogonal matrix of size $\rho \times q$ and $\tilde{R}_f$ is square. Moreover, $Q_\perp$ exists such that with

$$\tilde{Q} = \begin{bmatrix} Q_\perp & Q \end{bmatrix}, \quad T_2 = \begin{bmatrix} I_{r_g} & \mathbf{0} \\ \mathbf{0} & \tilde{Q}^* \end{bmatrix},$$

$\tilde{Q}$ is a unitary matrix and T_2 is nonsingular. Setting $T = T_2T_1$ yields

$$GT^{-1} = \begin{bmatrix} R_g & \mathbf{0}_{p\times(n-r_g)} \end{bmatrix} T_2^{-1} = \begin{bmatrix} R_g & \mathbf{0}_{p\times(\rho-q)} & \mathbf{0}_{p\times q} \end{bmatrix},$$

$$TF = T_2 \begin{bmatrix} \mathbf{0}_{r_g\times r_f} \\ V_\perp^* U R_f \end{bmatrix} = \begin{bmatrix} I_{r_g} & \mathbf{0} \\ \mathbf{0} & \tilde{Q}^* \end{bmatrix} \begin{bmatrix} \mathbf{0}_{r_g\times q} \\ Q\tilde{R}_f \end{bmatrix} = \begin{bmatrix} \mathbf{0}_{(n-q)\times q} \\ \tilde{R}_f \end{bmatrix},$$

where (3.64), (3.65), and $V_\perp^* U R_f = Q\tilde{R}_f$ are used. The above expressions are the same as in (3.63), thereby concluding the proof. □

In Lemma 3.3, the hypothesis $r_g + q \leq n$ is indispensable which ensures the existence of some nonsingular matrix T for (3.63) to hold. The number of rows for G is irrelevant. Lemma 3.3 is now used to prove the Kalman decomposition for LTV systems.

Theorem 3.33. *Let (A_t, B_t, C_t, D_t) be a realization of order n.* (i) *Suppose that (A_t, B_t) is unreachable over $[t_0, t_f)$. Then there exist an integer $r_c < n$ and similarity transforms $\{T_t\}_{t=t_0}^{t_f}$ such that*

$$\tilde{A}_t = T_{t+1}A_tT_t^{-1} = \begin{bmatrix} \tilde{A}_c(t) & \tilde{A}_{c\bar{c}}(t) \\ \mathbf{0} & \tilde{A}_{\bar{c}}(t) \end{bmatrix}, \quad \tilde{B}_t = T_{t+1}B_t = \begin{bmatrix} \tilde{B}_c(t) \\ \mathbf{0} \end{bmatrix} \tag{3.66}$$

for $t_0 \leq t < t_f$ where the pair $\{\tilde{A}_c(t), \tilde{B}_c(t)\}$ is reachable over $[t_0, t_f)$ and $\tilde{A}_c(t)$ and $\tilde{B}_c(t)$ have r_c rows, provided that rank$\{B_t\} \leq r_c$ *for $t_0 \leq t < t_f$.* (ii) *Suppose that (C_t, A_t) is unobservable over $[t_0, t_f)$. Then there exist an integer $r_o < n$ and similarity transforms $\{S_t\}_{t=t_0}^{t_f}$ such that*

$$\tilde{C}_t = C_tS_t^{-1} = \begin{bmatrix} \tilde{C}_o(t) & \mathbf{0} \end{bmatrix}, \quad \tilde{A}_t = S_{t+1}A_tS_t^{-1} = \begin{bmatrix} \tilde{A}_o(t) & \mathbf{0} \\ \tilde{A}_{o\bar{o}}(t) & \tilde{A}_{\bar{o}}(t) \end{bmatrix} \tag{3.67}$$

for $t_0 \leq t < t_f$ where the pair $\{\tilde{C}_o(t), \tilde{A}_o(t)\}$ is observable over $[t_0, t_f)$ and $\tilde{A}_o(t)$ and $\tilde{C}_o(t)$ have r_o columns, provided that rank$\{C_t\} \leq r_o$ *for $t_0 \leq t < t_f$.*

Proof. Because (i) and (ii) are dual to each other, only (ii) will be shown. By the hypothesis on unobservability over $[t_0, t_f)$, the observability matrix $\mathscr{O}_{t_0,t_f}(C_t, A_t)$ as in (3.53) has rank $r < n$. That is, an elementary matrix E exists such that

$$\mathscr{O}_{t_0,t_f}(C_t, A_t)E = \mathscr{O}_{t_0,t_f}(\tilde{C}_t, \tilde{A}_t) = \begin{bmatrix} \tilde{\mathscr{O}}_{t_0,t_f} & \mathbf{0}_{pn\times(n-r)} \end{bmatrix} \tag{3.68}$$

for some full column rank matrix $\tilde{\mathscr{O}}_{t_0,t_f}$ where p is the dimension of the output. The induction process is used to show the existence of $\{S_t\}_{t=t_0}^{t_f}$ such that (3.67) holds. It is noted that (3.68) implies

$$C_k \Phi_A(k,t_0) E \begin{bmatrix} \mathbf{0}_{r\times(n-r)} \\ I_{n-r} \end{bmatrix} = \mathbf{0}_{p\times(n-r)} \quad \text{for} \quad t_0 \le k < t_f. \tag{3.69}$$

For $k = t_0$, take $S_{t_0} = E^{-1}$ and note that $\Phi_A(t_0,t_0) = I_n$ leading to

$$\tilde{C}_{t_0} = C_{t_0} S_{t_0}^{-1} = \begin{bmatrix} \tilde{C}_{\mathrm{o}}(t) & \mathbf{0}_{p\times(n-r)} \end{bmatrix}. \tag{3.70}$$

For $k = t_0 + 1$, $\Phi_A(t_0+1,t_0) = A_{t_0}$. Let

$$G = C_{t_0+1}, \quad F = A_{t_0} S_{t_0}^{-1} \begin{bmatrix} \mathbf{0}_{r\times(n-r)} \\ I_{n-r} \end{bmatrix}.$$

Then (3.69) implies that $GF = \mathbf{0}_{p\times q}$ with $q = n-r$. Since $\mathrm{rank}\{C_t\} \le r = n-q$ for $t_0 \le t < t_f$ by taking $r_{\mathrm{o}} = r$, Lemma 3.3 can be employed to conclude the existence of S_{t_0+1} such that

$$\tilde{C}_{t_0+1} = C_{t_0+1} S_{t_0+1}^{-1} = \begin{bmatrix} \tilde{C}_{\mathrm{o}}(t_0+1) & \mathbf{0}_{p\times(n-r)} \end{bmatrix}, \tag{3.71}$$

$$\tilde{A}_{t_0} = S_{t_0+1} A_{t_0} S_{t_0}^{-1} = \begin{bmatrix} \tilde{A}_{\mathrm{o}}(t_0) & \mathbf{0}_{r\times(n-r)} \\ \tilde{A}_{\mathrm{o}\bar{\mathrm{o}}}(t_0) & \tilde{A}_{\bar{\mathrm{o}}}(t_0) \end{bmatrix}. \tag{3.72}$$

To complete the induction process assume that $\{S_t\}_{t=t_0}^{t_k-1}$ exist such that

$$\tilde{C}_i = C_i S_i^{-1} = \begin{bmatrix} \tilde{C}_{\mathrm{o}}(i) & \mathbf{0}_{p\times(n-r)} \end{bmatrix}, \tag{3.73}$$

$$\tilde{A}_{i-1} = S_i A_{i-1} S_{i-1}^{-1} = \begin{bmatrix} \tilde{A}_{\mathrm{o}}(i-1) & \mathbf{0}_{r\times(n-r)} \\ \tilde{A}_{\mathrm{o}\bar{\mathrm{o}}}(i-1) & \tilde{A}_{\bar{\mathrm{o}}}(i-1) \end{bmatrix}, \tag{3.74}$$

where $t_0 < i < t_k$ for some $t_k < t_f$. At $k = t_k$, (3.74) implies that

$$\Phi_{\tilde{A}}(t_k-1,t_0) = \begin{bmatrix} \tilde{\Phi}_{1,1} & \mathbf{0}_{r\times(n-r)} \\ \tilde{\Phi}_{1,2} & \tilde{\Phi}_{2,2} \end{bmatrix}$$

by the fact that the product of lower block triangular matrices of the same partition is again lower block triangular. Thus, (3.69) can be written as

$$C_{t_k} A_{t_k-1} S_{t_k-1}^{-1} \Phi_{\tilde{A}}(t_k-1,t_0) \begin{bmatrix} \mathbf{0}_{r\times(n-r)} \\ I_{n-r} \end{bmatrix} = \mathbf{0}_{p\times(n-r)}$$

by $k = t_k$. Recall the expression of $\Phi_{\tilde{A}}(t_k - 1, t_0)$. Taking $G = C_{t_k}$ and

$$F = A_{t_k-1}S_{t_k-1}^{-1}\Phi_{\tilde{A}}(t_k - 1, t_0)\begin{bmatrix} \mathbf{0}_{r\times(n-r)} \\ I_{n-r} \end{bmatrix} = A_{t_k-1}S_{t_k-1}^{-1}\begin{bmatrix} \mathbf{0}_{r\times(n-r)} \\ I_{n-r} \end{bmatrix}\tilde{\Phi}_{2,2}$$

implies that $GF = \mathbf{0}_{p\times q}$ with $q = n - r$. Since $\text{rank}\{C_t\} \leq r = n - q$ for $t_0 \leq t < t_f$ with $r_{\text{o}} = r$, Lemma 3.3 can be used again to conclude the existence of nonsingular S_{t_k} such that

$$\tilde{C}_{t_k} = C_{t_k}S_{t_k}^{-1} = \begin{bmatrix} \tilde{C}_{\text{o}}(t_k) & \mathbf{0}_{p\times(n-r)} \end{bmatrix}, \tag{3.75}$$

$$S_{t_k}A_{t_k-1}S_{t_k-1}^{-1}\begin{bmatrix} \mathbf{0}_{r\times(n-r)} \\ I_{n-r} \end{bmatrix}\tilde{\Phi}_{2,2} = \begin{bmatrix} \mathbf{0}_{r\times(n-r)} \\ \tilde{A}_{\bar{o}}(t_k - 1) \end{bmatrix}\tilde{\Phi}_{2,2}. \tag{3.76}$$

That is, a nonsingular S_{t_k} exists such that $\tilde{C}_{t_k-1} = C_{t_k-1}S_{t_k-1}^{-1}$ and $\tilde{A}_{t_k-1} = S_{t_k}A_{t_k-1}S_{t_k-1}^{-1}$ have the same form as in (3.67), concluding the proof for the observability Kalman decomposition. □

Complete Kalman decomposition can also be carried out for time-varying systems. The proof of the following result is left as an exercise (Problem 3.34).

Theorem 3.34. *Let (A_t, B_t, C_t, D_t) be a time-varying state-space realization which is unreachable and unobservable over $[t_0, t_f)$ with $t_f > t_0$. Then there exists a set of similarity transforms $\{T_t\}_{t=t_0}^{t_f}$ such that*

$$\tilde{A}_t = T_{t+1}A_tT_t^{-1} = \begin{bmatrix} \tilde{A}_{\text{co}}(t) & \mathbf{0} & \tilde{A}_{13}(t) & \mathbf{0} \\ \tilde{A}_{21}(t) & \tilde{A}_{\text{c}\bar{\text{o}}}(t) & \tilde{A}_{23}(t) & \tilde{A}_{24}(t) \\ \mathbf{0} & \mathbf{0} & \tilde{A}_{\bar{\text{c}}\text{o}}(t) & \mathbf{0} \\ \mathbf{0} & \mathbf{0} & \tilde{A}_{43}(t) & \tilde{A}_{\bar{\text{c}}\bar{\text{o}}}(t) \end{bmatrix},$$

$$\tilde{B}_t = T_{t+1}B_t = \begin{bmatrix} \tilde{B}_{\text{co}}(t)' & \tilde{B}_{\text{c}\bar{\text{o}}}(t)' & \mathbf{0} & \mathbf{0} \end{bmatrix}',$$

$$\tilde{C}_t = C_tT_t^{-1} = \begin{bmatrix} \tilde{C}_{\text{co}}(t) & \mathbf{0} & \tilde{C}_{\bar{\text{c}}\text{o}}(t) & \mathbf{0} \end{bmatrix},$$

where $\{\tilde{A}_{\text{co}}(t), \tilde{B}_{\text{co}}(t)\}$ is reachable and $\{\tilde{C}_{\text{co}}(t), \tilde{A}_{\text{co}}(t)\}$ is observable over the time interval $[t_0, t_f)$.

Theorem 3.34 does not indicate whether a reachable and observable realization is minimal over the time interval $[t_0, t_f)$. In addition, there are difficulties in extending the reachability and observability from the local time interval to the infinity time horizon. Partition of the time axis into a collection of nonoverlapping local time intervals does not work due to two reasons. The first is that the conventional realization theory requires implicitly that $\{A_t\}$ be square. Hence, if the minimal possible order of (A_t, B_t, C_t, D_t) is different over each local time interval, minimal realization is not possible over all time t. The second is the computational difficulty. It is noted that if the kth local time interval is $[t_0(k), t_f(k))$, then $T_{t_f(k)} = T_{t_0(k+1)}$ needs hold, which is not possible in general by the fact that T_{t_0} and T_{t_f} accomplish different goals in Kalman decomposition as shown in the proof of Theorem 3.32. The only known exception is the periodic time-varying systems with $\ell = t_f - t_0$ the period. See Problem 3.35 in Exercises. As such, it is not possible to use Kalman decomposition to search for minimal realizations over the infinite time horizon. Instead, the emphasis is placed on characterizations of minimal realizations. For this purpose, a new notion is needed.

Definition 3.7. (i) The pair (A_t, B_t) is uniformly reachable, if there exists some integer $\ell_c > 0$ and some real number $\delta_c > 0$ such that

$$Q_{t,t+\ell_c} = \mathscr{R}_{t,t+\ell_c}\mathscr{R}^*_{t,t+\ell_c} \geq \delta_c I_n \quad \forall t.$$

(ii) The pair (C_t, A_t) is uniformly observable, if there exists some integer $\ell_o > 0$ and some real number $\delta_o > 0$ such that

$$P_{t,t+\ell_o} = \mathscr{O}^*_{t,t+\ell_o}\mathscr{O}_{t,t+\ell_o} \geq \delta_o I_n \quad \forall t.$$

The uniform reachability will be called ℓ-step reachability, if $\ell_c = \ell$. Similarly, the uniform observability will be called ℓ-step observability, if $\ell_o = \ell$. The Definition 3.7 is much stronger than the local reachability and observability over $[t_0, t_f)$ because now the local time interval $[t_0, t_f)$ with $t_f = t_0 + \ell$ is "moving" by changing t_0 so that the time axis is covered by the collection of all such local time intervals. The next result follows.

Theorem 3.35. *Let (A_t, B_t, C_t, D_t) be a realization for the given time-varying system whose impulse response is $\{H(t,k)\}$. Suppose that (A_t, B_t) is ℓ-step reachable and (C_t, A_t) is ℓ-step observable for some $\ell > 0$. Then (A_t, B_t, C_t, D_t) is a minimal realization.*

Proof. The proof is similar to that for time-invariant systems. Assume that (A_t, B_t, C_t, D_t) is both ℓ-step reachable and observable, but its order n is not minimal. Then there exists a different realization $(\tilde{A}_t, \tilde{B}_t, \tilde{C}_t, D_t)$ with an order $\tilde{n}$ smaller than n for the same impulse response $\{H(t,k)\}$. Hence, for $t > k$,

$$H(t,k) = C_t \Phi_A(t, k+1) B_k = \tilde{C}_t \Phi_{\tilde{A}}(t, k+1)\tilde{B}_k.$$

By the definition of the transition matrix, there holds

$$\Phi(t_k + k, t_k)\Phi(t_k, t_k - i) = \Phi(t_k + k, t_k - i) \tag{3.77}$$

for all positive integers i, k, and t_k. It follows that

$$\begin{aligned}
\mathscr{H}_{t_0,t_f} &= \begin{bmatrix} H(t_k+1,t_i-1) & H(t_k+1,t_i-2) & \cdots & H(t_k+1,t_i-\ell) \\ H(t_k+2,t_i-1) & H(t_k+2,t_i-2) & \cdots & H(t_k+2,t_k-\ell) \\ \vdots & \cdots & \cdots & \vdots \\ H(t_k+\ell,t_i-1) & H(t_k+\ell,t_i-2) & \cdots & H(t_k+\ell,t_i-\ell) \end{bmatrix} \\
&= \mathscr{O}_{t_k+1,t_f+1}(C_t,A_t)\Phi_A(t_k+1,t_i)\mathscr{R}_{t_0,t_i}(A_t,B_t) \\
&= \mathscr{O}_{t_k+1,t_f+1}(\tilde{C}_t,\tilde{A}_t)\Phi_{\tilde{A}}(t_k+1,t_i)\mathscr{R}_{t_0,t_i}(\tilde{A}_t,\tilde{B}_t),
\end{aligned}$$

where $t_f - t_k = t_i - t_0 = \ell$. Hence for $t_i = t_k + 1$,

$$\Phi_A(t_k+1,t_i) = I_n \neq \Phi_{\tilde{A}}(t_k+1,t_i) = I_{\tilde{n}}.$$

As a result, $\text{rank}\{\mathscr{H}_{t_0,t_f}\} < n$. Thus, at least one of the ranks of $\mathscr{O}_{t_k+1,t_f+1}(C_t,A_t)$ and $\mathscr{R}_{t_0,t_i}(A_t,B_t)$ is smaller than n, implying that either (A_t,B_t) is not ℓ-step reachable, (C_t,A_t) is not ℓ-step observable, or both. This contradicts the assumption that (A_t,B_t,C_t,D_t) is both ℓ-step reachable and observable. The proof is now completed. □

As a concluding remark for this subsection, it needs to be pointed out that the uniform reachability/observability and ℓ-step reachability/observability play the same role as reachability/observability for LTI systems. This point will become more clear in the next subsection where uniform stabilizability and detectability will be introduced. It also needs to be pointed out that the realization theory for LTV systems is a difficult subject in general due to the time-varying nature of the corresponding realizations. The time-dependent similarity transforms are prohibitive in computing reachable and observable realizations over the infinite time horizon. For this and other reasons, the results presented in this subsection are not as complete as one wishes them to be but do capture the essentials of the realization theory for LTV systems.

3.3.2 Stability

The notion of stability for LTV state-space systems is similar to the one for the asymptotic stability as investigated in Sect. 3.2 for LTI systems. The following is adapted from Definition 3.3 with a minor modification.

Definition 3.8. The state-space system (3.45) is asymptotically stable, or stable in the Lyapunov sense, if the solution to

$$\mathbf{x}(t+1) = A_t\mathbf{x}(t), \quad \mathbf{x}(t_0) = \mathbf{x}_0, \tag{3.78}$$

satisfies $\|\mathbf{x}(t)\| \to 0$ as $t \to \infty$ for arbitrary $\mathbf{x}_0$ and any $t_0 \geq 0$.

It is noted that if $A_t = \mathbf{0}$ for some $t > t_0$ with t_0 a fixed integer, then $\mathbf{x}(t_f) = \Phi(t_f, t_0)\mathbf{x}_0 = \mathbf{0}_n$ for all $t_f > t$ regardless of $\mathbf{x}_0$. Thus, the initial time t_0 cannot be fixed in defining stability for LTV state-space systems in order for it to be inclusive. For this reason, stability as in Definition 3.8 is also termed *uniform asymptotic stability*. The next result is true.

Lemma 3.4. *Consider the difference Lyapunov equation*

$$X_{t_0,t+1} = A_t^* X_{t_0,t} A_t + Z_t, \quad X_{t_0,t_0} = \mathbf{0}, \tag{3.79}$$

where Z_t is positive definite for all t. Suppose that the limiting solution

$$X_{t_0} := \lim_{t \to \infty} X_{t_0,t} = \sum_{k=t_0}^{\infty} \Phi(k,t_0)^* Z_k \Phi(k,t_0)$$

is bounded for any $t_0 \geq 0$. Then the state-space system (3.78) is asymptotically stable.

Proof. The solutions to the difference Lyapunov equation (3.79) are given by

$$X_{t_0,t} = \sum_{k=t_0}^{t-1} \Phi(k,t_0)^* Z_k \Phi(k,t_0).$$

Each of the terms in the above summation is nonnegative. The hypothesis on Z_t and $X_{t_0,t}$ in the limit implies that X_{t_0} is bounded and positive definite. By the fact that $Z_t > 0$ for all $t \geq t_0$,

$$\lim_{k \to \infty} \Phi(k,t_0) = \mathbf{0} \implies \lim_{k \to \infty} \Phi(k,t_0)\mathbf{x}_0 = \mathbf{0}_n.$$

That is, $\|\mathbf{x}(t)\| \to 0$ asymptotically for any initial condition $\mathbf{x}_0$ and any initial time $t_0 \geq 0$. As a result, the state-space system (3.78) is asymptotically stable, thereby concluding the proof. □

Different from LTI systems, asymptotic stability does not imply that the difference Lyapunov equation (3.79) has bounded solutions asymptotically.

Example 3.36. Consider the unforced state-space system of order 1 with

$$A_t = \sqrt{\left|\frac{t+k_0}{t+k_0+1}\right|}, \quad t = 0, \pm 1, \pm 2, \ldots. \tag{3.80}$$

Assume that k_0 is real but not an integer. Then for $t_f > t_0$,

$$\Phi(t_f,t_0) = A_{t_f-1} \cdots A_{t_0+1} A_{t_0} = \sqrt{\left|\frac{t_0+k_0}{t_f+k_0}\right|} \to 0 \text{ as } t_f \to \infty.$$

Thus, the asymptotic stability holds. Let $Z_t = 1$ for all t. It follows that the limiting solution to (3.79) is given by

$$X_{t_0} = \lim_{t\to\infty} X_{t_0,t} = \sum_{k=t_0}^{\infty} \Phi(k,t_0)^* Z_k \Phi(k,t_0) = \sum_{k=t_0}^{\infty} \sqrt{\left(\frac{t_0+k_0}{k+k_0}\right)^2} \to \infty.$$

That is, asymptotic stability does not necessarily imply that the limiting solution to (3.79) is bounded. Now consider the case when

$$A_t = \left|\frac{t+k_0}{t+k_0+1}\right|, \quad t = 0, \pm 1, \pm 2, \ldots, \quad k_0 > 0. \tag{3.81}$$

The asymptotic stability can be concluded similarly. Let $Z_t = 1$ for all t. Then the limiting solution to (3.79) is given by

$$X_{t_0} = \lim_{t\to\infty} X_{t_0,t} = \sum_{k=t_0}^{\infty} \Phi(k,t_0)^* Z_k \Phi(k,t_0) = \sum_{k=t_0}^{\infty} \left(\frac{t_0+k_0}{k+k_0}\right)^2 < \infty.$$

That is, there exist bounded solutions $\{X_{t_0,t}\}$ to (3.79) as $t \to \infty$.

A moment of reflection indicates that the deficiency in stability testing for LTV systems lies in the lack of equivalence of the asymptotic stability and the exponential stability (refer to Sect. 2.2.3). For convenience, the exponential stability is next defined again with some suitable modification on its statement.

Definition 3.9. The system $\mathbf{x}(t+1) = A_t\mathbf{x}(t)$ is exponentially stable, if with any initial time t_0,

$$\|\mathbf{x}(t)\| \le \alpha\beta^{(t-t_0)} \quad \forall\, t \ge t_0$$

for some $\alpha > 0$ and $0 < \beta < 1$, where α may depend on $\mathbf{x}(t_0)$.

For LTI state-space systems, exponential stability and asymptotic stability are equivalent to each other. On the other hand, the exponential stability is not a necessary condition for the difference Lyapunov equation (3.79) to have bounded solutions, which is the case when A_t is given in (3.81). Hence, it is the decay rate of $\|\Phi(t,t_0)\|$ to zero as $t \to \infty$ that determines whether the Lyapunov stability criterion in Lemma 3.4 can be used as a necessary and sufficient condition to test the asymptotic stability of the LTV systems. The next result is more interesting.

Theorem 3.37. *Consider the difference Lyapunov equation*

$$Y_{t,t_f} = A_t^* Y_{t+1,t_f} A_t + Z_t, \quad Y_{t_f,t_f} = \mathbf{0}, \tag{3.82}$$

where $t < t_f$ and $\varepsilon I_n \le Z_t \le \delta I_n$ for all t and some $\delta > \varepsilon > 0$. Suppose that there exists $\alpha > 0$ such that for all finite t,

$$Y_t := \lim_{t_f \to \infty} Y_{t,t_f} = \lim_{t_f \to \infty} \sum_{k=t}^{t_f-1} \Phi(t_f,k+1)Z_k\Phi(t_f,k+1)^* \geq \alpha I_n.$$

Then the state-space system (3.78) is exponentially stable, if and only if the limiting solutions $\{Y_t\}$ are bounded.

Proof. Suppose that the limiting solutions $\{Y_t\}$ are bounded. Then

$$\alpha I_n \leq Y_t \leq \beta I_n \quad \forall\, t \tag{3.83}$$

for some $\beta > \alpha > 0$, which in turn implies that

$$A_t^* Y_{t+1} A_t - Y_t = -Z_t \leq -\varepsilon I_n \tag{3.84}$$

by taking the limit $t_f \to \infty$ in (3.82). Thus for any given t_0 and $t \geq t_0$, multiplying (3.84) by $\mathbf{x}(t)^*$ from left and by $\mathbf{x}(t)$ from right yields

$$\mathbf{x}(t+1)^* Y_{t+1}\mathbf{x}(t+1) - \mathbf{x}(t)^* Y_t \mathbf{x}(t) \leq -\varepsilon \|\mathbf{x}(t)\|^2,$$

where $\mathbf{x}(t+1) = A_t\mathbf{x}(t)$ is used. The inequality (3.83) leads to

$$\mathbf{x}(t)^* Y_t \mathbf{x}(t) \leq \beta \|\mathbf{x}(t)\|^2$$

which is equivalent to

$$-\|\mathbf{x}(t)\|^2 \leq -\beta^{-1}\mathbf{x}(t)^* Y_t \mathbf{x}(t).$$

Therefore for all $t \geq t_0$ there holds

$$\mathbf{x}(t+1)^* Y_{t+1}\mathbf{x}(t+1) - \mathbf{x}(t)^* Y_t \mathbf{x}(t) \leq -\varepsilon\beta^{-1}\mathbf{x}(t)^* Y_t \mathbf{x}(t).$$

The above can be rewritten as

$$\mathbf{x}(t+1)^* Y_{t+1}\mathbf{x}(t+1) \leq \left(1 - \frac{\varepsilon}{\beta}\right)\mathbf{x}(t)^* Y_t \mathbf{x}(t). \tag{3.85}$$

Since ε is a lower bound for Z_t, it can be taken such that

$$0 \leq \gamma^2 = 1 - \frac{\varepsilon}{\beta} < 1.$$

It follows from (3.85) that

$$\mathbf{x}(t+1)^* Y_{t+1}\mathbf{x}(t+1) \leq \gamma^{2(t+1-t_0)}\mathbf{x}(t_0)^* Y_{t_0}\mathbf{x}(t_0)$$

for all $t \geq t_0$. The bounds in (3.83) can then be used to conclude that

$$\|\mathbf{x}(t+1)\| \leq \gamma^{t+1-t_0}\sqrt{\beta/\alpha}\|\mathbf{x}(t_0)\|$$

from which the ex potential stability for (3.78) is proven. Now suppose that (3.78) is exponentially stable. Then for all $t \geq t_0$, there holds

$$\|\Phi(t,t_0)\| \leq \xi\gamma^{t-t_0}, \quad \xi > 0, \quad 0 < \gamma < 1.$$

The hypothesis on Z_t and the expression of Y_t yield

$$\|Y_t\| \leq \delta\xi^2 \lim_{t_f\to\infty} \sum_{k=t}^{t_f-1} \gamma^{2(t_f-k-1)} = \delta\xi^2 \sum_{i=0}^{\infty} \gamma^{2i} = \frac{\delta\xi^2}{1-\gamma^2}$$

which is bounded for all $t \geq t_0$, establishing the theorem. □

It is easy to see that the two difference Lyapunov equations (3.79) and (3.82) are different from each other in that one is forward recursion, and the other is backward recursion. While the boundedness of X_{t_0} can only conclude the asymptotic stability, the boundedness of Y_t can in fact establish the exponential stability with the aid of some additional mild assumptions. In fact, the lower bound $Y_t \geq \alpha I_n > 0$ can be removed as shown in the next result which improves the Lyapunov stability criterion in Theorem 3.37.

Theorem 3.38. *Consider the difference Lyapunov equation*

$$Y_{t,t_f} = A_t^* Y_{t+1,t_f} A_t + C_t^* C_t, \quad Y_{t_f,t_f} = \mathbf{0}, \tag{3.86}$$

where $t < t_f$ and (C_t, A_t) is ℓ-step observable. Let $\{Y_t\}$ be the limits of $\{Y_{t,t_f}\}$ as $t_f \to \infty$. Then the state-space system (3.78) is exponentially stable, if and only if $\{Y_t\}$ are bounded for all finite t.

Proof. It is easy to show that exponential stability of (3.78) implies that the limiting solution to (3.86) satisfies

$$Y_t = \lim_{t_f\to\infty} Y_{t,t_f} = \lim_{t_f\to\infty} \sum_{k=t}^{t_f-1} \Phi(t_f,k+1)C_k^*C_k\Phi(t_f,k+1)^* \leq \beta I_n$$

for some $\beta > 0$ by the same argument as that in the proof of Theorem 3.37. Now suppose that $\{Y_t\}$ are bounded. That is, $Y_t \leq \beta I_n$ for some $\beta > 0$. Let $V[t,\mathbf{x}(t)]$ be the Lyapunov function, defined by

$$V[t,\mathbf{x}(t)] = \mathbf{x}(t)^*[Y_t + \varepsilon I_n]\mathbf{x}(t), \quad \varepsilon > 0.$$

Denote $V_t = V[t, \mathbf{x}(t)]$. Then direct calculations show

$$V_{t+1} = V_t - \mathbf{x}(t)^*[\varepsilon I_n + C_t^* C_t - \varepsilon A_t^* A_t]\mathbf{x}(t),$$
$$V_{t+\ell} = V_t - \mathbf{x}(t)^*[\varepsilon I_n + P_{t,t+\ell} - \varepsilon \Phi(t+\ell,t)^* \Phi(t+\ell,t)]\mathbf{x}(t),$$

where $P_{t,t+\ell} = \mathcal{O}_{t,t+\ell}^* \mathcal{O}_{t,t+\ell}$ is the observability gramian as defined in (3.58). The hypothesis on ℓ-step observability implies that

$$P_{t,t+\ell} = \mathcal{O}_{t,t+\ell}^* \mathcal{O}_{t,t+\ell} \geq \alpha I_n$$

for some $\alpha > 0$. It follows that

$$\begin{aligned} V_t - V_{t+\ell} &= \mathbf{x}^*(t)[\varepsilon I_n + P_{t,t+\ell} - \varepsilon \Phi(t+\ell,t)^* \Phi(t+\ell,t)]\mathbf{x}(t) \\ &\geq (\alpha + \varepsilon[1-r]) \|\mathbf{x}(t)\|^2, \end{aligned}$$

where $r > 0$ satisfies $\Phi(t+\ell,t)^* \Phi(t+\ell,t) \leq r I_n$. Hence, $\varepsilon > 0$ in the Lyapunov function $V_t = V[t, \mathbf{x}(t)]$ can be taken sufficiently small such that

$$\rho = \alpha + \varepsilon[1-r] > 0.$$

Because $Y_t \leq \beta I_n$, $\|\mathbf{x}\|^2 \geq (\beta + \varepsilon)^{-1} V_t$, which establishes

$$V_t - V_{t+\ell} \geq \rho \|\mathbf{x}(t)\|^2 \geq \rho(\beta + \varepsilon)^{-1} V_t.$$

The above implies that $V_{t+\ell} \leq \gamma V_t$ with

$$0 < \gamma = 1 - \rho(\beta + \varepsilon)^{-1} < 1$$

by $0 < V_{t+\ell} < V_t$ whenever $\|\mathbf{x}(t)\| \neq 0$. As a result,

$$V_{t_0 + k\ell} \leq \gamma^k \|\mathbf{x}(t_0)\|^2 \to 0$$

exponentially as $k \to 0$. The fact that

$$\varepsilon \|\mathbf{x}(t)\|^2 \leq V[\mathbf{x}(t)] \leq (\beta + \varepsilon) \|\mathbf{x}(t)\|^2$$

establishes that $\|\mathbf{x}(t)\| \to 0$ exponentially as well. □

In the above proof, the boundedness of A_t and C_t for all t is assumed implicitly which holds for the section. Careful readers may notice that the proofs for Theorems 3.37 and 3.38 are quite similar. In fact, these two proofs can be adapted to prove further results in Lyapunov stability criteria parallel to that for LTI systems. However, the detectability notion needs be extended first from that for LTI systems.

Definition 3.10. The pair (C_t, A_t) is *uniformly detectable*, if there exist positive integers $\ell, k \geq 0$ and constants α, β with $0 < \beta < 1$, $0 < \alpha < \infty$ such that whenever

$$\|\Phi(t+k,t)\mathbf{v}\| \geq \beta \|\mathbf{v}\| \tag{3.87}$$

for some column vector $\mathbf{v}$ and time t, then

$$\mathbf{v}^* P_{t,t+\ell}\mathbf{v} \geq \alpha \mathbf{v}^*\mathbf{v}, \tag{3.88}$$

where $P_{t,t+\ell}$ is the observability gramian as defined in (3.58).

It is interesting to note how detectability is extended from that of LTI systems without aid of modal observability. It is also interesting to observe how uniform observability, i.e., $P_{t,t+\ell} \geq \delta I_n > 0$, is weakened to (3.88) subject to (3.87). The next theorem is the last result of this section. Because the proof is similar to that for Theorem 3.38, it is left as an exercise (Problem 3.38).

Theorem 3.39. *Consider the difference Lyapunov equation (3.86) where $t < t_f$, and (C_t, A_t) is uniformly detectable. Let $\{Y_t\}$ be the limits of $\{Y_{t,t_f}\}$ as $t_f \to \infty$. Then the state-space system (3.78) is exponentially stable, if and only if $\{Y_t\}$ are bounded for all finite t.*

Uniform stabilizability can be defined similarly.

Definition 3.11. The pair (A_t, B_t) is *uniformly stabilizable*, if there exist positive integers $\ell, k \geq 0$ and constants α, β with $0 < \beta < 1$, $0 < \alpha < \infty$ such that whenever

$$\|\Phi(t+k,t)\mathbf{v}\| \geq \beta \|\mathbf{v}\| \tag{3.89}$$

for some column vector $\mathbf{v}$ and time t, then

$$\mathbf{v}^* Q_{t,t+\ell}\mathbf{v} \geq \alpha \mathbf{v}^*\mathbf{v}, \tag{3.90}$$

where $Q_{t,t+\ell}$ is the reachability gramian as defined in (3.57).

The next result is basically the same as Theorem 3.39 by replacing A_t with A_t^*, and C_t by B_t^*. Hence, the proof is skipped.

Corollary 3.1. *Consider the difference Lyapunov equation*

$$X_{t,t_f} = A_t X_{t+1,t_f} A_t^* + B_t B_t^*, \quad X_{t_f,t_f} = \mathbf{0}, \tag{3.91}$$

where $t < t_f$, and (A_t, B_t) are uniformly stabilizable. Let $\{X_t\}$ be the limits of $\left\{X_{t,t_f}\right\}$ as $t_f \to \infty$. Then the state-space system (3.78) is exponentially stable, if and only if $\{X_t\}$ are bounded for all finite t.

It is noted that if the backward recursions in the Lyapunov equations (3.86) and (3.91) are changed into forward recursions, then the exponential stability in

Theorem 3.39 and Corollary 3.1 cannot be claimed, and the boundedness of the corresponding solutions is sufficient (not necessary) to ensure only asymptotic stability. The details are omitted.

Notes and References

Realization and stability theory for MIMO system are largely owe to Kalman [55, 56, 61]. Other people also contributed to the system theory. A sample of references includes [4, 34, 62, 65, 82, 89]. Several textbooks on linear system theory are favorite of this author, including [21, 53, 94]. Several other books [6, 18, 29, 30, 33] are listed here for further reading.

Exercises

3.1. Given the pair (C,A) with C dimension $p \times n$ and A $n \times n$, derive an algorithm for computing the observability matrix $\mathcal{O}_n$ such that its computational complexity is in the order of pn^3. In addition, program the algorithm using Matlab and test it with several numerical examples.

3.2. Find realizations in block canonical observer and controller form for both

$$\mathbf{H}_1(z) = \frac{1}{A(z)}\begin{bmatrix} B_1(z) & B_2(z) \end{bmatrix}, \quad \mathbf{H}_2(z) = \begin{bmatrix} B_1(z) \\ B_2(z) \end{bmatrix}\frac{1}{A(z)},$$

where $A(z) = z^n + a_1 z^{n-1} + \cdots + a_n$ and $B_k(z) = b_{k,1}z^{n-1} + \cdots + b_{k,n}$ for $k = 1,2$. Answer the following: If $\mathbf{H}(z)$ has size $p \times m$, which (observer or controller form) realization has smaller order for the case $p > m$ and $p < m$?

3.3. The matrix A_o in Example 3.4 for the SISO system $H(z)$ is called *left* companion matrix. (i) Show that

$$\det(\lambda I_n - A_o) = \lambda^n - \mu_1\lambda^{n-1} - \cdots - \mu_{n-1}\lambda - \mu_n.$$

(ii) Let $\mathbf{c}_o$ be the same as in Example 3.4. Show that

$$\mathbf{c}_o\,(zI_n - A_o)^{-1} = \begin{bmatrix} z^{n-1} & \cdots & z & 1 \end{bmatrix} / \det(zI_n - A_o).$$

(iii) If all eigenvalues $\{\lambda_i\}_{i=1}^n$ of A_o are distinct, show that

$$\mathbf{q}_i = \begin{bmatrix} \lambda_i^{n-1} & \cdots & \lambda_i & 1 \end{bmatrix}$$

is a left eigenvector of A_o associated with eigenvalue λ_i. That is, $\mathbf{q}_i A_o = \lambda_i \mathbf{q}_i$. (iv) Suppose that all eigenvalues $\{\lambda_i\}_{i=1}^n$ are distinct. Denote $\mathbf{e}_i$ as a column vector of size n with all zero elements except a one in the ith position. Let Vandermonde matrix be defined as

$$Q = \begin{bmatrix} \mathbf{q}_1' & \mathbf{q}_2' & \cdots & \mathbf{q}_n' \end{bmatrix}'. \tag{3.92}$$

Show that Q^{-1} exists, and $\mathbf{v}_i = Q^{-1}\mathbf{e}_i$ is the ith eigenvector corresponding eigenvalue λ_i.

3.4. Let $\mathbf{c}_o$ and A_o be as in Example 3.4 which have an observer form. Let $\Gamma_n(\mu)$ be a lower triangular Toeplitz matrix defined by

$$\Gamma_n(\mu) = \begin{bmatrix} 1 & 0 & \cdots & 0 \\ -\mu_1 & 1 & \ddots & \vdots \\ \vdots & \ddots & \ddots & 0 \\ -\mu_{n-1} & \cdots & -\mu_1 & 1 \end{bmatrix}.$$

Show that $\Gamma_n(\mu)^{-1} = \mathcal{O}_n$ is the observability matrix of $(\mathbf{c}_o, A_o)$.

3.5. (Observability form) For the SISO transfer function $H(z)$ in Example 3.4, denote $\{h(t)\}$ as its impulse response. (i) Show that

$$\mathbf{b}_{ob} = \begin{bmatrix} h(1) \\ h(2) \\ \vdots \\ h(n) \end{bmatrix} = \begin{bmatrix} 1 & 0 & \cdots & 0 \\ -\mu_1 & 1 & \ddots & \vdots \\ \vdots & \ddots & \ddots & 0 \\ -\mu_{n-1} & \cdots & -\mu_1 & 1 \end{bmatrix}^{-1} \begin{bmatrix} v_1 \\ v_2 \\ \vdots \\ v_n \end{bmatrix}.$$

(ii) Show that with $\mathbf{b}_{ob}$ as above, $\mathbf{c}_{ob} = \begin{bmatrix} 1 & \mathbf{0}_{n-1}^* \end{bmatrix}$, and

$$A_{ob} = \begin{bmatrix} \mathbf{0}_{n-1} & I_{n-1} \\ \mu_n & \tilde{\mathbf{v}}_{n-1}' \end{bmatrix}, \quad \tilde{\mathbf{v}}_{n-1}' = \begin{bmatrix} \mu_{n-1} & \cdots & \mu_1 \end{bmatrix},$$

$(A_{ob}, \mathbf{b}_{ob}, \mathbf{c}_{ob}, d)$ is a realization of $H(z)$. (iii) Show that the observability matrix $\mathcal{O}_n = I_n$, which is why it is called *observability form.* (iv) Find the block observability form for the transfer matrix $\mathbf{H}(z)$ as in Example 3.4.

3.6. Let (A, B, C, D) be a realization of some $m \times m$ transfer matrix with $n \times n$ the dimension of A. Show that the pair

$$\left(\begin{bmatrix} C & D \end{bmatrix}, \begin{bmatrix} A & B \\ \mathbf{0}_{m\times n} & \mathbf{0}_{m\times m} \end{bmatrix} \right)$$

is observable, if and only if (C,A) is observable and

$$\det\left(\begin{bmatrix} A & B \\ C & D \end{bmatrix}\right) \neq 0.$$

3.7. Prove Theorem 3.7. (*Hint:* Show first that

$$\mathbf{x}(n) = A^n \mathbf{x}_0 + \mathscr{R}_n(A,B)\mathscr{U}_n$$

is the solution to $\mathbf{x}(t+1) = A\mathbf{x}(t) + B\mathbf{u}(t)$, where

$$\mathscr{U}_n = \mathrm{vec}\left(\begin{bmatrix} \mathbf{u}(n-1) & \cdots & \mathbf{u}(1) & \mathbf{u}(0) \end{bmatrix}\right), \quad \mathbf{x}(0) = \mathbf{x}_0$$

and then conclude that $\mathscr{R}_n(A,B)$ needs have full row rank.)

3.8. Prove Theorem 3.9.

3.9. (i) Show that PBH test in Theorem 3.6 is equivalent to

$$\mathrm{rank}\left\{\begin{bmatrix} zI_n - A \\ C \end{bmatrix}\right\} = n \ \forall\ z \in \mathbb{C}.$$

(ii) Show that PBH test in Theorem 3.10 is equivalent to

$$\mathrm{rank}\left\{\begin{bmatrix} zI_n - A & B \end{bmatrix}\right\} = n \ \forall\ z \in \mathbb{C}.$$

3.10. (i) Show that (C,A) is unobservable, if and only if there exists a square matrix $P \neq \mathbf{0}$ such that $PA = AP$ and $CP = \mathbf{0}$. (ii) Show that (A,B) is unreachable, if and only if there exists a square matrix $Q \neq \mathbf{0}$ such that $AQ = QA$ and $QB = \mathbf{0}$.

3.11. (Modal observability and reachability) Let (A,B,C,D) be a realization of order n. (i) Show that a mode corresponding to eigenvalue λ of A is unobservable, if and only if

$$\mathrm{rank}\left\{\begin{bmatrix} \lambda I_n - A \\ C \end{bmatrix}\right\} < n.$$

(ii) Show that a mode corresponding to eigenvalue λ of A is unreachable, if and only if

$$\mathrm{rank}\left\{\begin{bmatrix} \lambda I_n - A & B \end{bmatrix}\right\} < n.$$

3.12. Find a minimal realization for

$$\text{(i) } \mathbf{G}(z) = \begin{bmatrix} \dfrac{z+1}{z^2 - z + \frac{2}{9}} & \dfrac{1}{z - \frac{1}{3}} \\ \dfrac{-z^2 + z + 1}{z^3 + \frac{2}{3}z^2 - \frac{1}{9}z - \frac{2}{27}} & \dfrac{z - \frac{1}{4}}{z^2 + z + \frac{2}{9}} \end{bmatrix},$$

$$\text{(ii) } \mathbf{G}(z) = \begin{bmatrix} \dfrac{z+1}{z^2 - z + \frac{1}{4}} & \dfrac{1}{z - \frac{1}{2}} \\ \dfrac{-z^2 + z + 1}{z^3 + \frac{1}{2}z^2 - \frac{1}{4}z - \frac{1}{8}} & \dfrac{z - \frac{1}{4}}{z^2 + z + \frac{1}{4}} \end{bmatrix}.$$

3.13. (Transmission zeros) Let (A,B,C,D) be a realization for the transfer matrix $\mathbf{H}(z)$ of size $p \times m$. Suppose that $\det(z_0 I_n - A) \neq 0$.

(i) For the case $p \geq m$, show that $z = z_0$ is an input zero of $\mathbf{H}(z)$, if and only if there is a nonzero solution to

$$\begin{bmatrix} z_0 I_n - A & B \\ -C & D \end{bmatrix} \begin{bmatrix} \mathbf{x}_0 \\ \mathbf{u}_0 \end{bmatrix} = \begin{bmatrix} \mathbf{0}_n \\ \mathbf{0}_p \end{bmatrix}.$$

Show that in this case there exists an initial condition $\mathbf{x}(0) = \mathbf{x}_0$ and an input $\mathbf{u}(t) = \mathbf{u}_0 z_0^t$ for $t \geq 0$ such that $\mathbf{y}(t) \equiv \mathbf{0}$ for all $t \geq 0$.

(ii) For the case $p \leq m$, show that $z = z_0$ is an output zero of $\mathbf{H}(z)$, if and only if there is a nonzero solution to

$$\begin{bmatrix} \tilde{\mathbf{x}}_0^* & \tilde{\mathbf{y}}_0^* \end{bmatrix} \begin{bmatrix} z_0 I_n - A & B \\ -C & D \end{bmatrix} = \begin{bmatrix} \mathbf{0}_n^* & \mathbf{0}_m^* \end{bmatrix}.$$

What is the interpretation of the transmission zero for the case $p < m$?

3.14. (Poles) Suppose that p_0 is a pole of $\mathbf{H}(z) = D + C(zI - A)^{-1}B$. Show that there exits an initial condition such that with input identically zero for each time sample, the output $\mathbf{y}(t) = \mathbf{v}p_0^t$ for some nonzero column vector $\mathbf{v}$ and each $t \geq 0$.

3.15. (Parallel realization) Suppose that $\{A_i\}_{i=1}^{\ell}$ are square and (A_i, A_k) have no common eigenvalues whenever $i \neq k$. Show that realization (A,B,C,D), given by

$$A = \operatorname{diag}(A_1, \ldots, A_\ell), \quad B^* = \begin{bmatrix} B_1^* & \cdots & B_\ell^* \end{bmatrix}^*, \quad C = \begin{bmatrix} C_1 & \cdots & C_\ell \end{bmatrix},$$

is minimal, if and only if (A_i, B_i) is reachable and (C_i, A_i) is observable for all i.

3.16. Consider a SISO system with realization $(A, \mathbf{b}, \mathbf{c})$. (i) Let A be a stability matrix and X be a solution to

$$X = AXA + \mathbf{bc}.$$

Show that if $(A, \mathbf{b}, \mathbf{c})$ is not a minimal realization, then X is singular. (ii) Suppose that Y is a unique solution to

$$AY + YA + \mathbf{bc} = \mathbf{0}.$$

Show that if Y is singular, then $(A, \mathbf{b}, \mathbf{c})$ is not a minimal realization.

Fig. 3.7 Cascade connection

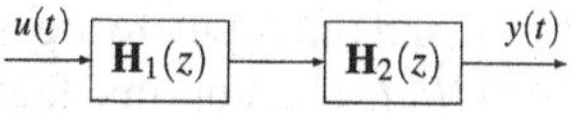

3.17. Show that any two minimal realizations for the same transfer matrix are equivalent to each other in the sense that a similarity transform exists which connects the two realizations.

3.18. Let $(A, \mathbf{b}, \mathbf{c}, d)$ be a realization of a SISO system. (i) Show that, if $(\mathbf{c}, A)$ is observable, then there exists a similarity transform T_o such that $\left(\mathbf{c}T_o^{-1}, T_oAT_o^{-1}\right)$ is in the observer form. (ii) Show that, if $(A, \mathbf{b})$ is reachable, then there exists a similarity transform T_c such that $\left(T_cAT_c^{-1}, T_c\mathbf{b}\right)$ is in the controller form.

3.19. Show that the Lyapunov equation (3.18) can be rewritten as

$$\left[I - A' \otimes A^*\right] \mathrm{vec}(P) = \mathrm{vec}(Z),$$

where $\otimes$ denotes the Kroneker product, and $\mathrm{vec}(P)$ and $\mathrm{vec}(Z)$ denote column vectors obtained by stacking columns of P and Z, respectively. Show also that the Lyapunov equation (3.18) has a unique solution, if and only $\lambda_i\bar{\lambda}_k \neq 1$ for all $i \neq k$, where $\{\lambda_k\}$ are eigenvalues of A.

3.20. Let (A, B, C, D) be a realization of order n, and A be stable. Let P and Q be the observability and reachability gramians, respectively. That is,

$$P = A^*PA + C^*C, \quad Q = AQA^* + BB^*.$$

(i) Let $P = U\Sigma_oU^*$ be SVD where U is unitary, and Σ_o is diagonal and of rank $r_o < n$. Show that similarity transform $T = U^*$ yields observability Kalman decomposition.
(ii) Let $Q = V\Sigma_cV^*$ be SVD where V is unitary, and Σ_c is diagonal and of rank $r_c < n$. Show that similarity transform $T = V^*$ yields reachability Kalman decomposition.

3.21. (Cascade realization) Consider cascade connection in Fig. 3.7. Let (A_1, B_1, C_1, D_1) be realization of $\mathbf{H}_1(z)$, and (A_2, B_2, C_2, D_2) be realization of $\mathbf{H}_2(z)$.

(i) Show that $\mathbf{H}(z) = \mathbf{H}_2(z)\mathbf{H}_1(z)$ has realization (A, B, C, D), given by

$$A = \begin{bmatrix} A_1 & \mathbf{0} \\ B_2C_1 & A_2 \end{bmatrix}, \quad B = \begin{bmatrix} B_1 \\ B_2D_1 \end{bmatrix}, \quad C = \begin{bmatrix} D_2C_1 & C_2 \end{bmatrix}, \quad D = D_2D_1.$$

(ii) Assume that both (A_1, B_1, C_1, D_1) and (A_2, B_2, C_2, D_2) are minimal realizations. Show that (C, A) is unobservable, if and only if there are cancellations between poles of $\mathbf{H}_1(z)$ and zeros of $\mathbf{H}_2(z)$. Similarly, show that (A, B) is unreachable, if and only if there are cancellations between poles of $\mathbf{H}_2(z)$ and zeros of $\mathbf{H}_1(z)$. (*Hint:* Use coprime factorizations in Sect. 3.2.3.)

3.22. (i) Let $\mathbf{H}(z)$ be an $m \times m$ transfer matrix with realization (A,B,C,D), and D nonsingular. Let $\tilde{A} = A - BD^{-1}C$. Show that

$$\mathbf{H}(z)^{-1} = \left[D + C(zI - A)^{-1}B\right]^{-1} = D^{-1} - D^{-1}C\left(zI - \tilde{A}\right)^{-1}BD^{-1}.$$

(ii) Show that the double Bezout identity (3.37) in Theorem 3.28 is true.

3.23. Suppose that $(A+BF)$ and $(A+LC)$ are stability matrices. Show that the observer-based controller

$$\mathbf{K}(z) = -F(zI - A - BF - LC - LDF)^{-1}L$$

admits coprime factorizations as in (3.40).

3.24. Consider the feedback system in Fig. 3.6 where $\mathbf{P}(z)$ has realization (A,B,C,D) and $\mathbf{K}(z)$ is an observer-based controller as in Problem 3.23. Show that the closed-loop transfer matrix from $\mathbf{r}(t)$ to $\mathbf{y}(t)$ is given by

$$-\left[(C+DF)(zI - A - BF)^{-1}B + D\right]F(zI - A - LC)^{-1}L.$$

3.25. Consider Theorem 3.28. Show that the realizations for

$$\text{(i)}\ \ \begin{bmatrix}\mathbf{N}(z) & \mathbf{M}(z)\end{bmatrix} = \left[\begin{array}{c|cc} A+LC & (B+LD) & L \\ \hline C & D & I_p \end{array}\right],$$

$$\text{(ii)}\ \ \begin{bmatrix}\tilde{\mathbf{M}}(z) \\ \tilde{\mathbf{N}}(z)\end{bmatrix} = \left[\begin{array}{c|c} A+BF & B \\ \hline F & I_m \\ (C+DF) & D \end{array}\right]$$

are minimal, if and only if the realization (A,B,C,D) is minimal.

3.26. Suppose that (A,B,C,D) is a minimal realization for $\mathbf{H}(z)$ with size $p \times m$. Show that (3.38) and (3.39) hold true, and thus for all $z \in \mathbb{C}$, except those eigenvalues of $(A+BF)/(A+LC)$,

$$\text{rank}\left\{\begin{bmatrix}\mathbf{N}(z) & \mathbf{M}(z)\end{bmatrix}\right\} = p, \quad \text{rank}\left\{\begin{bmatrix}\tilde{\mathbf{M}}(z) \\ \tilde{\mathbf{N}}(z)\end{bmatrix}\right\} = m.$$

3.27. (i) Let $G(z) = z/(1+z)$ and $K(z) = 1$ in the feedback system of Fig. 3.6. Show that $y(t) = r(t+1)$, and the condition (3.29) fails. (ii) Assume that both $\mathbf{K}(z)$ and $\mathbf{G}(z)$ are proper. Show that if (3.29) holds, then all signals in Fig. 3.6 are causal.

3.28. Let $\mathbf{T}(z) = D + C(zI - A)^{-1}B$. Suppose that (A,B) is stabilizable, and (C,A) is detectable. Show that the state-space system with realization (A,B,C,D) is

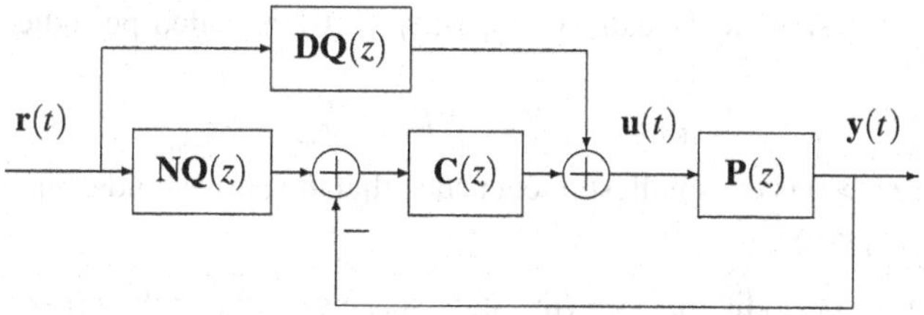

Fig. 3.8 Two-degree-of-freedom control system

internally stable, if and only if the transfer matrix $\mathbf{T}(z)$ has all its poles strictly inside the unit circle.

3.29. Suppose that the feedback system in Fig. 3.6 is well posed, and the realizations of $\mathbf{K}(z)$ and $\mathbf{G}(z)$ are both stabilizable and detectable. Show that the feedback system is internally stable, if and only if

$$\mathbf{T}(z) = \begin{bmatrix} I & -\mathbf{K}(z) \\ -\mathbf{G}(z) & I \end{bmatrix}^{-1}$$

is proper and stable. (*Hint:* Apply the result in Problem 3.28 to realization of $\mathbf{T}(z)$, by assuming realizations of $\mathbf{K}(z)$ and $\mathbf{G}(z)$.)

3.30. Consider the two-degree-of-freedom control system in Fig. 3.8 where $\mathbf{P}(z) = \mathbf{N}(z)\mathbf{D}(z)^{-1}$ is right coprime factorization. Show that

$$\mathbf{T}_{ry}(z) = \mathbf{N}(z)\mathbf{Q}(z), \quad \mathbf{T}_{ru}(z) = \mathbf{D}(z)\mathbf{Q}(z),$$

where $\mathbf{T}_{ry}(z)$ and $\mathbf{T}_{ru}(z)$ are transfer matrices from $\mathbf{r}(t)$ to $\mathbf{y}(t)$ and $\mathbf{u}(t)$, respectively.

3.31. Refer to the proof of Theorem 3.29, show that $\mathbf{Q}(z) = \tilde{\mathbf{Q}}(z)$ if and only if $\mathbf{X}(z)\tilde{\mathbf{Y}}_0(z) - \mathbf{Y}(z)\tilde{\mathbf{X}}_0(z) = \mathbf{Y}_0(z)\tilde{\mathbf{X}}(z) - \mathbf{X}_0(z)\tilde{\mathbf{Y}}(z)$ where

$$\mathbf{X}_0(z) = \mathbf{X}(z) - \mathbf{Q}(z)\mathbf{N}(z), \quad \mathbf{Y}_0(z) = \mathbf{Y}(z) + \mathbf{Q}(z)\mathbf{M}(z),$$
$$\tilde{\mathbf{Y}}_0(z) = \tilde{\mathbf{Y}}(z) + \tilde{\mathbf{M}}(z)\tilde{\mathbf{Q}}(z), \quad \tilde{\mathbf{X}}_0(z) = \tilde{\mathbf{X}}(z) - \tilde{\mathbf{N}}(z)\tilde{\mathbf{Q}}(z).$$

3.32. Consider the double Bezout identity in (3.37). Suppose that $\{\mathbf{M}(z), \mathbf{N}(z)\}$ and $\{\tilde{\mathbf{M}}(z), \tilde{\mathbf{N}}(z)\}$ are normalized coprime factors, i.e.,

$$\mathbf{M}(z)\mathbf{M}(z)^\sim + \mathbf{N}(z)\mathbf{N}(z)^\sim = I_p, \quad \tilde{\mathbf{M}}(z)^\sim\tilde{\mathbf{M}}(z) + \tilde{\mathbf{N}}(z)^\sim\tilde{\mathbf{N}}(z) = I_m.$$

Show that $\mathbf{Y}(z)\mathbf{M}(z)^\sim - \mathbf{X}(z)\mathbf{N}(z)^\sim = \tilde{\mathbf{M}}(z)^\sim\tilde{\mathbf{Y}}(z) - \tilde{\mathbf{N}}(z)^\sim\tilde{\mathbf{X}}(z)$.

3.33. Prove (3.61) and (3.62) by direct verifications.

3.34. Prove Theorem 3.34.

3.35. An LTV system with realization $\{A_t, B_t, C_t, \mathbf{0}\}$ is called periodic time varying, if

$$A_{t+\ell} = A_t, \quad B_{t+\ell} = B_t, \quad C_{t+\ell} = C_t$$

for all t where $\ell > 0$ is the smallest integer such that the above holds. Such an integer ℓ is termed the period.

1. Show that for periodic time-varying state-space systems with period $\ell > 0$, local reachability over $[t_0, t_0 + \ell)$ is equivalent to ℓ-step reachability.
2. Let $t_f = t_0 + \ell$. Suppose that (C_t, A_t) is unobservable over $[t_0, t_f)$. Show that the similarity transforms $\{S_t\}_{t=t_0}^{t_f-1}$ exist such that (3.67) holds for $t_0 \le t \le t_f$. (*Hint:* Use the proof of Theorem 3.32.)
3. Show that $(A_t, B_t, C_t, \mathbf{0})$ is an ℓ-step minimal realization, if and only if (A_t, B_t) is reachable and (C_t, A_t) is observable over $[t_0, t_f)$ with $t_f = t_0 + \ell$.
4. Suppose that $(A_t, B_t, C_t, \mathbf{0})$ and $(\tilde{A}_t, \tilde{B}_t, \tilde{C}_t, \mathbf{0})$ are both ℓ-step reachable and observable realizations for the same LTV system. Show that there exist similarity transforms $\{T_i\}_{i=t_0}^{\infty}$ such that

$$\left(\tilde{A}_t, \tilde{B}_t, \tilde{C}_t, \mathbf{0}\right) = \left(T_{t+1} A_t T_t^{-1}, T_{t+1} B_t, C_t T_t^{-1}, \mathbf{0}\right) \ \forall t.$$

3.36. Show that $\mathbf{x}(t+1) = A_t \mathbf{x}(t)$ is exponentially stable, if there exists $\rho > 1$ such that $\mathbf{x}(t+1) = \rho A_t \mathbf{x}(t)$ is asymptotically stable.

3.37. Suppose that (C_t, A_t) is detectable. Let

$$\mathbf{x}(t+1) = A_t \mathbf{x}(t), \quad \mathbf{y}(t) = C_t \mathbf{x}(t).$$

Show that $\lim_{t\to\infty} \|\mathbf{y}(t)\| = 0 \implies \lim_{t\to\infty} \|\mathbf{x}(t)\| = 0$.

3.38. Prove Theorem 3.39. (*Hint:* Use the same method as in the proof of Theorem 3.38.)

Chapter 4
Model Reduction

Because real systems are highly complex and may involve physical phenomena beyond mathematical description, plant models are likely to be of high order. In addition, system models are obtained either from identification based on experimental data or from modeling based on physics principles which add complexity to state-space descriptions. Even in the case when such models do not involve pole and zero cancellation which can be removed anyway with Kalman decomposition, redundancies may still exist in the state-space model owing to its being nearly unreachable or nearly unobservable or both. Hence, being minimal for realization is not adequate. High order state-space models will lead to high order controllers and increase overhead to analysis, design, and implementation of the feedback control systems. For this reason, there is a strong incentive to reduce the order of the system model. The real issue is how to quantify and remove redundancies in the original high order model so that the reduced order model admits high fidelity in representation of the physical process.

In the past a few decades, several techniques are emerged for order reduction of state-space models. In this chapter, methods of balanced realization and optimal Hankel-norm approximation will be presented. For a given state-space realization, its Hankel singular values will be shown to provide a suitable measure of the model redundancy. Specifically under the balanced realization, the subsystem corresponding to small Hankel singular values contributes little to the system behavior which can thus be truncated directly. This method can be further improved to obtain reduced order models of higher fidelity. More importantly, upper bounds will be derived to quantify the approximation error between the reduced model and the original high order model. The contents of this chapter include error measures, balanced truncation, and optimal Hankel-norm approximation, which rely heavily on basic concepts and mathematical analysis from system theory in the previous chapter.

G. Gu, *Discrete-Time Linear Systems: Theory and Design with Applications*,
DOI 10.1007/978-1-4614-2281-5_4,

4.1 Performance Measures

Approximation errors are inevitable when low order models are employed to represent the high order models. Performance of model reduction is measured by the corresponding approximation error. The two most frequently used ones are $\mathscr{H}_2$ and $\mathscr{H}_\infty$-norms.

For a given causal and rational plant $\mathbf{P}(z)$, it can be decomposed to

$$\mathbf{P}(z) = \mathbf{H}(z) + \mathbf{U}(z), \tag{4.1}$$

where $\mathbf{H}(z)$ is stable with all poles strictly inside the unit circle and all $\mathbf{U}(z)$ is anti-stable in the sense that none of its poles is stable. Because feedback stabilization in control requires the full knowledge of $\mathbf{U}(z)$, model reduction is normally carried out for the stable component of $\mathbf{P}(z)$. The $\mathscr{H}_2$ norm of $\mathbf{H}(z)$ is defined by

$$\|\mathbf{H}\|_2 = \sqrt{\mathrm{Tr}\left\{\frac{1}{2\pi}\int_{-\pi}^{\pi} \mathbf{H}\left(e^{j\omega}\right)\mathbf{H}\left(e^{j\omega}\right)^* \, d\omega\right\}} \tag{4.2}$$

that is the same as (2.51) in Chap. 2 induced by power norm. This norm is often encountered in feedback control and signal processing due to the white nature of noises and disturbances. Collection of all stable transfer matrices with bounded $\mathscr{H}_2$-norm forms a Hardy space on the unit circle which is in fact a Hilbert space. On the other hand, the $\mathscr{H}_\infty$-norm of $\mathbf{H}(z)$ is defined by

$$\|\mathbf{H}\|_\infty = \sup_{|z|>1} \overline{\sigma}\left[\mathbf{H}(z)\right]. \tag{4.3}$$

Collection of all stable transfer matrices with bounded $\mathscr{H}_\infty$-norm is also a Hardy space on the unit circle which is now a Banach space. Readers are referred to other texts for more complete knowledge of Hardy spaces. For rational $\mathbf{H}(z)$, its $\mathscr{H}_\infty$-norm has a simpler form:

$$\|\mathbf{H}\|_\infty = \max_{\omega\in\mathbf{R}} \overline{\sigma}\left[\mathbf{H}\left(e^{j\omega}\right)\right]. \tag{4.4}$$

Basically, $\mathscr{H}_\infty$-norm is the maximum "magnitude response."

The mathematical expressions in (4.2) and (4.4) suggest that transfer matrices can be regarded as matrix-valued functions of real variable ω. Thus, Hardy spaces of $\mathscr{H}_2$ and $\mathscr{H}_\infty$ can be extended to Lebesgue spaces of $\mathscr{L}_2[0, 2\pi]$ and $\mathscr{L}_\infty[0, 2\pi]$, respectively. Indeed, let $\mathscr{L}_2[0, 2\pi]$ be the collection of all transfer matrices $\mathbf{H}(z)$ such that the integration in (4.2) is bounded in the sense of Lebesgue. It includes $\mathscr{H}_2$ as a subspace. Another subspace of $\mathscr{L}_2[0, 2\pi]$ consists of those transfer matrices whose impulse responses are anticausal that is the complement of $\mathscr{H}_2$ and denoted by $\mathscr{H}_2^{\perp}$. In fact, each $\mathbf{H}\left(e^{\omega}\right) \in \mathscr{L}_2[0, 2\pi]$ can be written as

$$\mathbf{H}(z) = \mathbf{H}_{\mathrm{s}}(z) + \mathbf{H}_{\mathrm{as}}, \quad \mathbf{H}_{\mathrm{s}}(z) \in \mathscr{H}_2, \ \mathbf{H}_{\mathrm{as}}(z) \in \mathscr{H}_{2\perp}, \tag{4.5}$$

owing to the nice geometry property of the Hilbert space. Similarly, $\mathscr{H}_{\infty\perp}$ can be defined as the collection of all anticausal transfer matrices whose frequency responses are essentially bounded. Both $\mathscr{H}_\infty$ and $\mathscr{H}_{\infty\perp}$ are subspaces of $\mathscr{L}_\infty[0, 2\pi]$. However, each $\mathbf{H}(e^{\omega}) \in \mathscr{L}_\infty[0, 2\pi]$ does not admit a similar decomposition to that in (4.5) because of the geometry complexity of the Banach space.

In engineering practice white noises and disturbances are abundant. See many examples presented in Chap. 2. As a result, $\mathscr{H}_2$ norm is often adopted to measure the model reduction error. Let $\widehat{\mathbf{H}}(z)$ be a reduced order model. Then small value of $\left\|\mathbf{H} - \widehat{\mathbf{H}}\right\|_2$ implies good quality of the reduced order model in terms of approximation to the high order model $\mathbf{H}(z)$. It turns out that $\mathscr{H}_2$ norm is not a good measure for applications to control systems because of the consideration for feedback stability. Recall Nyquist criterion in classic control: It is the shortest distance between the frequency response of the loop transfer function and the critical point of -1 on complex plane that measures the stability margin. In other words, if $\widehat{\mathbf{H}}(z)$ is the approximate loop transfer function, and results in stable feedback system, then the closed-loop system with loop transfer function $\mathbf{H}(z)$ remains stable, if $\left\|\mathbf{H} - \widehat{\mathbf{H}}\right\|_\infty$ is strictly smaller than the stability margin. Clearly, $\mathscr{H}_\infty$-norm is the worst-case measure that can be conservative. However, because at which frequency the shortest distance takes place is unknown prior to controller design, the $\mathscr{H}_\infty$-norm measure is indispensable for model reduction in feedback control systems, if the controller is synthesized based on the reduced order model.

For a stable SISO system $\mathbf{H}(z)$, there holds $\|\mathbf{H}\|_2 \le \|\mathbf{H}\|_\infty$. In fact, the difference between $\|\mathbf{H}\|_2$ and $\|\mathbf{H}\|_\infty$ can be substantial, if the magnitude response of $\mathbf{H}(z)$ varies substantially implying that the $\mathscr{H}_\infty$-norm measure can be very conservative. In addition, optimal model reduction under either $\mathscr{H}_2$ or $\mathscr{H}_\infty$-norm involves nonlinear optimization and is thus not tractable. A different norm, termed Hankel norm, is introduced to aid model reduction.

Let a causal signal $\{\mathbf{u}(t)\}_{t=0}^{\infty}$ of dimension d be energy bounded. The collection of all such signals is denoted by ℓ_{2+}^d. Its complement, ℓ_{2-}^d, is the collection of all anticausal and energy-bounded signals of dimension d. As such, each d-dimensional signal $\{\mathbf{s}(t)\}_{t=-\infty}^{\infty}$ with bounded energy admits the following decomposition:

$$\mathbf{s}(t) = \mathbf{s}_c(t) + \mathbf{s}_a(t), \quad \mathbf{s}_c(t) \in \ell_{2+}^d, \quad \mathbf{s}_a(t) \in \ell_{2-}^d.$$

It follows that $\ell_2^d := \ell_{2+}^d \oplus \ell_{2-}^d$ is the collection of energy-bounded signals of dimension d which is also a Hilbert space with ℓ_{2+}^d and ℓ_{2-}^d, two complementary complete subspaces. For each $\mathbf{s}(t) \in \ell_2^d$, its DTFT is defined as

$$\mathbf{S}\left(e^{j\omega}\right) = \sum_{t=-\infty}^{\infty} \mathbf{s}(t)e^{-j\omega t}, \quad \omega \in \mathbb{R}.$$

In light of Parseval's theorem, there holds $\|\mathbf{s}\|_2 = \|\mathbf{S}\|_2$ where

$$\|\mathbf{s}\|_2 = \sqrt{\sum_{t=-\infty}^{\infty} \|\mathbf{s}(t)\|^2}, \quad \|\mathbf{S}\|_2 = \sqrt{\frac{1}{2\pi}\int_{-\pi}^{\pi} \|\mathbf{S}(e^{j\omega})\|^2 \, d\omega}$$

are called 2-norms with $\|\cdot\|$, the Euclidean norm.

The transfer matrix $\mathbf{H}(z)$ of size $p \times m$ can be viewed as a mapping from ℓ_{2+}^m to ℓ_{2+}^p. Let $\delta(t)$ be Kroneker delta function and $\mathbf{e}_k$ be a vector of size m with 1 at the kth entry and rest zeros. Denote $\{H(t)\}_{t=0}^{\infty}$ the impulse response of $\mathbf{H}(z)$. Then

$$\mathbf{y}_k(t) = H(q)\delta(t)\mathbf{e}_k = H(t)\mathbf{e}_k$$

is the impulse response with impulse excited at the kth input channel. It can be verified that the $\mathscr{H}_2$ norm has the following time-domain expression:

$$\|\mathbf{H}\|_2 = \sqrt{\sum_{k=1}^{m} \|\mathbf{y}_k\|_2^2} = \sqrt{\sum_{t=-\infty}^{\infty} \mathrm{Tr}\{H(t)H^*(t)\}}.$$

However, the time-domain interpretation for $\mathscr{H}_\infty$-norm is more intriguing that is presented in the next result.

Theorem 4.1. *Let $\{\mathbf{u}(t)\} \in \ell_{2+}^m$ and $\{\mathbf{y}(t)\} \in \ell_{2+}^p$ be input and output of a causal and stable system represented by its transfer matrix $\mathbf{H}(z)$ under zero initial condition. Then*

$$\|\mathbf{H}\|_\infty = \sup_{0 \neq \mathbf{u} \in \ell_{2+}^m} \frac{\|\mathbf{y}\|_2}{\|\mathbf{u}\|_2}. \tag{4.6}$$

Proof. Applying DTFT under zero initial condition yields

$$\mathbf{Y}\left(e^{j\omega}\right) = \mathbf{H}\left(e^{j\omega}\right)\mathbf{U}\left(e^{j\omega}\right) \;\; \forall\, \omega \in \mathbb{R}.$$

By the definition of $\mathscr{H}_\infty$-norm, there holds

$$\begin{aligned}
\|\mathbf{y}\|_2^2 = \|\mathbf{Y}\|_2^2 &= \frac{1}{2\pi}\int_{-\pi}^{\pi} \mathbf{U}\left(e^{j\omega}\right)^* \mathbf{H}\left(e^{j\omega}\right)^* \mathbf{H}\left(e^{j\omega}\right)\mathbf{U}\left(e^{j\omega}\right) d\omega \\
&\leq \frac{1}{2\pi}\int_{-\pi}^{\pi} \overline{\sigma}^2\left[\mathbf{H}\left(e^{j\omega}\right)\right]\mathbf{U}\left(e^{j\omega}\right)^*\mathbf{U}\left(e^{j\omega}\right) d\omega \\
&\leq \|\mathbf{H}\|_\infty^2 \|\mathbf{U}\|_2^2 = \|\mathbf{H}\|_\infty^2 \|\mathbf{u}\|_2^2,
\end{aligned}$$

where Parseval's theorem is used twice. Hence, $\|\mathbf{H}\|_\infty$ is an upper bound for $\|\mathbf{y}\|_2/\|\mathbf{u}\|_2$. This upper bound can be asymptotically achieved. Consider the first case when $\mathbf{H}(z)$ admits continuous frequency response. Specifically, there exists ω_m

at which $\overline{\sigma}\left[\mathbf{H}\left(e^{j\omega_m}\right)\right] = \|\mathbf{H}\|_\infty$. Hence, there exists a pair of complex-valued unit vectors $\{\mathbf{u}_m, \mathbf{y}_m\}$ of appropriate dimensions such that

$$\mathbf{H}\left(e^{j\omega_m}\right)\mathbf{u}_m = \|\mathbf{H}\|_\infty \mathbf{y}_m. \tag{4.7}$$

If a causal input $\mathbf{u}(t)$ with DTFT $\mathbf{U}\left(e^{j\omega}\right) = \mathbf{u}_m \psi_\varepsilon(\omega)$ is taken where

$$|\psi_\varepsilon(\omega)|^2 = \begin{cases} \frac{\pi}{\varepsilon}, & \omega \in [\omega_m - \varepsilon,\ \omega_m + \varepsilon], \\ 0, & \text{elsewhere}, \end{cases} \tag{4.8}$$

and $\varepsilon > 0$, then $\mathbf{u}(t) \in \ell_{2+}^m$ and $\|\mathbf{u}\|_2 = \|\mathbf{U}\|_2 = 1$. For this input signal,

$$\|\mathbf{y}\|_2^2 = \|\mathbf{H}\mathbf{u}\|_2^2 = \frac{1}{2\pi}\int_{-\pi}^{\pi} |\psi_\varepsilon(\omega)|^2 \mathbf{u}_m^* \mathbf{H}\left(e^{j\omega}\right)^* \mathbf{H}\left(e^{j\omega}\right)\mathbf{u}_m \, d\omega \ \to\ \|\mathbf{H}\|_\infty^2$$

as $\varepsilon \to 0$ in light of (4.7) and (4.8). The proof for the case when $\mathbf{H}(z)$ does not admit continuous frequency response is similar but involves more sophisticated mathematics which is skipped. □

It is interesting to observe that $\mathscr{H}_\infty$-norm, defined in frequency domain, has an interpretation as the induced ℓ_2-norm in time domain. Next consider an input $\{\mathbf{u}(t)\}_{t=-\infty}^{-1} \in \ell_{2-}^m$. The corresponding output is given by

$$\mathbf{y}(t) = \mathbf{H}(q)\mathbf{u}(t) = \sum_{k=t+1}^{\infty} H(k)\mathbf{u}(t-k) \ \in \ell_2^p. \tag{4.9}$$

Let $\Pi_+[\cdot]$ be the orthogonal projection from ℓ_2 to ℓ_{2+}. Then the Hankel operator Γ_{H}, associated with $\mathbf{H}(z)$ of dimension $p \times m$, is defined via

$$\mathbf{y}(t) = \Gamma_{\mathrm{H}}\mathbf{u} = \Pi_+[\mathbf{H}(q)\mathbf{u}(t)] \tag{4.10}$$

that is a mapping from ℓ_{2-}^m to ℓ_{2+}^p. The Hankel norm is defined via

$$\|\mathbf{H}\|_{\mathrm{H}} := \sup_{0 \neq \mathbf{u} \in \ell_{2-}^m} \frac{\|\Pi_+[\mathbf{y}]\|_2}{\|\mathbf{u}\|_2}.$$

Hankel matrices of finite size are used in the previous chapter in studying minimal realizations. Let (A,B,C,D) be a realization of $\mathbf{H}(z)$. Its impulse response is given by $H(t) = CA^{t-1}B$ for $t \geq 1$ and $H(0) = D$. The Hankel matrix of infinite size associated with $\mathbf{H}(z)$ is given by

$$\mathscr{H}_H = \lim_{\ell \to \infty} \begin{bmatrix} H(1) & H(2) & \cdots & H(\ell) \\ H(2) & H(3) & \cdots & H(\ell+1) \\ \vdots & \cdots & \cdots & \vdots \\ H(\ell) & \cdots & \cdots & H(2\ell-1) \end{bmatrix} = \lim_{\ell \to \infty} \mathscr{O}_\ell \mathscr{R}_\ell, \tag{4.11}$$

where $\mathscr{O}_\ell$ and $\mathscr{R}_\ell$ are the observability and reachability matrices of size ℓ, respectively. For anticausal input $\{\mathbf{u}(t)\}_{t=-\infty}^{-1} \in \ell_{2-}^m$ and the projected output $\{\mathbf{y}(t)\}_{t=0}^{\infty} \in \ell_{2+}^m$, there holds $\underline{\mathbf{y}} = \mathscr{H}_H \underline{\mathbf{u}}$ where

$$\underline{\mathbf{u}} = \lim_{\ell\to\infty} \begin{bmatrix} \mathbf{u}(-1) \\ \mathbf{u}(-2) \\ \vdots \\ \mathbf{u}(-\ell) \end{bmatrix}, \quad \underline{\mathbf{y}} = \lim_{\ell\to\infty} \begin{bmatrix} \mathbf{y}(0) \\ \mathbf{y}(1) \\ \vdots \\ \mathbf{y}(\ell) \end{bmatrix}.$$

Indeed, it can be verified rather easily using (4.9) that

$$\begin{bmatrix} \mathbf{y}(0) \\ \mathbf{y}(1) \\ \vdots \\ \mathbf{y}(t) \end{bmatrix} = \lim_{\ell\to\infty} \begin{bmatrix} H(1) & H(2) & \cdots & H(\ell) \\ H(2) & H(3) & \cdots & H(\ell+1) \\ \vdots & \cdots & \cdots & \vdots \\ H(t) & \cdots & \cdots & H(2\ell-1) \end{bmatrix} \begin{bmatrix} \mathbf{u}(-1) \\ \mathbf{u}(-2) \\ \vdots \\ \mathbf{u}(-\ell) \end{bmatrix}$$

leading to $\underline{\mathbf{y}} = \mathscr{H}_H \underline{\mathbf{u}}$. As a result, $\|\mathbf{H}\|_{\mathrm{H}} = \overline{\sigma}(\mathscr{H}_H)$.

A salient feature of the Hankel matrix is its finite rank being the same as the McMillan degree of the corresponding transfer matrix. Hence, model reduction can be carried out via approximation of the Hankel matrix with a lower rank Hankel matrix. For $\mathbf{H}(z) = D + C(zI_n - A)^{-1}B$ of order n, its observability and reachability gramians are solutions to the following two Lyapunov equations

$$P = A^*PA + C^*C, \quad Q = AQA^* + BB^*, \tag{4.12}$$

respectively. Stability of A ensures that

$$P = \lim_{\ell\to\infty} \mathscr{O}_\ell^* \mathscr{O}_\ell = \sum_{k=0}^{\infty} \left(CA^k\right)^* \left(CA^k\right), \tag{4.13}$$

$$Q = \lim_{\ell\to\infty} \mathscr{R}_\ell \mathscr{R}_\ell^* = \sum_{k=0}^{\infty} \left(A^kB\right) \left(A^kB\right)^*. \tag{4.14}$$

Properties of eigenvalues imply

$$\sigma_i(\mathscr{H}_H) = \sqrt{\lambda_i(\mathscr{H}_H \mathscr{H}_H^*)} = \sqrt{\lambda_i(PQ)}, \quad i = 1, 2, \ldots, n.$$

For this reason, square roots of eigenvalues of PQ are referred to as Hankel singular values, which depend only on the impulse response of $\mathbf{H}(z)$. It is noted that neither Hankel operator nor Hankel matrix involves D.

Lemma 4.1. *For* $\mathbf{H}(z) = C(zI - A)^{-1}B$ *of dimension* $p \times m$ *with* A *a stability matrix, there holds*

$$\frac{1}{\min\{p,m\}}\|\mathbf{H}\|_2 \le \|\mathbf{H}\|_{\mathrm{H}} \le \|\mathbf{H}\|_\infty.$$

Proof. By stability of A and definition of the Hankel and $\mathscr{H}_\infty$-norms,

$$\|\mathbf{H}\|_{\mathrm{H}} = \sup_{\|\mathbf{u}\|_2=1, \mathbf{u}\in\ell_{2-}} \|\Pi_+[\mathbf{H}\mathbf{u}]\|_2 \le \sup_{\|\mathbf{u}\|_2=1, \mathbf{u}\in\ell_{2+}} \|\mathbf{H}\mathbf{u}\|_2 = \|\mathbf{H}\|_\infty.$$

For $\mathscr{H}_2$ norm, assuming $m \le p$ has no loss of generality. It is claimed that

$$\|\mathbf{H}\|_2^2 = \mathrm{Tr}\{B^*PB\}, \quad P = A^*PA + C^*C. \tag{4.15}$$

Stability of A ensures that $P \ge \mathbf{0}$. In addition, there holds

$$\begin{aligned}(z^{-1}I - A^*)P(zI - A) &= P - zA^*P - z^{-1}PA + A^*PA \\ &= 2A^*PA + C^*C - zA^*P - z^{-1}PA \\ &= C^*C - A^*P(zI - A) - (z^{-1}I - A^*)PA \end{aligned} \tag{4.16}$$

in which the Lyapunov equation from (4.15) is used. It follows that

$$C^*C = A^*P(zI - A) + (z^{-1}I - A^*)PA + (z^{-1}I - A^*)P(zI - A).$$

Recall $\mathbf{H}(z) = C(zI - A)^{-1}B$. Multiplying the above equality by $(zI - A)^{-1}B$ from right and by $B^*(z^{-1}I - A^*)^{-1}$ from left yields

$$\mathbf{H}(z)^*\mathbf{H}(z) = B^*(z^{-1}I - A^*)^{-1}A^*PB + B^*PA(zI - A)^{-1}B + B^*PB \tag{4.17}$$

for each z on the unit circle. By Cauchy's integral theorem, contour integration of the above matrix function on the unit circle counterclockwise with trace operation verifies (4.15). On the other hand, the Lyapunov equation $Q = AQA^* + BB^*$ shows that $BB^* \le Q$ by again stability of A. Consequently,

$$\begin{aligned}\frac{1}{m}\|\mathbf{H}\|_2^2 &= \frac{1}{m}\mathrm{Tr}\{B^*PB\} = \frac{1}{m}\mathrm{Tr}\{PBB^*\} \\ &= \frac{1}{m}\sum_{i=1}^{m}\lambda_i(PBB^*) \le \lambda_{\max}(PBB^*) \\ &\le \lambda_{\max}(PQ) = \|\mathbf{H}\|_{\mathrm{H}}^2\end{aligned}$$

by positivity of all the eigenvalues that completes the proof. □

Example 4.2. Let us examine a specific $\mathbf{H}(z)$ from Problem 3.12 given by

$$\mathbf{H}(z) = \begin{bmatrix} \dfrac{z+1}{z^2 - z + \frac{1}{4}} & \dfrac{1}{z - \frac{1}{2}} \\ \dfrac{-z^2 + z + 1}{z^3 + \frac{1}{2}z^2 - \frac{1}{4}z - \frac{1}{8}} & \dfrac{z - \frac{1}{4}}{z^2 + z + \frac{1}{4}} \end{bmatrix}. \tag{4.18}$$

Its partial fraction is obtained as

$$\mathbf{H}(z) = \frac{\begin{bmatrix} 1 & 1 \\ 1.25 & 0 \end{bmatrix} z - \begin{bmatrix} -1 & 0.5 \\ 0.675 & 0 \end{bmatrix}}{\left(z - \frac{1}{2}\right)^2} + \frac{\begin{bmatrix} 0 & 0 \\ -2.25 & 1 \end{bmatrix} z - \begin{bmatrix} 0 & 0 \\ 1.375 & 0.25 \end{bmatrix}}{\left(z + \frac{1}{2}\right)^2}.$$

The second column of the first term has only one pole at $\frac{1}{2}$, and the second term is a rank 1 matrix and thus has McMillan degree 2. Hence, a minimal realization (A, B, C, D) is constructed to be

$$A = \begin{bmatrix} 1 & -0.25 & 0 & 0 & 0 \\ 1 & 0 & 0 & 0 & 0 \\ 0 & 0 & 0.5 & 0 & 0 \\ 0 & 0 & 0 & -1 & 1 \\ 0 & 0 & 0 & -0.25 & 0 \end{bmatrix}, \quad B = \begin{bmatrix} 1 & 0 \\ 0 & 0 \\ 0 & 1 \\ -2.25 & 1 \\ -1.375 & -0.25 \end{bmatrix},$$

$$C = \begin{bmatrix} 1 & 1 & 1 & 0 & 0 \\ 1.25 & -0.675 & 0 & 1 & 0 \end{bmatrix}, \quad D = \begin{bmatrix} 0 & 0 \\ 0 & 0 \end{bmatrix}.$$

The observability and reachability gramians are given by

$$P = \begin{bmatrix} 12.65 & -2.404 & 2.667 & 0.826 & 0.381 \\ -2.404 & 2.246 & 0.667 & -0.445 & -0.206 \\ 2.667 & 0.667 & 1.333 & 0 & 0 \\ 0.826 & -0.445 & 0 & 2.963 & -2.370 \\ 0.381 & -0.206 & 0 & -2.370 & 2.963 \end{bmatrix},$$

$$Q = \begin{bmatrix} 2.963 & 2.370 & 0 & -1.568 & -0.944 \\ 2.370 & 2.963 & 0 & 0.624 & 0.392 \\ 0 & 0 & 1.333 & 0.560 & -0.320 \\ -1.568 & 0.624 & 0.560 & 10.27 & 4.329 \\ -0.944 & 0.392 & -0.320 & 4.329 & 2.595 \end{bmatrix}.$$

As a result, $\|\mathbf{H}\|_2 = \sqrt{\text{Tr}\{B^*PB\}} = 4.414$, $\|\mathbf{H}\|_{\text{H}} = \sqrt{\lambda_{\max}(PQ)} = 5.539$, and $\|\mathbf{H}\|_\infty = 8.28$ that takes place at $\omega = 0$.

Lemma 4.1 and Example 4.2 show that the Hankel norm is between $\mathscr{H}_2$ and $\mathscr{H}_\infty$-norms that can be an appropriate measure for model reduction. More importantly, the Hankel matrix has finite rank. Model reduction can thus be converted to low rank matrix approximation that is a tractable problem. In the next two sections, two different techniques will be developed for model reduction. Surprisingly, both admit good approximation error in $\mathscr{H}_\infty$ norm.

4.2 Balanced Truncation

Algorithms for model reduction in this section are based on approximation of gramian matrices with the lower rank ones via direct truncation. However, two gramians exist for each transfer matrix. How to approximate them with lower rank ones becomes a problem. A moment of reflection suggests that the two gramians need to be balanced prior to truncation in order to help minimization of the approximation error. There are two different ways for balancing the gramian matrices leading to two different reduction methods based on balanced truncation which will be studied in this section.

4.2.1 Balanced Realization

Given a state-space realization (A,B,C,D) with A stable, both observability gramian P and reachability gramian Q are positive semidefinite, which are the unique solutions to the two Lyapunov equations in (4.12), respectively. Moreover, P is positive definite, if and only if (C,A) is observable, and Q is positive definite, if and only if (A,B) is reachable. The next result illustrates that a similarity transform can be applied to obtain a minimal realization, which admits equal and diagonal observability and reachability gramians. Such a realization is called balanced realization.

Proposition 4.1. *Let P and Q be respective observability and reachability gramians associated with realization (A,B,C,D) where A is a stability matrix. There exists a similarity transform T such that the new realization*

$$(A_b,B_b,C_b,D) = \left(TAT^{-1},TB,CT^{-1},D\right) \tag{4.19}$$

admits observability and reachability gramians given by

$$P_b = Q_b = \operatorname{diag}(\sigma_1,\sigma_2,\ldots,\sigma_n) =: \Sigma \geq 0. \tag{4.20}$$

Proof. Let $QP = S\Lambda S^{-1}$ be eigenvalue decomposition. Since both P and Q are nonnegative matrices, Λ is diagonal and nonnegative. It follows that $\Lambda = \Sigma^2$ where

Σ is diagonal with $\{\sigma_i\}$ on the diagonal. In fact, P and Q are simultaneously diagonalizable (refer to Appendix A). Hence, a nonsingular matrix S exists such that $\Sigma = S^{-1}Q(S^*)^{-1} = S^*PS$ and

$$\Sigma^2 = \left[S^{-1}Q(S^*)^{-1}\right]\left[S^*PS\right] = S^{-1}QPS.$$

Let (A_b, B_b, C_b) be as in (4.19) and $P_b = Q_b$ as in (4.20). Multiplying the second Lyapunov equation in (4.12) by S^{-1} from left and $(S^*)^{-1}$ from right yields

$$Q_b = A_b Q A_b^* + B_b B_b^*,$$

if $T = S^{-1}$. Similarly, multiplying the first Lyapunov equation in (4.12) by S^* from left and S from right yields

$$P_b = A_b^* P_b A_b + C_b^* C_b,$$

if $T = S^{-1}$. The proposition is thus true by taking $T = S^{-1}$. □

The sequence $\{\sigma_i\}$ is called *Hankel singular values* which can be arranged in descending order:

$$\sigma_1 \geq \sigma_2 \geq \cdots \geq \sigma_n \geq 0.$$

It is noted that the required similarity transform T in Proposition 4.1 may not be the same as the S^{-1} matrix in eigenvalue decomposition of PQ. In this case, it is suggested to first eliminate the unobservable and unreachable modes separately as in Problem 3.20 and then find the similarity transform to obtain the balanced realization. When P and Q are nonsingular, it can be verified that $T = \Sigma^{-1/2}U^*R$ is the required similarity transform in obtaining the balanced realization where $P = R^*R$ is the Cholesky factorization, and

$$RQR^* = U\Sigma^2U^*$$

is the SVD with U a unitary matrix.

There are two reasons for studying balanced realizations. The first is their insensitivity to rounding off errors, compared with other realizations, subsuming the ARMA models. Recall that discrete-time signal processing requires digital implementations. The second is model reduction which is studied in this chapter. Even for minimal realizations, not all the modes contribute equally to the dynamic behavior of the system. If insignificant modes are truncated directly, then the reduced order dynamic system approximates the original system with little noticeable error. However, direct truncation of less significant modes is not the right approach in terms of minimization of the approximate error. A reflection on Kalman decomposition indicates that the approximation error is suitably small, if the truncated subsystems or state variables are nearly unobservable and unreachable. It needs to be emphasized that for minimal realizations, a state variable that is nearly unobservable does not imply that it has an insignificant contribution to the input/output behavior of the system. In fact, because (A,B,C,D) and $(A,\rho B,\rho^{-1}C,D)$ describe the same input/output system with $\rho \neq 0$, the second realization can be made nearly unobservable by making $|\rho|$ sufficiently large. Hence, there needs to be some balance between the observability and reachability. A right measure is the observability and reachability gramians, which represent the energy functions in the Lyapunov stability criteria, leading to balanced realizations.

Suppose that (A,B,C,D) is a balanced realization with A stable. Then

$$\Sigma = A^*\Sigma A + C^*C, \quad \Sigma = A\Sigma A^* + BB^*, \tag{4.21}$$

where Σ is both the observability and reachability gramian, given by

$$\Sigma = \text{diag}(\sigma_1, \sigma_2, \ldots, \sigma_n), \quad \sigma_1 \geq \sigma_2 \geq \cdots \geq \sigma_n \tag{4.22}$$

with n the order of (A,B,C,D). Assume that there is a large gap between σ_r and σ_{r+1} with $1 \leq r < n$, i.e., $\sigma_r \gg \sigma_{r+1}$. Then Σ can be partitioned into $\Sigma = \text{diag}(\Sigma_1, \Sigma_2)$ with Σ_1 containing the first r significant Hankel singular values and Σ_2 containing the less significant or tail of the n Hankel singular values. Partition the realization compatibly as

$$A = \begin{bmatrix} A_{11} & A_{12} \\ A_{21} & A_{22} \end{bmatrix}, \quad B = \begin{bmatrix} B_1 \\ B_2 \end{bmatrix}, \quad C = \begin{bmatrix} C_1 & C_2 \end{bmatrix}, \tag{4.23}$$

where (A_{11}, B_1, C_1, D) has order r, and is called balance truncated model. It follows that the state vector can be partitioned conformally as

$$\mathbf{x}(t) = \begin{bmatrix} \mathbf{x}_1^*(t) & \mathbf{x}_2^*(t) \end{bmatrix}^*$$

with $\mathbf{x}_2(t)$ much less observable and reachable as compared with $\mathbf{x}_1(t)$ which has size r. Kalman decomposition suggests that (A_{11}, B_1, C_1, D) should be a good approximation to (A,B,C,D). This is indeed true. The following result provides an a priori error bound in $\mathscr{H}_\infty$ norm for the balance truncated model.

Theorem 4.3. *Let (A,B,C,D) be a balanced realization of order n for $\mathbf{H}(z)$, satisfying (4.21) and partitioned as in (4.23). Suppose that $\sigma_r > \sigma_{r+1}$ with Σ as given in (4.22). Then the balance truncated model (A_{11}, B_1, C_1, D) of order r is internally stable, and there holds error bound*

$$\left\| \mathbf{H} - \widehat{\mathbf{H}} \right\|_\infty \leq 2 \sum_{i=r+1}^{n} \sigma_i, \tag{4.24}$$

where $\widehat{\mathbf{H}}(z) = D + C_1(zI_r - A_{11})^{-1}B_1$.

Proof. It is noted that the Lyapunov equations in (4.22) can be written into the following two inequalities:

$$\text{(i) } A^*\Sigma A + C^*C \leq \Sigma, \quad \text{(ii) } A\Sigma A^* + BB^* \leq \Sigma, \tag{4.25}$$

where $X \le Y$ stands for $Y - X \ge \mathbf{0}$, i.e., positive semidefinite. By the partitions in (4.23) and $\Sigma = \text{diag}(\Sigma_1, \Sigma_2)$,

$$A_{11}^* \Sigma_1 A_{11} + A_{21}^* \Sigma_2 A_{21} - \Sigma_1 \le C_1^* C_1. \tag{4.26}$$

Let λ be an eigenvalue of A_{11} with eigenvector $\hat{\mathbf{v}}$. Multiplying the above inequality by $\hat{\mathbf{v}}^*$ from left and $\hat{\mathbf{v}}$ from right yields

$$\left(|\lambda|^2 - 1\right)\left(\hat{\mathbf{v}}^* \Sigma_1 \hat{\mathbf{v}}\right) + \hat{\mathbf{v}}^* A_{21}^* \Sigma_2 A_{21} \hat{\mathbf{v}} \le -\|C_1 \hat{\mathbf{v}}\|^2.$$

Since $\Sigma > \mathbf{0}$, $|\lambda| \le 1$ is true. Moreover, if $|\lambda| = 1$, then

$$A_{21}\hat{\mathbf{v}} = \mathbf{0}, \quad C_1 \hat{\mathbf{v}} = \mathbf{0}$$

by $\Sigma_2 > 0$, which implies that λ is also an eigenvalue of A by

$$A \begin{bmatrix} \hat{\mathbf{v}} \\ \mathbf{0} \end{bmatrix} = \begin{bmatrix} A_{11}\hat{\mathbf{v}} \\ A_{21}\hat{\mathbf{v}} \end{bmatrix} = \lambda \begin{bmatrix} \hat{\mathbf{v}} \\ \mathbf{0} \end{bmatrix}.$$

Because A is a stability matrix, λ with $|\lambda| = 1$ cannot be an eigenvalue of A_{11} that concludes stability of A_{11}. Define

$$\begin{aligned} \tilde{A}(z) &= A_{22} + A_{21}(zI_r - A_{11})^{-1} A_{21}, \\ \tilde{B}(z) &= B_2 + A_{21}(zI_r - A_{11})^{-1} B_1, \\ \tilde{C}(z) &= C_2 + C_1(zI_r - A_{11})^{-1} A_{21}. \end{aligned} \tag{4.27}$$

It is left as an exercise (Problem 4.3) to show that

$$\mathbf{H}(z) - \widehat{\mathbf{H}}(z) =: \mathbf{E}(z) = \tilde{C}(z)\left[zI_r - \tilde{A}(z)\right]^{-1} \tilde{B}(z), \tag{4.28}$$

$$\tilde{B}\left(e^{j\omega}\right)\tilde{B}\left(e^{j\omega}\right)^* \le \Phi\left(e^{j\omega}\right)\Sigma_2 e^{-j\omega} + e^{j\omega}\Sigma_2 \Phi\left(e^{j\omega}\right)^*, \tag{4.29}$$

$$\tilde{C}\left(e^{j\omega}\right)^* \tilde{C}\left(e^{j\omega}\right) \le \Phi\left(e^{j\omega}\right)^* \Sigma_2 e^{j\omega} + e^{-j\omega}\Sigma_2 \Phi\left(e^{j\omega}\right), \tag{4.30}$$

for each ω where $\Phi(z) = \left[zI_{n-r} - \tilde{A}(z)\right]$. To prove the error bound (4.24), assume first that $\Sigma_2 = \sigma I_{n-r}$. Then multiplying (4.29) by $\tilde{C}\left(e^{j\omega}\right)\left[\Phi\left(e^{j\omega}\right)\right]^{-1}$ from left and by $\left(\tilde{C}\left(e^{j\omega}\right)\left[\Phi\left(e^{j\omega}\right)\right]^{-1}\right)^*$ from right leads to

$$\mathbf{E}\left(e^{j\omega}\right)\mathbf{E}\left(e^{j\omega}\right)^* \le \sigma \tilde{C}\left(e^{j\omega}\right)\left(e^{-j\omega}\left[\Phi\left(e^{j\omega}\right)^*\right]^{-1} + e^{j\omega}\left[\Phi\left(e^{j\omega}\right)\right]^{-1}\right)\tilde{C}\left(e^{j\omega}\right)^*.$$

Therefore, at each frequency ω,

$$\left\|\mathbf{E}\left(e^{j\omega}\right)\right\|^2 \le 2\sigma \left\|\tilde{C}\left(e^{j\omega}\right)\left[e^{j\omega}I_{n-r} - \tilde{A}\left(e^{j\omega}\right)\right]^{-1} \tilde{C}\left(e^{j\omega}\right)^*\right\|.$$

Denote $\Pi\left(e^{j\omega}\right) = \tilde{C}\left(e^{j\omega}\right)\left[\Phi\left(e^{j\omega}\right)\right]^{-1}\tilde{C}\left(e^{j\omega}\right)^*$. Then

$$\left\|\mathbf{E}\left(e^{j\omega}\right)\right\| = \sqrt{2\sigma\left\|\Pi\left(e^{j\omega}\right)\right\|} \tag{4.31}$$

at each frequency ω. Now multiplying (4.30) by $\left[\Phi\left(e^{j\omega}\right)\right]^{-1}\tilde{C}\left(e^{j\omega}\right)^*$ from right and by $\left(\left[\Phi\left(e^{j\omega}\right)\right]^{-1}\tilde{C}\left(e^{j\omega}\right)^*\right)^*$ from left leads to

$$\Pi\left(e^{j\omega}\right)^*\Pi\left(e^{j\omega}\right) \le \sigma\left[e^{j\omega}\Pi\left(e^{j\omega}\right) + e^{-j\omega}\Pi\left(e^{j\omega}\right)^*\right].$$

The above is equivalent to

$$\left[\Pi\left(e^{j\omega}\right) - \sigma e^{j\omega}I_{n-r}\right]^*\left[\Pi\left(e^{j\omega}\right) - \sigma e^{j\omega}I_{n-r}\right] \le \sigma^2 I_{n-r}.$$

That is, $\left\|\Pi\left(e^{j\omega}\right) - \sigma e^{j\omega}I_{n-r}\right\| \le \sigma$ from which $\left\|\Pi\left(e^{j\omega}\right)\right\| \le 2\sigma$ follows. Combined with (4.31) yields $\left\|E\left(e^{j\omega}\right)\right\| \le 2\sigma$ for all ω. The error bound (4.24) is thus true for the case $\Sigma_2 = \sigma I_{n-r}$. Note that the balance truncated model satisfies the same inequalities as in (4.25):

$$A_{11}^*\Sigma_1 A_{11} + C_1^*C_1 \le \Sigma_1, \quad A_{11}\Sigma_1 A_{11}^* + B_1B_1^* \le \Sigma_1$$

by (4.26) and its dual inequality. Hence, if $\Sigma_2 \ne \sigma I_{n-r}$ and Hankel singular values are all distinct, then balanced truncation can be applied repeatedly to each individual Hankel singular value from σ_n to σ_{r+1}. Because each balance truncated model satisfies the same inequalities as in (4.25), the truncation error is $2\sigma_{n-i+1}$ at the ith truncation stage with $i = 1, \ldots, n-r$. The error bound (4.24) is thus proven, which can be improved if there are repeated Hankel singular values. See Problem 4.4 in Exercises. □

The error bound in $\mathcal{H}_\infty$ norm is a pleasant surprise. It will be shown later that optimal Hankel-norm approximation admits a better error bound but involves higher computational complexity. The next result follows from the proof of Theorem 4.3 that will be useful in the next subsection. Its proof is left as an exercise (Problem 4.5).

Corollary 4.1. *Suppose that A_{11} is a stability matrix. Let $\tilde{A}(z)$, $\tilde{B}(z)$, $\tilde{C}(z)$ be defined in (4.27), and $\tilde{\mathbf{T}}(z) = \tilde{C}(z)\left[zI - \tilde{A}(z)\right]^{-1}\tilde{B}(z)$. If*

$$\tilde{B}(z)\tilde{B}(z)^* \le z^{-1}\left[zI - \tilde{A}(z)\right]U_2 + zU_2\left[zI - \tilde{A}(z)\right]^*,$$
$$\tilde{C}(z)^*\tilde{C}(z) \le z\left[zI - \tilde{A}(z)\right]^*V_2 + z^{-1}V_2\left[zI - \tilde{A}(z)\right],$$

for each z on the unit circle and $U_2V_2 = \sigma^2 I$ with $\sigma > 0$, then $\left\|\tilde{\mathbf{T}}\right\|_\infty \le 2\sigma$.

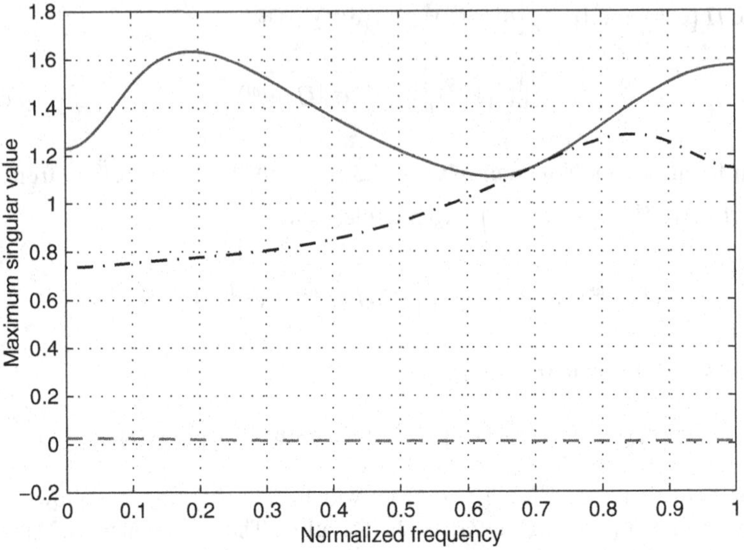

Fig. 4.1 Maximum error frequency response plots

Example 4.4. Consider the same transfer matrix as in Example 4.2. A similarity transformation matrix can be computed to yield a balanced realization for $\mathbf{H}(z)$ so that

$$\Sigma = \text{diag}(5.5386, 3.8150, 1.3345, 1.0299, 0.0176) \tag{4.32}$$

is both the observability and reachability gramians. The maximum singular values of the error frequency responses are plotted in the following figure for approximation with the second-order (upper solid curve), third-order (middle dot–dash curve), and fourth-order (lower dashed curve) models based on balanced truncation (see Fig. 4.1).

It is seen that the error bound in Theorem 4.3 holds. In fact, the error bound is quite conservative for small r.

A SISO plant is considered in the next example.

Example 4.5. The transfer function in consideration is given by

$$H(z) = \frac{1}{z-0.9}\left(\frac{1+0.8z}{z+0.8}\right)^6$$

consisting of one subsystem having the pole at 0.9 and the subsystem with Blaschke product[1] of order 6. It can be shown that all the Hankel singular values of the

[1]Blaschke product is a transfer function whose poles and zeros are in mirror pattern with respect to the unit circle.

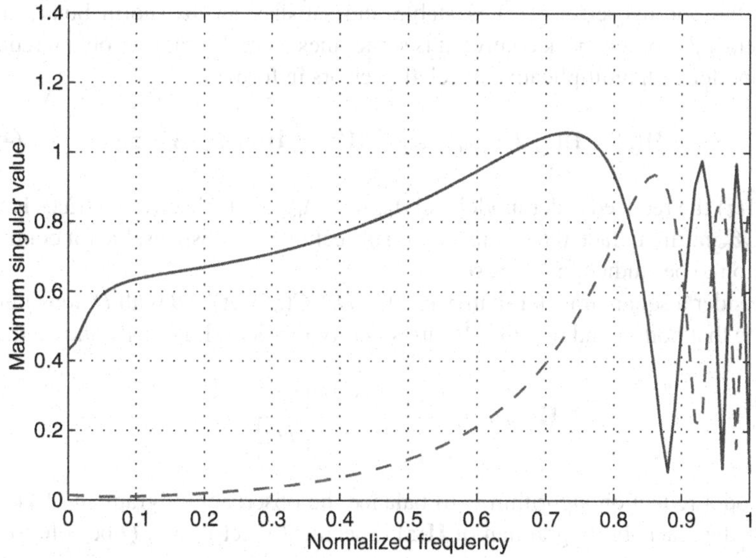

Fig. 4.2 Frequency response error plot

Blaschke product are 1 (Problem 4.10 in Exercises), and thus, the order of $H(z)$ cannot be reduced without the additional pole at 0.9. However, with the pole at 0.9 included, its Hankel singular values are given by

$$5.5419,\ 0.8323,\ 0.5601,\ 0.5352,\ 0.5294,\ 0.5274,\ 0.5265.$$

Hence, except the first Hankel singular value σ_1, all others are very close to each other and are considerably smaller than σ_1. The first- and second-order balanced truncations provide pretty good approximation with maximum error roughly 10% of the $\mathcal{H}_\infty$-norm of $H(z)$ that is 10. See the frequency response error plots in the next figure with solid line for the first-order and dashed line for the second-order approximations. It can be seen that the error bound is again very conservative. In fact, the actual $\mathcal{H}_\infty$-norm error is no more than $2\sigma_7$ (see Fig. 4.2).

4.2.2 *Inverse Balanced Realization*

Balanced truncation yields an additive representation of the plant given by

$$\mathbf{H}(z) = \widehat{\mathbf{H}}(z) + \Delta(z), \tag{4.33}$$

where the additive error $\Delta(z)$ is stable and satisfies an $\mathcal{H}_\infty$-norm bound as in Theorem 4.3. For feedback control, it is sometimes more desirable to obtain reduced order model with multiplicative or relation errors in form of

$$\mathbf{H}(z) = \widehat{\mathbf{H}}(z)\,[I + \Delta_{\rm mul}(z)], \quad \widehat{\mathbf{H}}(z) = \mathbf{H}(z)\,[I + \Delta_{\rm rel}(z)] \tag{4.34}$$

with $\widehat{\mathbf{H}}(z)$ the reduced order model and $\Delta_{\rm mul}(z)/\Delta_{\rm rel}(z)$ stable error matrices having small $\mathcal{H}_\infty$ norm. In fact, the results in this subsection are indispensable for controller reduction to be studied in Chap. 6.

Consider a square transfer matrix $\mathbf{H}(z) = D + C(zI - A)^{-1}B$ with D nonsingular. Assume that both A and $A - BD^{-1}C$ are stability matrices. It is easily shown that

$$\mathbf{H}(z)^{-1} = \left[\begin{array}{c|c} A - BD^{-1}C & BD^{-1} \\ \hline -D^{-1}C & D^{-1} \end{array}\right].$$

The model reduction algorithm is to balance the observability gramian of $\mathbf{H}(z)^{-1}$ against the reachability gramian of $\mathbf{H}(z)$. Specifically, let W and Q be solutions to the following two Lyapunov equations:

$$W = \left(A - BD^{-1}C\right)^* W \left(A - BD^{-1}C\right) + C^*R^{-1}C, \tag{4.35}$$

$$Q = AQA^* + BB^*, \quad R = DD^*. \tag{4.36}$$

Stability of A and $A - BD^{-1}C$ implies that $W \geq \mathbf{0}$ and $Q \geq \mathbf{0}$. If

$$W = Q = S := \mathrm{diag}\left(s_1 I_{i_1}, s_2 I_{i_2}, \ldots, s_\eta I_{i_\eta}\right), \tag{4.37}$$

with $s_1 > s_2 > \cdots > s_\eta \geq 0$ and $i_1 + i_2 + \cdots + i_\eta = n$, then the realization of $\mathbf{H}(z)$ is termed inverse balanced. Let $i_1 + i_2 + \cdots + i_\rho = r$ for some $\rho < \eta$. Partition the realization $\{A, B, C\}$ compatibly as in (4.23) that results in

$$A = \begin{bmatrix} A_{11} & A_{12} \\ A_{21} & A_{22} \end{bmatrix}, \quad B = \begin{bmatrix} B_1 \\ B_2 \end{bmatrix}, \quad C = \begin{bmatrix} C_1 & C_2 \end{bmatrix}, \tag{4.38}$$

where (A_{11}, B_1, C_1, D) has order r yielding the reduced order model

$$\widehat{\mathbf{H}}(z) = D + C_1\left(zI_r - A_{11}\right)^{-1} B_1.$$

Although same notations are used, the reduced order model is different from that in the previous subsection due to the use of inverse balanced truncation (IBT). A different error bound in $\mathcal{H}_\infty$ norm holds.

Theorem 4.6. *Consider square transfer matrix* $\mathbf{H}(z) = D + C(zI_n - A)^{-1}B$ *of which* $\det(D) \neq 0$. *Assume that both* A *and* $A - BD^{-1}C$ *are stability matrices. Let* $\widehat{\mathbf{H}}(z)$ *of order* $r < n$ *be obtained through the* IBT *procedure where* $\{s_k\}$ *in (4.37) are* *all distinct. Then the relations in (4.34) hold for some stable $\Delta_{\rm mul}(z)$ and $\Delta_{\rm rel}(z)$ satisfying*

$$\|\Delta_{\rm mul}\|_\infty \le \prod_{k=\rho+1}^{\eta}\left(1+2s_k\sqrt{1+s_k^2}+2s_k^2\right)-1, \tag{4.39}$$

$$\|\Delta_{\rm rel}\|_\infty \le \prod_{k=\rho+1}^{\eta}\left(1+2s_k\sqrt{1+s_k^2}+2s_k^2\right)-1. \tag{4.40}$$

Proof. The proof is similar to that for Theorem 4.3 by first recognizing that the Lyapunov functions associated with IBT can be written as

$$\left(A-BD^{-1}C\right)^* S\left(A-BD^{-1}C\right)+C^*R^{-1}C \le S, \tag{4.41}$$

$$ASA^*+BB^* \le S, \tag{4.42}$$

where S in (4.37) is diagonal. Next, the error matrices $\Delta_{\rm mul}(z)$ and $\Delta_{\rm rel}(z)$ in (4.34) can be rewritten as

$$\Delta_{\rm mul} = \widehat{\mathbf{H}}^{-1}\left[\mathbf{H}-\widehat{\mathbf{H}}\right], \quad \Delta_{\rm rel} = \mathbf{H}^{-1}\left[\mathbf{H}-\widehat{\mathbf{H}}\right].$$

Recall (4.27) for $C(zI-A)^{-1}B$ which is now modified to

$$\tilde{A}_{\rm m}(z) = A_{22}+A_{21}(zI_r-A_{11})^{-1}A_{12} = \left[\begin{array}{c|c} A_{11} & A_{12} \\ \hline A_{21} & A_{22}\end{array}\right],$$

$$\tilde{B}_{\rm m}(z) = B_2+A_{21}(zI_r-A_{11})^{-1}B_1 = \left[\begin{array}{c|c} A_{11} & B_1 \\ \hline A_{21} & B_2\end{array}\right],$$

$$\tilde{C}_{\rm m}(z) = D^{-1}\left[C_2+C_1(zI_r-A_{11})^{-1}A_{12}\right] = D^{-1}\left[\begin{array}{c|c} A_{11} & A_{12} \\ \hline C_1 & C_2\end{array}\right], \tag{4.43}$$

for $D^{-1}\mathbf{H}(z)-I = D^{-1}C(zI-A)^{-1}B$. It follows from (4.28) that

$$D^{-1}\left[\mathbf{H}(z)-\widehat{\mathbf{H}}(z)\right] = \tilde{C}_{\rm m}(z)\left[zI_r-\tilde{A}_{\rm m}(z)\right]^{-1}\tilde{B}_{\rm m}(z). \tag{4.44}$$

The following transfer matrices

$$\tilde{A}_r(z) = \left[\begin{array}{c|c} A_{11}-B_1D^{-1}C_1 & A_{12}-B_1D^{-1}C_2 \\ \hline A_{21}-B_2D^{-1}C_1 & A_{22}-B_2D^{-1}C_2\end{array}\right],$$

$$\tilde{B}_r(z) = \left[\begin{array}{c|c} A_{11}-B_1D^{-1}C_1 & B_1 \\ \hline A_{21}-B_2D^{-1}C_1 & B_2\end{array}\right],$$

$$\tilde{C}_r(z) = \left[\begin{array}{c|c} A_{11}-B_1D^{-1}C_1 & A_{12}-B_1D^{-1}C_2 \\ \hline D^{-1}C_1 & D^{-1}C_2\end{array}\right], \tag{4.45}$$

are then defined based on $I - \mathbf{H}(z)^{-1}D = D^{-1}C\left(zI - A + BD^{-1}C\right)^{-1}B$ by appropriate modification of (4.27). The expression in (4.28) can be adapted to suite the error matrix $\mathbf{E}_1(z) = \left[I - \mathbf{H}(z)^{-1}D\right] - \left[I - \widehat{\mathbf{H}}(z)^{-1}D\right]$ to arrive at

$$\mathbf{E}_1(z) = \widehat{\mathbf{H}}(z)^{-1}D - \mathbf{H}(z)^{-1}D = \tilde{C}_{\mathrm{r}}(z)\left[zI_r - \tilde{A}_{\mathrm{r}}(z)\right]^{-1}\tilde{B}_{\mathrm{r}}(z). \tag{4.46}$$

It is left as an exercise (Problem 4.7) to show that

$$\begin{aligned} &\text{(i)} \quad \tilde{A}_{\mathrm{r}}(z) = \tilde{A}_{\mathrm{m}}(z) - \tilde{B}_{\mathrm{m}}(z)\tilde{C}_{\mathrm{r}}(z), \\ &\text{(ii)} \ \tilde{B}_{\mathrm{m}}(z) = \tilde{B}_{\mathrm{r}}(z)D^{-1}\widehat{\mathbf{H}}(z), \\ &\text{(iii)} \ \tilde{C}_{\mathrm{r}}(z) = \widehat{\mathbf{H}}(z)^{-1}D\tilde{C}_{\mathrm{m}}(z). \end{aligned} \tag{4.47}$$

Multiplying (4.44) by $\widehat{\mathbf{H}}(z)^{-1}D$ from left yields

$$\begin{aligned} \Delta_{\mathrm{mul}}(z) &= \widehat{\mathbf{H}}(z)^{-1}D\tilde{C}_{\mathrm{m}}(z)\left[zI - \tilde{A}_{\mathrm{m}}(z)\right]^{-1}\tilde{B}_{\mathrm{m}}(z) \\ &= \tilde{C}_{\mathrm{r}}(z)\left[zI - \tilde{A}_{\mathrm{m}}(z)\right]^{-1}\tilde{B}_{\mathrm{m}}(z). \end{aligned} \tag{4.48}$$

by (iii) of (4.47). Since

$$\begin{aligned} \Delta_{\mathrm{rel}}(z) &= \mathbf{H}(z)^{-1}\left[\mathbf{H}(z) - \widehat{\mathbf{H}}(z)\right] \\ &= \left[\widehat{\mathbf{H}}(z)^{-1} - \mathbf{H}(z)^{-1}\right]\widehat{\mathbf{H}}(z) = \mathbf{E}_1(z)D^{-1}\widehat{\mathbf{H}}(z), \end{aligned}$$

multiplying (4.46) by $D^{-1}\widehat{\mathbf{H}}(z)$ from right and using (ii) of (4.47) yield

$$\begin{aligned} \Delta_{\mathrm{rel}}(z) &= \tilde{C}_{\mathrm{r}}(z)\left[zI_{n-r} - \tilde{A}_{\mathrm{r}}(z)\right]^{-1}\tilde{B}_{\mathrm{r}}(z)D^{-1}\widehat{\mathbf{H}}(z) \\ &= \tilde{C}_{\mathrm{r}}(z)\left[zI_{n-r} - \tilde{A}_{\mathrm{r}}(z)\right]^{-1}\tilde{B}_{\mathrm{m}}(z). \end{aligned} \tag{4.49}$$

The Lyapunov inequalities in (4.41) and (4.42) lead to

$$\tilde{B}_{\mathrm{m}}\left(e^{j\omega}\right)\tilde{B}_{\mathrm{m}}\left(e^{j\omega}\right)^* \le \Phi_{\mathrm{m}}\left(e^{j\omega}\right)S_2e^{-j\omega} + e^{j\omega}S_2\Phi_{\mathrm{m}}\left(e^{j\omega}\right)^*, \tag{4.50}$$

$$\tilde{C}_{\mathrm{r}}\left(e^{j\omega}\right)^*\tilde{C}_{\mathrm{r}}\left(e^{j\omega}\right) \le \Phi_{\mathrm{r}}\left(e^{j\omega}\right)^*S_2e^{j\omega} + e^{-j\omega}S_2\Phi_{\mathrm{r}}\left(e^{j\omega}\right), \tag{4.51}$$

for each ω where $\Phi_{\mathrm{m}}(z) = \left[zI_{n-r} - \tilde{A}_{\mathrm{m}}(z)\right]$ and $\Phi_{\mathrm{r}}(z) = \left[zI_{n-r} - \tilde{A}_{\mathrm{r}}(z)\right]$. The above two inequalities are similar to (4.29) and (4.30), respectively. Suppose that $S_2 = s_\eta I_{n-r}$. Substituting (i) of (4.47) into inequality (4.51) gives

$$s_\eta^{-1}\tilde{C}_{\mathrm{r}}^*\tilde{C}_{\mathrm{r}} \le z^{-1}\Phi_{\mathrm{m}} + z\Phi_{\mathrm{m}}^* + z^{-1}\tilde{B}_{\mathrm{m}}\tilde{C}_{\mathrm{r}} + z\tilde{C}_{\mathrm{r}}^*\tilde{B}_{\mathrm{m}}^*,$$

where the argument $e^{j\omega}$ is suppressed and $z = e^{j\omega}$. Applying the inequality from Problem 4.8 in Exercises with $U = z^{-1}\tilde{B}_m$ and $V = \tilde{C}_r$ leads to

$$\begin{aligned} s_\eta^{-1}\tilde{C}_r^*\tilde{C}_r &\le z^{-1}\Phi_m + z\Phi_m^* + \alpha^{-1}\tilde{B}_m\tilde{B}_m^* + \alpha\tilde{C}_r^*\tilde{C}_r \\ &\le \left(1+s_\eta\alpha^{-1}\right)\left(z^{-1}\Phi_m + z\Phi_m^*\right) + \alpha\tilde{C}_r^*\tilde{C}_r \end{aligned}$$

in which (4.50) is used and $\alpha > 0$. For $\alpha s_\eta < 1$, the above inequality implies

$$\tilde{C}_r^*\tilde{C}_r \le \left(\frac{1+s_\eta\alpha^{-1}}{s_\eta^{-1}-\alpha}\right)\left(z^{-1}\Phi_m + z\Phi_m^*\right).$$

Recall that $|z| = 1$ is arbitrary. Together with (4.50) yields

$$\|\Delta_{\text{mul}}\|_\infty \le 2s_\eta\sqrt{(\alpha+s_\eta)/[\alpha(1-\alpha s_\eta)]}$$

in light of Corollary 4.1 and $S_2 = s_\eta I$. Noting that the right-hand side is minimized by $\alpha = \sqrt{1-s_\eta^2} - s_\eta$ that satisfies $\alpha s_\eta < 1$, there holds

$$\|\Delta_{\text{mul}}\|_\infty \le 2s_\eta\left(\sqrt{1+s_\eta^2} + s_\eta\right)$$

that agrees with (4.39) in the special case $\rho + 1 = \eta$. To prove the multiplicative error bound for the general case, denote

$$\Delta_m(k) = \widehat{\mathbf{H}}_{k-1}^{-1}\left(\widehat{\mathbf{H}}_k - \widehat{\mathbf{H}}_{k-1}\right), \quad \rho+1 \le k \le \eta,$$

where $\mathbf{H} = \widehat{\mathbf{H}}_\eta$. Then $\|\Delta_m(k)\|_\infty \le 2s_k\left(\sqrt{1+s_k^2} + s_k\right)$ and

$$\widehat{\mathbf{H}}_k = \widehat{\mathbf{H}}_{k-1}[I + \Delta_m(k)]. \tag{4.52}$$

Let $\widehat{\mathbf{H}}_\rho = \widehat{\mathbf{H}}$. Repeated use of the above equation leads to

$$\begin{aligned} \mathbf{H} = \widehat{\mathbf{H}}_\eta &= \widehat{\mathbf{H}}_{\eta-1}[I + \Delta_m(\eta)] = \cdots \\ &= \widehat{\mathbf{H}}[I + \Delta_m(\rho+1)]\cdots[I + \Delta_m(\eta)]. \end{aligned}$$

Since $\Delta_{\text{mul}} = \widehat{\mathbf{H}}^{-1}\mathbf{H} - I$, there holds

$$\Delta_{\text{mul}} = [I + \Delta_m(\rho+1)]\cdots[I + \Delta_m(\eta)] - I. \tag{4.53}$$

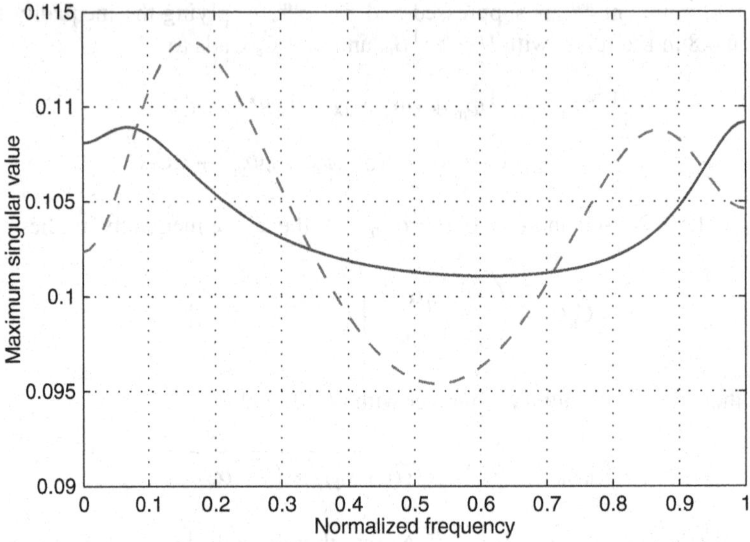

Fig. 4.3 Maximum error frequency response plots

Applying the inequality in Problem 4.11 in Exercises yields

$$\|\Delta_{\text{mul}}\|_\infty \le \prod_{i=\rho+1}^{\eta}\left(1+2s_k\sqrt{1+s_k^2}+2s_k^2\right)-1$$

that completes the proof for (4.39). The proof for the relative error bound in (4.40) is similar and is left as an exercise (Problem 4.12). □

Example 4.7. Consider $\mathbf{H}(z) = C(zI-A)^{-1}B$ with (A,B,C) given in Example 4.2. Since inverse balanced realization requires proper inverse, this example examines

$$\tilde{\mathbf{H}}(z) = z\mathbf{H}(z) = \left[\begin{array}{c|c} A & B \\ \hline CA & CB \end{array}\right], \quad CB = \begin{bmatrix} 1 & 1 \\ -1 & 1 \end{bmatrix},$$

for IBT. Clearly, $\det(CB) \ne 0$. By applying the IBT algorithm to the above $\tilde{\mathbf{H}}(z)$, the reduced plant with order 3 is obtained. The following figure shows the maximum singular value plots for both the relative (solid line) and multiplicative (dashed line) errors in frequency domain (see Fig. 4.3).

It is seen that the model reduction error based on IBT does not exceed 11.35% in the case of $r = 3$. The IB singular values are given by

$$S = \text{diag}(3.3882, 2.6863, 0.3614, 0.1041, 0.0150).$$

The error bound in the case of $r = 3$ is 0.2686 that is considerably greater than the actual error. On the other hand, if $r = 2$ is taken, then the error bound is 1.5752 exceeding 1, even though the actual error is 54.51%, which is not plotted. So the error bound is conservative again, but it does provide some indication on the true error.

Although square plants are assumed, the IBT procedure is actually applicable to nonsquare transfer matrix $\mathbf{G}(z) = D + C(zI_n - A)^{-1}B$, provided that it does not have transmission zeros on and outside the unit circle, including at ∞. Without loss of generality assume that $\mathbf{G}(z)$ is wide or its dimension is $p \times m$ with $p < m$. Then D has full row rank and there exists a right inverse D^+ and a matrix $D_\perp$ of size $(m-p) \times m$ such that

$$\begin{bmatrix} D \\ D_\perp \end{bmatrix} \begin{bmatrix} D^+ & D_\perp^+ \end{bmatrix} = I_m. \tag{4.54}$$

It is left as an exercise to show that (Problem 4.14) $\mathbf{G}(z)$ is strictly minimum phase, if and only if $\{A - BD^+C, BD_\perp\}$ is stabilizable. Hence, there exists a stabilizing state-feedback gain F such that

$$A_F = A - BD^+C + BD_\perp F = A + BC_F \tag{4.55}$$

is a stability matrix with $C_F = D_\perp F - D^+C$. Rewrite

$$\mathbf{G}(z) = D\left[\begin{array}{c|c} A & B \\ \hline -C_F & I \end{array}\right] \implies \mathbf{G}(z)^+ = \left[\begin{array}{c|c} A_F & B \\ \hline C_F & I \end{array}\right] D^+. \tag{4.56}$$

The IBT balances the reachability gramian against the observability gramian of $\mathbf{G}(z)^+$ which are given respectively by

$$Q = AQA^* + BB^*, \quad P = A_F^* P A_F + C_F^* C_F.$$

The following result is true.

Corollary 4.2. *Consider $\mathbf{G}(z) = D + C(zI_n - A)^{-1}B$ of size $p \times m$ with $p < m$ and D full row rank. Suppose that both A and A_F in (4.55) are stability matrices for some F. Then $\mathbf{G}(z)^+$ in (4.56) is a stable right inverse of $\mathbf{G}(z)$. If realizations of $\mathbf{G}(z)$ and $\mathbf{G}(z)^+$ are inverse balanced or*

$$P = Q = S := \operatorname{diag}(s_1 I_{i_1}, s_2 I_{i_2}, \ldots, s_\eta I_{i_\eta}),$$

with $s_1 > s_2 > \cdots > s_\eta \geq 0$ and $i_1 + i_2 + \cdots + i_\eta = n$, and $\widehat{\mathbf{G}}(z)$ is obtained through direct truncation of the states associated with $\{s_k\}_{k=\rho+1}^{\eta}$, then

$$\mathbf{G}(z) = \widehat{\mathbf{G}}(z)[I + \Delta_{\text{mul}}(z)], \quad \widehat{\mathbf{G}}(z) = \mathbf{G}(z)[I + \Delta_{\text{rel}}(z)] \tag{4.57}$$

hold true for some stable $\Delta_{\text{mul}}(z)$ and $\Delta_{\text{rel}}(z)$, satisfying the error bounds in (4.39) and (4.40), respectively.

Proof. By $C_F = D_\perp F - D^+C$, $D_\perp$ in (4.54), and $\mathbf{G}(z)$ in (4.56),

$$\mathbf{H}(z) = \begin{bmatrix} \mathbf{G}(z) \\ \mathbf{G}_\perp(z) \end{bmatrix} := \begin{bmatrix} D \\ D_\perp \end{bmatrix} \left[I - C_F(zI - A)^{-1}B\right]$$

is square. Its inverse is given by

$$\mathbf{H}(z)^{-1} = \left[I + C_F(zI - A_F)^{-1}B\right]\left[D^+ \; D_\perp^+\right]$$

that is also stable. More importantly, inverse balanced realization for $\mathbf{G}(z)$ and $\mathbf{G}(z)^+$ implies that of $\mathbf{H}(z)$ and $\mathbf{H}(z)^+$ as well. An application of Theorem 4.6 leads to the existence of $\Delta_{\text{mul}}(z)$ and $\Delta_{\text{rel}}(z)$ such that

$$\mathbf{H}(z) = \widehat{\mathbf{H}}(z)[I + \Delta_{\text{mul}}(z)], \quad \widehat{\mathbf{H}}(z) = \mathbf{H}(z)[I + \Delta_{\text{rel}}(z)]$$

from which (4.57) follows, and the error bounds in (4.39) and (4.40) hold. □

The error bounds clearly depend on P and thus the stabilizing gain F. Using the procedure in proof of Lemma 4.1, it can be shown that

$$\left\|\mathbf{G}^+\right\|_2^2 = \text{Tr}\left\{\left(D^+\right)^* (I + B^*PB)D^+\right\}.$$

See also Problem 4.1 in Exercises. Hence, minimization of the error bound is hinged to find a stable right inverse of $\mathbf{G}(z)$ with the minimum $\mathscr{H}_2$ norm. This problem will be studied in Chap. 7 in connection with the precoder design for wireless transceivers.

4.3 Optimal Hankel-Norm Approximation

Model reduction based on optimal Hankel-norm approximation is much more involved than that based on balanced realizations. Recall that the rank of the Hankel-matrix $\mathscr{H}_H$ in (4.11) is equals to the McMillan degree of $\mathbf{H}(z)$, the stable and strictly causal part of the underlying plant model $\mathbf{P}(z)$ in (4.1). Optimal Hankel-norm approximation seeks a lower rank Hankel-matrix $\widehat{\mathscr{H}_H}$ such that $\|\mathbf{H} - \widehat{\mathbf{H}}\|_{\text{H}} = \overline{\sigma}(\mathscr{H}_H - \widehat{\mathscr{H}_H})$ is minimized.

For an arbitrary matrix without structural constraint, the optimal approximation is easy to compute via an SVD procedure. Specifically, consider matrix M of size $n \times m$ with rank q. Let $M = U\Sigma V^*$ be its SVD where U has dimension $n \times q$, V dimension $m \times q$, given by

$$U = \left[\mathbf{u}_1 \; \mathbf{u}_2 \; \cdots \; \mathbf{u}_q\right], \quad V = \left[\mathbf{v}_1 \; \mathbf{v}_2 \; \cdots \; \mathbf{v}_q\right],$$

respectively, and $\Sigma = \text{diag}(\sigma_1, \sigma_2, \ldots, \sigma_q)$. Then

$$\widehat{M} = \begin{bmatrix} \mathbf{u}_1 & \cdots & \mathbf{u}_r \end{bmatrix} \text{diag}(\sigma_1, \ldots, \sigma_r) \begin{bmatrix} \mathbf{v}_1 & \cdots & \mathbf{v}_r \end{bmatrix}^*$$

with $r < q$ is the optimal approximation to M in the sense that

$$\overline{\sigma}\left(M - \widehat{M}\right) = \min_{\text{rank}\{M_r\}=r} \overline{\sigma}(M - M_r) = \sigma_{r+1}. \tag{4.58}$$

The proof is left as an exercise (Problem 4.17). However, if such an SVD procedure is applied to the Hankel-matrix $\mathcal{H}_H$, the resultant $\widehat{\mathcal{H}_H}$ may not admit Hankel structure and thus results in no reduced order transfer matrix $\widehat{\mathbf{H}}(z)$. It turns out that the SVD procedure is not applicable directly, and optimal Hankel-norm approximation needs to be developed in operator's framework. By allowing unstable components in $\widehat{\mathbf{H}}(z)$, optimal Hankel-norm approximation admits "allpass" characterization. More specifically, for optimal Hankel-norm approximation of rth order, $\widehat{\mathbf{H}}(z) = \widehat{\mathbf{H}}_s(z) + \widehat{\mathbf{H}}_{as}(z)$ can be employed where $\widehat{\mathbf{H}}_s(z)$ is stable and has McMillan degree r, and $\widehat{\mathbf{H}}_{as}(z)$ is antistable, i.e., it has all its poles strictly outside the unit circle. Since the Hankel matrix depends only on the strictly causal part of the impulse response, the Hankel matrix associated with $\widehat{\mathbf{H}}(z)$ indeed has rank r, independent of $\widehat{\mathbf{H}}_{as}(z)$. The key is to come up with $\widehat{\mathbf{H}}_s(z)$ and $\widehat{\mathbf{H}}_{as}(z)$ such that $\mathbf{H}(z) - \widehat{\mathbf{H}}(z)$ is "allpass" with constant amplitude. It will be shown that a similar procedure to SVD can be developed. More importantly, an $\mathcal{H}_\infty$-norm error bound exists for optimal Hankel-norm approximation that beats the error bound for balanced truncation by a factor of two.

4.3.1 Optimal Model Reduction

Let σ_i be the ith singular value of the Hankel matrix $\mathcal{H}_H$ arranged in descending order: $\sigma_1 \geq \sigma_2 \geq \cdots \geq \sigma_n$. Let $\mathbf{u}_i$ and $\mathbf{v}_i$ be the left and right singular vectors of $\mathcal{H}_H$ corresponding to σ_i. Then for positive integer $k = r + 1 < n$,

$$\mathcal{H}_H \underline{\mathbf{u}}_k = \sigma_k \underline{\mathbf{v}}_k \quad \mathcal{H}_H^* \underline{\mathbf{v}}_k = \sigma_k \underline{\mathbf{u}}_k, \tag{4.59}$$

where $\underline{\mathbf{u}}_k = \text{vec}\left(\begin{bmatrix} \mathbf{u}_k(-1) \; \mathbf{u}_k(-2) \; \cdots \; \mathbf{u}_k(-t) \; \cdots \end{bmatrix}\right)$ corresponds to the past input and $\underline{\mathbf{v}}_k = \text{vec}\left(\begin{bmatrix} \mathbf{v}_k(0) \; \mathbf{v}_k(1) \; \cdots \; \mathbf{v}_k(t) \; \cdots \end{bmatrix}\right)$ corresponds to the present and future output. Recall that vec$(\cdot)$ stacks all its columns into a single vector sequentially. Then $\mathbf{Y}(z) = \sigma_k \mathbf{V}_k(z)$ and $\mathbf{X}(z) = \mathbf{U}_k(z)$ satisfy the Hankel operator equation

$$\mathbf{Y}(z) = \Gamma_{\mathrm{H}} \mathbf{X}(z) := \Pi_+[\mathbf{H}(z)\mathbf{X}(z)] \tag{4.60}$$

with $\varGamma_{\mathrm{H}}$ the Hankel operator associated with $\mathbf{H}(z)$, and

$$\mathbf{U}_k(z) = \sum_{t=1}^{\infty} \mathbf{u}_k(-t)z^t, \quad \mathbf{V}_k(z) = \sum_{t=0}^{\infty} \mathbf{v}_k(t)z^{-t} \tag{4.61}$$

as $\mathscr{Z}$ transforms of the input and output, respectively.

Suppose that σ_{r+1} has multiplicity $\rho > 1$. Then $\sigma_{r+1} = \cdots = \sigma_{r+\rho}$. Each σ_i with its associated Schmidt pair $(\underline{\mathbf{u}}_i, \underline{\mathbf{v}}_i)$ satisfies (4.59) for $k \le i < k+\rho-1$. Recall $k = r+1$. There thus holds

$$\varGamma_{\mathrm{H}}\left[\mathbf{U}_{r+1}\ \mathbf{U}_{r+2}\ \cdots\ \mathbf{U}_{r+\rho}\right] = \sigma_k\left[\mathbf{V}_{r+1}\ \mathbf{V}_{r+2}\ \cdots\ \mathbf{V}_{r+\rho}\right]$$

by the equivalence of (4.60) and (4.59). The remarkable result in optimal Hankel-norm approximation is the following identity:

$$\left(\mathbf{H} - \widehat{\mathbf{H}}\right)\left[\mathbf{U}_{r+1}\ \cdots\ \mathbf{U}_{r+\rho}\right] = \sigma_k\left[\mathbf{V}_{r+1}\ \cdots\ \mathbf{V}_{r+\rho}\right], \tag{4.62}$$

where $\widehat{\mathbf{H}}(z)$ has exactly r poles strictly inside the unit circle plus those outside the unit circle. Since $\mathbf{HU}_i = \varPi_+[\mathbf{HU}_i] + \varPi_-[\mathbf{HU}_i]$ for $r+1 \le i \le r+\rho$, the identity in (4.62) yields

$$\varPi_+\left[\widehat{\mathbf{H}}\mathbf{U}_i\right] = \mathbf{0}, \quad \widehat{\mathbf{H}}\mathbf{U}_i = \varPi_-\left[\widehat{\mathbf{H}}\mathbf{U}_i\right] = \varPi_-\left[\mathbf{HU}_i\right],$$

for the same range of index i. Although σ_i and its corresponding Schmidt pair $(\underline{\mathbf{u}}_i, \underline{\mathbf{v}}_i)$ involve SVD for $\mathscr{H}_H$ of infinite size, they can all be obtained through computations of finite size matrices.

Consider again $\mathbf{H}(z) = C(zI_n - A)^{-1}B$ with A a stability matrix. Let $\mathscr{O}_\infty$ and $\mathscr{R}_\infty$ be the corresponding observability and reachability matrices. Then

$$P = \mathscr{O}_\infty^* \mathscr{O}_\infty, \quad Q = \mathscr{R}_\infty \mathscr{R}_\infty^*$$

are bounded, and satisfy the Lyapunov equations

$$\text{(i)}\ P - A^*PA = C^*C, \quad \text{(ii)}\ Q - AQA^* = BB^*, \tag{4.63}$$

respectively. Denote $\lambda_i(\cdot)$ as the ith eigenvalue. Then $\mathscr{H}_H = \mathscr{O}_\infty \mathscr{R}_\infty$ implies

$$\sigma_i = \sqrt{\lambda_i(QP)} = \sqrt{\lambda_i(PQ)}.$$

Let $v_i \in \mathbb{R}^n$ be the eigenvector of QP corresponding to σ_i:

$$QPv_i = \sigma_i^2 v_i. \tag{4.64}$$

Define $\mu_i = \sigma_i^{-1} P v_i$. Then for $1 \le i \le n$,

$$Q\mu_i = \sigma_i v_i, \quad P v_i = \sigma_i \mu_i, \tag{4.65}$$

by (4.64). Hence, a dual equation to (4.64)

$$PQ\mu_i = \sigma_i^2 \mu_i \tag{4.66}$$

holds. Multiplying (4.66) by $\mathscr{R}_\infty^*$ from left yields

$$\mathscr{R}_\infty^* PQ\mu_i = \mathscr{H}_H^* \mathscr{H}_H \left(\mathscr{R}_\infty^* \mu_i\right) = \sigma_i^2 \left(\mathscr{R}_\infty^* \mu_i\right).$$

The above yields the following expression for the input vector:

$$\underline{\mathbf{u}}_i = \begin{bmatrix} \mathbf{u}_i(-1) \\ \mathbf{u}_i(-2) \\ \vdots \end{bmatrix} = \mathscr{R}_\infty^* \mu_i = \begin{bmatrix} B^* \mu_i \\ (AB)^* \mu_i \\ \vdots \end{bmatrix}.$$

Similarly, by

$$\mathscr{O}_\infty PQ v_i = \mathscr{H}_H \mathscr{H}_H^* \left(\mathscr{O}_\infty v_i\right) = \sigma_i^2 \left(\mathscr{O}_\infty v_i\right),$$

the output vector has the following expression:

$$\underline{\mathbf{v}}_i = \begin{bmatrix} \mathbf{v}_i(0) \\ \mathbf{v}_i(1) \\ \vdots \end{bmatrix} = \mathscr{O}_\infty v_i = \begin{bmatrix} C v_i \\ CA v_i \\ \vdots \end{bmatrix}.$$

It follows that for $r+1 \le i \le r+\rho$, there hold

$$\mathbf{U}_i(z) = \sum_{t=1}^{\infty} B^* (A^*)^{t-1} z^t \mu_i = B^* \left(z^{-1} I - A^*\right)^{-1} \mu_i,$$

$$\mathbf{V}_i(z) = \sum_{t=0}^{\infty} C A^* z^{-t} v_i = zC(zI - A)^{-1} v_i.$$

Denote $\tilde{U} = \begin{bmatrix} \mu_{r+1} & \cdots & \mu_{r+\rho} \end{bmatrix}$ and $\tilde{V} = \begin{bmatrix} v_{r+1} & \cdots & v_{r+\rho} \end{bmatrix}$. Then

$$P\tilde{V} = \tilde{U}\sigma_k, \quad Q\tilde{U} = \tilde{V}\sigma_k. \tag{4.67}$$

The optimal Hankel-norm approximation in (4.62) satisfies

$$\left[\mathbf{H}(z) - \widehat{\mathbf{H}}(z)\right] \tilde{\mathbf{X}}(z) = \sigma_k \tilde{\mathbf{Y}}(z), \tag{4.68}$$

where $\tilde{\mathbf{X}}(z) = B^* \left(z^{-1} I - A^*\right)^{-1} \tilde{U}$, and $\tilde{\mathbf{Y}}(z) = zC(zI - A)^{-1} \tilde{V}$.

Lemma 4.2. *For* $\mathbf{H}(z) = C(zI - A)^{-1}B$ *of size* $p \times m$*, its rth or* $(k-1)$*st order optimal Hankel-norm approximation* $\widehat{\mathbf{H}}(z)$ *satisfies*

$$\widehat{\mathbf{H}}(z)B^*\left(z^{-1}I - A^*\right)^{-1}\tilde{U} = CQA^*\left(z^{-1}I - A^*\right)^{-1}\tilde{U}. \tag{4.69}$$

If the kth Hankel singular value σ_k *has multiplicity* $\rho \geq m$ *and* $p = m$*, then the optimal error function* $\Delta(z) = \mathbf{H}(z) - \widehat{\mathbf{H}}(z)$ *is "allpass" in the sense that* $\Delta(z)^*\Delta(z) = \sigma_k^2 I_m \; \forall |z| = 1$.

Proof. Denote $\mathbf{G}(z) = \mathbf{H}(z)B^*\left(z^{-1}I - A^*\right)^{-1}\tilde{U}$. The equality (4.68) can be rewritten as

$$\widehat{\mathbf{H}}(z)B^*\left(z^{-1}I - A^*\right)^{-1}\tilde{U} = \mathbf{G}(z) - \sigma_k z C(zI - A)^{-1}\tilde{V}. \tag{4.70}$$

A dual form to (4.17) in the proof of Lemma 4.1 is given by

$$(zI - A)Q\left(z^{-1}I - A^*\right) = BB^* - AQ\left(z^{-1}I - A^*\right) - (zI - A)QA^*.$$

in light of (i) in (4.63). The above can be rewritten as

$$BB^* = (zI - A)Q\left(z^{-1}I - A^*\right) + AQ\left(z^{-1}I - A^*\right) + (zI - A)QA^* \tag{4.71}$$

Recall the expression of $\mathbf{G}(z)$. Multiplying (4.71) from left by $C(zI - A)^{-1}$ and from right by $\left(z^{-1}I - A^*\right)^{-1}\tilde{U}$ yields

$$\begin{aligned}
\mathbf{G}(z) &= C(zI - A)^{-1}BB^*\left(z^{-1}I - A^*\right)^{-1}\tilde{U} \\
&= C\left[I + (zI - A)^{-1}A\right]Q\tilde{U} + CQA^*\left(z^{-1}I - A^*\right)^{-1}\tilde{U} \\
&= \sigma_k z C(zI - A)^{-1}\tilde{V} + CQA^*\left(z^{-1}I - A^*\right)^{-1}\tilde{U},
\end{aligned} \tag{4.72}$$

where $Q\tilde{U} = \sigma_k\tilde{V}$ is used. Substituting the above into (4.70) gives

$$\widehat{\mathbf{H}}(z)B^*\left(z^{-1}I - A^*\right)^{-1}\tilde{U} = CQA^*\left(z^{-1}I - A^*\right)^{-1}\tilde{U}$$

that is the same as (4.69). In other words, the optimal Hankel-norm approximation $\widehat{\mathbf{H}}(z)$ has to satisfy (4.69). Let

$$\Psi_x(z) = \tilde{\mathbf{X}}^*\left(z^{-1}\right)\tilde{\mathbf{X}}(z) = \tilde{U}^*(zI - A)^{-1}BB^*\left(z^{-1}I - A^*\right)^{-1}\tilde{U}.$$

Then a similar derivation to (4.71) and (4.72) also gives

$$\Psi_x(z) = \tilde{U}^*Q\tilde{U} + \tilde{U}^*A(zI - A)^{-1}Q\tilde{U} + \tilde{U}^*Q\left(z^{-1}I - A^*\right)^{-1}A^*\tilde{U}. \tag{4.73}$$

Let $\Psi_y(z) = \tilde{\mathbf{Y}}^*\left(z^{-1}\right)\tilde{\mathbf{Y}}(z) = \tilde{V}^*\left(z^{-1}I - A^*\right)^{-1}C^*C(zI-A)^{-1}\tilde{V}$. Then (ii) of (4.63) gives the rise of the dual equation to (4.73):

$$\Psi_y(z) = \tilde{V}^*P\tilde{V} + \tilde{V}^*A^*\left(z^{-1}I - A^*\right)^{-1}P\tilde{V} + \tilde{V}^*P(zI-A)^{-1}A\tilde{V}. \qquad (4.74)$$

By (4.67), $\tilde{V}^*P\tilde{V} = \sigma_k\tilde{V}^*\tilde{U}$ and $\tilde{U}^*Q\tilde{U} = \sigma_k\tilde{U}^*\tilde{V}$. Thus, both $\tilde{V}^*\tilde{U}$ and $\tilde{U}^*\tilde{V}$ are hermitian. Consequently,

$$\tilde{V}^*P\tilde{V} = \sigma_k\tilde{V}^*\tilde{U} = \left(\sigma_k\tilde{V}^*\tilde{U}\right)^* = \sigma_k\tilde{U}^*\tilde{V} = \tilde{U}^*Q\tilde{U}.$$

Hence, the constant term of (4.73) is the same as the one in (4.74). Furthermore, the strictly causal term in (4.73) is, by using (4.67) again,

$$\tilde{U}^*A\left(z^{-1}I - A\right)^{-1}Q\tilde{U} = \sigma_k\tilde{U}^*A(zI-A)^{-1}\tilde{V} = \tilde{V}^*P(zI-A)^{-1}A\tilde{V}$$

which is the same as the strictly causal term in (4.74). It follows that $\Psi_x(z) = \Psi_y(z)$ for all z. Hence, with $\Delta(z) = \widehat{\mathbf{H}}(z) - \mathbf{H}(z)$ and $\Psi_\delta(z) = \Delta^*\left(z^{-1}\right)\Delta(z)$, (4.68) can be manipulated to yield

$$\tilde{\mathbf{X}}^*\left(z^{-1}\right)\Psi_\delta(z)\tilde{\mathbf{X}}(z) = \sigma_k^2\tilde{\mathbf{Y}}^*\left(z^{-1}\right)\tilde{\mathbf{Y}}(z) = \sigma_k^2\tilde{\mathbf{X}}^*\left(z^{-1}\right)\tilde{\mathbf{X}}(z)$$

in light of $\Psi_x(z) = \Psi_y(z)$. It follows that

$$\tilde{\mathbf{X}}^*\left(z^{-1}\right)\left[\sigma_k^2 I_m - \Psi_\delta(z)\right]\tilde{\mathbf{X}}(z) = 0.$$

For the case $\rho \geq m$, it can be concluded that $\Psi_\delta(z) = \sigma_k^2 I_m \ \forall |z| = 1$. □

The result in Lemma 4.2 indicates the allpass feature of the optimal Hankel-norm approximation in the case of $\rho \geq m$ and $p = m$. It will be shown later that

$$\widehat{\mathbf{H}}(z) = \widehat{\mathbf{H}}_s(z) + \widehat{\mathbf{H}}_{as}(z),$$

where $\widehat{\mathbf{H}}_{rms}(z)$ is stable and has the McMillan degree r and $\widehat{\mathbf{H}}_{as}(z)$ is antistable. However, (4.69) may have more than one solution in general. In fact, the optimal Hankel-norm problem has more than one optimal solution if $m > 1$. Our goal in this subsection aims to derive a particular optimal Hankel-norm approximation as a solution to (4.69) for the case $p = m = \rho$.

An Optimal Solution

Several assumptions will be made in order to numerically compute the optimal Hankel-norm approximation. The first two are:

- A1: The realization (A,B,C) of $\mathbf{H}(z)$ is minimal, and balanced in the sense that with P and Q the unique solutions to (i) and (ii) of (4.63), respectively, $P = \text{diag}(\Sigma_2, \sigma_k I_\rho)$ and $Q = \text{diag}(\Sigma_1, \sigma_k I_\rho)$ where Σ_1 and Σ_2 are diagonal and positive.
- A2: $m = p = \rho$ and $\det\left(\begin{bmatrix} A_{11} & B_1 \\ A_{21} & B_2 \end{bmatrix}\right) \neq 0$.

Clearly, A1 can be made true and thus has no loss of generality. The nonsingular assumption in A2 holds generically, if $m = p = \rho$ is true. However, $m = p = \rho$ in A2 does not hold in general. An extension theorem will be introduced in the late part of the subsection to address this issue.

Under A1, the two Lyapunov equations in (4.63) can be decomposed into:

$$\Sigma_1 - A_{11}\Sigma_1 A_{11}^* - \sigma_k A_{12} A_{12}^* = B_1 B_1^*, \tag{4.75}$$

$$-A_{11}\Sigma_1 A_{21}^* - \sigma_k A_{12} A_{22}^* = B_1 B_2^*, \tag{4.76}$$

$$\sigma_k I_p - A_{21}\Sigma_1 A_{21}^* - \sigma_k A_{22} A_{22}^* = B_2 B_2^*, \tag{4.77}$$

$$\Sigma_2 - A_{11}^*\Sigma_2 A_{11} - \sigma_k A_{21}^* A_{21} = C_1^* C_1, \tag{4.78}$$

$$\sigma_k I_p - A_{12}^*\Sigma_2 A_{12} - \sigma_k A_{22}^* A_{22} = C_2^* C_2 \tag{4.79}$$

$$-A_{11}^*\Sigma_2 A_{12} - \sigma_k A_{21}^* A_{22} = C_1^* C_2, \tag{4.80}$$

where A_{ij}, B_i, and C_i for $i,j = 1,2$ are partitions of A, B, and C, such that they are compatible with partitions of P and Q. In addition, $\tilde{U} = \tilde{V} = \begin{bmatrix} 0 & I_p \end{bmatrix}^*$. Thus, with $\mathbf{G}_1(z) = \left[(zI_p - A_{22}^*) - A_{12}^*(zI_{n-p} - A_{11}^*)^{-1} A_{21}^*\right]^{-1}$,

$$\begin{aligned}
\left(z^{-1}I_n - A^*\right)^{-1}\tilde{U} &= \begin{bmatrix} z^{-1}I_{n-p} - A_{11}^* & -A_{21}^* \\ -A_{12}^* & zI_p - A_{22}^* \end{bmatrix}^{-1} \begin{bmatrix} 0 \\ I_r \end{bmatrix} \\
&= \begin{bmatrix} \left(z^{-1}I_{n-p} - A_{11}^*\right)^{-1} A_{21}^* \\ I_p \end{bmatrix} \mathbf{G}_1(z)
\end{aligned}$$

holds. As a result, (4.69) in Lemma 4.2 is equivalent to

$$\widehat{\mathbf{H}}(z)\left[B_2^* + B_1^*\left(z^{-1}I - A_{11}^*\right)^{-1} A_{21}^*\right] = \Omega_2 + \Omega_1\left(z^{-1}I - A_{11}^*\right)^{-1} A_{21}^*, \tag{4.81}$$

where $\begin{bmatrix} \Omega_1 & \Omega_2 \end{bmatrix} = CQA^*$ given by

$$\Omega_1 = C_1\Sigma_1 A_{11}^* + \sigma_k C_2 A_{12}^*, \quad \Omega_2 = C_1\Sigma_1 A_{21}^* + \sigma_k C_2 A_{22}^*. \tag{4.82}$$

Lemma 4.3. *Let $(\hat{A}, \hat{B}, \hat{C}, \hat{D})$ be a minimal realization for $\widehat{\mathbf{H}}(z)$ that is the optimal Hankel-norm approximation to $\mathbf{H}(z) = C(zI - A)^{-1}B$. Under the* Assumption A1–A2, *there exist a nonsingular solution W to*

$$W = \hat{A} W A_{11}^* + \hat{B} B_1^*. \tag{4.83}$$

Proof. The equality (4.81) can be rewritten as

$$\widehat{\mathbf{H}}(z)B_1^*\left(z^{-1}I-A_{11}^*\right)^{-1}A_{21}^* = \Omega_2-\widehat{\mathbf{H}}(z)B_2^*+\Omega_1\left(z^{-1}I-A_{11}^*\right)^{-1}A_{21}^*.$$

The left-hand side involves poles of $\widehat{\mathbf{H}}(z)$ (which are eigenvalues of $\hat{A}$) and poles of $B_1^*\left(z^{-1}I-A_{11}^*\right)^{-1}A_{21}^*$ (which are eigenvalues of $(A_{11}^*)^{-1}$) in light of the hypothesis on minimal realizations. These two sets of poles are decomposed into the summation form as shown on the right-hand side, implying that (4.83) has a solution W. It follows that the next equality

$$\hat{B}B_1^* = \left(zI-\hat{A}\right)W\left(z^{-1}I-A_{11}^*\right)+\hat{A}W\left(z^{-1}I-A_{11}^*\right)+\left(zI-\hat{A}\right)WA_{11}^* \tag{4.84}$$

holds that can be derived in a similar way to that in (4.71). Multiplying (4.84) by $\hat{C}\left(zI-\hat{A}\right)^{-1}$ from left and by $\left(z^{-1}I-A_{11}^*\right)^{-1}A_{21}^*$ from right yields

$$\begin{aligned}\widehat{\mathbf{H}}(z)B_1^*\left(z^{-1}I-A_{11}^*\right)^{-1}A_{21}^* &= \hat{C}WA_{21}^*+\hat{C}\left(zI-\hat{A}\right)^{-1}\hat{A}WA_{21}^*\\ &\quad+\hat{C}WA_{11}^*\left(z^{-1}I-A_{11}^*\right)^{-1}A_{21}^*,\end{aligned}$$

if $\hat{D}=\mathbf{0}$. Otherwise, the above can be modified to yield

$$\begin{aligned}\mathbf{T}_{\text{tmp}}(z) &:= \widehat{\mathbf{H}}(z)\left[B_2^*+B_1^*(z^{-1}I_{n-r}-A_{11}^*)^{-1}A_{21}^*\right]\\ &= \hat{C}WA_{21}^*+\hat{D}B_2^*+\hat{C}\left(zI-\hat{A}\right)^{-1}\left(\hat{A}WA_{21}^*+\hat{B}B_2^*\right)\\ &\quad+\left(\hat{C}WA_{11}^*+\hat{D}B_1^*\right)\left(z^{-1}I-A_{11}^*\right)^{-1}A_{21}^*\\ &= \Omega_2+\Omega_1\left(z^{-1}I_{n-r}-A_{11}^*\right)^{-1}A_{21}^*\end{aligned}$$

in light of (4.81). Therefore, there hold

$$\begin{aligned}\hat{A}WA_{21}^*+\hat{B}B_2^* &= \mathbf{0},\\ \hat{C}WA_{11}^*+\hat{D}B_1^* &= \Omega_1,\\ \hat{C}WA_{21}^*+\hat{D}B_2^* &= \Omega_2.\end{aligned}$$

In conjunction with $W=\hat{A}WA_{11}^*+\hat{B}B_1^*$ leads to

$$\begin{bmatrix}\hat{A}W & \hat{B}\\ \hat{C}W & \hat{D}\end{bmatrix}\begin{bmatrix}A_{11} & B_1\\ A_{21} & B_2\end{bmatrix}^* = \begin{bmatrix}W & \mathbf{0}\\ \Omega_1 & \Omega_2\end{bmatrix}. \tag{4.85}$$

Assume on the contrary that W is singular. Then there exists a nonzero row vector $\mathbf{q}$ such that $\mathbf{q}W = \mathbf{0}$. Multiplying the augmented vector $\begin{bmatrix} \mathbf{q} & \mathbf{0} \end{bmatrix}$ to the equality (4.85) from left yields

$$\mathbf{q}\begin{bmatrix} \hat{A}W & \hat{B} \end{bmatrix} \begin{bmatrix} A_{11} & B_1 \\ A_{21} & B_2 \end{bmatrix}^* = \mathbf{0}.$$

The nonsingular assumption in A2 implies that

$$\mathbf{q}\hat{A}W = \mathbf{0}, \quad \mathbf{q}\hat{B} = \mathbf{0} \implies \mathbf{q}_1 W = \mathbf{0}$$

by taking $\mathbf{q}_1 = \mathbf{q}\hat{A}$. Applying the same process with $\mathbf{q}$ replaced by $\mathbf{q}_1$ yields

$$\mathbf{q}_1\hat{A}W = \mathbf{0}, \quad \mathbf{q}_1\hat{B} = \mathbf{q}\hat{A}\hat{B} = \mathbf{0} \implies \mathbf{q}_2 W = \mathbf{0}$$

by taking $\mathbf{q}_2 = \mathbf{q}_1\hat{A} = \mathbf{q}\hat{A}^2$. By induction, $\mathbf{q}\hat{A}^i\hat{B} = \mathbf{0}$ holds for $i = 1, 2, \ldots$, leading to nonreachability of $(\hat{A}, \hat{B})$ that contradicts to the hypothesis on minimal realization for $\widehat{\mathbf{H}}(z)$. The lemma is thus true. □

By the proof of Lemma 4.3, $\operatorname{diag}\left(W^{-1}, I\right)$ can be multiplied from left to both sides of (4.85). Because $\left\{W^{-1}\hat{A}W, W^{-1}\hat{B}, \hat{C}W, \hat{D}\right\}$ is also a realization of $\widehat{\mathbf{H}}(z)$, $W = I$ can be assumed that has no loss of generality, yielding the new equality:

$$\begin{bmatrix} \hat{A} & \hat{B} \\ \hat{C} & \hat{D} \end{bmatrix} \begin{bmatrix} A_{11} & B_1 \\ A_{21} & B_2 \end{bmatrix}^* = \begin{bmatrix} I & \mathbf{0} \\ \Omega_1 & \Omega_2 \end{bmatrix}. \tag{4.86}$$

Hence, a realization of the optimal Hankel-norm approximation $\widehat{\mathbf{H}}(z)$ satisfies (4.86) under Assumptions A1 and A2, and can be easily computed by inverting the second square matrix on left. The following illustrates the use of the optimal Hankel-norm approximation in model reduction.

Example 4.8. Recall the transfer function in Example 4.5:

$$H(z) = \frac{1}{z - 0.9}\left(\frac{1 + 0.8z}{z + 0.8}\right)^6.$$

This is a SISO model, and its minimal realization can be easily obtained. Assumptions A1 and A2 can thus be made true by noticing that $p = m = \rho = 1$ and by the fact that all the Hankel singular values of $H(z)$ are different. A minimal realization of the optimal Hankel-norm approximation $\widehat{H}(z)$ can be computed based on the relation derived in (4.86). For each positive integer r smaller than 6, allpass frequency responses are observed for $\left|H\left(e^{j\omega}\right) - \widehat{H}\left(e^{j\omega}\right)\right|$ with magnitude equal to $\sigma_k = \sigma_{r+1}$.

It is noted that for $1 < r < 6$, the Hankel-norm approximation $\widehat{H}(z)$ has $(n - \rho)$ poles with exactly r stable poles and $(n - \rho - r)$ unstable poles where $\rho = 1$. This fact will be proven later. The extraction of the stable part of $\widehat{H}(z)$ can be carried out in two steps. The first step uses Schur decomposition of $\hat{A}$. The Matlab command "$[U_T, A_T] = \text{schur}(\hat{A})$" can be used to obtain $\hat{A} = U_T A_T U_T^*$ with A_T upper triangle and U_A unitary. Unfortunately, the eigenvalues of A_T are not ordered. Thus, Matlab command "$[U_{T_D}, A_{T_D}] = \text{ordschur}(U_T, A_T, \text{"udi"})$" needs to be used to obtain the upper triangular A_{T_D} in the following ordered form:

$$A_{T_D} = \begin{bmatrix} A_{\rm s} & A_{\rm x} \\ \mathbf{0} & A_{\rm as} \end{bmatrix},$$

where $A_{\rm s}$ and $A_{\rm as}^{-1}$ are both stability matrices. The keyword "udi" stands for "interior of unit disk" and specifies the ordering of the matrix A_{T_D} with respect to the unit circle. There holds $A = U_{T_D} A_{T_D} U_{T_D}^*$. In the second step, the solution to the following Sylvester equation

$$-A_{\rm s} Z + Z A_{\rm x} + A_{\rm as} = 0$$

needs to be computed with the Matlab command "$Z = \text{lyap}(-A_{\rm s}, A_{\rm as}, A_{\rm x})$." It is left as an exercise (Problem 4.16) to show that

$$S = \begin{bmatrix} I_r & Z \\ \mathbf{0} & I_{n-k} \end{bmatrix} U_{T_D}^*$$

is the required similarity transform to decompose $\widehat{H}(z) = \widehat{H}_{\rm s}(z) + \widehat{H}_{\rm as}(z)$ with $\widehat{H}_{\rm s}(z)$ the stable and strictly causal part, and $\widehat{H}_{\rm as}(z)$ the anti-causal part of $\widehat{H}(z)$. The next figure shows the magnitude response of $H(z) - \widehat{H}_{\rm s}(z)$:

The solid line shows the error response based on the first-order approximation, while the dashed line shows the error response based on the second-order approximation. Both are smaller than their respective counterpart based on balanced truncation in Example 4.5. In fact there exits an error bound for the optimal Hankel-norm approximation which is only half of that for balanced truncation (see Fig. 4.4).

To facilitate the derivation of the error bound associated with the optimal Hankel-norm approximation, the last Assumption A3: $\det(B_2) \neq 0$ is made temporarily. This assumption implies A2 and has no loss of generality. If A3 fails, $B_{2\varepsilon} = B_2 + \varepsilon I_p$ can be used by taking $\varepsilon > 0$ a small value such that $\det(B_{2\varepsilon}) \neq 0$. After the error bound is proven, $\varepsilon \to 0$ can be taken.

Under Assumption A3, the equation

$$\begin{bmatrix} A_{21}^* \\ C_2 A_{22}^* \end{bmatrix} = \begin{bmatrix} \Theta \\ \Phi \end{bmatrix} B_2^* \tag{4.87}$$

has unique solutions (Θ, Φ). It follows that

$$\Omega_2 = C_1 \Sigma_1 A_{21}^* + \sigma_k C_2 A_{22}^* = (C_1 \Sigma_1 \Theta + \sigma_k \Phi) B_2^*$$

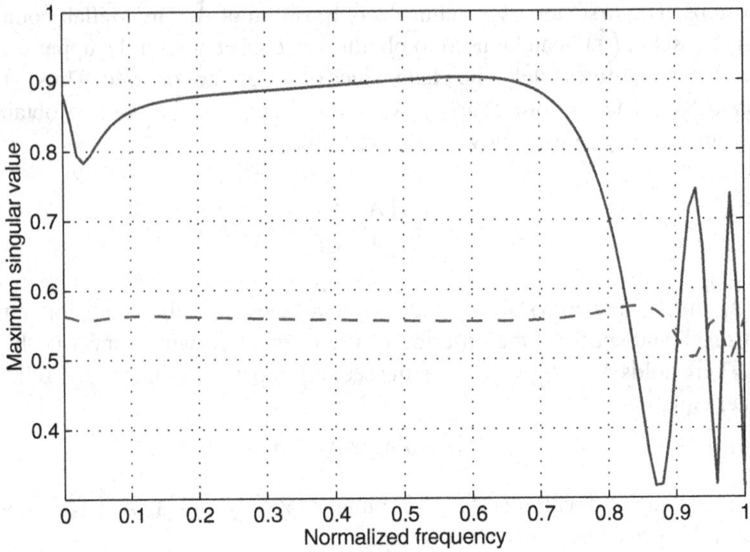

Fig. 4.4 Frequency response error plot

by (4.82). Multiplying (4.86) by $\begin{bmatrix} I & \mathbf{0} \end{bmatrix}^*$ from right leads to

$$\begin{bmatrix} \hat{A} & \hat{B} \\ \hat{C} & \hat{D} \end{bmatrix} \begin{bmatrix} A_{21}^* \\ B_2^* \end{bmatrix} = \begin{bmatrix} \hat{A}\Theta + \hat{B} \\ \hat{C}\Theta + \hat{D} \end{bmatrix} B_2^* = \begin{bmatrix} \mathbf{0} \\ C_1\Sigma_1\Theta + \sigma_k\Phi \end{bmatrix} B_2^*$$

which holds, if $\hat{A}\Theta + \hat{B} = \mathbf{0}$ and $\hat{C}\Theta + \hat{D} = C_1\Sigma_1\Theta + \sigma\Phi$. The former gives

$$\hat{B} = -\hat{A}\Theta \implies \hat{A}\,(A_{11}^* - \Theta B_1^*) = I \tag{4.88}$$

by substituting $\hat{B} = -\hat{A}\Theta$ into $\hat{A}A_{11}^* + \hat{B}B_1 = I$. The latter yields

$$\hat{D} = C_1\Sigma_1\Theta + \sigma_k\Phi - \hat{C}\Theta. \tag{4.89}$$

Substituting the above into $\hat{C}A_{11}^* + \hat{D}B_1^* = \Omega_1$ gives

$$\hat{C}\,(A_{11}^* - \Theta B_1^*) = \Omega_1 - (C_1\Sigma_1\Theta + \sigma_k\Phi)\,B_1^*.$$

In light of (4.88) and the expression of Ω_1 in (4.82), $\hat{C}$ can be solved as

$$\hat{C} = C_1\Sigma_1 + \sigma_k\,(C_2A_{12}^* - \Phi B_1^*)\,\hat{A}. \tag{4.90}$$

Substituting the expression of $\hat{C}$ into (4.89) and using the relation $\hat{A}\Theta = -\hat{B}$ yield

$$\hat{D} = \sigma_k \Phi + \sigma_k \left(C_2 A_{12}^* - \Phi B_1^*\right) \hat{B}. \tag{4.91}$$

Thus, under Assumption A3, the state-space realization of $\widehat{\mathbf{H}}(z)$ has more appealing expressions in (4.88), (4.90), and (4.91).

Characterization of Allpass

The next two lemmas will be useful.

Lemma 4.4. *Let Q be a solution to $Q - AQA^* = BB^*$. (a) If (A,B) is reachable, then A has no eigenvalues on the unit circle. (b) If A has no eigenvalues on the unit circle, and Q is nonsingular, then (A,B) is reachable.*

Proof. If A has an eigenvalue $e^{j\omega}$ for some real valued ω with eigenvector $\mathbf{v}$, then multiplying $\mathbf{v}$ from right, and $\mathbf{v}^*$ from left to $Q - AQA^* = BB^*$ yields

$$\left(1 - |\lambda|^2\right) \mathbf{v}^* Q \mathbf{v} = \|B\mathbf{v}\|^2. \tag{4.92}$$

Since $\lambda = e^{j\omega}$, $B\mathbf{v} = \mathbf{0}$ implying that (A,B) is not reachable that proves (a) by the contrapositive argument. For (b), let λ be any eigenvalue of A with eigenvector $\mathbf{v}$ for which (4.92) again holds. Since $|\lambda|^2 \neq 1$, $B\mathbf{v} \neq \mathbf{0}$, and thus, (A,B) is reachable, in light of the PBH test. □

Lemma 4.5. *Let a square transfer matrix $\mathbf{S}(z) = D + C(zI - A)^{-1}B$ have a minimal realization with nonsingular A. If $\mathbf{S}(z)$ is "allpass" or $\mathbf{S}^*\left(z^{-1}\right)\mathbf{S}(z) = I$, then there exist X and Y such that*

$$\begin{aligned}
&\text{(i)}\ \ Y - A^*YA = C^*C, \quad \text{(ii)}\ X - AXA^* = BB^*, \quad \text{(iii)}\ XY = I,\\
&\text{(iv)}\ BD^* + AXC^* = 0, \quad \text{(v)}\ D^*C + B^*YA = 0,\\
&\text{(vi)}\ D^*D + B^*YB = DD^* + CXC^* = I.
\end{aligned}$$

Proof. By $\mathbf{S}^*\left(z^{-1}\right) = D^* + B^*\left(z^{-1}I - A^*\right)^{-1} C^*$, direct calculation gives

$$\mathbf{S}^*\left(z^{-1}\right) = D^* - B^*A^{*-1}C^* - B^*A^{*-1}\left(zI - A^*\right)^{-1}A^{*-1}C^*.$$

By $\mathbf{S}^*\left(z^{-1}\right)\mathbf{S}(z) = I$, $\mathbf{S}^*\left(z^{-1}\right) = \mathbf{S}(z)^{-1}$. Thus

$$\lim_{z\to\infty} \mathbf{S}^*\left(z^{-1}\right) = D^* - B^*A^{*-1}C^* = \lim_{z\to\infty} \mathbf{S}(z)^{-1} = D^{-1}.$$

That is, A being nonsingular implies D being nonsingular. The above yields

$$D^*D - B^*A^{*-1}C^*D = I, \tag{4.93}$$

$$DD^* - DB^*A^{*-1}C^* = I. \tag{4.94}$$

The first equality is obtained by multiplying $D^{-1} = D^* - B^*A^{*-1}C^*$ by D from right, and the second by D from left. Since D is invertible,

$$\mathbf{S}(z)^{-1} = D^{-1} - D^{-1}C\left(zI - A + BD^{-1}C\right)^{-1}BD^{-1}.$$

The minimality of the realization and $\mathbf{S}^*\left(z^{-1}\right) = \mathbf{S}(z)^{-1}$ implies the existence of a nonsingular matrix T such that

$$(\alpha)\;\; T^{-1}A^{*-1}T = A - BD^{-1}C, \tag{4.95}$$

$$(\beta)\;\; T^{-1}A^{*-1}C^* = BD^{-1}, \tag{4.96}$$

$$(\gamma)\;\; B^*A^{*-1}T = D^{-1}C. \tag{4.97}$$

The equality (α) gives rise to

$$T = A^*TA - A^*TBD^{-1}C, \quad T^{-1} = AT^{-1}A^* - BD^{-1}CT^{-1}A^* \tag{4.98}$$

by multiplying both sides of (α) by A^*T from left and by $T^{-1}A^*$ from right, respectively. Multiplying both sides of (β) by A^*T from left brings to

$$C^* = A^*TBD^{-1} \implies C^*D = A^*TB. \tag{4.99}$$

On the other hand, multiplying both sides of (γ) by $T^{-1}A^*$ from right yields

$$B^* = D^{-1}CT^{-1}A^* \implies DB^* = CT^{-1}A^*. \tag{4.100}$$

Setting $Y = -T$ and substituting $C^*D = A^*TB$ of (4.99) into (4.93) shows

$$I = D^*D - B^*A^{*-1}C^*D = D^*D - B^*TB = D^*D + B^*YB.$$

With $X = -T^{-1}$ and $DB^* = CT^{-1}A^*$ in (4.100), (4.94) can be rewritten as

$$I = DD^* - DB^*A^{*-1}C^* = DD^* - CT^{-1}C^* = DD^* + CXC^*.$$

Thus, (vi) holds. Since $XY = \left(-T^{-1}\right)(-T) = I$, (iii) holds as well. Substituting $A^*TB = C^*D$ of (4.99) into the first equality of (4.98) yields

$$A^*TA - A^*TBD^{-1}C = A^*TA - C^*C = T$$

which is the same as (i) by $Y = -T$. Substituting $D^{-1}CT^{-1}A^* = B^*$ of (4.100) into the second equality of (4.98) shows

$$AT^{-1}A^* - BD^{-1}CT^{-1}A^* = AT^{-1}A^* - BB^* = T^{-1}$$

which is the same as (ii) by $X = -T^{-1}$. Finally, (iv) is identical to (4.100) and (v) is identical to (4.99) thereby concluding the proof. □

Under Assumptions A1–A3, the allpass property for the error matrix $\Delta(z) = \mathbf{H}(z) - \widehat{\mathbf{H}}(z)$ holds. It can be represented by

$$\Delta(z) = \left[\begin{array}{c|c} A_e & B_e \\ \hline C_e & D_e \end{array}\right] = \left[\begin{array}{cc|c} A & \mathbf{0} & B \\ \mathbf{0} & \hat{A} & \hat{B} \\ \hline C & -\hat{C} & -\hat{D} \end{array}\right].$$

Consider Lyapunov equations for the error system:

$$\text{(a)}\ P_e - A_e^* P_e A_e = C_e^* C_e, \quad \text{(b)}\ Q_e - A_e Q_e A_e^* = B_e B_e^*. \tag{4.101}$$

Partition P_e and Q_e compatibly with $\{A_e, B_e, C_e\}$ as

$$P_e = \begin{bmatrix} P_{11} & P_{12} \\ P_{21} & P_{22} \end{bmatrix}, \quad Q_e = \begin{bmatrix} Q_{11} & Q_{12} \\ Q_{21} & Q_{22} \end{bmatrix}.$$

Then (a) is equivalent to the following three equations:

$$P_{11} - A^* P_{11} A = C^* C \implies P_{11} = P, \tag{4.102}$$

$$P_{22} - \hat{A}^* P_{22} \hat{A} = \hat{C}^* \hat{C}, \tag{4.103}$$

$$P_{21} = \hat{A}^* P_{21} A - \hat{C}^* C \implies P_{21} = \begin{bmatrix} \Gamma_1 & \Gamma_2 \end{bmatrix} \tag{4.104}$$

with Γ_1 of size $(n-p) \times (n-p)$ and Γ_2 of size $(n-p) \times p$. For (b), there hold

$$Q_{11} - A Q_{11} A^* = BB^* \implies Q_{11} = Q \tag{4.105}$$

$$Q_{22} - \hat{A} Q_{22} \hat{A}^* = \hat{B}\hat{B}^*, \tag{4.106}$$

$$Q_{12} = A Q_{12} \hat{A}^* + B\hat{B}^* \implies Q_{12} = \begin{bmatrix} I_{n-p} \\ 0 \end{bmatrix} \tag{4.107}$$

The last equation can be verified by (4.88) that leads to

$$Q_{12} = A Q_{12} \hat{A}^* - B\Theta^* \hat{A}^* \iff Q_{12}(A_{11}^* - B_1 \Theta^*) = A Q_{12} - B\Theta^*,$$

and thus, Q_{12} in (4.107) is true in light of $A_{21} = B_2\Theta^*$ in (4.87). It is left as an exercise to show that (Problem 4.20)

$$\left(\hat{D}+\hat{C}\Theta\right)B^* + \hat{C}\hat{A}^{-1}Q_{21} = CQA^* = \begin{bmatrix}\Omega_1 & \Omega_2\end{bmatrix}. \tag{4.108}$$

Lemma 4.6. *Let* $\widehat{\mathbf{H}}(z)$ *be an optimal Hankel-norm approximate to* $\mathbf{H}(z) = C(zI - A)^{-1}B$ *with its realization computed from (4.86). Under* Assumptions A_1–A_3, *there hold*

$$\Gamma_1 = \sigma_k^2 I_{n-p} - \Sigma_1\Sigma_2, \quad \Gamma_2 = 0, \quad Q_{22} = -\Sigma_2\Gamma_1^{-1}, \quad P_{22} = -\Sigma_1\Gamma_1. \tag{4.109}$$

Proof. Since $p = m = \rho$, the error function $\Delta(z)$ is indeed "allpass" with amplitude σ_k, implying that the solutions P_e and Q_e to the two Lyapunov equations in (4.101), respectively, satisfy $Q_eP_e = \sigma_k^2 I_{2n-p}$, or equivalently

$$\begin{aligned} Q_eP_e &= \begin{bmatrix} \Sigma_1 & 0 & I_{n-p} \\ 0 & \sigma_k I_p & 0 \\ I_{n-p} & 0 & Q_{22} \end{bmatrix} \begin{bmatrix} \Sigma_2 & 0 & \Gamma_1^* \\ 0 & \sigma_k I_p & \Gamma_2^* \\ \Gamma_1 & \Gamma_2 & P_{22} \end{bmatrix} \\ &= \begin{bmatrix} \Gamma_1 + \Sigma_1\Sigma_2 & 0 & P_{22} + \Sigma_1\Gamma_1^* \\ 0 & \sigma_k^2 I_p & \sigma_k\Gamma_2^* \\ Q_{22}\Gamma_1 + \Sigma_2 & Q_{22}\Gamma_2 & Q_{22}P_{22} + \Gamma_1^* \end{bmatrix} = \sigma_k^2 I_{2n-p}, \end{aligned}$$

in light of Lemma 4.5. Recall that $Q = \text{diag}(\Sigma_1, \sigma_k I)$ and $P = \text{diag}(\Sigma_2, \sigma_k I)$. Hence, (4.109) can be easily verified. □

Since $\Sigma_1\Sigma_2$ is diagonal consisting of Hankel singular values other than σ_k, both P_{22} and Q_{22} have exactly $(k-1) = r$ positive eigenvalues for the square "allpass" $\Delta(z)$ (Problem 4.21 in Exercises). As a result, $\widehat{\mathbf{H}}(z)$ has no more than r poles inside the unit circle as claimed earlier. Because the minimal achievable Hankel-norm error is the same as $\overline{\sigma}\left(\mathscr{H}_H - \widehat{\mathscr{H}}_H\right)$ that is no smaller than the kth Hankel singular value σ_k over all possible $\widehat{\mathbf{H}}(z)$ with $(k-1)$ poles inside the unit circle, the allpass property does yield an optimal Hankel-norm approximation, which in turn implies that $\widehat{\mathbf{H}}(z)$ has exactly r poles strictly inside the unit circle. It is noted that $\Gamma_2 = 0$ as in Lemma 4.6 converts equivalently (4.104) into

$$\text{(i) } A_{12}^*\Gamma_1\hat{A} = C_2^*\hat{C}, \quad \text{(ii) } \Gamma_1 + C_1^*\hat{C} = A_{11}^*\Gamma_1\hat{A}. \tag{4.110}$$

4.3.2 Error Bound in $\mathscr{H}_\infty$ Norm

The most salient feature of optimal Hankel-norm approximation is the $\mathscr{H}_\infty$-norm error bound. Specifically, for $\mathbf{H}(z) = C\left(zI_n - A\right)^{-1}B$ with A a stability matrix, its optimal Hankel-norm approximation $\widehat{\mathbf{H}}(z)$ can be decomposed into

$$\widehat{\mathbf{H}}(z) = \widehat{\mathbf{H}}_s(z) + \widehat{\mathbf{H}}_{as}(z)$$

with $\widehat{\mathbf{H}}_s(z)$ and $\widehat{\mathbf{H}}_{as}\left(z^{-1}\right)$ both stable transfer matrices. Suppose that $\mathbf{H}(z)$ has η distinct Hankel singular values given by $\{\sigma_{i_j}\}_{j=1}^{\eta}$ with $i_\kappa = k$ and $i_\eta = n$. Let $\widehat{\mathbf{H}}(z)$ be the optimal Hankel-norm approximation obtained in the previous subsection. Then there holds:

$$\left\|\mathbf{H} - \widehat{\mathbf{H}}_s\right\|_\infty \le \sum_{j=\kappa}^{\eta} \sigma_{i_j}. \tag{4.111}$$

This subsection is devoted to the proof of the above error bound.

Several issues need to be resolved first, including the McMillan degree of $\widehat{\mathbf{H}}_s(z)$ and Hankel singular values of $\widehat{\mathbf{H}}_{as}\left(z^{-1}\right)$.

Lemma 4.7. *Let* $\widehat{\mathbf{H}}(z) = \hat{D} + \hat{C}\left(zI - \hat{A}\right)^{-1}\hat{B}$ *be the* $(k-1) = r$*th order optimal Hankel-norm approximation under* Assumption A$_1$–A$_3$. *If* $\det(\Gamma_1) = \det(\sigma_k^2 I - \Sigma_1\Sigma_2) \neq 0$, *then* $\hat{A}$ *has exactly* r *and* $(n-p-r)$ *eigenvalues inside and outside the unit circle, respectively, and zero eigenvalue on the unit circle.*

Proof. Suppose on the contrary that $\hat{A}$ has an eigenvalue λ on the unit circle with eigenvector $\mathbf{x}$. Then multiplying $\mathbf{x}^*$ from left and $\mathbf{x}$ from right to $P_{22} = \hat{A}^* P_{22}\hat{A} + \hat{C}^*\hat{C}$ yields

$$\left(1 - |\lambda|^2\right)\mathbf{x}^* P_{22}\mathbf{x} = \left\|\hat{C}\mathbf{x}\right\|^2 = 0, \tag{4.112}$$

concluding $\hat{C}\mathbf{x} = \mathbf{0}$. Since $\left\|\mathbf{H} - \widehat{\mathbf{H}}\right\|_\infty = \sigma_k$ by the allpass property, the assumption on $p = m = \rho$ reveals that $\widehat{\mathbf{H}}(z)$ has no pole on the unit circle. Hence, any eigenvalue of $\hat{A}$ on the unit circle is unobservable by (4.112). Multiplying (4.110) by $\mathbf{x}$ from right gives

$$\text{(i) } \lambda A_{12}^*\Gamma_1\mathbf{x} = \mathbf{0}, \quad \text{(ii) } \lambda A_{11}^*\Gamma_1\mathbf{x} = \Gamma\mathbf{x}$$

The hypothesis on nonsingular Γ_1 leads to

$$A^*\begin{bmatrix}\Gamma_1\mathbf{x}\\ \mathbf{0}\end{bmatrix} = \lambda^{-1}\begin{bmatrix}\Gamma_1\mathbf{x}\\ \mathbf{0}\end{bmatrix}, \quad |\lambda| = 1,$$

that contradicts stability of A. It is thus concluded that $\hat{A}$ has no eigenvalue on the unit circle. Recall that Hankel-norm of $\mathbf{H}(z)$ is dependent only on the strictly causal part of $\mathbf{H}(z)$. There thus holds

$$\overline{\sigma}\left(\mathscr{H}_H - \widehat{\mathscr{H}}_H\right) = \left\|\mathbf{H} - \widehat{\mathbf{H}}\right\|_H = \left\|\mathbf{H} - \widehat{\mathbf{H}}_s\right\|_H = \sigma_k.$$

It follows that the McMillan degree of $\widehat{\mathbf{H}}_s$ is $(k-1)$. A similarity transform S can be applied so that

$$S\hat{A}S^{-1} = \begin{bmatrix} \hat{A}_s & \mathbf{0} \\ \mathbf{0} & \hat{A}_{as} \end{bmatrix}, \quad SP_{22}S^* = \begin{bmatrix} X_{11} & X_{12} \\ X_{21} & X_{22} \end{bmatrix}$$

with $\hat{A}_s$ and $\hat{A}_{as}^{-1}$ being stability matrices. It follows that $\hat{A}_s$ has dimension no smaller than $(k-1)$. The Lyapunov equation $P_{22} = \hat{A}^* P_{22}\hat{A} + \hat{C}^*\hat{C}$ is now converted to

$$\begin{aligned} X_{11} &= \hat{A}_s^* X_{11}\hat{A}_s + \hat{C}_s^*\hat{C}_s, \\ X_{22} &= \hat{A}_{as}^* X_{11}\hat{A}_{as} + \hat{C}_{as}^*\hat{C}_{as}, \\ X_{12} &= \hat{A}_s^* X_{12}\hat{A}_{as} + \hat{C}_s^*\hat{C}_{as}, \end{aligned}$$

where $X_{11} \geq \mathbf{0}$ and $X_{22} \leq \mathbf{0}$. The fact that $P_{22} = -\Sigma_1\Gamma_1$ has $(k-1)$ positive eigenvalues implies that X_{11} has dimension $(k-1)\times(k-1)$ and so does $\hat{A}_s$ that concludes the proof. □

A crucial observation is made: For any antistable transfer matrix $\mathbf{J}(z)$, there holds

$$\left\| \mathbf{H} - \widehat{\mathbf{H}}_s - \mathbf{J} \right\|_\infty \geq \left\| \mathbf{H} - \widehat{\mathbf{H}} \right\|_\infty = \sigma_k.$$

That is, the lower bound is achieved by taking $\mathbf{J}(z) = \widehat{\mathbf{H}}_{as}$. Optimal Hankel-norm approximation takes $\widehat{\mathbf{H}}_s(z)$ as the reduced order model that is stable. Since

$$\left\| \mathbf{H} - \widehat{\mathbf{H}}_s \right\|_\infty \leq \left\| \mathbf{H} - \widehat{\mathbf{H}} \right\|_\infty + \left\| \widehat{\mathbf{H}}_{as} \right\|_\infty, \tag{4.113}$$

the error bound in (4.111) is hinged to bounding $\left\| \widehat{\mathbf{H}}_{as} \right\|_\infty$.

Lemma 4.8. *Let* $\mathbf{H}(z) = \widehat{\mathbf{H}}_s(z) + \widehat{\mathbf{H}}_{as}(z)$ *be the optimal Hankel-norm approximation in Lemma 4.7 where both* $\widehat{\mathbf{H}}_s(z)$ *and* $\widehat{\mathbf{H}}_{as}(z^{-1})$ *have all poles inside the unit circle. Then* $\sigma_i(\mathbf{J}) = \sigma_{i+k+p-1}(\mathbf{H})$ *with* $\mathbf{J}(z) = \widehat{\mathbf{H}}_{as}(z^{-1})$.

Proof. Let P_e and Q_e be the solutions to the two Lyapunov equations in (4.101), respectively. Since $\mathbf{H}(z)$ is square and $\widehat{\mathbf{H}}(z)$ is an optimal Hankel-norm approximation, the solutions P_e and Q_e are given as in (4.102)–(4.107), and as in Lemma 4.6. By $Q_eP_e = P_eQ_e = \sigma_k^2 I_{2n-p}$, $Q_{11}P_{11} = \sigma_k^2 I_n - Q_{12}P_{12}$, and $P_{21}Q_{12} = \Gamma_1 = \sigma_k^2 I_{n-p} - P_{22}Q_{22}$, there holds

$$\begin{aligned} \det(\lambda I_n - Q_{11}P_{11}) &= \det\left(\lambda I_n - \left(\sigma_k^2 I_n - Q_{12}P_{12}\right)\right) \\ &- \det\left(\left(\lambda - \sigma_k^2\right) I_n + Q_{12}P_{12}\right) \\ &= \left(\lambda - \sigma_k^2\right)^p \det\left[\left(\lambda - \sigma_k^2\right) I_{n-p} + \Gamma_1\right] \\ &= \left(\lambda - \sigma_k^2\right)^p \det\left(\lambda I_{n-p} - P_{22}Q_{22}\right). \end{aligned}$$

In light of $Q_{22}P_{22} = \Sigma_1\Sigma_2$ and $Q_{11}P_{11} = QP = \text{diag}\left(\Sigma_1\Sigma_2, \sigma_k^2 I_p\right)$,

$$\lambda_i(Q_{11}P_{11}) = \begin{cases} \lambda_i(Q_{22}P_{22}), & i = 1,2,\ldots,k-1, \\ \sigma_k^2, & i = k,k+1,\ldots,k+p-1, \\ \lambda_{i-p}(Q_{22}P_{22}), & i = k+p,k+p-1,\ldots,n. \end{cases}$$

Because the first $(k-1)$ eigenvalues of Q_{22} and P_{22} are positive, $\sqrt{\lambda_i(Q_{22}P_{22})}$ is the ith Hankel singular values of $\widehat{\mathbf{H}}_s(z)$, for $i = 1,2,\ldots,k-1$. Similarly, because the last $(n-k-p+1)$ eigenvalues of Q_{22} and P_{22} are both negative,

$$\sqrt{\lambda_{i+k-1}(Q_{22}P_{22})} = \sqrt{\lambda_{i+k+p-1}(QP)}$$

is the ith Hankel singular values of $\mathbf{J}(z) = \widehat{\mathbf{H}}_{\text{as}}\left(z^{-1}\right)$ for $i = 1,2,\ldots,n-k-p+1$. The fact that $\sqrt{\lambda_i(Q_{11}P_{11})} = \sqrt{\lambda_i(QP)}$ is the ith Hankel singular value of $\mathbf{H}(z)$ concludes the proof. □

Although the allpass property holds for the case $p = m = \rho$, an allpass embedding procedure is available for the case when $p \neq m$ and for $\rho \geq 1$. Due to the complexity of the procedure, it is not presented in this textbook but will be assumed to hold in proving the error bound.

Corollary 4.3. *Let* $\mathbf{H}(z) = C(zI_n - A)^{-1}B$ *be a square transfer matrix with minimal realization, and* A *a stability matrix. Suppose that* $\mathbf{H}(z)$ *has* $\eta \leq n$ *distinct Hankel singular values* σ_{i_j}*, where* $j = 1,2,\ldots,\eta$*. Then there exists a constant matrix* D_0 *such that*

$$\|\mathbf{H} - D_0\|_\infty \leq \sum_{j=1}^{\eta} \sigma_{i_j}. \tag{4.114}$$

Proof. Taking optimal Hankel-norm approximation $\widehat{\mathbf{H}}(z) = \widehat{\mathbf{H}}_\eta(z)$ with $k = n$ implies that $\widehat{\mathbf{H}}_\eta(z)$ has all poles inside the unit circle. By Lemma 4.6 and Lemma 4.8, $\widehat{\mathbf{H}}_\eta(z)$ has the same first $(\eta - 1)$ distinct Hankel singular values as $\mathbf{H}(z)$. Moreover, $\left\|\mathbf{H} - \widehat{\mathbf{H}}_\eta\right\|_\infty = \sigma_n = \sigma_{i_\eta}$. Repeating the same optimal Hankel-norm approximation to $\widehat{\mathbf{H}}_\eta(z)$ yields $\widehat{\mathbf{H}}_{\eta-1}(z)$ with the same first $(\eta - 2)$ distinct Hankel singular values as $\mathbf{H}(z)$, and

$$\left\|\mathbf{H} - \widehat{\mathbf{H}}_{\eta-1}\right\|_\infty \leq \left\|\mathbf{H} - \widehat{\mathbf{H}}_\eta\right\|_\infty + \left\|\mathbf{H}_\eta - \widehat{\mathbf{H}}_{\eta-1}\right\|_\infty = \sigma_{i_\eta} + \sigma_{i_{\eta-1}}.$$

By induction, (4.114) holds for some constant matrix D_0, that is the sum of all the constant matrices from each one step optimal Hankel-norm approximation. □

By the result and proof of Corollary (4.3), the following identity:

$$\mathbf{H}(z) = D_0 + \sum_{j=1}^{\eta} \sigma_{i_j}\mathbf{E}_{i_j}(z)$$

exists with $\mathbf{E}_{i_j}(z)$ allpass. Since

$$\overline{\sigma}(D_0) \leq \sup_{|z|>1} \overline{\sigma}\left[\mathbf{H}(z) - D_0\right] = \|\mathbf{H} - D_0\|_\infty \leq \sum_{j=1}^{\eta} \sigma_{i_j}$$

for the same D_0 as in (4.114), there holds inequality

$$\|\mathbf{H}\|_\infty \leq \overline{\sigma}(D_0) + \|\mathbf{H} - D_0\|_\infty \leq 2\sum_{j=1}^{\eta} \sigma_{i_j}$$

that coincides with the one deduced from balanced truncation.

Theorem 4.9. *Let $\widehat{\mathbf{H}}(z) = \widehat{\mathbf{H}}_s(z) + \widehat{\mathbf{H}}_{as}(z)$ be an optimal Hankel-norm approximation to $\mathbf{H}(z)$, where both $\widehat{\mathbf{H}}_s(z)$ and $\widehat{\mathbf{H}}_{as}\left(z^{-1}\right)$ have all its poles inside the unit circle, and $\widehat{\mathbf{H}}_{as}\left(z^{-1}\right)$ is proper. Assume that $\mathbf{H}(z)$ has η distinct Hankel singular values with $\sigma_k = \sigma_{i_\kappa}$, and $\widehat{\mathbf{H}}_s(z)$ has McMillan degree $(k-1)$. Then the error bound in (4.111) holds.*

Proof. The triangle inequality (4.113) can be used. Since $\Delta(z) = \mathbf{H}(z) - \widehat{\mathbf{H}}(z)$ is "allpass" with modulo σ_k, and since $\left\|\widehat{\mathbf{H}}_{as}\right\|_\infty$ is no larger than the sum of its Hankel singular values in light of Corollary 4.3 (note that the constant term is included in $\widehat{\mathbf{H}}_{as}(z)$, the error bound in (4.111) is thus true, in light of Lemma 4.8. □

4.3.3 Extension Theorem

In the previous two subsections, $p = m = \rho$ is assumed (Assumption A2). This subsection presents an extension theorem to fulfill A2 when it is violated. Without loss of generality, $p = m$ is assumed for $\mathbf{H}(z)$ of size $p \times m$ by adding zero rows to C or zero columns to B. Generically, all Hankel singular values are distinct. The extension theorem asserts the existence of H_0 such that

$$\tilde{\mathbf{H}}(z) = z^{-1}\left[\mathbf{H}(z) + H_0\right]$$

admits p identical Hankel singular values at $\sigma \in (\sigma_k,\ \sigma_{k-1})$. There are $r = (k-1)$ Hankel singular values greater than σ, and the rest smaller than σ. Hence, the rth order optimal Hankel-norm approximation from the previous subsections can be applied to $\tilde{\mathbf{H}}(z)$ from which the rth order (sub)optimal Hankel-norm approximation to $\mathbf{H}(z)$ can be obtained.

Let (A, B, C) be a minimal realization for the square $\mathbf{H}(z)$. Then

$$\tilde{\mathbf{H}}(z) = \left[\begin{array}{c|c} \tilde{A} & \tilde{B} \\ \hline \tilde{C} & \tilde{D} \end{array}\right] := \left[\begin{array}{cc|c} A & \mathbf{0} & B \\ C & \mathbf{0} & H_0 \\ \hline \mathbf{0} & I_p & \mathbf{0} \end{array}\right].$$

With P and Q the observability and reachability gramians for $\mathbf{H}(z)$, respectively, the corresponding two gramians for $\tilde{\mathbf{H}}(z)$ are given respectively by

$$\tilde{P} = \begin{bmatrix} P & \mathbf{0} \\ \mathbf{0} & I_p \end{bmatrix}, \quad \tilde{Q} = \begin{bmatrix} Q & (H_0B^* + CQA^*)^* \\ H_0B^* + CQA^* & H_0H_0^* + CQC^* \end{bmatrix}.$$

By definition, the Hankel singular values of $\tilde{\mathbf{H}}(z)$, denoted by $\{\tilde{\sigma}_k\}$, are the square roots of the eigenvalues of

$$\tilde{P}^{1/2}\tilde{Q}\tilde{P}^{1/2} = \begin{bmatrix} P^{1/2}QP^{1/2} & P^{1/2}(H_0B^* + CQA^*)^* \\ (H_0B^* + CQA^*)P^{1/2} & H_0H_0^* + CQC^* \end{bmatrix}$$

It follows that $\tilde{\sigma}_r > \tilde{\sigma}_{r+1} = \cdots \tilde{\sigma}_{r+p} > \tilde{\sigma}_{r+p+1}$, if and only if

$$\Psi = \tilde{P}^{1/2}\tilde{Q}\tilde{P}^{1/2} - \sigma^2 I = \begin{bmatrix} P^{1/2}QP^{1/2} - \sigma^2 I & P^{1/2}(H_0B^* + CQA^*)^* \\ (H_0B^* + CQA^*)P^{1/2} & H_0H_0^* + CQC^* - \sigma^2 I \end{bmatrix}$$

has r positive eigenvalues and p zero eigenvalues. Denote

$$R = \left(\sigma^2 I - P^{1/2}QP^{1/2}\right)^{-1} \tag{4.115}$$

that exists by $\sigma \in (\sigma_k, \sigma_{k-1})$. Define

$$L = \begin{bmatrix} I_n & \mathbf{0} \\ (H_0B^* + CQA^*)P^{1/2} & I_p \end{bmatrix}.$$

It can be verified by direct calculations that

$$L\Psi L^* = \begin{bmatrix} P^{1/2}QP^{1/2} - \sigma^2 I & \mathbf{0} \\ \mathbf{0} & V \end{bmatrix},$$

where V is given by, using (4.115),

$$V = H_0H_0^* + CQC^* - \sigma^2 I - (H_0B^* + CQA^*)P^{1/2}RP^{1/2}(H_0B^* + CQA^*)^*.$$

The extension theorem holds for some H_0, if and only if $V = \mathbf{0}$, in light of the fact that Ψ and $L\Psi L^*$ have the same inertia, i.e., their respective eigenvalues have the same signs. This *congruent* relation is represented by the mathematical expression of $\Psi \sim L\Psi L^*$. Hence, the central issue in proving the extension theorem is hinged to the existence of H_0 that zeros V. Define

$$\begin{aligned} E &= CQA^*P^{1/2}RP^{1/2}AQC^* + CQC^* - \sigma^2 I, \\ F &= CQA^*P^{1/2}RP^{1/2}B, \\ G &= I_p + B^*P^{1/2}RP^{1/2}B. \end{aligned} \tag{4.116}$$

Then by the expression of V, the condition for $V = \mathbf{0}$ is equivalent to

$$V = FH_0^* + H_0F^* + H_0GH_0^* + E = \mathbf{0}. \tag{4.117}$$

The above is in the form of continuous-time ARE. The extension theorem amounts to the existence of a solution H_0.

Theorem 4.10. *Assume that G is nonsingular. Then G and $FG^{-1}F^* - E$ are congruent, and thus, there exists at least one solution H_0 to (4.117).*

Proof. The proof is quite long. First, it is noted that the continuous-time ARE in (4.117) can be written as

$$\left(H_0 + FG^{-1}\right) G \left(H_0 + FG^{-1}\right)^* = FG^{-1}F^* - E.$$

Hence, the existence of a solution H_0 to (4.117) is equivalent to the congruent relation between G and $FG^{-1}F^* - E$. If the congruent relation holds, then there exists a nonsingular matrix T such that

$$TGT^* = FG^{-1}F^* - E \implies T = H_0 + FG^{-1}$$

from which H_0 can be obtained. Next, it will be shown that

$$(\alpha)\ FG^{-1}F^* - E \sim \tilde{G} = I + CQ^{1/2}\left(\sigma^2 I - Q^{1/2}PQ^{1/2}\right)^{-1} Q^{1/2}C^*$$

that is dual to G and $(\beta)\ \tilde{G} \sim G$, thereby concluding the congruent relation between G and $FG^{-1}F^* - E$.

To prove (α), matrix inversion formula:

$$\left(X + YZ^{-1}W\right)^{-1} = X^{-1} - X^{-1}Y\left(Z + WX^{-1}Y\right)^{-1} WX^{-1} \tag{4.118}$$

is useful. An application of the above formula yields

$$\begin{aligned}(\tilde{G})^{-1} &= \left[I + CQ^{1/2}(\sigma^2 I - Q^{1/2}PQ^{1/2})^{-1}Q^{1/2}C^*\right]^{-1} \\ &= I - CQ^{1/2}\left[\left(\sigma^2 I - Q^{1/2}PQ^{1/2}\right) + Q^{1/2}C^*CQ^{1/2}\right]^{-1} Q^{1/2}C^*.\end{aligned}$$

Recalling $P - C^*C = A^*PA$ leads to

$$\begin{aligned}(\tilde{G})^{-1} &= I - CQ^{1/2}\left(\sigma^2 I - Q^{1/2}A^*PAQ^{1/2}\right)^{-1} Q^{1/2}C^* \\ &= I - \sigma^{-2}CQ^{1/2}\Big[I + Q^{1/2}A^*P^{1/2} \\ &\qquad \times \left(\sigma^2 I - P^{1/2}AQA^*P^{1/2}\right)^{-1} P^{1/2}A^*Q^{1/2}\Big] Q^{1/2}C^*,\end{aligned}$$

where $\left(\sigma^2 I - YY^*\right)^{-1} = \sigma^{-2}\left[I + Y\left(\sigma^2 I - Y^*Y\right)^{-1}Y^*\right]$ is used. Note that

$$\left(\sigma^2 I - P^{1/2}AQA^*P^{1/2}\right)^{-1} = \left[R^{-1} + P^{1/2}BB^*P^{1/2}\right]^{-1}$$

by $AQA^* = Q - BB^*$ and the expression of R in (4.115). Applying the matrix inversion formula again leads to

$$\begin{aligned}\left(\sigma^2 I - P^{\frac{1}{2}}AQA^*P^{\frac{1}{2}}\right)^{-1} &= R - RP^{\frac{1}{2}}B\left(I + B^*P^{\frac{1}{2}}RP^{\frac{1}{2}}B\right)^{-1}B^*P^{\frac{1}{2}}R \\ &= R - RP^{1/2}BG^{-1}B^*P^{1/2}R\end{aligned}$$

by the definition of G in (4.116). It follows that

$$\begin{aligned}(\tilde{G})^{-1} &= I - \sigma^{-2}CQC^* - \sigma^{-2}CQAP^{\frac{1}{2}}\left[R - RP^{\frac{1}{2}}BG^{-1}B^*P^{\frac{1}{2}}R\right]P^{\frac{1}{2}}AQC^* \\ &= \sigma^{-2}\left(-E + FG^{-1}F^*\right)\end{aligned}$$

that concludes (α). For (β), it is claimed that

$$\begin{aligned}\left(-\sigma^2 I + P^{1/2}AQA^*P^{1/2}\right) &\sim \left(-\sigma^2 I + Q^{1/2}A^*PAQ^{1/2}\right), \\ \left(-\sigma^2 I + P^{1/2}QP^{1/2}\right) &\sim \left(-\sigma^2 I + Q^{1/2}PQ^{1/2}\right), \\ \begin{bmatrix} -\sigma^2 I + Q^{1/2}PQ^{1/2} & \mathbf{0} \\ \mathbf{0} & G^* \end{bmatrix} &\sim \begin{bmatrix} -\sigma^2 I + Q^{1/2}A^*PAQ^{1/2} & \mathbf{0} \\ \mathbf{0} & I_p \end{bmatrix}, \\ M_{\text{tmp}} := \begin{bmatrix} -\sigma^2 I + P^{1/2}QP^{1/2} & \mathbf{0} \\ \mathbf{0} & G \end{bmatrix} &\sim \begin{bmatrix} -\sigma^2 I + P^{1/2}AQA^*P^{1/2} & \mathbf{0} \\ \mathbf{0} & I_p \end{bmatrix}.\end{aligned}$$

Only the last congruent relation will be proven. The rest is left as an exercise (Problem 4.23). To verify the last congruent relation, denote

$$L_1 = \begin{bmatrix} I_n & \mathbf{0} \\ B^*P^{1/2}R & I_p \end{bmatrix}, \quad L_2 = \begin{bmatrix} I_n & P^{1/2}B \\ \mathbf{0} & I_p \end{bmatrix}.$$

Direct calculation shows that

$$\begin{aligned}L_2L_1M_{\text{tmp}}L_1^*L_2^* &= L_2\begin{bmatrix} -\sigma^2 I + P^{1/2}QP^{1/2} & -P^{1/2}B \\ B^*P^{1/2} & I_p \end{bmatrix}L_2^* \\ &= \begin{bmatrix} -\sigma^2 I + P^{1/2}QP^{1/2} - P^{1/2}BB^*P^{1/2} & \mathbf{0} \\ \mathbf{0} & I_p \end{bmatrix} \\ &= \begin{bmatrix} -\sigma^2 I + P^{1/2}AQA^*P^{1/2} & \mathbf{0} \\ \mathbf{0} & I_p \end{bmatrix}\end{aligned}$$

in which definitions of G and $Q - BB^* = AQA^*$ are used. □

An example is worked out to illustrate the use of the extension theorem and the optimal Hankel-norm approximation for MIMO models.

Example 4.11. Consider the same transfer matrix as in Example 4.2 with the second- and third-order approximations ($r = 2,3$) based on the optimal Hankel-norm approximation. For $r = 2$, $\sigma = \sigma_3 + 0.1 (< \sigma_2)$ is used. Applying the extension theorem yields

$$H_0 = \begin{bmatrix} 3.56194 & -0.84895 \\ 2.37756 & -3.51080 \end{bmatrix}.$$

For $r = 3$, $\sigma = \sigma_4 + 0.001 (< \sigma_3)$ is used. The extension theorem yields

$$H_0 = \begin{bmatrix} 5.68751 & -1.69553 \\ 2.49880 & -2.58960 \end{bmatrix}.$$

Trial and errors are used to determine the value of σ. Normally, the closer σ is to σ_{r+1}, the better the approximation error, but numerical problems may arise if σ is too close to σ_{r+1}.

Once H_0 is available, the optimal Hankel-norm approximation $\widehat{\tilde{\mathbf{H}}}(z)$ to $\tilde{\mathbf{H}}(z) = z^{-1}[\mathbf{H}(z) + H_0]$ can be computed rather easily based on the relation derived in (4.86). The stable and strictly proper part of $\widehat{\tilde{\mathbf{H}}}(z)$, denoted by $\widehat{\tilde{\mathbf{H}}}_s(z)$, can then be extracted following the same steps as those in Example 4.8. The error responses of $\tilde{\mathbf{H}}(z) - \widehat{\tilde{\mathbf{H}}}_s$ are plotted in Fig. 4.5 that shows smaller frequency response errors than the corresponding ones computed by the BT method. This error response plot validates the smaller error bound for the Hankel-norm method, although it is not as much as by half. Recall $\tilde{\mathbf{H}}(z) = z^{-1}[\mathbf{H}(z) + H_0]$. Setting

$$\widehat{\mathbf{H}}_s(z) = z\widehat{\tilde{\mathbf{H}}}_s - H_0 \implies \mathbf{H}(z) - \widehat{\mathbf{H}}_s(z) = z\left[\tilde{\mathbf{H}}(z) - \widehat{\tilde{\mathbf{H}}}_s\right].$$

Hence, the magnitude error responses do not change. Since $\widehat{\tilde{\mathbf{H}}}_s(z)$ does not include the constant term, $\widehat{\mathbf{H}}_s(z)$ is both stable and proper. In fact, for our example, the constant term is zero for both the second- and third-order optimal Hankel-norm approximations.

Notes and References

Balanced truncation to model reduction was first introduced in [86]. See also [28, 36]. It also has applications to digital filters design with minimum roundoff noises [3,88]. Optimal Hankel-norm approximation was initiated in [2] and was later introduced to the control literature for model reduction in [70], which was further explored in [13, 36, 40, 42, 71, 72]. Inverse balanced truncation (IBT) can be traced

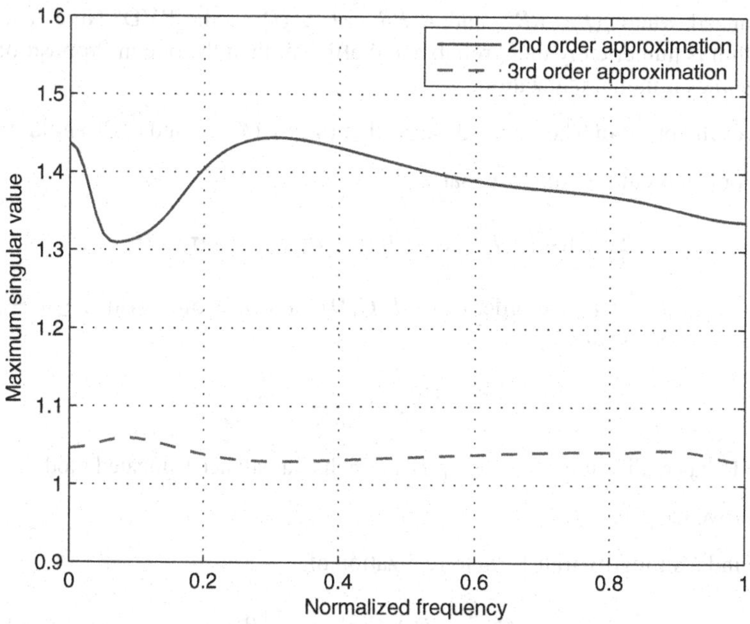

Fig. 4.5 Frequency response error $\sigma_i\left(\mathbf{H}\left(e^{j\omega}\right) - \hat{\mathbf{H}}\left(e^{j\omega}\right)\right)$ for $i = 1,2$

back to [28]. The algorithm presented in this chapter is generalized from [125] that is derived for continuous-time systems. The paper [41, 76] considers application of IBT to controller reduction.

Exercises

4.1. For $\mathbf{H}(z) = D + C(zI - A)^{-1}B$ with A a stability matrix, show that

$$\|\mathbf{H}\|_2^2 = \mathrm{Tr}\{D^*D + B^*PB\},$$

where $P \geq \mathbf{0}$ solves the Lyapunov equation $P = A^*PA + C^*C$. (*Hint:* Modify the procedure in the proof of Lemma 4.1.)

4.2. (Balanced realization) Suppose that (A, B) is reachable, (C, A) is observable, and A is a stability matrix. Let

$$P = A^*PA + C^*C, \quad Q = AQA^* + BB^*.$$

Show that under the similarity transform $T = \Sigma^{1/2}U^*R^{-1}$,

$$(A_b, B_b, C_b) = \left(TAT^{-1}, TB, CT^{-1}\right)$$

is balanced where $Q = RR^*$, and $R^*PR = U\Sigma^2U^*$ is the SVD. Note that this algorithm is numerically more reliable and efficient than the one in Proposition 4.1 but assumes minimal realization.

4.3. For the proof of Theorem 4.3, show that (4.28), (4.29), and (4.30) hold true.

4.4. For Theorem 4.3, suppose that

$$\Sigma = \text{diag}(\sigma_1 I_{\mu_1}, \ldots, \sigma_\ell I_{\mu_\ell}), \quad \sigma_1 > \cdots > \sigma_\ell > 0,$$

with $n = \mu_1 + \cdots + \mu_\ell$ the order of (A,B,C,D), and $\sigma_\kappa \gg \sigma_{\kappa+1}$ with $1 \leq r = \mu_1 + \cdots + \mu_\kappa < n$. Show that

$$\left\|\mathbf{H} - \widehat{\mathbf{H}}\right\|_\infty \leq 2 \sum_{i=\kappa+1}^{\ell} \sigma_i,$$

where $\widehat{\mathbf{H}}$ has realization (A_{11}, B_1, C_1, D), which is a balance truncated model.

4.5. Prove Corollary 4.1.

4.6. Find a similarity transform to realization of

$$\mathbf{H}(z) = D + C(zI - A)^{-1}B \tag{4.119}$$

that is inverse balanced, i.e., the solutions to Lyapunov equations in (4.35) and (4.36) satisfy (4.37).

4.7. Prove the relations in (4.47).

4.8. For possibly complex valued matrices U and V such that UV is square, show that for each $\alpha > 0$, there holds

$$UV + V^*U^* \leq \frac{1}{\alpha}UU^* + \alpha V^*V.$$

4.9. In the proof of Theorem 4.6, show that

$$\|\Delta_\text{m}\|^2 \leq 2\sigma \left\|\tilde{C}_\text{r}\left(zI - \tilde{A}_\text{m}\right)^{-1}\tilde{C}_\text{r}^*\right\|, \tag{4.120}$$

$$\|\Delta_\text{r}\|^2 \leq 2\sigma \left\|\tilde{B}_\text{m}^*\left(zI - \tilde{A}_\text{r}\right)^{-1}\tilde{B}_\text{m}\right\|. \tag{4.121}$$

4.10. Consider the Blaschke product

$$B(z) = \prod_{i=1}^{n} \frac{1 + \bar{p}_k z}{z - p_k}, \quad |p_k| < 1.$$

(i) Show that $\left|B(e^{j\omega})\right| = 1$ for all real ω. Such a transfer function admits the "allpass" property.

(ii) Consider a transfer matrix $\mathbf{T}(z) = D + C(zI - A)^{-1}B$ with A a stability matrix and (A, B, C) a minimal realization for $\mathbf{T}(z)$. Let P and Q be the observability, and reachability gramian, respectively. Show that $\mathbf{T}\left(e^{j\omega}\right)^* \mathbf{T}\left(e^{j\omega}\right) = I$ for all ω, if and only if

$$\text{(a) } D^*D + B^*PB = I, \quad \text{(b) } D^*C + B^*PA = \mathbf{0},$$

and $\mathbf{T}\left(e^{j\omega}\right) \mathbf{T}\left(e^{j\omega}\right)^* = I$ for all ω, if and only if

$$\text{(a) } DD^* + CQC^* = I, \quad \text{(b) } BD^* + AQB = \mathbf{0}.$$

(iii) If $\mathbf{T}(z)$ is square, show that $PQ = I$, and thus all the Hankel singular values of $\mathbf{T}(z)$ are 1. (*Hint:* Use similar derivation in the proof of Lemma 4.5.)

4.11. Suppose that the square transfer matrices $\{\Delta_i(z)\}_{i=1}^n$ are all stable satisfying $\|\Delta_i\|_\infty \le \Delta_i < 1$ for each i. Show that

$$\|(I + \Delta_1) \cdots (I + \Delta_n) - I\|_\infty \le (1 + \delta_1) \cdots (1 + \delta_n) - 1. \tag{4.122}$$

4.12. Complete the proof for relative error bound in Theorem 4.6.

4.13. Let D be of size $p \times m$ with $m > p$. If D has rank p, show that there exists a right inverse D^+ and a matrix $D_\perp$ of size $(m-p) \times m$ such that

$$\begin{bmatrix} D \\ D_\perp \end{bmatrix} \begin{bmatrix} D^+ & D_\perp^+ \end{bmatrix} = I_m, \tag{4.123}$$

where $D^+ = D^*(DD^*)^{-1}$ can be chosen. Show that all right inverses of D are given by $D^\dagger = D^+ + D_\perp^+ \Theta$ with Θ of size $(m-p) \times p$ a free parameter.

4.14. (Continuation of Problem 4.13) Consider the transfer matrix $\mathbf{G}(z) = D + C(zI - A)^{-1}B$ of size $p \times m$ with D the same as in Problem 4.13:

(i) Show that $\mathbf{G}(z)$ is strictly minimum phase, if and only if $(A - BD^+C, BD_\perp)$ is stabilizable.

(ii) Let F be stabilizing or $(A - BD^+C + BD_\perp F)$ is a stability matrix. Show that $\mathbf{G}(z)$ admits a stable right inverse given by

$$\mathbf{G}^+(z) = \left[\begin{array}{c|c} A - BD^+C + BD_\perp^* F & BD^+ \\ \hline D_\perp^* F - D^+C & D^+ \end{array}\right].$$

(*Hint:* $\mathbf{G}(z) = D\left[I - (D_\perp^* F - D^+C)(zI - A)^{-1}B\right]$.)

(iii) Let $\mathbf{G}_\perp(z) = D_\perp - F(zI - A)^{-1}B$. Show that $\mathbf{G}_\perp(z)$ admits a stable right inverse given by

$$\mathbf{G}_\perp^+(z) = \left[\begin{array}{c|c} A - BD^+C + BD_\perp^* F & BD_\perp^* \\ \hline D_\perp^* F - D^+C & D_\perp^* \end{array}\right].$$

(iv) Let A be a stability matrix. Show that for all $|z| \geq 1$, there holds

$$\begin{bmatrix} \mathbf{G}(z) \\ \mathbf{G}_{\perp}(z) \end{bmatrix} \begin{bmatrix} \mathbf{G}^{+}(z) & \mathbf{G}_{\perp}^{+}(z) \end{bmatrix} = I_m.$$

(v) Show that each stable right inverse of $\mathbf{G}(z)$ is given by

$$\mathbf{G}^{\text{inv}}(z) = \mathbf{G}^{+}(z) + \mathbf{G}_{\perp}^{+}(z)\mathbf{Q}(z)$$

for some stable transfer matrix $\mathbf{Q}(z)$.

4.15. Let D and $D_{\perp}$ be in Problem (4.13) satisfying (4.123). Show that $D^{+}D + D_{\perp}^{+}D_{\perp} = I$ and

$$\overline{\sigma}\left(D^{+}D\right) = \overline{\sigma}\left(D_{\perp}^{+}D_{\perp}\right).$$

(*Hint:* Both $D^{+}D$ and $D_{\perp}^{+}D_{\perp}$ are projection matrices.)

4.16. Let $\mathbf{G}(z) = C(zI - A)^{-1}B$ where A has eigenvalues strictly inside or outside the unit circle. Show that there exists a similarity transform S such that the transformed realization $\left(SAS^{-1}, SB, CS^{-1}\right)$ can be partitioned compatibly into

$$SAS^{-1} = \begin{bmatrix} A_{\text{s}} & \mathbf{0} \\ \mathbf{0} & A_{\text{u}} \end{bmatrix}, \quad SB = \begin{bmatrix} B_{\text{s}} \\ B_{\text{u}} \end{bmatrix}, \quad CS^{-1} = \begin{bmatrix} C_{\text{s}} & C_{\text{u}} \end{bmatrix},$$

where A_{s} and A_{u}^{-1} are both stability matrices.

4.17. Show that (4.58) holds.

4.18. Consider Assumption A2 in Sect. 4.3.1. Suppose that $m = p = \rho$ holds. Show that

$$\det\left(\begin{bmatrix} A_{11} & B_1 \\ A_{21} & B_2 \end{bmatrix}\right) \neq 0,$$

if and only if $A_{22}(I - \sigma_k A_{12}^{*}\Sigma_1^{-1}A_{12})^{-1}A_{22}^{*}$ has no eigenvalue at 1.

4.19. Consider $\mathbf{G}(z) = D_g + C_g(zI - A_g)^{-1}B_g$ of size $p \times m$ with (A_g, B_g) reachable and $p \geq m$. Let Y_g be a solution to $Y_g - A_g^{*}Y_gA_g = C_g^{*}C_g$. Show that $\mathbf{G}^{*}\left(z^{-1}\right)\mathbf{G}(z) = I_m$, if and only if

$$\text{(i)}\ D_g^{*}C_g + B_g^{*}Y_gA_g = 0, \quad \text{(ii)}\ D_g^{*}D_g + B_g^{*}Y_gB_g = I_m.$$

(*Hint*: Use (4.17) by setting $\mathbf{G}(z) = D + \mathbf{H}(z)$.)

4.20. Show that (4.108) holds, if $Q_{21} = \begin{bmatrix} I_{n-p} & \mathbf{0} \end{bmatrix}$.

4.21. Suppose that (A, B) is reachable and there exists a solution $X = X^{*}$ such that $X = AXA^{*} + BB^{*}$. Show that (i) X is nonsingular and A has no eigenvalue on the unit circle, (ii) number of eigenvalues of A inside the unit circle is equal to the number of positive eigenvalues of X, and number of eigenvalues of A outside the unit circle is equal to the number of negative eigenvalues of X.

4.22. Let A and B be square matrices of the same dimension. Suppose that A and B are congruent. Find a nonsingular matrix T such that $TAT^* = B$. (*Hint*: Use Schur decomposition with the Matlab command "$[U_{\mathrm{a}}, T_{\mathrm{a}}] = \mathrm{schur}(A)$" and "$[U_{\mathrm{a}}, T_{\mathrm{a}}] = \mathrm{ordschur}(U_{\mathrm{a}}, T_{\mathrm{a}},$"lhp"$)$", do the same for B to figure out the required matrix T.)

4.23. Consider the proof of Theorem 4.10. Show that

$$\left(-\sigma^2 I + P^{1/2} A Q A^* P^{1/2}\right) \sim \left(-\sigma^2 I + Q^{1/2} A^* P A Q^{1/2}\right),$$

$$\left(-\sigma^2 I + P^{1/2} Q P^{1/2}\right) \sim \left(-\sigma^2 I + Q^{1/2} P Q^{1/2}\right),$$

$$\begin{bmatrix} -\sigma^2 I + Q^{1/2} P Q^{1/2} & \mathbf{0} \\ \mathbf{0} & \tilde{G} \end{bmatrix} \sim \begin{bmatrix} -\sigma^2 I + Q^{1/2} A^* P A Q^{1/2} & \mathbf{0} \\ \mathbf{0} & I \end{bmatrix}.$$

(*Hint*: For the first two equalities, use the property $\lambda_i(rI + M) = r + \lambda_i(M)$ and $\lambda_i(NM) = r + \lambda_i(MN)$ for eigenvalues; For, the last equality, it is dual to the congruent relation for M_{tmp} in the proof of Theorem 4.10.)

circle, (b) number of eigenvalues of A inside the unit circle is equal to the number of positive eigenvalues of X, and number of eigenvalues of A outside the unit circle is equal to the number of negative eigenvalues of X.

4. Let A and B be square matrices of the same dimension. Suppose that A and B commute. Find a nonsingular matrix T such that [illegible] [illegible] with the Matlab command [illegible] [illegible] and [illegible] the same for [illegible] out the [illegible] matrix, P.

5. Consider the proof of Theorem 8.10. Show that

$$[illegible]$$

(Hint: [illegible] is equivalent to the property of [illegible] and [illegible] for [illegible]. The last equality [illegible] is equal to the [illegible].)

Chapter 5
Optimal Estimation and Control

Estimation and control are the two fundamental problems for dynamical systems. Many engineering design tasks can be formulated into either an estimation or control problem associated with some appropriate performance index. In order to simplify the design issues in practice, dynamical systems are usually assumed to be linear and finite-dimensional. Otherwise, various approximation methods can be applied to derive linear and finite-dimensional models with negligible modeling errors. As such, state-space representations are made possible providing computational tools for optimal design and enabling design of optimal estimators and controllers.

Estimation aims at design of state estimators that reconstruct the state vector based on measurements of the past and present input and output data. Due to the unknown and random nature of the possible disturbance at the input and the corrupting noise at the output, it is impossible to reconstruct the true state vector in real-time. Therefore, the design objective for state estimators will be minimization of the estimation error variance by assuming white noises for input disturbances and output measurement errors. The focus will be on design of optimal linear estimators.

Disturbance rejection has been the primary objective in feedback control system design in which white noises are the main concern. The emphasis will be placed on the design of state-feedback controllers to not only stabilize the feedback control system but also minimize the adverse effect due to white noise disturbances. With the variance as the performance measure, optimal control leads to linear feedback controllers that are dual to optimal linear estimators.

Without exaggeration, optimal estimation and control are the two most celebrated results in engineering system design. They have brought in not only the design algorithms but also the new design methodology that has had far reaching impacts as evidenced by the wide use of Kalman filtering and feedback control in almost every aspect of the system engineering. Nevertheless, it is the conceptual notions from linear system theory that empower the state-of-the-art design algorithms and allow applications of optimal estimation and control in engineering practice. This chapter will cover the well-known results in Kalman filtering and quadratic regulators that have enriched engineering system design. It also covers optimal output estimators and full information control that are developed more recently.

G. Gu, *Discrete-Time Linear Systems: Theory and Design with Applications*,
DOI 10.1007/978-1-4614-2281-5_5, © Springer Science+Business Media, LLC 2012

5.1 Minimum Variance Estimation

5.1.1 Preliminaries

As a prelude to optimal estimation for state-space systems, a simpler and more intuitive estimation problem will be investigated. Let X and Y be two random vectors with the PDFs $p_X(\mathbf{x})$ and $p_Y(\mathbf{y})$, respectively. A natural question is how knowledge of the value taken by Y can provide information about the value taken by X. In other words, how an *estimate* of $X = \hat{\mathbf{x}}$ can be made based on the observation of $Y = \mathbf{y}$? Clearly, with $Y = \mathbf{y}$ being observed the PDF of X is modified into the *conditional PDF* given by Bayes' rule:

$$p_{X|Y}(\mathbf{x}|\mathbf{y}) = \frac{p_{X,Y}(\mathbf{x},\mathbf{y})}{p_Y(\mathbf{y})} \tag{5.1}$$

assuming that $p_Y(\mathbf{y}) \neq 0$.

The quality of estimation is better measured by *maximum a posteriori* (MAP). That is, $\mathbf{x} = \hat{\mathbf{x}}$ should maximize $p_{X|Y}(\mathbf{x}|\mathbf{y})$. As a result, computation of the MAP estimate involves nonlinear optimization procedures, which is not tractable in general due to the existence of multiple peaks in $p_{X|Y}(\mathbf{x}|\mathbf{y})$ or high dimension of $\mathbf{x}$. An alternate measure is the conditional error variance $\mathrm{E}\left\{\|X - \hat{\mathbf{x}}\|^2 |Y = \mathbf{y}\right\}$. The minimum variance estimate $X = \hat{\mathbf{x}}$ satisfies

$$\mathrm{E}\left\{\|X - \hat{\mathbf{x}}\|^2 |Y = \mathbf{y}\right\} \leq \mathrm{E}\left\{\|X - \mathbf{x}\|^2 |Y = \mathbf{y}\right\} \quad \forall\, \mathbf{x}. \tag{5.2}$$

The left-hand side of (5.2) is often termed the *minimum mean-squared error* (MMSE). The MMSE estimate or the minimum variance estimate has the closed-form solution which is a contrast to the MAP estimate, as shown next.

Theorem 5.1. *Let X and Y be two jointly distributed random vectors. The* MMSE *estimate $\hat{\mathbf{x}}$ of X given observation $Y = \mathbf{y}$ is uniquely specified as the conditional mean (by an abuse of the notation for integration)*

$$\hat{\mathbf{x}} = \mathrm{E}\{X|Y = \mathbf{y}\} = \int_{-\infty}^{\infty} \mathbf{x} p_{X|Y}(\mathbf{x}|\mathbf{y})d\mathbf{x}. \tag{5.3}$$

Proof. Let $h(\mathbf{z}) = \mathrm{E}\left\{\|X - \mathbf{z}\|^2 |Y = \mathbf{y}\right\}$ with $\mathbf{z}$ to be chosen. Then

$$\begin{aligned} h(\mathbf{z}) &= \int_{-\infty}^{\infty} \|\mathbf{x} - \mathbf{z}\|^2 p_{X|Y}(\mathbf{x}|\mathbf{y})d\mathbf{x} \\ &= \int_{-\infty}^{\infty} \|\mathbf{x}\|^2 p_{X|Y}(\mathbf{x}|\mathbf{y})d\mathbf{x} + \|\mathbf{z}\|^2 - 2\mathrm{Re}\{\mathbf{z}^* \mathrm{E}[X|Y = \mathbf{y}]\} \end{aligned}$$

$$= \|\mathbf{z} - \mathrm{E}\{X|Y = \mathbf{y}\}\|^2 + \int_{-\infty}^{\infty} \|\mathbf{x}\|^2 p_{X|Y}(\mathbf{x}|\mathbf{y})\, d\mathbf{x} - \|\mathrm{E}\{X|Y = \mathbf{y}\}\|^2$$

$$\geq \int_{-\infty}^{\infty} \|\mathbf{x}\|^2 p_{X|Y}(\mathbf{x}|\mathbf{y})\, d\mathbf{x} - \|\mathrm{E}\{X|Y = \mathbf{y}\}\|^2.$$

The minimum is achieved uniquely with $\mathbf{z} = \hat{\mathbf{x}}$ in (5.3). □

Theorem 5.1 indicates that the MMSE estimate is the same as the conditional mean. Its closed-form offers a great advantage in its computation compared with the MAP estimate. In some cases, the conditional mean can be experimentally determined which can be extremely valuable if the joint PDF of X and Y is unavailable. Clearly, the MMSE estimate is different from the MAP estimate in general unless the global maximum of $p_{X|Y}(\mathbf{x}|\mathbf{y})$ takes place at the conditional mean $\mathbf{x} = \mathrm{E}\{X|Y = \mathbf{y}\}$. The next example is instrumental.

Example 5.2. Let the two random vectors X and Y be jointly Gaussian.Then the random vector $Z = \begin{bmatrix} X^* & Y^* \end{bmatrix}^*$ is Gaussian distributed with

$$\mathbf{m_z} = \mathrm{E}\{Z\} = \begin{bmatrix} \mathbf{m_x} \\ \mathbf{m_y} \end{bmatrix}, \quad \Sigma_{\mathbf{zz}} = \mathrm{cov}\{Z\} = \begin{bmatrix} \Sigma_{\mathbf{xx}} & \Sigma_{\mathbf{xy}} \\ \Sigma_{\mathbf{yx}} & \Sigma_{\mathbf{yy}} \end{bmatrix}.$$

Clearly, $\Sigma_{\mathbf{xy}} = \mathrm{E}\{(\mathbf{x} - \mathbf{m_x})(\mathbf{y} - \mathbf{m_y})^*\}$ and

$$\mathbf{m_x} = \mathrm{E}\{X\}, \ \Sigma_{\mathbf{xx}} = \mathrm{cov}\{X\} := \mathrm{E}\{(\mathbf{x} - \mathbf{m_x})(\mathbf{x} - \mathbf{m_x})^*\},$$
$$\mathbf{m_y} = \mathrm{E}\{Y\}, \ \Sigma_{\mathbf{yy}} = \mathrm{cov}\{Y\} := \mathrm{E}\{(\mathbf{y} - \mathbf{m_y})(\mathbf{y} - \mathbf{m_y})^*\}.$$

Suppose that the covariance matrices $\Sigma_{\mathbf{xx}}$ and $\Sigma_{\mathbf{yy}}$ are nonsingular. Then X and Y have marginal PDFs

$$p_X(\mathbf{x}) = \frac{1}{\sqrt{(2\pi)^n \det(\Sigma_{\mathbf{xx}})}} \exp\left\{ -\frac{1}{2}(\mathbf{x} - \mathbf{m_x})^* \Sigma_{\mathbf{xx}}^{-1} (\mathbf{x} - \mathbf{m_x}) \right\}, \tag{5.4}$$

$$p_Y(\mathbf{y}) = \frac{1}{\sqrt{(2\pi)^n \det(\Sigma_{\mathbf{yy}})}} \exp\left\{ -\frac{1}{2}(\mathbf{y} - \mathbf{m_y})^* \Sigma_{\mathbf{yy}}^{-1} (\mathbf{y} - \mathbf{m_y}) \right\}, \tag{5.5}$$

respectively. It is left as an exercise to show that the conditional PDF of X, given $Y = \mathbf{y}$, is

$$p_{X|Y}(\mathbf{x}|\mathbf{y}) = \frac{p_{X,Y}(\mathbf{x},\mathbf{y})}{p_Y(\mathbf{y})} = \frac{p_Z(\mathbf{z})}{p_Y(\mathbf{y})}$$

$$= \frac{1}{\sqrt{(2\pi)^n \det\left(\tilde{\Sigma}_{\mathbf{xx}}\right)}} \exp\left\{ -\frac{1}{2}(\mathbf{x} - \tilde{\mathbf{m}}_{\mathbf{x}})^* \tilde{\Sigma}_{\mathbf{xx}}^{-1} (\mathbf{x} - \tilde{\mathbf{m}}_{\mathbf{x}}) \right\}, \tag{5.6}$$

where $\tilde{\mathbf{m}}_{\mathbf{x}} = \mathbf{m}_{\mathbf{x}} + \Sigma_{\mathbf{xy}}\Sigma_{\mathbf{yy}}^{-1}(\mathbf{y} - \mathbf{m}_{\mathbf{y}})$ and $\tilde{\Sigma}_{\mathbf{xx}} = \Sigma_{\mathbf{xx}} - \Sigma_{\mathbf{xy}}\Sigma_{\mathbf{yy}}^{-1}\Sigma_{\mathbf{yx}}$. Hence, the conditional PDF in (5.6) is also Gaussian. Its MAP estimate is identical to the MMSE estimate given by

$$\hat{\mathbf{x}} = \tilde{\mathbf{m}}_{\mathbf{x}} = \mathbf{m}_{\mathbf{x}} + \Sigma_{\mathbf{xy}}\Sigma_{\mathbf{yy}}^{-1}(\mathbf{y} - \mathbf{m}_{\mathbf{y}}). \tag{5.7}$$

Suppose that X and Y have the same dimension and are related by

$$Y = X + N,$$

where N is Gaussian independent of X with zero vector mean and $\Sigma_{\mathbf{nn}}$ the covariance. Then by the independence of X and N,

$$\mathbf{m}_{\mathbf{y}} = \mathbf{m}_{\mathbf{x}}, \quad \Sigma_{\mathbf{xy}} = \Sigma_{\mathbf{xx}}, \quad \Sigma_{\mathbf{yy}} = \Sigma_{\mathbf{xx}} + \Sigma_{\mathbf{nn}}.$$

The optimal estimate in (5.7) reduces to

$$\hat{\mathbf{x}} = \mathbf{m}_{\mathbf{x}} + \Sigma_{\mathbf{xx}}(\Sigma_{\mathbf{xx}} + \Sigma_{\mathbf{nn}})^{-1}(\mathbf{y} - \mathbf{m}_{\mathbf{y}}). \tag{5.8}$$

The linear form (strictly speaking it is the affine form) of the estimate in terms of the observed data $Y = \mathbf{y}$ is due to the Gaussian distribution which does not hold in general.

Example 5.2 reveals several nice properties about the Gaussian random vectors. First, one needs know only the mean and covariance matrix in order to have the complete knowledge of the PDF. Often such statistical quantities can be experimentally determined. Second, if X and Y are jointly Gaussian, then each marginal and conditional distribution is also Gaussian. Moreover, a linear combination of Gaussian random vectors is Gaussian as well. Finally, the Gaussian assumption leads to the linear form of the optimal estimate for both the MMSE and MAP criteria. Because the observed data $\mathbf{y}$ in (5.7) can be any value and is in fact random, (5.7) actually gives the expression of the optimal *estimator* (a function of the observation) for jointly Gaussian random vectors. More generally, the following result on the MMSE estimator holds.

Theorem 5.3. *Let X and Y be two jointly distributed random vectors. Then the* MMSE *estimator $\hat{X}$ of X in terms of Y is given by* $\hat{X} = \mathrm{E}\{X|Y\}$.

Proof. The difference between $\mathrm{E}\{X|Y\}$ and $\mathrm{E}\{X|Y = \mathbf{y}\}$ lies in that $\mathrm{E}\{X|Y\}$ takes the expectation over all possible values of X and Y. Hence, the MMSE estimator is more difficult to prove than the MMSE estimate. However, the following two properties of the conditional expectation are helpful:

$$\mathrm{E}_{X|Y}\{f(X,Y)|Y = \mathbf{y}\} = \mathrm{E}_{X|Y}\{f(X,\mathbf{y})|Y = \mathbf{y}\}, \tag{5.9}$$

$$\mathrm{E}_{Y}\left\{\mathrm{E}_{X|Y}[f(X,Y)|Y = \mathbf{y}]\right\} = \mathrm{E}_{X,Y}\{f(X,Y)\}, \tag{5.10}$$

where the subscripts indicate the variables with respect to which expectation is being taken. Hence, by the MMSE estimate in (5.3),

$$\mathrm{E}_{X|Y}\left\{\|X - \hat{X}(\mathbf{y})\|^2 |Y = \mathbf{y}\right\} \leq \mathrm{E}_{X|Y}\left\{\|X - \tilde{X}(\mathbf{y})\|^2 |Y = \mathbf{y}\right\}$$

for any other estimator $\tilde{X}(\cdot)$. On the other hand, (5.9) implies that

$$\mathrm{E}_{X|Y}\left\{\|X - \hat{X}(\mathbf{y})\|^2 |Y = \mathbf{y}\right\} \leq \mathrm{E}_{X|Y}\left\{\|X - \tilde{X}(\mathbf{Y})\|^2 |Y = \mathbf{y}\right\}.$$

The above inequality is preserved with expectation being taken with respect to Y. Now with the aid of (5.10), there holds

$$\mathrm{E}_{X,Y}\left\{\|X - \hat{X}(Y)\|^2\right\} \leq \mathrm{E}_{X,Y}\left\{\|X - \tilde{X}(Y)\|^2\right\}$$

which establishes the desired result. □

The next example shows that the MAP and MMSE estimators are nonlinear in general. For convenience, the estimator $\hat{X}(Y)$ is still denoted by $\hat{\mathbf{x}}$.

Example 5.4. A typical case in digital communications is when the random variables X and Y are related as $Y = X + N$. Suppose that the random variable X is binary and equiprobable with the probability

$$P_X[X = 1] = 0.5, \quad P_X[X = -1] = 0.5.$$

The random variable N represents the additive noise which is assumed to be Gaussian distributed with zero mean and the variance σ_n^2. Suppose that X and N are independent. Then X and Y are jointly distributed. If $X = x$ (x only takes values ± 1) is transmitted, then the PDF of $Y = y$ is given by

$$p_{Y|X}(y|x) = \frac{1}{\sqrt{2\pi}\sigma_n}\exp\left\{-\frac{(y-x)^2}{2\sigma_n^2}\right\}. \tag{5.11}$$

It is easy to see that the marginal PDF for $Y = y$ is given by

$$\begin{aligned} p_Y(y) &= P_X[X = 1]p_{Y|X}(y|x = 1) + P_X[X = -1]p_{Y|X}(y|x = -1) \\ &= 0.5p_{Y|X}(y|x = 1) + 0.5p_{Y|X}(y|x = -1). \end{aligned} \tag{5.12}$$

By an abuse of notation, the conditional probability for $X = 1$, given $Y = y$ is received, is given by

$$
\begin{aligned}
P_{X|Y}[X=1|Y=y] &= \frac{p_{Y|X}(y|x=1)P_X[X=1]}{p_Y(y)} \\
&= \frac{0.5p_{Y|X}(y|x=1)}{0.5p_{Y|X}(y|x=1)+0.5p_{Y|X}(y|x=-1)} \\
&= \left(1+\frac{p_{Y|X}(y|x=-1)}{p_{Y|X}(y|x=1)}\right)^{-1} = \left(1+\exp\left\{-\frac{2y}{\sigma_n^2}\right\}\right)^{-1}
\end{aligned}
$$

in light of (5.11) and (5.12). Similarly, the conditional probability for $X=-1$, given $Y=y$ is received, is given by

$$
P_{X|Y}[X=-1|Y=y] = \left(1+\frac{p_{Y|X}(y|x=1)}{p_{Y|X}(y|x=-1)}\right)^{-1} = \left(1+\exp\left\{\frac{2y}{\sigma_n^2}\right\}\right)^{-1}.
$$

It is easy to verify that $P_{X|Y}[X=1|Y=y]+P_{X|Y}[X=-1|Y=y]=1$. If $y>0$, then

$$
P_{X|Y}[X=1|Y=y] > 0.5, \quad P_{X|Y}[X=-1|Y=y] < 0.5.
$$

Because X is binary, the maximum of $P_{X|Y}[X=x|Y=y]$ for $y>0$ takes place at $X=1$. If $y<0$, then

$$
P_{X|Y}[X=1|Y=y] < 0.5, \quad P_{X|Y}[X=-1|Y=y] > 0.5,
$$

and thus, the maximum of $P_{X|Y}[X=x|Y=y]$ takes place at $X=-1$. Consequently, the MAP estimator is obtained as

$$
\hat{x}_{\text{MAP}} = \begin{cases} 1, & \text{for } y>0, \\ -1, & \text{for } y<0. \end{cases} \tag{5.13}
$$

This is identical to the optimal decision rule as discussed in Sect. 2.3 in the sense that the BER is minimized.

On the other hand, given received data $Y=y$ the conditional mean for X is given by (recall that X is binary valued):

$$
\hat{x}_{\text{MMSE}} = \text{E}\{X|Y=y\} = \left(1+\exp\left\{-\frac{2y}{\sigma_n^2}\right\}\right)^{-1} - \left(1+\exp\left\{\frac{2y}{\sigma_n^2}\right\}\right)^{-1}.
$$

Different from the MAP estimate, $\hat{x}_{\text{MMSE}}$ is not binary valued. Its values as function of the received y are plotted in the following figure where the dotted line is for the case $\sigma_n^2=1$, the dash–dotted line for $\sigma_n^2=0.1$, and the solid line for $\sigma_n^2=0.01$ (see Fig. 5.1).

As the variance σ_n^2 decreases (which corresponds to increase of the SNR), the MMSE estimate approaches the MAP estimate. It is noted that both estimators are nonlinear functions of the received data y.

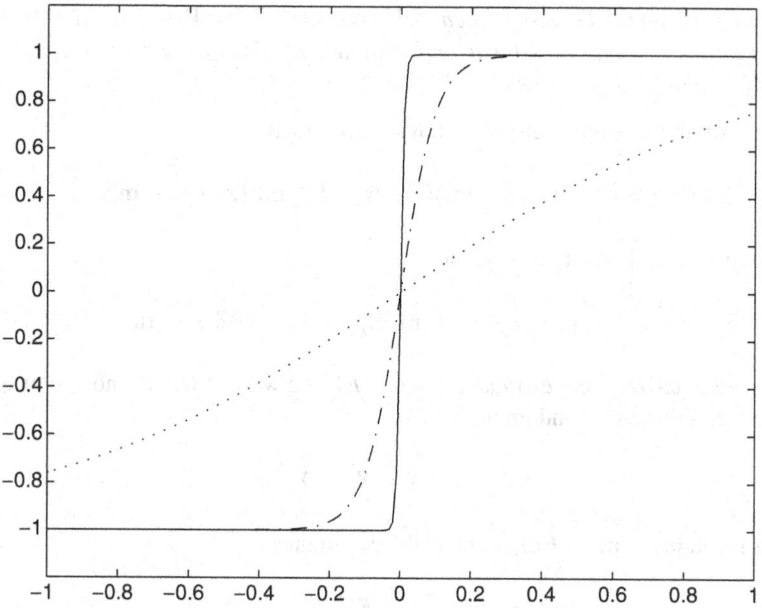

Fig. 5.1 The conditional mean estimates $\hat{x}$ as function of y

In spite of the fact that optimal estimators are nonlinear in general, the linear estimator (strictly speaking it is the affine estimator) has its appeal owing to its simplicity and mathematical tractability. Moreover, in the special case of Gaussian random vectors, both MMSE and MAP estimators are linear. Hence, there is an incentive to focus on linear estimators and search for the optimal estimator among all the linear estimators. The next theorem contains the complete result for MMSE estimators.

Theorem 5.5. *Let X and Y be two jointly distributed random vectors with*

$$\mathrm{E}\left\{\begin{bmatrix} X \\ Y \end{bmatrix}\right\} = \begin{bmatrix} \mathbf{m_x} \\ \mathbf{m_y} \end{bmatrix}, \quad \mathrm{cov}\left\{\begin{bmatrix} X \\ Y \end{bmatrix}\right\} = \begin{bmatrix} \Sigma_{\mathbf{xx}} & \Sigma_{\mathbf{xy}} \\ \Sigma_{\mathbf{yx}} & \Sigma_{\mathbf{yy}} \end{bmatrix}. \tag{5.14}$$

Then the linear MMSE *estimator for X in terms of Y is given by*

$$\hat{X} = \mathbf{m_x} + \Sigma_{\mathbf{xy}}\Sigma_{\mathbf{yy}}^{-1}(Y - \mathbf{m_y}), \tag{5.15}$$

where $\Sigma_{\mathbf{yy}}^{+}$ can be used if $\Sigma_{\mathbf{yy}}$ is singular (Problem 5.1 in Exercises). *The error covariance associated with $\hat{X}$ is unconditioned and given by*

$$\mathrm{E}\left\{\left(X - \hat{X}\right)\left(X - \hat{X}\right)^*\right\} = \Sigma_{\mathbf{xx}} - \Sigma_{\mathbf{xy}}\Sigma_{\mathbf{yy}}^{-1}\Sigma_{\mathbf{yx}}. \tag{5.16}$$

If X *and* Y *are jointly Gaussian, then (5.15) is also the* MMSE *estimate* $\hat{X} = \mathrm{E}\{X|Y\}$ *and is optimal among all (linear and nonlinear) estimators. Its error covariance conditioned on* Y *is the same as in (5.16).*

Proof. For any random vector Z its covariance satisfies

$$\mathrm{cov}\{Z\} = \mathrm{E}\{(Z - \mathbf{m_z})(Z - \mathbf{m_z})^*\} = \mathrm{E}\{ZZ^*\} - \mathbf{m_z}\mathbf{m_z}^*,$$

where $\mathbf{m_z} = \mathrm{E}\{Z\}$. As a consequence,

$$\mathrm{E}\left\{\|Z\|^2\right\} = \mathrm{Tr}\left\{\mathrm{cov}[Z]\right\} + \mathrm{Tr}\left\{\mathbf{m_z}\mathbf{m_z}^*\right\} = \mathrm{Tr}\left\{\mathrm{cov}[Z]\right\} + \|\mathbf{m_z}\|^2. \tag{5.17}$$

Now parameterize linear estimators as $\tilde{X} = FY + \mathbf{g}$ with matrix F and vector $\mathbf{g}$ free to choose. Setting the random vector

$$Z = X - \tilde{X} = X - FY - \mathbf{g}$$

yields mean $\mathbf{m_z} = \mathbf{m_x} - F\mathbf{m_y} - \mathbf{g}$ and the covariance

$$\mathrm{cov}\{Z\} = \Sigma_{\mathbf{xx}} + F\Sigma_{\mathbf{yy}}F^* - F\Sigma_{\mathbf{yx}} - \Sigma_{\mathbf{xy}}F^*.$$

Hence, the error variance $\mathrm{E}\left\{\|Z\|^2\right\} = \mathrm{E}\{\|X - FY - \mathbf{g}\|^2\}$ is given by

$$\begin{aligned}
\mathrm{E}\left\{\|Z\|^2\right\} &= \mathrm{Tr}\left\{\mathrm{cov}[Z]\right\} + \|\mathbf{m_z}\|^2 \geq \mathrm{Tr}\left\{\mathrm{cov}[Z]\right\} \\
&= \mathrm{Tr}\left\{\Sigma_{\mathbf{xx}} + F\Sigma_{\mathbf{yy}}F^* - F\Sigma_{\mathbf{yx}} - \Sigma_{\mathbf{xy}}F^*\right\} \\
&= \mathrm{Tr}\left\{\left[F - \Sigma_{\mathbf{xy}}\Sigma_{\mathbf{yy}}^{-1}\right]\Sigma_{\mathbf{yy}}\left[F - \Sigma_{\mathbf{xy}}\Sigma_{\mathbf{yy}}^{-1}\right]^* + \Sigma_{\mathbf{xx}} - \Sigma_{\mathbf{xy}}\Sigma_{\mathbf{yy}}^{-1}\Sigma_{\mathbf{yx}}\right\} \\
&\geq \mathrm{Tr}\left\{\Sigma_{\mathbf{xx}} - \Sigma_{\mathbf{xy}}\Sigma_{\mathbf{yy}}^{-1}\Sigma_{\mathbf{yx}}\right\}
\end{aligned}$$

for any F and $\mathbf{g}$ where (5.17) is used. By taking

$$F = F_{\mathrm{opt}} = \Sigma_{\mathbf{xy}}\Sigma_{\mathbf{yy}}^{-1} \quad \text{and} \quad \mathbf{g} = \mathbf{g}_{\mathrm{opt}} = \mathbf{m_x} - F_{\mathrm{opt}}\mathbf{m_y},$$

$\mathbf{m_z} = \mathbf{0}$ and $\mathrm{E}\left\{\|Z\|^2\right\} = \mathrm{Tr}\left\{\Sigma_{\mathbf{xx}} - \Sigma_{\mathbf{xy}}\Sigma_{\mathbf{yy}}^{-1}\Sigma_{\mathbf{yx}}\right\} = \mathrm{E}\left\{\|Z\|^2\right\}$ which is the unconditional error variance. Therefore, the error covariance in (5.16) holds and the linear MMSE estimator is given by

$$\hat{X} = F_{\mathrm{opt}}Y + \mathbf{g}_{\mathrm{opt}} = \mathbf{m_x} + F_{\mathrm{opt}}(Y - \mathbf{m_y})$$

which is identical to (5.15) by $F_{\mathrm{opt}} = \Sigma_{\mathbf{xy}}\Sigma_{\mathbf{yy}}^{-1}$. It is noted that the linear MMSE estimator is identical to (5.7), if $Y = \mathbf{y}$, due to the fact that the linear MMSE estimator coincides with the overall MMSE estimator (among all linear and nonlinear estimators) for the Gaussian case. □

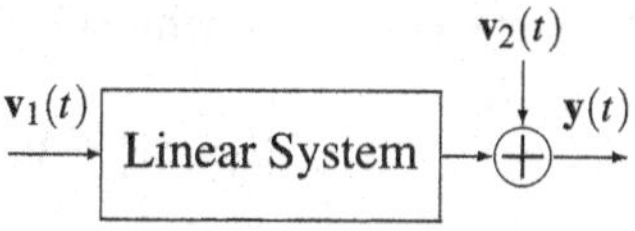

Fig. 5.2 Signal model for Kalman filtering

It should be emphasized that (5.16) is the unconditional error covariance in general. Only when X and Y are jointly Gaussian is it also the conditional error covariance. In the case when $\Sigma_{\mathbf{yy}}$ is singular, $\Sigma_{\mathbf{yy}}^{-1}$ needs to be replaced by its pseudo-inverse $\Sigma_{\mathbf{yy}}^{+}$ for which the results in Theorem 5.5 still hold. See Problem 5.1 in Exercises.

5.1.2 Kalman Filters

The design of state estimators is considerably more difficult than that of the estimator in the previous subsection due to the dynamical model for the random processes.

Consider the time-varying state-space system

$$\mathbf{x}(t+1) = A_t\mathbf{x}(t) + \mathbf{v}_1(t), \quad \mathbf{y}(t) = C_t\mathbf{x}(t) + \mathbf{v}_2(t), \tag{5.18}$$

where both $\{\mathbf{v}_1(t)\}$ and $\{\mathbf{v}_2(t)\}$ are random processes. Traditionally, $\{\mathbf{v}_1(t)\}$ is called the *process noise* and $\{\mathbf{v}_2(t)\}$ the *observation noise*. See the signal model in Fig. 5.2.

It is assumed that both $\{\mathbf{v}_1(t)\}$ and $\{\mathbf{v}_2(t)\}$ are white random processes with Gaussian distributions for each t and with zero means:

$$\mathrm{E}\{\mathbf{v}_1(t)\} = \mathbf{0}, \quad \mathrm{E}\{\mathbf{v}_2(t)\} = \mathbf{0}.$$

Since $\{\mathbf{v}_1(t)\}$ and $\{\mathbf{v}_2(t)\}$ are white, the covariance matrices are given by

$$\begin{aligned}
\mathrm{E}\{\mathbf{v}_1(t+k)\mathbf{v}_1(t)^*\} &= B_tB_t^*\delta(k), \\
\mathrm{E}\{\mathbf{v}_2(t+k)\mathbf{v}_2(t)^*\} &= D_tD_t^*\delta(k), \\
\mathrm{E}\{\mathbf{v}_2(t+k)\mathbf{v}_1(t)^*\} &= D_tB_t^*\delta(k),
\end{aligned}$$

for some matrices B_t of size $n \times m$ and D_t of size $p \times m$ where n and p are the sizes of the state vector $\mathbf{x}(t)$ and the observed output $\mathbf{y}(t)$, respectively. For this reason, the state-space model (5.18) can be equivalently written as

$$\mathbf{x}(t+1) = A_t\mathbf{x}(t) + B_t\mathbf{v}(t), \quad \mathbf{y}(t) = C_t\mathbf{x}(t) + D_t\mathbf{v}(t) \tag{5.19}$$

for some equivalent white Gaussian random process $\{\mathbf{v}(t)\}$ where

$$\mathrm{E}\{\mathbf{v}(t)\} = \mathbf{0}_m \text{ and } \mathrm{E}\{\mathbf{v}(t+k)\mathbf{v}(t)^*\} = I_m\delta(k). \tag{5.20}$$

Basically, substitutions of $\mathbf{v}_1(t) = B_t\mathbf{v}(t)$ and $\mathbf{v}_2(t) = D_t\mathbf{v}(t)$ are employed in arriving at the state-space model (5.19).

Suppose that the initial condition $\mathbf{x}_0 = \mathbf{x}(0)$ is also random, and Gaussian distributed with mean and covariance

$$\mathrm{E}\{\mathbf{x}_0\} = \overline{\mathbf{x}}_0 \text{ and } \mathrm{cov}\{\mathbf{x}_0\} = P_0, \tag{5.21}$$

respectively. Assume further that $\mathbf{x}_0$ is independent of $\{\mathbf{v}(t)\}$, and $\mathbf{x}_0$ and $\mathbf{v}(0)$ are jointly Gaussian. Then

$$\mathbf{x}(1) = A_0\mathbf{x}_0 + B_0\mathbf{v}(0)$$

is a linear combination of two jointly distributed Gaussian random vectors. Thus, $\mathbf{x}(1)$ is Gaussian distributed as well. By the white assumption on $\{\mathbf{v}(t)\}$ and independence of $\mathbf{x}_0$ to $\{\mathbf{v}(t)\}$, $\mathbf{x}(1)$ and $\mathbf{v}(t)$ are independent random vectors for all $t \geq 1$. As a result, $\mathbf{x}(1)$ and $\mathbf{v}(1)$ are jointly Gaussian. Hence, by the induction process, the state vectors $\{\mathbf{x}(t)\}$ are Gaussian random processes. In fact, $\mathbf{x}(t)$ and $\mathbf{v}(t)$ are jointly Gaussian for each $t \geq 0$. Optimal state estimators are concerned with the MMSE estimate of $\mathbf{x}(t+1)$ for $t \geq 0$, based on the observation data $\{\mathbf{y}(k)\}_{k=0}^{t}$. Due to the Gaussian property, such an MMSE estimator is also a MAP estimator. The solution to the optimal state estimator is the well-publicized Kalman filtering which will be studied in this subsection.

Under the Gaussian assumption, the MMSE estimator is easy to derive for $\mathbf{x}(t)$ and $\mathbf{x}(t+1)$ based on $\{\mathbf{y}(k)\}_{k=0}^{t}$ using the basic result in Theorem 5.5. Specifically in the case of $\mathbf{x}(t+1)$, denote

$$\overline{\mathbf{x}}(t+1) = \mathrm{E}\{\mathbf{x}(t+1)\}, \quad \overline{\mathscr{Y}}_t = \mathrm{E}\{\mathscr{Y}_t\},$$

where $\mathscr{Y}_t$ is the observation up to time $t \geq 0$ with an expression

$$\mathscr{Y}_t = \mathrm{vec}\left(\left[\mathbf{y}(0)\ \mathbf{y}(1) \cdots \mathbf{y}(t)\right]\right). \tag{5.22}$$

Then $\overline{\mathbf{x}}(t+1) = A_t\overline{\mathbf{x}}(t)$. The state vector $\mathbf{x}(t+1)$ and the observed data $\mathscr{Y}_t$ are jointly Gaussian with mean $\{A_t\overline{\mathbf{x}}(t), \overline{\mathscr{Y}}_t\}$ and covariance

$$\mathrm{cov}\left\{\begin{bmatrix} \mathbf{x}(t+1) \\ \mathscr{Y}_t \end{bmatrix}\right\} = \begin{bmatrix} P_{t+1} & Z_t \\ Z_t^* & \Psi_t \end{bmatrix}. \tag{5.23}$$

It is easy to see that $\Psi_t = \mathrm{cov}\{\mathscr{Y}_t\}$, $P_{t+1} = \mathrm{cov}\{\mathbf{x}(t+1)\}$, and

$$Z_t = \mathrm{E}\left\{\left[\mathbf{x}(t+1) - \overline{\mathbf{x}}_{t+1}\right]\left[\mathscr{Y}_t - \overline{\mathscr{Y}}_t\right]^*\right\}. \tag{5.24}$$

Note that Ψ_t is nonsingular provided that $\det\left(D_k D_k^*\right) \neq 0$ for $0 \leq k \leq t$. Hence, Theorem 5.5 can be applied to compute the MMSE estimate for $\mathbf{x}(t+1)$ based on $\mathscr{Y}_t$ which is given by

$$\hat{\mathbf{x}}_{t+1|t} = A_t \overline{\mathbf{x}}_t + Z_t \Psi_t^{-1} \left(\mathscr{Y}_t - \overline{\mathscr{Y}}_t\right) \tag{5.25}$$

by $\overline{\mathbf{x}}(t+1) = A_t \overline{\mathbf{x}}(t)$. The associated error covariance according to (5.16) is

$$\Sigma_{t+1|t} = \mathrm{E}\left\{\left[\mathbf{x}(t+1) - \hat{\mathbf{x}}_{t+1|t}\right]\left[\mathbf{x}(t+1) - \hat{\mathbf{x}}_{t+1|t}\right]^*\right\} = P_{t+1} - Z_t \Psi_t^{-1} Z_t^*. \tag{5.26}$$

However, the MMSE estimator as described in (5.25) and (5.26) has no value in practice because the associated computational complexity grow with respect to the time index t. A remarkable feature of the Kalman filter is its recursive computation of the MMSE estimate $\hat{\mathbf{x}}_{k+1|k}$ and recursive update of the optimal error covariance Σ_{k+1} with complexity dependent only on the order of the state-space model in (5.19) rather than the time index.

Theorem 5.6. *Consider the state-space model in (5.19) where $\{\mathbf{v}(t)\}$ is the white Gaussian random process satisfying (5.20) and the initial condition $\mathbf{x}(0) = \mathbf{x}_0$ is also Gaussian distributed, independent of $\{\mathbf{v}(t)\}$, with the mean $\overline{\mathbf{x}}_0$ and the covariance P_0. Suppose*

$$B_t D_t^* = \mathbf{0}, \quad R_t = D_t D_t^* > \mathbf{0}, \ \forall\, t \geq 0. \tag{5.27}$$

Denote $\hat{\mathbf{x}}_{k|i}$ as the MMSE *estimate of $\mathbf{x}(k)$ based on $\mathscr{Y}_i$, and $\Sigma_{k|i}$ as the corresponding error covariance where $k \geq i \geq 0$. Then*

$$\hat{\mathbf{x}}_{t|t} = \hat{\mathbf{x}}_{t|t-1} + L_t\left[\mathbf{y}(t) - C_t \hat{\mathbf{x}}_{t|t-1}\right], \tag{5.28}$$

$$L_t = \Sigma_{t|t-1} C_t^* \left(R_t + C_t \Sigma_{t|t-1} C_t^*\right)^{-1}, \tag{5.29}$$

$$\Sigma_{t|t} = \Sigma_{t|t-1} - \Sigma_{t|t-1} C_t^* \left(R_t + C_t \Sigma_{t|t-1} C_t^*\right)^{-1} C_t \Sigma_{t|t-1}, \tag{5.30}$$

$$\hat{\mathbf{x}}_{t+1|t} = A_t \hat{\mathbf{x}}_{t|t}, \quad \Sigma_{t+1|t} = A_t \Sigma_{t|t} A_t^* + B_t B_t^*, \tag{5.31}$$

initialized by $\hat{\mathbf{x}}_{0|-1} = \overline{\mathbf{x}}_0$ and $\Sigma_{0|-1} = P_0$.

Proof. Given $\hat{\mathbf{x}}_{t|t-1}$, $\Sigma_{t|t-1}$, and observation $\mathscr{Y}_t$, it can be verified that

$$\mathrm{E}\left\{\begin{bmatrix}\mathbf{x}(t)\\ \mathbf{y}(t)\end{bmatrix}\Bigg|\mathscr{Y}_{t-1}\right\} = \begin{bmatrix}\hat{\mathbf{x}}_{t|t-1}\\ C_t \mathbf{x}_{t|t-1}\end{bmatrix}, \tag{5.32}$$

$$\mathrm{cov}\left\{\begin{bmatrix}\mathbf{x}(t)\\ \mathbf{y}(t)\end{bmatrix}\Bigg|\mathscr{Y}_{t-1}\right\} = \begin{bmatrix}\Sigma_{t|t-1} & \Sigma_{t|t-1} C_t^*\\ C_t \Sigma_{t|t-1} & R_t + C_t \Sigma_{t|t-1} C_t^*\end{bmatrix}. \tag{5.33}$$

Applying Theorem 5.5 with $X = \mathbf{x}(t)$ and $Y = \mathscr{Y}_t$ leads to the MMSE estimate $\hat{\mathbf{x}}_{t|t}$ and error covariance $\Sigma_{t|t}$ in (5.28)–(5.30) which are referred to as *measurement update*. Because of (5.27), the random vectors $B_t\mathbf{v}(t)$ and $D_t\mathbf{v}_t$ are uncorrelated or $\mathrm{E}\{B_t\mathbf{v}(t)[D_t\mathbf{v}_t]^*\} = B_tD_t^* = \mathbf{0}$. In fact, $B_t\mathbf{v}(t)$ and $D_t\mathbf{v}_t$ are independent of each other due to the Gauss assumption. It follows that $B_t\mathbf{v}(t)$ and $\mathbf{y}(t)$ are independent of each other. Hence

$$\mathrm{E}\{B_t\mathbf{v}(t)|\mathscr{Y}_t\} = \mathrm{E}\{B_t\mathbf{v}(t)|\mathbf{y}_t\} = \mathbf{0}.$$

The above leads to $\mathrm{E}\{\mathbf{x}(t+1)|\mathscr{Y}_t\} = A_t\hat{\mathbf{x}}_{t|t} + \mathrm{E}\{B_t\mathbf{v}(t)|\mathscr{Y}_t\} = A_t\hat{\mathbf{x}}_{t|t}$ and

$$\Sigma_{t+1|t} = A_t\Sigma_{t|t}A_t^* + B_tB_t^*$$

or (5.31) that is referred to as *time update*. The proof is now complete. □

Theorem 5.6 indicates that Kalman filtering is basically an efficient and recursive algorithm for implementing the MMSE estimator. Recall that the computational complexity for those in (5.25)–(5.26) grows with respect to the time index. It is surprising that the MMSE estimator is linear and finite-dimensional with the same order as that of the signal model in (5.19), rather than nonlinear and infinite-dimensional as one might have speculated at the beginning. Of course, such linear and finite-dimensional properties of the MMSE estimator are owing to the Gaussian assumption. If the noise process $\{\mathbf{v}(t)\}$ is not Gaussian, then the Kalman filter can only be claimed to be optimal among all linear filters of arbitrary orders in light of Theorem 5.5. In addition, its property of being an MAP estimator is lost in general.

It is observed that the Kalman filter in Theorem 5.6 actually consists of two MMSE estimators: One is the *measurement update* as described in (5.28)–(5.30), and the other is *time update* as described in (5.31). While Theorem 5.6 is the main result of Kalman filtering, Kalman filter is often referred to the recursive algorithm for computing $\hat{\mathbf{x}}_{t+1|t}$ based on $\hat{\mathbf{x}}_{t|t-1}$. The next result shows the structure of such an optimal one-step predictor. The proof is left as an exercise (Problem 5.9).

Theorem 5.7. *Denote $\Sigma_k = \Sigma_{k|k-1}$ for each integer $k \geq 1$. Under the same hypotheses of Theorem 5.6, the* MMSE *estimate $\hat{\mathbf{x}}_{t+1|t}$ for $\mathbf{x}(t+1)$ based on the observation $\mathscr{Y}_t = \{\mathbf{y}(k)\}_{k=0}^t$ is given recursively as*

$$\hat{\mathbf{x}}_{t+1|t} = [A_t + K_tC_t]\hat{\mathbf{x}}_{t|t-1} - K_t\mathbf{y}(t), \quad \hat{\mathbf{x}}_{0|-1} = \bar{\mathbf{x}}_0, \tag{5.34}$$

$$K_t = -A_t\Sigma_tC_t^*\left(R_t + C_t\Sigma_tC_t^*\right)^{-1}, \tag{5.35}$$

$$\Sigma_{t+1} = A_t\Sigma_tA_t^* + B_tB_t^* + K_tC_t\Sigma_tA_t^*, \quad \Sigma_0 = P_0. \tag{5.36}$$

It is interesting to observe that (5.34) can be written as

$$\hat{\mathbf{x}}_{t+1|t} = A_t\hat{\mathbf{x}}_{t|t-1} - K_t\left[\hat{\mathbf{y}}_{t|t-1} - \mathbf{y}(t)\right]$$

with $\hat{\mathbf{y}}_{t|t-1} = C_t\hat{\mathbf{x}}_{t|t-1}$. So, it is similar to (5.19) with only one difference in that $B_t\mathbf{v}(t)$ is replaced by $-K_t\left[\mathbf{y}(t) - C_t\hat{\mathbf{x}}_{t|t-1}\right]$. A reflection on this indicates that the vector $\left(\mathbf{y}(t) - C_t\hat{\mathbf{x}}_{t|t-1}\right)$ provides new information that is not contained in $\mathscr{Y}_{t-1}$.

For this reason, $\left\{\left(\mathbf{y}(t) - C_t\hat{\mathbf{x}}_{t|t-1}\right)\right\}$ is called *innovation sequence* which is in fact a white process (refer to Problem 5.12 in Exercises). It is also interesting to observe that the error covariance $\Sigma_{t+1|t}$ is independent of the observation $\mathscr{Y}_t$. That is, no one set of measurements helps any more than any other set to eliminate the uncertainty in $\mathbf{x}_t$. For convenience, $\Sigma_{t+1} := \Sigma_{t+1|t}$ will be used in the rest of the text. Equation (5.36) governing the error covariance is called the difference Riccati equation (DRE).

The initial covariance $\Sigma_0 = P_0$ measures the confidence of the a priori information on the initial estimate $\mathbf{x}_{0|-1} = \overline{\mathbf{x}}_0$. Small P_0 means high confidence whereas large P_0 means low confidence. In practice, the knowledge on the a priori information of $\overline{\mathbf{x}}_0$ and P_0 may not be available. In this case, $\overline{\mathbf{x}}_0 = \mathbf{0}$ and $P_0 = \rho I_n$ are often taken with $\rho > 0$ sufficiently large. However, so long as the Kalman filter is stable (to be investigated in the next subsection), the impact of $\overline{\mathbf{x}}_0$ and P_0 to the MMSE estimate $\hat{\mathbf{x}}_{t+1|t}$ will fade away as t gets large. The next result is obtained for time-invariant systems.

Proposition 5.1. *Suppose that $(A_t, B_t, C_t, D_t) = (A, B, C, D)$ for all t and the hypotheses of Theorem 5.6 hold. If $\Sigma_0 = \mathbf{0}$, then the solution to the* DRE *(5.36) is monotonically increasing, i.e., $\Sigma_{t+1} \geq \Sigma_t$ for all $t \geq 0$.*

Proof. For $t = 0$, the DRE (5.36) gives $\Sigma_1 = BB^* \geq \Sigma_0 = \mathbf{0}$ in light of the time-invariance hypothesis. Using the induction, assume that $\Sigma_k \geq \Sigma_{k-1}$ for $k > 1$. The proof can be completed by showing $\Sigma_{k+1} \geq \Sigma_k$. Denote $\Delta_t = \Sigma_t - \Sigma_{t-1}$ for $t = k$ and $k+1$. The DRE (5.36) is equivalent to

$$\Sigma_{t+1} = A\left(I + \Sigma_t C^* R^{-1} C\right)^{-1} \Sigma_t A^* + BB^*, \quad R = DD^*.$$

See Problem 5.10 in Exercises. Taking the difference $\Delta_{k+1} = \Sigma_{k+1} - \Sigma_k$ gives

$$\begin{aligned}
\Delta_{k+1} &= A\left[\left(I + \Sigma_k C^* R^{-1} C\right)^{-1} \Sigma_k - \Sigma_{k-1}\left(I + C^* R C \Sigma_{k-1}\right)^{-1}\right] A^* \\
&= A\left(I + \Sigma_k C^* R^{-1} C\right)^{-1} \Delta_k \left(I + C^* R^{-1} C \Sigma_{k-1}\right)^{-1} A^* \\
&= A\left(I + \Sigma_{k-1} C^* R^{-1} C + \Delta_k C^* R^{-1} C\right)^{-1} \Delta_k \left(I + C^* R C \Sigma_{k-1}\right)^{-1} A^* \\
&= \overline{A}_{k-1}\left[I + \Delta_k C^* \left(R + C\Sigma_k C^*\right)^{-1} C\right]^{-1} \Delta_k \overline{A}_{k-1}^* \geq 0
\end{aligned}$$

by $\Delta_k = \Sigma_k - \Sigma_{k-1} \geq 0$ where $\overline{A}_{k-1} = A\left(I + \Sigma_{k-1} C^* R^{-1} C\right)^{-1}$. □

Before ending this subsection, the removal of the assumption (5.27) needs to be addressed. It is noted that the difference between the estimated and the true state vectors satisfies the difference equation

$$\hat{\mathbf{e}}(t+1) = (A_t + K_t C_t)\hat{\mathbf{e}}(t) + (B_t + K_t D_t)\mathbf{v}(t) \tag{5.37}$$

by taking the difference of (5.19) and (5.34) where $\hat{\mathbf{e}}(t) = \mathbf{x}(t) - \hat{\mathbf{x}}_{t|t-1}$. This is the error equation for the associated Kalman filter under the assumption (5.27) or $B_t D_t^* = \mathbf{0}$ for all t. For the case $B_t D_t^* \neq \mathbf{0}$, it is claimed that the error equation associated with the MMSE estimate has the form

$$\hat{\mathbf{e}}(t+1) = \left(\tilde{A}_t + \tilde{K}_t C_t\right)\hat{\mathbf{e}}(t) + \left(\tilde{B}_t + \tilde{K}_t D_t\right)\mathbf{v}(t), \tag{5.38}$$

$$\tilde{A}_t = A_t - B_t D_t^* R_t^{-1} C_t, \quad \tilde{B}_t = B_t\left[I_m - D_t^* R_t^{-1} D_t\right], \tag{5.39}$$

where $\tilde{B}_t D_t^* = \mathbf{0}$. Specifically, the state-space system (5.19) can be written as

$$\mathbf{x}(t+1) = \tilde{A}_t \mathbf{x}(t) + \tilde{B}_t \mathbf{v}(t) + B_t D_t^* R_t^{-1}\mathbf{y}(t). \tag{5.40}$$

Because $\tilde{B}_t D_t^* = \mathbf{0}$, and $\mathbf{y}(t)$ is the measured output, the Kalman filter can be adapted to compute the MMSE estimate for $\mathbf{x}(t+1)$ in accordance with

$$\hat{\mathbf{x}}_{t+1|t} = \left[\tilde{A}_t + \tilde{K}_t C_t\right]\hat{\mathbf{x}}_{t|t-1} - \tilde{K}_t \mathbf{y}(t) + B_t D_t^* R_t^{-1}\mathbf{y}(t), \tag{5.41}$$

where the Kalman gain and the error covariance are given by

$$\tilde{K}_t = -\tilde{A}_t \Sigma_t C_t^* \left(R_t + C_t \Sigma_t C_t^*\right)^{-1}, \tag{5.42}$$

$$\Sigma_{t+1} = \tilde{A}_t \Sigma_t \tilde{A}_t^* + \tilde{B}_t \tilde{B}_t^* - \tilde{A}_t \Sigma_t C_t^* \left(R_t + C_t \Sigma_t C_t^*\right)^{-1} C_t \Sigma_t \tilde{A}_t^*, \tag{5.43}$$

respectively. The Kalman gain $\tilde{K}_t$ is associated with $\left(\tilde{A}_t, \tilde{B}_t\right)$, but Σ_t is the same error covariance as before. Subtracting (5.41) from (5.40) yields (5.38) as claimed earlier. On the other hand, (5.41) can be equivalently written as

$$\hat{\mathbf{x}}_{t+1|t} = (A_t + K_t C_t)\hat{\mathbf{x}}_{t|t-1} - K_t \mathbf{y}(t) \quad K_t = \tilde{K}_t - B_t D_t^* R_t^{-1}, \tag{5.44}$$

where K_t is the Kalman gain associated with (A_t, B_t). There holds

$$K_t = \tilde{K}_t - B_t D_t^* R_t^{-1} = -\left(A_t \Sigma_t C_t^* + B_t D_t^*\right)\left(R_t + C_t \Sigma_t C_t^*\right)^{-1}. \tag{5.45}$$

Therefore, the Kalman filter has the same form with a slight increase in the complexity of computing the Kalman gain and the associated DRE for the error covariance. The next result summarizes the above discussion.

Corollary 5.1. *Let $\tilde{A}_t$ and $\tilde{B}_t$ be as in (5.39). Under the same hypotheses of Theorem 5.7, except that $B_t D_t^* \neq \mathbf{0}$, the Kalman filter for $\mathbf{x}(t+1)$ based on the observation $\mathscr{Y}_t = \{\mathbf{y}(k)\}_{k=0}^t$ is given recursively by (5.44), (5.45), and (5.43) which collapse to those in Theorem 5.7 for the case $B_t D_t^* = \mathbf{0}$.*

Corollary 5.1 indicates that there is no loss of generality in focusing on the case $B_t D_t^* = \mathbf{0}$ for Kalman filtering. The case $B_t D_t^* \neq \mathbf{0}$ causes only some minor computational modifications for (linear) MMSE estimators.

5.1.3 Stability

An immediate question regarding the Kalman filter is its stability. Are Kalman filters always stable? What stability properties do Kalman filters possess? Such questions will be answered first for time-varying systems and then for time-invariant systems. The following stability result holds.

Theorem 5.8. *Let the state-space model be as in (5.19) with $\{\mathbf{v}(t)\}$ the white noise satisfying (5.20). For the case $B_tD_t^* = \mathbf{0}$, assume that (A_t, B_t) is uniformly stabilizable and (C_t, A_t) is uniformly detectable. Then the Kalman filter as described in Theorem 5.7 is asymptotically stable. For the case $B_tD_t^* \neq \mathbf{0}$, assume that $(\tilde{A}_t, \tilde{B}_t)$ is uniformly stabilizable, and (C_t, A_t) is uniformly detectable where $\tilde{A}_t$ and $\tilde{B}_t$ are as in (5.39). Then the Kalman filter as described in Corollary 5.1 is asymptotically stable.*

Proof. If $B_tD_t^* = \mathbf{0}$, then the stabilizability and detectability assumptions imply the existence of linear state estimation gains $\{L_t\}$ such that

$$\mathbf{x}(t+1) = (A_t + L_tC_t)\mathbf{x}(t)$$

is asymptotically stable. Hence, the difference Lyapunov equation

$$Q_{t+1} = (A_t + L_tC_t)Q_t(A_t + L_tC_t)^* + L_tR_tL_t^* + B_tB_t^*$$

has bounded nonnegative solutions $\{Q_t\}$. On the other hand, the DRE (5.36) which governs the error covariance of the Kalman filter can be written into the same form as the above difference Lyapunov equation:

$$\Sigma_{t+1} = (A_t + K_tC_t)\Sigma_t(A_t + K_tC_t)^* + B_tB_t^* + K_tR_tK_t^*.$$

See Problem 5.10 in Exercises. It is noted that $\left(A + K_tC_t, \begin{bmatrix} B_t & K_tR_t^{1/2} \end{bmatrix}\right)$ is stabilizable. In light of the discussions at the end of Chap. 3, stability of the Kalman filter is hinged to the boundedness of Σ_t as $t \to \infty$. But the Kalman filter is optimal among all linear estimators. It follows that $\text{Tr}\{\Sigma_t\} \leq \text{Tr}\{Q_t\}$. The Kalman filter is thus asymptotically stable. The proof for the case $B_tD_t^* \neq \mathbf{0}$ is similar, and is skipped. □

For asymptotically exponentially stable systems, the assumptions of uniform stabilizability and detectability hold. Hence, the Kalman filter preserves the stability property that is owing to the fact $\Sigma_t \leq P_t$ or the error covariance for the estimated state vector is no larger than the covariance of the state vector to be estimated, which is in turn owing to the optimality of the Kalman filter. The hypothesis on stabilizability of (A_t, B_t) in Theorem 5.8 might seem unnecessary by the argument on the existence of the stable linear estimators. However, it cannot be removed for the stability result in Theorem 5.8 to hold true. Nevertheless, this hypothesis can be weakened for stability of Kalman filters if the underlying state-space system is time invariant and the additive noise process is stationary. For this purpose, consider the following signal model

$$\mathbf{x}(t+1) = A\mathbf{x}(t) + B\mathbf{v}(t), \quad \mathbf{y}(t) = C\mathbf{x}(t) + D\mathbf{v}(t), \tag{5.46}$$

where $\mathbf{x}(0) = \mathbf{x}_0$ with mean $\bar{\mathbf{x}}_0$ and covariance P_0. Suppose that $\mathbf{x}_0$ and $\{\mathbf{v}(t)\}$ are independently distributed and $\{\mathbf{v}(t)\}$ satisfies (5.20). Then $\{B\mathbf{v}(t)\}$ and $\{D\mathbf{v}(t)\}$ are both WSS white processes. Stability of A and $\mathrm{E}\{\mathbf{x}(t)\} = \mathbf{0}$ imply that $\{\mathbf{x}(t)\}$ can be made a stationary process, provided that $P = P_0$ satisfies the Lyapunov equation

$$P = APA^* + BB^*. \tag{5.47}$$

Assume that A is stable. Then there is a unique solution $P \geq 0$ to the above Lyapunov equation. If $P_0 \neq P$, then $\{\mathbf{x}(t)\}$ is not a WSS process in general. But, it is WSS asymptotically. Indeed, let $P_k = \mathrm{cov}\{\mathbf{x}(k)\}$ for $k \geq 0$. Then

$$P_{t+1} = AP_tA^* + BB^* = A^{t+1}P_0(A^*)^{t+1} + \sum_{k=0}^{t} A^kBB^*(A^*)^k$$

by $P_0 = \mathrm{cov}\{\mathbf{x}_0\}$. Stability of A implies that

$$P = \lim_{t\to\infty} P_{t+1} = \sum_{k=0}^{\infty} A^kBB^*(A^*)^k$$

exists and is bounded that is the unique solution to (5.47). However, if A is not a stability matrix, then $\{\mathbf{x}(t)\}$ may diverge and is thus not a WSS process in general. A somewhat surprising fact is that the error state vectors $\hat{\mathbf{e}}(t) = \mathbf{x}(t) - \hat{\mathbf{x}}_{t|t-1}$ associated with the Kalman filter are WSS process asymptotically, provided that the Kalman filter is asymptotically stable. Stability of A is not required. Consider the following nth order LTI estimator

$$\hat{\mathbf{x}}_{t+1} = (A + KC)\hat{\mathbf{x}}_t - K\mathbf{y}(t), \quad \hat{\mathbf{x}}_0 = \bar{\mathbf{x}}(0), \tag{5.48}$$

$$K = -A\Sigma C^*\left(R + C\Sigma C^*\right)^{-1}, \tag{5.49}$$

$$\Sigma = A\left(I_n + \Sigma C^* R^{-1} C\right)^{-1}\Sigma A^* + BB^*. \tag{5.50}$$

This is the same as the Kalman filter in (5.34)–(5.36) after removing the time indexes of the matrices. The equation for the error covariance in (5.50) is called the algebraic Riccati equation (ARE).

Example 5.9. Consider the inverted pendulum system. Its state-space realization after discretization with sampling period $T_s = 0.25$ is obtained in Example 3.16 of Chap. 3. Suppose that $0.1BB^*$ is the covariance for the process noise and

$\text{diag}(0.1, 0.01)$ is the covariance for the measurement noise, assuming that the noises for measurements of position and angle are uncorrelated. The Matlab command "dare" can be used to compute the solution Σ to the ARE in (5.50) and the stationary estimation gain K in (5.49). The numerical results are given by

$$\Sigma = \begin{bmatrix} 21.762 & 5.5171 & 1.6675 & 0.7219 \\ 5.5171 & 1.6269 & 0.6957 & 0.4560 \\ 1.6675 & 0.6957 & 0.4602 & 0.3910 \\ 0.7219 & 0.4560 & 0.3910 & 0.3699 \end{bmatrix}, \quad K = \begin{bmatrix} 0.4017 & -44.5166 \\ 0.4402 & -10.6004 \\ 0.4451 & -2.5400 \\ 0.4330 & -0.6398 \end{bmatrix}.$$

It can be easily verified with Matlab that the eigenvalues of $(A + KC)$ are all positive real, and strictly smaller than 1. Hence, $(A + KC)$ is a stability matrix, implying that the error for estimation of the state vector approaches zero asymptotically.

The ARE solution computed in Matlab, if it exists, is called stabilizing solution that is defined next.

Definition 5.1. The solution Σ to the ARE (5.50) is said to be stabilizing, if K as in (5.49) is stabilizing. That is, $(A + KC)$ is a stability matrix.

The stabilizing solution to ARE (5.50), if it exists, is unique (refer to Problem 5.13 in Exercises). The next result regards stability of the Kalman filter.

Theorem 5.10. *Suppose that* (A, B) *is stabilizable,* $BD^* = \mathbf{0}$*, and* $R = DD^* > 0$ *for the random process in (5.46) where* $\{\mathbf{v}(t)\}$ *is* WSS *with zero mean and identity covariance. If the* ARE *(5.50) admits a stabilizing solution, then the Kalman filter for (5.46) is asymptotically stable and its associated state estimation error vector* $\hat{\mathbf{e}}(t) = \mathbf{x}(t) - \hat{\mathbf{x}}_{t|t-1}$ *is* WSS *asymptotically.*

Proof. Let Σ_t be the solution to the DRE (5.36). Since the time-invariant estimator described in (5.48)–(5.49) is a special case of the linear estimator in Problem 5.11, its error covariance $\Sigma \geq \Sigma_t \geq 0$ for all $t \geq 0$, provided that $\Sigma \geq P_0 \geq \mathbf{0}$. In this case, $\{\Sigma_t\}$ is monotonically increasing in light of Proposition 5.1 and uniformly bounded above by Σ. Hence, its limit $\overline{\Sigma}$ exists. Since the limit of the DRE (5.36) is identical to the ARE (5.50), it can be written as

$$\overline{\Sigma} = (A + \overline{K}C)\,\overline{\Sigma}\,(A + \overline{K}C)^* + \overline{K}R\overline{K}^* + BB^*,$$

where $\overline{K} = -A\overline{\Sigma}C^*\left(R + C\overline{\Sigma}C^*\right)^{-1}$. Stabilizability of (A, B) implies that $\overline{\Sigma}$ is stabilizing, and thus $\overline{\Sigma} = \Sigma$ by its uniqueness. As such the Kalman filter converges to the linear estimator as described in (5.48)–(5.49) which is stable. The fact that the Kalman filter is linear implies that its stability is independent of the initial condition $\hat{\mathbf{x}}_{0|-1} = \overline{\mathbf{x}}_0$ which in turn implies that the convergence of Σ_t to Σ is independent of the boundary condition $\Sigma_0 = P_0$. Moreover, the conditional mean and covariance associated with $\hat{\mathbf{e}}(t)$ are zero and Σ (asymptotically), respectively. So, the state error vector $\hat{\mathbf{e}}(t)$ is WSS asymptotically. The proof is completed. □

The LTI estimator in (5.48)–(5.49) is referred to as stationary Kalman filter. It is the state-space version of the Wiener filter. In lieu of the optimality properties of the Kalman filter, the stationary Kalman filter outperforms all LTI estimators of arbitrary orders. If, in addition, the noise process is Gaussian, then the stationary Kalman filter outperforms all (linear or nonlinear) time-invariant estimators. The premise is the existence of the stabilizing solution to the ARE (5.50) for which the following result provides the necessary and sufficient condition.

Theorem 5.11. *There exists a stabilizing solution to the* ARE *in (5.50), if and only if* (C,A) *is detectable and*

$$\text{rank}\left\{\left[A - e^{j\omega}I_n \; B\right]\right\} = n \quad \forall\ \omega \in \mathbb{R}.$$

Different from time-varying systems with nonstationary noises, stabilizability of (A,B) is not required so long as (A,B) does not have unreachable modes on the unit circle. The proof is delayed to the next section. After the brain-storm of materials on Kalman filtering in the style of theorem/proof, it will be wise to pause for a while with readings of a few examples. It does need to be pointed out though that the results on time-invariant systems and WSS noises are established under the assumption that $BD^* = \mathbf{0}$. If the assumption does not hold, then the stationary Kalman filter is still the same as in (5.48) but the Kalman gain in (5.49) and the ARE in (5.50) need to be replaced by

$$K = -(A\Sigma C^* + BD^*)\left(R + C\Sigma C^{-1}\right)^{-1}, \tag{5.51}$$

$$\Sigma = \tilde{A}\left(I_n + \Sigma C^* R^{-1} C\right)^{-1} \Sigma \tilde{A}^* + B\left(I - D^* R^{-1} D\right) B^*, \tag{5.52}$$

respectively, where $\tilde{A} = A - BD^*R^{-1}C$. Moreover, the ARE (5.52) has a stabilizing solution, if and only if (C,A) is detectable, and

$$\text{rank}\left\{\begin{bmatrix} A - e^{j\omega}I_n & B \\ C & D \end{bmatrix}\right\} = n + p \quad \forall\, \omega \in \mathbb{R}$$

with p the number of rows of C. Its proof is again delayed to the next section.

Two examples will be presented which are designed to help digest the theoretical results in this section. The first example is modified from digital communications as illustrated in the following Fig. 5.3 and discussed next.

Example 5.12. Consider estimation of the symbol $\mathbf{s}(t)$ in multiuser wireless data communications. The multipath channel is described by

$$\mathbf{r}(t) = \sum_{k=1}^{\ell} H_k(t)\mathbf{s}(t-k) \tag{5.53}$$

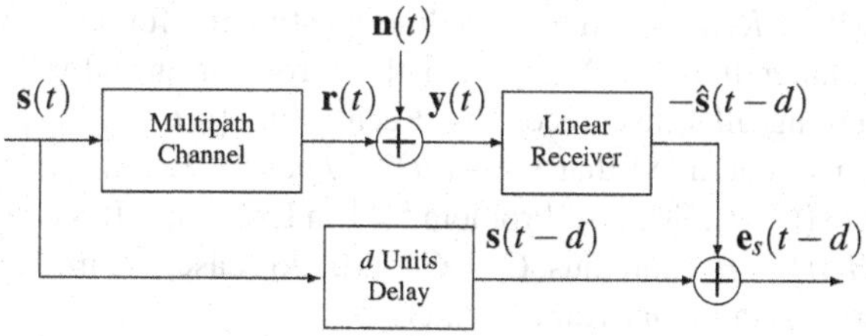

Fig. 5.3 Estimation of the symbol inputs

which is an MA model. It is assumed that the channel information or the impulse response $\{H_k(t)\}$ is known at time t and has dimension $p \times m$. The objective is to design linear receivers that estimate the symbol $\mathbf{s}(t-d)$ for some d satisfying $1 < d \leq \ell$ with the minimum error variance. The design problem seems to be different from Kalman filtering but is intimately related to the estimation problem in this subsection.

First, the input symbols are assumed to be independent and have the same (equiprobable) distributions. As such, $\{\mathbf{s}(t)\}$ is white with the zero mean and covariance $\sigma_s^2 I$ where σ_s^2 is the transmission power for each symbol. For simplicity, $\sigma_s = 1$ is taken via some suitable normalization. Secondly, the channel model can be associated with a realization with the state vector

$$\mathbf{x}(t) = \operatorname{vec}\left(\left[\, \mathbf{s}(t-1)\ \mathbf{s}(t-2)\ \cdots\ \mathbf{s}(t-\ell)\,\right]\right). \tag{5.54}$$

Denote $\mathbf{v}(t) = \left[\, \mathbf{s}(t)^*\ \mathbf{n}(t)^*\,\right]^*$. The observed signal $\mathbf{y}(t)$ at the receiver site can be described by the same state-space model in (5.19) with

$$\begin{aligned} A_t = A &= \begin{bmatrix} \mathbf{0} & \mathbf{0}_{m\times m} \\ I_{m(\ell-1)} & \mathbf{0} \end{bmatrix}, B_t = B = \begin{bmatrix} I_m & \mathbf{0} \\ \mathbf{0} & \mathbf{0}_{m(\ell-1)\times p} \end{bmatrix} \\ C_t &= \left[\, H_1(t) \cdots H_\ell(t)\,\right], D_t = \left[\, \mathbf{0}\ \Sigma_{\mathbf{n}}^{\frac{1}{2}}\,\right], \end{aligned} \tag{5.55}$$

where $\Sigma_{\mathbf{n}} > 0$ is the covariance of $\mathbf{n}$ assumed to be white and WSS. Hence, the signal to be estimated is given by

$$\mathbf{s}(t-d) = J_d\mathbf{x}(t), \quad J_d = \left[\, \mathbf{0} \cdots \mathbf{0}\ I_m\ \mathbf{0} \cdots \mathbf{0}\,\right], \tag{5.56}$$

where $1 < d \leq \ell$ and I_m is the dth block of J_d. Finally, it is noted that $\mathbf{v}(t)$ is white but not Gaussian. If $H_0(t) \equiv \mathbf{0}$, i.e., there is a pure delay in the multipath channel, then $BD_t^* = \mathbf{0}$. An application of Kalman filtering yields the optimal linear estimator for $\mathbf{s}(t-d) = J_d\mathbf{x}(t) = J_{d+1}\mathbf{x}(t+1)$ based on observations $\{\mathbf{y}(k)\}_{k=0}^{t}$ given by

$$\begin{aligned} \hat{\mathbf{x}}_{t+1|t} &= (A + K_tC_t)\hat{\mathbf{x}}_{t|t-1} - K_t\mathbf{y}(t), \quad K_t = -A\Sigma_tC_t^*\left(R + C_t\Sigma_tC_t^*\right)^{-1}, \\ \Sigma_{t+1} &= A\Sigma_tA^* + BB^* - A\Sigma_tC_t^*\left(R + C_t\Sigma_tC_t^*\right)^{-1}C_t\Sigma_tA^*, \end{aligned}$$

where $R = \Sigma_{\mathbf{n}}$. Any other linear estimator for $\mathbf{x}(t+1)$ has an error variance no smaller than $\text{Tr}\{\Sigma_{t+1}\}$, and the error variance for $\hat{\mathbf{s}}(t-d)$ is also the smallest among all linear receivers. Recall that $\{\mathbf{s}(t-k)\}_{k=1}^{\ell}$ are subsumed in $\mathbf{x}(t)$. It is thus concluded that $\hat{\mathbf{s}}(t-d|t) = J_d\hat{\mathbf{x}}_{t|t} = J_{d+1}\hat{\mathbf{x}}_{t+1|t}$ is an optimal linear estimate of $\mathbf{s}(t-d)$. See also Problem 5.11 in Exercises. If the channel is time invariant, then $H_k(t) = H_k$, and thus, $C_t = C \;\forall\, t$. In this case, the linear MMSE estimator converges to the stationary Kalman filter:

$$\hat{\mathbf{x}}_{t+1|t} = (A+KC)\hat{\mathbf{x}}_{t|t-1} - K\mathbf{y}(t), \;\; \hat{\mathbf{s}}(t-d|t) = J_{d+1}\hat{\mathbf{x}}_{t+1|t},$$
$$\Sigma = A\left(I_n + \Sigma C^* R^{-1} C\right)^{-1}\Sigma A^* + BB^*, \;\; K = -A\Sigma C^*\left(R + C\Sigma C^*\right)^{-1}.$$

As A is a stability matrix, the solution $\Sigma \geq 0$ exists and is stabilizing.

Example 5.12 shows that if the signals to be estimated are output of the form $\mathbf{z}(t) = J\mathbf{x}(t+1)$, then the optimal output estimator is the same as the optimal state estimator. Hence, the Kalman filter serves as the optimal linear estimator for both state and output estimation, provided that both estimators are restricted to being strictly causal. The following example regards the application of Kalman filtering to system identification.

Example 5.13. Suppose that the system is described by an ARMA model

$$y(t) = \sum_{k=1}^{n} \alpha_k y(t-k) + \sum_{k=1}^{m} \beta_k u(t-k) + \eta(t),$$

where $\{\eta(t)\}$ is white and Gaussian. Suppose that

$$\mathbf{h} = \begin{bmatrix} \alpha_1 & \cdots & \alpha_n & \beta_1 & \cdots & \beta_m \end{bmatrix}^*$$

is also Gaussian with a priori mean $\overline{\mathbf{h}}$ and covariance P. It is reasonable to assume that $\{\eta(t)\}$ and $\mathbf{h}$ are independent. The goal of system identification is to estimate the true value of $\mathbf{h}$ based on measured data $\{y(t)\}$ and the deterministic input data $\{u(t)\}$. For this purpose, consider the fictitious state-space equation

$$\mathbf{x}(t+1) = \mathbf{x}(t) = \mathbf{h}, \quad y(t) = \mathbf{q}(t)\mathbf{x}(t) + \eta(t), \tag{5.57}$$

where $\mathbf{q}(t) = \begin{bmatrix} y(t-1) & \cdots & y(t-n) & u(t-1) & \cdots & u(t-m) \end{bmatrix}$ is a row vector. This corresponds to the random process model (5.19) with $A_t = I_{n+m}, B_t = \mathbf{0}_{n+m}, \mathbf{C}_t = \mathbf{q}(t)$ and $D_t = [\text{cov}\{\eta(t)\}]^{1/2}$. An application of the Kalman filtering with $\hat{\mathbf{h}}_t = \hat{\mathbf{x}}_{t+1|t}$ yields the estimator:

$$\hat{\mathbf{h}}_t = \hat{\mathbf{h}}_{t-1} + \Sigma_t\mathbf{q}(t)\left[R_t + \mathbf{q}(t)\Sigma_t\mathbf{q}^*(t)\right]^{-1}\left[y(t) - \mathbf{q}(t)\hat{\mathbf{h}}_{t-1}\right],$$
$$\Sigma_{t+1} = \Sigma_t - \Sigma_t\mathbf{q}(t)^*\left[R_t + \mathbf{q}(t)\Sigma_t\mathbf{q}(t)^*\right]^{-1}\mathbf{q}(t)\Sigma_t, \qquad \Sigma_0 = P, \tag{5.58}$$

where $\hat{\mathbf{h}}_{-1} = \bar{\mathbf{h}}$ and $R_t = \text{cov}\{\eta(t)\}$. It is noted that Σ_{t+1} is not truly the error covariance associated with $\hat{\mathbf{h}}_t$ by the fact that $C_t = \mathbf{q}(t)$ is random to which the Kalman filter in Theorem 5.7 does not apply. Hence, the estimator in (5.58) is not an MMSE estimator for $\mathbf{h}$. This also explains why the MMSE estimator in (5.58) is nonlinear in terms of the observed data $\{\mathbf{y}(k)\}_{k=t-n}^{t-1}$ in $C_t = \mathbf{q}(t)$, even though the Kalman filter is linear. On the other hand, if at time t, $\mathbf{y}(k)$ is treated as deterministic containing the realization of the noise process $\eta(k)$ for $k < t$, then the estimator in (5.58) can be interpreted as an MMSE estimator for $\mathbf{h}$. However, this interpretation is rather far-fetched. A more interesting case is when $n = 0$, i.e., the system is an MA model. In this case, the row vector $\mathbf{q}(t)$ does not contain any measured output data $\{y(k)\}_{k=0}^{t}$. Since the input $\{\mathbf{u}(t)\}$ is deterministic, the estimator in (5.58) becomes linear, and thus, the estimator in (5.58) is truly the MMSE estimator for $\mathbf{h}$ outperforming any other system identification algorithms for FIR models. Moreover, Σ_{t+1} is truly the error covariance associated with $\hat{\mathbf{h}}_t$. If the joint Gaussian assumption is dropped, then the estimator in (5.58) is the linear MMSE estimator outperforming any other linear algorithms for identification of FIR models. Nonetheless, such claims do not hold for the case $n > 0$ or the IIR models.

5.1.4 Output Estimators

The Kalman filter estimates $\mathbf{x}(t)$ or $\mathbf{x}(t+1)$ based on observations $\mathscr{Y}_t = \{\mathbf{y}(k)\}_{k=0}^{t}$ at time $t \geq 0$. A more practical problem is the output estimation or estimation of the linear combination of the state vector and the process noise at time t based on observation $\mathscr{Y}_t$. Such an estimation problem is described by the following state-space model:

$$\begin{bmatrix} \mathbf{x}(t+1) \\ \mathbf{z}(t) \\ \mathbf{y}(t) \end{bmatrix} = \begin{bmatrix} A_t & B_t \\ C_{1t} & D_{1t} \\ C_{2t} & D_{2t} \end{bmatrix} \begin{bmatrix} \mathbf{x}(t) \\ \mathbf{v}(t) \end{bmatrix}, \tag{5.59}$$

where $\mathbf{v}(t)$ is the white noise process as in (5.20), the initial condition $\mathbf{x}(0) = \mathbf{x}_0$ is a random vector, and $\mathbf{z}(t)$ is the signal to be estimated. The goal is design of a linear estimator represented by state-space realization $\left(\widehat{A}_t, \widehat{B}_t, \widehat{C}_t, \widehat{D}_t\right)$ such that $\hat{\mathbf{z}}_{t|t}$, the estimate of $\mathbf{z}(t)$ based on the observation $\{\mathbf{y}(k)\}_{k=0}^{t}$, minimizes the error variance of $\mathbf{e}_z(t) = \mathbf{z}(t) - \hat{\mathbf{z}}_{t|t}$. Figure 5.4 shows the schematic diagram for output estimation that is different from the state estimation problem in Kalman filtering. In the special case of $C_{1t} = I$ and $D_{1t} = \mathbf{0}$ for all t, it aims to estimate $\mathbf{x}(t)$, based on observation $\{\mathbf{y}(k)\}_{k=0}^{t}$. On the other hand, if $C_{1t} = \mathbf{0}$ and $D_{1t} = I$ for all t, then it is an estimator for the noise process $\mathbf{v}(t)$ based on observations $\{\mathbf{y}(k)\}_{k=0}^{t}$. Therefore, the output estimation problem is more versatile and more useful in engineering practice. It turns out that among all linear estimators, the MMSE estimator can be obtained from the Kalman filter with some minor modifications.

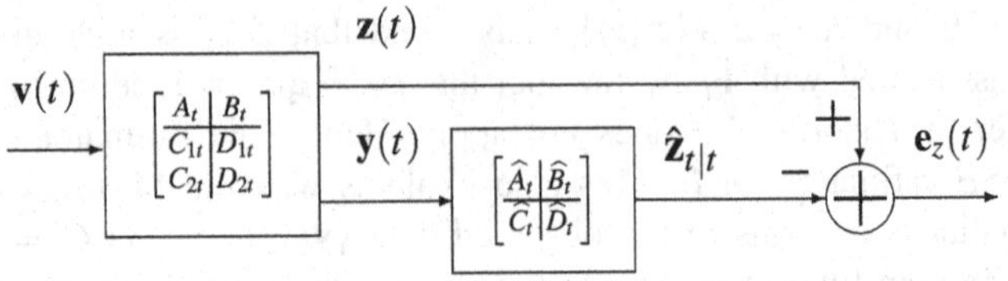

Fig. 5.4 Schematic diagram for linear output estimator

Theorem 5.14. *Let the state-space model be as in (5.59) with $\{\mathbf{v}(t)\}$ the white noise satisfying (5.20). Suppose that $R_t = D_{2t}D_{2t}^*$ is nonsingular and $\mathbf{x}(0) = \mathbf{x}_0$ has mean $\overline{\mathbf{x}}_0$ and covariance P_0 which is independent of $\{\mathbf{v}(t)\}$. Let $(\tilde{A}_t, \tilde{B}_t)$ be as in (5.39), i.e.,*

$$\tilde{A}_t = A_t - B_t D_{2t}^* R_t^{-1} C_{2t}, \quad \tilde{B}_t = B_t \left[I - D_{2t}^* R_t^{-1} D_{2t} \right]. \tag{5.60}$$

Then the linear MMSE estimation of $\mathbf{z}(t)$ based on observation $\{\mathbf{y}(k)\}_{k=0}^t$ is given recursively by

$$\begin{aligned} \hat{\mathbf{x}}_{t+1|t} &= [A_t + K_t C_{2t}]\,\hat{\mathbf{x}}_{t|t-1} - K_t \mathbf{y}(t), \quad \hat{\mathbf{x}}_{0|-1} = \overline{\mathbf{x}}_0, \\ \hat{\mathbf{z}}_{t|t} &= [C_{1t} + L_t C_{2t}]\,\hat{\mathbf{x}}_{t|t-1} - L_t \mathbf{y}(t), \end{aligned} \tag{5.61}$$

where the Kalman gains K_t and L_t are given by

$$\begin{bmatrix} K_t \\ L_t \end{bmatrix} = -\begin{bmatrix} A_t \Sigma_t C_{2t}^* + B_t D_{2t}^* \\ C_{1t} \Sigma_t C_{2t}^* + D_{1t} D_{2t}^* \end{bmatrix} (R_t + C_{2t} \Sigma_t C_{2t}^*)^{-1}, \tag{5.62}$$

$$\Sigma_{t+1} = \tilde{A}_t \left(I_n + \Sigma_t C_{2t}^* R_t^{-1} C_{2t} \right)^{-1} \Sigma_t \tilde{A}_t^* + \tilde{B}_t \tilde{B}_t^*, \quad \Sigma_0 = P_0. \tag{5.63}$$

Proof. The trick of the proof is to convert the output estimation to Kalman filtering. For simplicity, assume that $B_t D_{2t}^* = \mathbf{0}$ and $D_{1t} D_{2t}^* = \mathbf{0}$ for each t. Augment the state vector

$$\check{\mathbf{x}}(t) = \begin{bmatrix} \mathbf{x}(t) \\ \mathbf{z}(t-1) \end{bmatrix}, \quad \check{\mathbf{x}}(0) = \begin{bmatrix} \mathbf{x}_0 \\ \mathbf{0} \end{bmatrix}.$$

Its associated a priori covariance is $\check{P}_0 = \text{diag}(P_0, \mathbf{0})$. There holds

$$\check{\mathbf{x}}(t+1) = \check{A}_t \check{\mathbf{x}}(t) + \check{B}_t \mathbf{v}(t), \quad \mathbf{y}(t) = \check{C}_t \check{\mathbf{x}}(t) + \check{D}_t \mathbf{v}(t) \tag{5.64}$$

by straightforward calculation where $\check{D}_t = D_{2t}$ and

$$\check{A}_t = \begin{bmatrix} A_t & \mathbf{0} \\ C_{1t} & \mathbf{0} \end{bmatrix}, \quad \check{B}_t = \begin{bmatrix} B_t \\ D_{1t} \end{bmatrix}, \quad \check{C}_t = \begin{bmatrix} C_{2t} & \mathbf{0} \end{bmatrix}. \tag{5.65}$$

Since $\mathbf{x}(t+1)$ and $\mathbf{z}(t)$ are subsumed in $\breve{\mathbf{x}}(t+1)$, the MMSE output estimator for $\mathbf{z}(t)$ is equivalent to the MMSE estimation of $\breve{\mathbf{x}}(t+1)$, among all linear estimators, based on $\{\mathbf{y}(k)\}_{k=0}^{t}$. Recall the discussion after Example 5.12. Hence, the optimal solution is the Kalman filter in Theorem 5.7 for the random process in (5.64) due to $\breve{B}_t\breve{D}_t^* = \mathbf{0}$ by the hypothesis $B_tD_{2t}^* = \mathbf{0}$ and $D_{1t}D_{2t}^* = \mathbf{0}$ leading to the DRE and Kalman gain:

$$\breve{\Sigma}_{t+1} = \breve{A}_t\left(I + \breve{\Sigma}_t\breve{C}_t^*R_t^{-1}\breve{C}_t\right)^{-1}\breve{\Sigma}_t\breve{A}_t^* + \breve{B}_t\breve{B}_t^*, \tag{5.66}$$

$$\breve{K}_t = -\breve{A}_t\breve{\Sigma}_t\breve{C}_t^*\left(R_t + \breve{C}_t\breve{\Sigma}_t\breve{C}_t^*\right)^{-1}. \tag{5.67}$$

Partition $\breve{\Sigma}_k$ into a 2×2 block matrix with Σ_k the $(1,1)$ block which is the error covariance for $\hat{\mathbf{x}}_{k|k-1}$ at $k=t$ and $k=t+1$. Then the $(1,1)$ block of the DRE (5.66) and the Kalman gain in (5.67) are obtained as

$$\Sigma_{t+1} = A_t\left(I + \Sigma_t C_{2t}^* R_t^{-1} C_{2t}\right)^{-1}\Sigma_t A_t^* + B_t B_t^*,$$

$$\breve{K}_t = \begin{bmatrix} K_t \\ L_t \end{bmatrix} = -\begin{bmatrix} A_t\Sigma_t C_{2t}^*, \\ C_{1t}\Sigma_t C_{2t}^* \end{bmatrix}\left(R_t + C_{2t}\Sigma_t C_{2t}^*\right)^{-1}$$

which are the same as in (5.63) and (5.62), respectively, for the case $B_tD_{2t}^* = \mathbf{0}$. It follows that the Kalman filter for $\breve{\mathbf{x}}(t+1)$ in the system (5.64) is given by

$$\begin{bmatrix} \hat{\mathbf{x}}_{t+1|t} \\ \hat{\mathbf{z}}_{t|t} \end{bmatrix} = \left(\breve{A}_t + \breve{K}_t\breve{C}_t\right)\begin{bmatrix} \hat{\mathbf{x}}_{t|t-1} \\ \hat{\mathbf{z}}_{t-1|t-1} \end{bmatrix} - \breve{K}_t\mathbf{y}(t),$$

$$= \begin{bmatrix} A_t + K_tC_{2t} \\ C_{1t} + L_tC_{2t} \end{bmatrix}\hat{\mathbf{x}}_{t|t-1} - \begin{bmatrix} K_t \\ L_t \end{bmatrix}\mathbf{y}(t),$$

by substitution of the expressions in (5.65). The above are the same as the linear MMSE output estimator in (5.61) for the case $B_tD_{2t}^* = \mathbf{0}$. If $B_tD_{2t}^* \neq \mathbf{0}$ and $D_{1t}D_{2t}^* \neq \mathbf{0}$, the same procedure can be carried out using the Kalman filtering results in Corollary 5.1 that will lead to the linear MMSE estimator in (5.61) with the Kalman gains in (5.62). The details are omitted here and left as an exercise (Problem 5.15). □

Theorem 5.14 indicates that the optimal output estimate $\hat{\mathbf{z}}_{t|t}$ is a linear function of the optimal state estimate $\hat{\mathbf{x}}_{t|t-1}$ in light of (5.61) and the associated error covariance is irrelevant to C_{1t} and D_{1t}. In this sense, optimal output estimation is equivalent to optimal state estimation. The realization of the linear MMSE output estimator in (5.61) is given by

$$\left(\widehat{A}_t, \widehat{B}_t, \widehat{C}_t, \widehat{D}_t\right) = \left(A_t + K_tC_{2t}, -K_t, C_{1t} + L_tC_{2t}, -L_t\right)$$

which has the same order as the original state-space model in (5.59). Its input is $\mathbf{y}(t)$, and output is $\hat{\mathbf{z}}_{t|t}$ as shown in Fig. 5.4. In light of the Kalman filtering, the linear estimator in (5.61) is optimal among all linear estimators with arbitrary orders. If, in addition, the noise process $\{\mathbf{v}(t)\}$ and the initial condition $\mathbf{x}_0$ are independent, and jointly Gaussian, then the linear estimator in (5.61) is optimal among all possible output estimators, including those nonlinear ones. The next example is related to Example 5.12, and illustrates the utility of output estimators.

Example 5.15. A commonly seen state-space model in applications is

$$\mathbf{x}(t) = A_t\mathbf{x}(t-1) + B_t\mathbf{v}(t), \quad \mathbf{y}(t) = C_t\mathbf{x}(t) + D_t\mathbf{v}(t) \tag{5.68}$$

that appears differently from the ones discussed in this chapter thus far. It will be shown that the results on output estimation are applicable to derive the optimal state estimator for the model in (5.68).

For simplicity, assume that $B_{t+1}D_t^* = \mathbf{0}$ and $R_t = D_tD_t^* > \mathbf{0}$. Under the same hypotheses on white and Gaussian $\mathbf{v}(t)$, and on the initial state $\mathbf{x}(0)$ that has mean $\overline{\mathbf{x}}_0$ and covariance P_0, Theorem 5.14 can be used to derive the equations for measurement update:

$$\hat{\mathbf{x}}_{t|t} = [I + L_tC_t]\,\hat{\mathbf{x}}_{t|t-1} - L_t\mathbf{y}(t), \tag{5.69}$$

$$\Sigma_{t|t} = \Sigma_t - \Sigma_tC_t^*\left(R_t + C_t\Sigma_tC_t^*\right)^{-1}C_t\Sigma_t, \tag{5.70}$$

initialized by $\hat{\mathbf{x}}_{0|-1} = \overline{\mathbf{x}}_0$ and covariance $\Sigma_0 = P_0$ where

$$L_t = -\Sigma_tC_t^*\left(R_t + C_t\Sigma_tC_t^*\right)^{-1}. \tag{5.71}$$

Recall $\Sigma_t = \Sigma_{t|t-1}$. Moreover, the time update equations can be obtained as

$$\hat{\mathbf{x}}_{t+1|t} = A_{t+1}\hat{\mathbf{x}}_{t|t}, \quad \Sigma_{t+1} = A_{t+1}\Sigma_{t|t}A_{t+1}^* + B_{t+1}B_{t+1}^*. \tag{5.72}$$

Specifically, the state-space model in (5.68) can be rewritten as

$$\mathbf{x}(t+1) = A_{t+1}\mathbf{x}(t) + B_{t+1}\mathbf{v}(t+1), \quad \mathbf{y}(t) = C_t\mathbf{x}(t) + D_t\mathbf{v}(t). \tag{5.73}$$

Because $B_{t+1}\mathbf{v}(t+1)$ and $D_t\mathbf{v}(t)$ are uncorrelated, replacing A_t by A_{t+1} and B_t by B_{t+1} in Theorem 5.14 leads to the optimal state estimator or one-step predictor $\hat{\mathbf{x}}_{t+1|t} = \mathrm{E}\{\mathbf{x}(t+1)|\mathscr{Y}_t\}$:

$$\hat{\mathbf{x}}_{t+1|t} = [A_{t+1} + K_tC_t]\,\hat{\mathbf{x}}_{t|t-1} - K_t\mathbf{y}(t) \tag{5.74}$$

initialized by $\hat{\mathbf{x}}_{0|-1} = \overline{\mathbf{x}}_0$ where $K_t = A_{t+1}L_t$ is the Kalman gain. In addition with $\Sigma_0 = P_0$, the associated error covariance Σ_{t+1} for $t \geq 0$ can be computed according to the DRE

$$\Sigma_{t+1} = A_{t+1}\Sigma_t \left(I + C_t^* R_t^{-1} C_t \Sigma_t\right)^{-1} A_{t+1}^* + B_{t+1}B_{t+1}^*. \tag{5.75}$$

For $\hat{\mathbf{x}}_{t|t} = \mathrm{E}\{\mathbf{x}(t)|\mathscr{Y}_t\}$, setting $\mathbf{z}(t) = \mathbf{x}(t)$ leads to the optimal estimator in (5.69) by $C_{1t} = I$, $D_{1t} = \mathbf{0}$, and $C_{2t} = C_t$. The fact of $K_t = A_{t+1}L_t$ and optimal estimate in (5.69) yield time update equations in (5.72). A comparison of Σ_{t+1} in (5.72) with the one in (5.75) shows

$$\Sigma_{t|t} = \Sigma_t \left(I + C_t^* R_t^{-1} C_t \Sigma_t\right)^{-1} \tag{5.76}$$

that is the same error covariance in (5.70).

The linear MMSE output estimator as in Theorem 5.14 has the same stability properties as those of the Kalman filter by the fact that they have the identical covariance matrices for the state vectors. Hence, all the results in the previous subsection apply to the linear MMSE output estimators, which will not be repeated here except for the following. Suppose that the state-space realization in (5.59) is independent of time t. Let $\tilde{A} = A - BD_2^* R^{-1} C_2$, $R = D_2 D_2^*$, and $\tilde{B} = B\left(I - D_2^* R^{-1} D_2\right)$. If the ARE

$$\Sigma = \tilde{A}\left(I_n + \Sigma C_2^* R^{-1} C_2\right)^{-1} \Sigma \tilde{A}^* + \tilde{B}\tilde{B}^* \tag{5.77}$$

has a stabilizing solution $\Sigma \geq 0$, then the output estimator in (5.61) is asymptotically stable, in light of Theorem 5.10. In this case, the output estimator in (5.61) converges asymptotically to the following time-invariant system

$$\begin{aligned} \hat{\mathbf{x}}_{t+1|t} &= [A + KC_2]\,\hat{\mathbf{x}}_{t|t-1} - K\mathbf{y}(t), \quad \hat{\mathbf{x}}_{0|-1} = \bar{\mathbf{x}}_0, \\ \hat{\mathbf{z}}_{t|t} &= [C_1 + LC_2]\,\hat{\mathbf{x}}_{t|t-1} - L\mathbf{y}(t), \end{aligned} \tag{5.78}$$

where K and L have the same expressions as in (5.62) with the time index t removed. One may employ the time-invariant estimator (5.78) directly for output estimation which can be computed off-line in order to reduce the computational complexity in its implementation. Clearly, the estimator in (5.78) admits the transfer matrix, denoted by $\mathbf{F}(z)$ and given by

$$\mathbf{F}(z) = -\left[L + (C_1 + LC_2)(zI_n - A - KC_2)^{-1} K\right]. \tag{5.79}$$

This section is concluded with an example on Wiener filtering (see Fig. 5.5).

Example 5.16. (Wiener filtering) Consider the signal model as in Fig. 4.5 where $\mathbf{v}_1(t)$ and $\mathbf{v}_2(t)$ are independent white noises of zero mean and identity covariance, and $\mathbf{G}_1(z)$ and $\mathbf{G}_2(z)$ are causal and stable rational transfer matrices. Wiener filtering aims to design a LTI filter $\mathbf{W}(z)$ which estimates $\mathbf{z}(t-m)$, the output of $\mathbf{G}_1(z)$, based on observations $\mathbf{y}(k)$ for all $k \leq t$ and some integer m. It is termed as smoothing, if $m > 0$ (estimation of the past output), filtering, if $m = 0$ (estimation of the present output), and prediction, if $m < 0$ (estimation of the future output). It is claimed

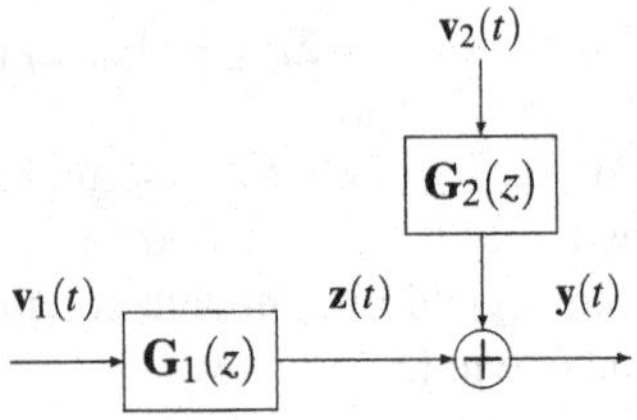

Fig. 5.5 Signal model for Wiener filtering

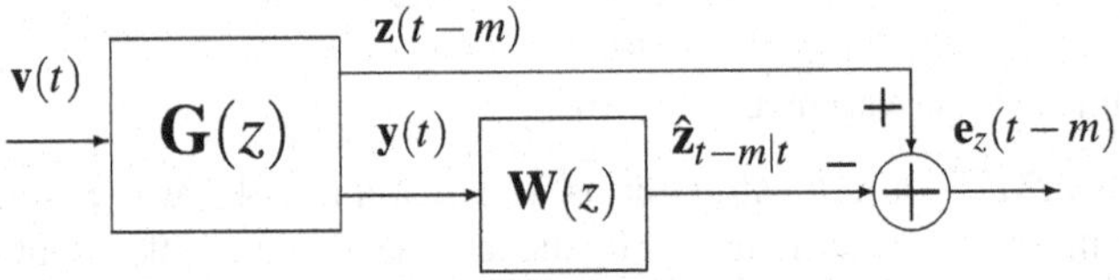

Fig. 5.6 Wiener filter as an output estimator

that these three estimation problems can all be cast into the output estimation as illustrated in the figure below, provided that $z^{-m}\mathbf{G}_1(z)$ is causal.

Indeed, if $z^{-m}\mathbf{G}_1(z)$ is causal, then

$$\mathbf{G}(z) = \begin{bmatrix} z^{-m}\mathbf{G}_1(z) & \mathbf{0} \\ \mathbf{G}_1(z) & \mathbf{G}_2(z) \end{bmatrix} = \left[\begin{array}{c|c} A & B \\ \hline C_1 & D_1 \\ C_2 & D_2 \end{array}\right]$$

for some realization matrices. Let $\mathbf{v}(t) = \begin{bmatrix} \mathbf{v}_1(t) & \mathbf{v}_2(t) \end{bmatrix}'$. Then Wiener filtering can be converted into output estimation as shown in Fig. 5.6. Consequently, the results on output estimation can be applied to design the stationary optimal estimator represented by $\mathbf{W}(z)$ which is the required Wiener filter. If $z^{-m}\mathbf{G}_1(z)$ is not causal, decompose

$$z^{-m}\mathbf{G}_1(z) = \mathbf{G}_A(z) + \mathbf{G}_C(z)$$

with $\mathbf{G}_C(z)$ causal and $\mathbf{G}_A(z)$ strictly anticausal. A state-space realization of $\mathbf{G}(z)$ can again be obtained with $z^{-m}\mathbf{G}_1(z)$ replaced by $\mathbf{G}_C(z)$. It can be shown (Problem 5.19 in Exercises) that the Wiener filter does not depend on $\mathbf{G}_A(z)$.

5.2 Minimum Variance Control

A common control problem is *disturbance rejection*. Engineering systems are designed to operate in various environments where unknown and random disturbances are unavoidable and detrimental to the system performances. Disturbance rejection aims to design effective control laws that suppress the devastating effects of the disturbances and ensure that the system operates as desired. Often it results in feedback control laws. This section investigates the case when disturbances are white noise processes and the variance is the performance measure for control system design.

The system under consideration is described by the state-space model

$$\mathbf{x}(t+1) = A_t\mathbf{x}(t) + B_{1t}\mathbf{v}(t) + B_{2t}\mathbf{u}(t), \quad \mathbf{z}(t) = C_t\mathbf{x}(t) + D_t\mathbf{u}(t), \tag{5.80}$$

where $\mathbf{v}(t)$ is the white noise disturbance with the same statistics as in (5.20), $\mathbf{u}(t)$ is the control input signal, and $\mathbf{z}(t)$ is the output to be controlled. The initial condition $\mathbf{x}(0) = \mathbf{x}_0$ is assumed to be random and has mean $\bar{\mathbf{x}}_0$ and covariance P_0. Due to the random nature and the dynamic impact of the initial condition, $\mathbf{x}_0$ is accounted as part of the disturbance. The objective is to design a control law $\mathscr{U}_T = \{\mathbf{u}(t)\}_{t=0}^{T-1}$ that minimizes

$$V_{[0,T)} = \sum_{t=0}^{T-1} V_t, \quad V_t = \mathrm{E}\left\{\|\mathbf{z}(t)\|^2 \,|\, \mathscr{U}_T\right\} = \mathrm{Tr}\left(\mathrm{E}\{\mathbf{z}(t)\mathbf{z}(t)^* | \, \mathscr{U}_T\}\right) \tag{5.81}$$

with V_t the variance of the controlled output.

The aforementioned control problem is very different from the estimation problem in the previous section, but the two are closely related. In fact, there exists a duality relation between the linear minimum variance control and the linear minimum variance estimation. As a fortiori, optimal disturbance rejection can be obtained from the Kalman filtering. However, such a derivation may blur out the distinctions between control and estimation and is thus not adopted in this text. Instead, linear minimum variance control will be derived independently. The duality will be interpreted at a later stage to deepen the understanding of the resultant optimal feedback control.

5.2.1 *Linear Quadratic Regulators*

Before tackling the problem of disturbance rejection, design of linear quadratic regulators (LQRs) will be studied. The LQR is a deterministic control problem. Yet, its solution coincides with that for disturbance rejection. Let

$$\mathbf{x}(t+1) = A_t\mathbf{x}(t) + B_t\mathbf{u}(t), \quad \mathbf{x}(0) = \mathbf{x}_0 \neq 0. \tag{5.82}$$

It is desirable to regulate the state vector $\mathbf{x}(T)$ to the origin $\mathbf{0}$ in finite time $T > 0$ through some suitable control action $\{\mathbf{u}(t)\}_{t=0}^{T-1}$. However, the exact regulation to $\mathbf{x}(T) = \mathbf{0}$ may not be feasible in some finite time T. Even if it is feasible, the cost of control input can be prohibitively high. Hence, it is appropriate to consider the quadratic performance index

$$J_T(t_0) = \mathbf{x}(T)^* Q_T \mathbf{x}(T) + \sum_{t=t_0}^{T-1} \mathbf{x}(t)^* Q_t \mathbf{x}(t) + \mathbf{u}(t)^* R_t \mathbf{u}(t) \tag{5.83}$$

for $t_0 = 0$ which provides the mechanism of trade-offs between the regulation of $\mathbf{x}(t)$ and the energy constraint on $\mathbf{u}(t)$. The weighting matrix $Q_t = Q_t^* \geq 0$ represents the penalty on the state vector and thus the quality of regulation, and $R_t = R_t^* > 0$ shows the penalty on the control input and thus the measure of the energy at time t. For convenience, $J_T = J_T(0)$ is used.

The hypothesis on the weighting matrices in J_T implies that

$$Q_t = C_t^* C_t, \quad R_t = D_t^* D_t, \quad D_t^* C_t = \mathbf{0} \tag{5.84}$$

for some C_t of dimension $p \times n$ and D_t of dimension $p \times m$ with n the order of the state-space model in (5.82) and m the dimension of the control input. Let $\mathbf{z}(t) = C_t\mathbf{x}(t) + D_t\mathbf{u}(t)$. Then together with the state-space equation (5.82),

$$\begin{bmatrix} \mathbf{x}(t+1) \\ \mathbf{z}(t) \end{bmatrix} = \begin{bmatrix} A_t & B_t \\ C_t & D_t \end{bmatrix} \begin{bmatrix} \mathbf{x}(t) \\ \mathbf{u}(t) \end{bmatrix} \tag{5.85}$$

which represents the system model. The decomposed LQR cost at time t is

$$\|\mathbf{z}(t)\|^2 = \mathbf{x}(t)^* Q_t \mathbf{x}(t) + \mathbf{u}(t)^* R_t \mathbf{u}(t).$$

For any square matrix $X_{t+1} = X_{t+1}^* \geq \mathbf{0}$ with size $n \times n$, let

$$W_{t+1} = \mathbf{x}(t+1)^* X_{t+1} \mathbf{x}(t+1) + \|\mathbf{z}(t)\|^2$$

be the candidate Lyapunov function for $0 \leq t < T$. Then

$$\begin{aligned} W_{t+1} &= \mathbf{x}(t+1)^* X_{t+1} \mathbf{x}(t+1) + \mathbf{z}(t)^* \mathbf{z}(t) \\ &= \begin{bmatrix} \mathbf{x}(t+1)^* & \mathbf{z}(t)^* \end{bmatrix} \begin{bmatrix} X_{t+1} & \mathbf{0} \\ \mathbf{0} & I \end{bmatrix} \begin{bmatrix} \mathbf{x}(t+1) \\ \mathbf{z}(t) \end{bmatrix} \\ &= \left(\begin{bmatrix} A_t & B_t \\ C_t & D_t \end{bmatrix} \begin{bmatrix} \mathbf{x}(t) \\ \mathbf{u}(t) \end{bmatrix} \right)^* \begin{bmatrix} X_{t+1} & \mathbf{0} \\ \mathbf{0} & I \end{bmatrix} \left(\begin{bmatrix} A_t & B_t \\ C_t & D_t \end{bmatrix} \begin{bmatrix} \mathbf{x}(t) \\ \mathbf{u}(t) \end{bmatrix} \right) \\ &= \begin{bmatrix} \mathbf{x}(t)^* & \mathbf{u}(t)^* \end{bmatrix} \begin{bmatrix} A_t^* X_{t+1} A_t + Q_t & A_t^* X_{t+1} B_t \\ B_t^* X_{t+1} A_t & R_t + B_t^* X_{t+1} B_t \end{bmatrix} \begin{bmatrix} \mathbf{x}(t) \\ \mathbf{u}(t) \end{bmatrix}, \end{aligned}$$

where (5.85) and the relations $R_t = D_t^* D_t$, $Q_t = C_t^* C_t$, and $D_t^* C_t = \mathbf{0}$ are used. Let $\Psi_t = A_t^* X_{t+1} A_t + Q_t, \Omega_t = B_t^* X_{t+1} A_t$, and $\Theta_t = R_t + B_t^* X_{t+1} B_t$. Because $\Theta_t > 0$ by $R_t > 0$, the Schur decomposition

$$\begin{bmatrix} \Psi_t & \Omega_t^* \\ \Omega_t & \Theta_t \end{bmatrix} = \begin{bmatrix} I & \Omega_t^*\Theta_t^{-1} \\ 0 & I \end{bmatrix} \begin{bmatrix} X_t & 0 \\ 0 & \Theta_t \end{bmatrix} \begin{bmatrix} I & 0 \\ \Theta_t^{-1}\Omega_t & I \end{bmatrix}$$

holds (refer to Problem 5.8 in Exercises) with $X_t = \left(\Psi_t - \Omega_t^*\Theta_t^{-1}\Omega_t\right)$ the Schur complement. The Lyapunov function candidate W_{t+1} can now be written into

$$W_{t+1} = \begin{bmatrix} \mathbf{x}(t)^* & (\mathbf{u}(t) - F_t\mathbf{x}(t))^* \end{bmatrix} \begin{bmatrix} X_t & 0 \\ 0 & \Theta_t \end{bmatrix} \begin{bmatrix} \mathbf{x}(t) \\ \mathbf{u}(t) - F_t\mathbf{x}(t) \end{bmatrix},$$

where $F_t = -\Theta_t^{-1}\Omega_t = -\left(R_t + B_t^*X_{t+1}B_t\right)^{-1} B_t^*X_{t+1}A_t$ and X_t satisfies

$$\begin{aligned} X_t &= A_t^*X_{t+1}A_t + Q_t - A_t^*X_{t+1}B_t\left(R_t + B_t^*X_{t+1}B_t\right)^{-1} B_t^*X_{t+1}A_t \\ &= A_t^*X_{t+1}\left(I_n + B_tR_t^{-1}B_t^*X_{t+1}\right)^{-1} A_t + Q_t \end{aligned} \tag{5.86}$$

which is a DRE dual to the estimation DRE as in Theorem 5.7. Because $\Theta_t > 0$, the minimum value of W_{t+1} is achieved by setting

$$\mathbf{u}(t) = \mathbf{u}_{\rm opt}(t) := F_t\mathbf{x}(t), \quad F_t = -\left(R_t + B_t^*X_{t+1}B_t\right)^{-1} B_t^*X_{t+1}A_t \tag{5.87}$$

for which $W_{t+1} = \mathbf{x}(t)^*X_t\mathbf{x}(t)$ is the minimum possible.

Let $X_T = Q_T$ be the boundary condition and $\{X_t\}_{t=0}^{T-1}$ be the solution to the DRE in (5.86). Then an induction process can be applied to W_{t+1} with $t = T-1, T-2, \ldots, 0$ in the performance index (5.83) yielding

$$\begin{aligned} J_T &= \mathbf{x}(T)^*X_T\mathbf{x}(T) + \sum_{k=0}^{T-1} \|\mathbf{z}(k)\|^2 = W_T + \sum_{k=0}^{T-2} \|\mathbf{z}(k)\|^2 \\ &\geq \mathbf{x}(T-1)^*X_{T-1}\mathbf{x}(T-1) + \sum_{k=0}^{T-2} \|\mathbf{z}(k)\|^2 \\ &= W_{T-1} + \sum_{k=0}^{T-3} \|\mathbf{z}(k)\|^2 \geq \cdots \geq \mathbf{x}(0)^*X_0\mathbf{x}(0), \end{aligned}$$

where the fact $W_{t+1} \geq \mathbf{x}(t)^*X_t\mathbf{x}(t)$ is used to arrive at the lower bound of J_T. It is noted that the lower bound $\mathbf{x}_0^*X_0\mathbf{x}_0$ for J_T is achievable by employing the optimal control law $\left\{\mathbf{u}_{\rm opt}(t)\right\}_{t=0}^{T-1}$ as in (5.87) which constitutes the optimal solution to the LQR control problem. The above derivations are summarized into the following result.

Theorem 5.17. *Suppose that $Q_t \geq 0$ and $R_t > 0$ for $0 \leq t < T$. Let $\{X_t\}_{t=0}^{T-1}$ be the solution to the* DRE *in (5.86) with the boundary condition $X_T = Q_T \geq 0$. Then the optimal control law minimizing the performance index J_T in (5.83) and subject* *to the dynamic equation (5.82) is given by* $\{\mathbf{u}_{\text{opt}}(t)\}_{t=0}^{T-1}$ *in (5.87). The associated minimum performance index of* J_T *is* $\mathbf{x}_0^* X_0 \mathbf{x}_0$ *with* $\mathbf{x}_0 \neq \mathbf{0}$ *the initial condition of the state vector* $\mathbf{x}(t)$.

It is surprising that the optimal control law for the LQR problem is linear and static. After all, one might have expected nonlinear and dynamic control laws as the possible optimal solution. On the other hand, the feedback structure of the control law is more or less expected. The optimal feedback gains $\{F_t\}_{t=0}^{T-1}$ as in (5.87) are functions of the solution to the DRE in (5.86) which has the form of backward recursion. Such a backward recursion is deeply rooted in the *principle of optimality* which states that an optimal control policy over the time interval $[k,\ T)$ for $0 < k < T$ constitutes the optimal control policy over the time interval $[0,\ T)$ regardless of the states and control inputs before the time k. Indeed, denote $\mathscr{U}_{[t_0,t_f)} = \{\mathbf{u}(t)\}_{t=t_0}^{t_f-1}$ with $t_f > t_0$. There holds

$$\min_{\mathscr{U}_{[0,T)}} J_T(0) = \min_{\mathscr{U}_{[0,T)}} \{J_k(0) + J_T(k)\} = \min_{\mathscr{U}_{[0,k)}} \left\{ J_k(0) + \min_{\mathscr{U}_{[k,T)}} J_T(k) \right\}$$

for $0 < k < T$ in light of the causality of the state-space equation. That is, minimization of $J_T(0)$ can be carried out in two stages with stage 1 for minimization of $J_T(k)$ over all possible $\mathscr{U}_{[k,T)}$ and stage 2 for minimization of $J_k(0) + \min\{J_T(k) : \mathscr{U}_{[k,T)}\}$ over all possible $\mathscr{U}_{[0,k)}$. The repeated use of this two-stage method for $k = T-1$, $T-2, \ldots$ is what is called *dynamic programming* and is employed in the derivation of the optimal control law for the LQR problem. It is worth emphasizing that, by again causality,

$$\min_{\mathscr{U}_{[0,T)}} J_T(0) \neq \min_{\mathscr{U}_{[k,T)}} \left\{ J_T(k) + \min_{\mathscr{U}_{[0,k)}} J_k(0) \right\} = \min_{\mathscr{U}_{[k,T)}} J_T(k) + \min_{\mathscr{U}_{[0,k)}} J_k(0).$$

One should also realize that the principle of optimality applies to more broad optimal control problems beyond the LQR for linear state-space systems.

Example 5.18. As an application of the LQR control, consider the tracking problem for the state-space system (5.82) with output $\mathbf{y}(t) = C_t\mathbf{x}(t)$. Given a desired output trajectory $\tilde{\mathbf{y}}(\cdot)$, how does one design a state-feedback control law which minimizes the tracking error and consumes the minimum energy? A reasonable measure is the quadratic performance index

$$J_T = \sum_{t=0}^{T-1} [\mathbf{y}(t) - \tilde{\mathbf{y}}(t)]^* Q_1(t) [\mathbf{y}(t) - \tilde{\mathbf{y}}(t)] + \mathbf{x}(t)^* Q_2(t)\mathbf{x}(t) + \mathbf{u}(t)^* R_t \mathbf{u}(t), \quad (5.88)$$

where $R_t > 0$, $Q_1(t) > 0$, and $Q_2(t) = I - C_t^* (C_t C_t^*)^{-1} C_t$ for all t assuming that rows of C_t are linearly independent. Let the dimension of $\mathbf{y}(t)$ be $p < n$ with n the dimension of the state vector $\mathbf{x}(t)$. A small tracking error imposes only the p

constraints on the state vector or in the subspace spanned by p columns of C_t^*. The weighting matrix $Q_2(t)$ regulates the state vectors with the remaining $(n-p)$ constraints or in the null space of C_t. The weighting factors can be changed through adjusting $Q_1(t)$ and R_t. The terminal penalty $Q_1(T)$ is omitted for convenience. The desired output trajectory $\tilde{\mathbf{y}}(\cdot)$ is assumed to be the output of a LTI model

$$\mathbf{w}(t+1) = A_w\mathbf{w}(t), \quad \tilde{\mathbf{y}}(t) = H\mathbf{w}(t).$$

Such a model includes step functions, sinusoidal signals, and their linear combinations. Together with the system model (5.82), there holds

$$\tilde{\mathbf{x}}(t+1) = \tilde{A}_t\mathbf{x}(t) + \tilde{B}_t\mathbf{u}(t), \quad \delta\mathbf{y}(t) = \tilde{C}_t\tilde{\mathbf{x}}(t),$$

where $\tilde{\mathbf{x}}(t) = \begin{bmatrix} \mathbf{x}(t)^* & \mathbf{w}(t)^* \end{bmatrix}^*, \delta\mathbf{y}(t) = \mathbf{y}(t) - \tilde{\mathbf{y}}(t)$, and thus,

$$\tilde{A}_t = \begin{bmatrix} A_t & \mathbf{0} \\ \mathbf{0} & A_w \end{bmatrix}, \quad \tilde{B}_t = \begin{bmatrix} B_t \\ \mathbf{0} \end{bmatrix}, \quad \delta C_t = \begin{bmatrix} C_t & -H \end{bmatrix}.$$

It follows that the performance index J_T in (5.88) can be written into the same form as in (5.83) with $\mathbf{x}(t)$ replaced by $\tilde{\mathbf{x}}(t)$ leading to

$$Q_t = \begin{bmatrix} Q_2(t) + C_t^*Q_1(t)C_t & -C_t^*Q_1(t)H \\ -H^*Q_1(t)C_t & H^*Q_1(t)H \end{bmatrix}.$$

Hence, the LQR control law in Theorem 5.17 can be readily applied.

In reference to the LQR control, a similar solution approach can be adopted to the minimum variance control. Recall the expression of V_t in (5.81) and the augmented performance index

$$V_{[t_0,T)} = V_{t_0} + V_{t_1} + \cdots + V_{T-1}, \quad 0 \le t_0 < T. \tag{5.89}$$

Then the control law $\{\mathbf{u}(t)\}_{t=0}^{T-1}$ minimizing $V_{[0,T)}$ is reminiscent of LQR control law as shown next.

Theorem 5.19. *Consider the state-space system (5.82) where $\{\mathbf{v}(t)\}$ is white satisfying (5.20) and $\mathbf{x}(0) = \mathbf{x}_0$ is a random vector independent of $\{\mathbf{v}(t)\}$ with mean $\overline{\mathbf{x}}_0$ and covariance P_0. Suppose that $R_t = D_t^*D_t > 0$ and $D_t^*C_t = \mathbf{0}$ for all t. Let $Q_t = C_t^*C_t$, and $\{X_t\}_{t=0}^{T-1}$ be the solution to the* DRE *in (5.86) with the boundary condition $X_T = \mathbf{0}$ and $B_t = B_{2t}$. Then the optimal control law minimizing $V_{[0,T)}$ in (5.89) is the same as $\mathbf{u}_{\text{opt}}(t)$ in (5.87). Denote $\mathscr{U}_T = \{\mathbf{u}(t)\}_{t=0}^{T-1}$. Let $\Sigma_t = \mathrm{E}\{\mathbf{x}(t)\mathbf{x}(t)^* \mid \mathscr{U}_T\}$. Then $\Sigma_0 = P_0$ and*

$$\Sigma_{t+1} = (A_t + B_{2t}F_t)\Sigma_t(A_t + B_{2t}F_t)^* + B_{1t}B_{1t}^*. \tag{5.90}$$

The minimum variance for the controlled output over $[0,\,T)$ *is given by*

$$\min_{\mathscr{U}_T} V_{[0,T)} = \sum_{t=0}^{T-1} \mathrm{Tr}\{(C_t + D_t F_t)\Sigma_t (C_t + D_t F_t)^*\}. \tag{5.91}$$

Proof. It is noted that the state-space equation in (5.82) can be written as

$$\begin{bmatrix} \mathbf{x}(t+1) \\ \mathbf{z}(t) \end{bmatrix} = \begin{bmatrix} A_t & B_{2t} \\ C_t & D_t \end{bmatrix} \begin{bmatrix} \mathbf{x}(t) \\ \mathbf{u}(t) \end{bmatrix} + \begin{bmatrix} B_{1t} \\ \mathbf{0} \end{bmatrix} \mathbf{v}(t) \tag{5.92}$$

which is similar to (5.85). Let $W_{k+1} = \mathrm{E}\{\mathbf{x}(k+1)^* X_{k+1} \mathbf{x}(k+1) | \mathscr{U}_T\} + V_k$. Since the solution to the DRE (5.86) with the boundary condition $X_T = \mathbf{0}$ satisfies $X_t \geq \mathbf{0}$ (refer to Problem 5.21 in Exercises), W_{k+1} is nonnegative for $0 \leq k < T$. Similar to the derivation for the LQR control, there holds

$$\begin{aligned} W_{k+1} &= \mathrm{E}\{\mathbf{x}(k+1)^* X_{k+1} \mathbf{x}(k+1) + \mathbf{z}(k)^* \mathbf{z}(k) | \mathscr{U}_T\} \\ &= \mathrm{E}\left\{ \begin{bmatrix} \mathbf{x}(k+1)^* & \mathbf{z}(k)^* \end{bmatrix} \begin{bmatrix} X_{k+1} & \mathbf{0} \\ \mathbf{0} & I \end{bmatrix} \begin{bmatrix} \mathbf{x}(k+1) \\ \mathbf{z}(k) \end{bmatrix} \middle| \mathscr{U}_T \right\} \\ &= \mathrm{Tr}\{B_{1k}^* X_{k+1} B_{1k}\} + \mathrm{E}\left\{ \begin{bmatrix} \mathbf{x}(k) \\ \delta\mathbf{u}(k) \end{bmatrix}^* \begin{bmatrix} X_k & 0 \\ 0 & \Theta_k \end{bmatrix} \begin{bmatrix} \mathbf{x}(k) \\ \delta\mathbf{u}(k) \end{bmatrix} \middle| \mathscr{U}_T \right\} \end{aligned}$$

by the independence of $\mathbf{u}(k)$ and $\mathbf{x}(k)$ to $\mathbf{v}(k)$ where $\Theta_k = R_k + B_{2k}^* X_k B_{2k}$, $\delta\mathbf{u}(k) = \mathbf{u}(k) - \mathbf{u}_{\mathrm{opt}}(k)$, and $\mathbf{u}_{\mathrm{opt}}(k)$ is defined as in (5.87) with $B_k = B_{2k}$. It follows that $\mathbf{u}(k) = \mathbf{u}_{\mathrm{opt}}(k)$ minimizes W_{k+1} for $k = T-1, T-2, \ldots, 0$ and thus $V_{[0,T)}$ in (5.89). Indeed, with $X_T = \mathbf{0}$,

$$\begin{aligned} V_{[0,T)} &= \sum_{k=0}^{T-1} V_k = W_T + \sum_{k=0}^{T-2} \mathrm{E}\{\mathbf{z}(k)^* \mathbf{z}(k) | \mathscr{U}_T\} \\ &\geq \mathrm{Tr}\left\{B_{1(T-1)}^* X_T B_{1(T-1)}\right\} + W_{T-1} + \sum_{k=0}^{T-3} \mathrm{E}\{\mathbf{z}(k)^* \mathbf{z}(k) | \mathscr{U}_T\} \\ &\geq \cdots \geq \sum_{k=0}^{T-1} \mathrm{Tr}\{B_{1k}^* X_{k+1} B_{1k}\} + \mathrm{E}\{\mathbf{x}(0)^* X_0 \mathbf{x}(0) | \mathscr{U}_T\}. \end{aligned}$$

The lower bound for $V_{[0,T)}$ is achieved by $\mathscr{U}_T = \{\mathbf{u}_{\mathrm{opt}}(t)\}_{t=0}^{T-1}$ which also minimizes V_t for all $t \in [0,\,T)$ in light of the principle of optimality. The expression in (5.91) can be easily verified by direct computation. □

It is observed that B_{1t} has no influence on the optimal feedback gain $\{F_t\}$ although it changes the performance index V_t. This feature is important and induces the duality between the minimum variance control and the Kalman filtering for which F_t is dual to the Kalman gain K_t, and the backward control DRE (5.86) is dual to the forward filtering DRE (5.36). For this reason, many properties of the Kalman filter also hold for the minimum variance control. In particular, the condition $D_t^* C_t = \mathbf{0}$ can be removed as shown next.

Corollary 5.2. *Under the same hypotheses as in Theorem 5.19 except that* $D_t^* C_t \neq \mathbf{0}$*, the optimal control law minimizing* V_t *in (5.89) is given by* $\mathbf{u}(t) = \mathbf{u}_{\text{opt}}(t) = F_t \mathbf{x}(t)$ *with*

$$F_t = -\left(R_t + B_{2t}^* X_{t+1} B_{2t}\right)^{-1} \left(B_{2t}^* X_{t+1} A_t + D_t^* C_t\right), \tag{5.93}$$

$$X_t = \tilde{A}_t^* X_{t+1} \left(I_n + B_{2t} R_t^{-1} B_{2t}^* X_{t+1}\right)^{-1} \tilde{A}_t + \tilde{C}_t^* \tilde{C}_t, \tag{5.94}$$

where $\tilde{A}_t = \left(A_t - B_{2t} R_t^{-1} D_t^* C_t\right)$, $\tilde{C}_t = \left(I - D_t R_t^{-1} D_t^*\right) C_t$*, and* $X_T = \mathbf{0}$.

Again the control gain F_t in (5.93) is dual to the filtering gain K_t in (5.45), and the control DRE in (5.94) is dual to the filtering DRE in (5.43).

Proof. Introduce the variable substitution $\mathbf{u}(t) = -R_t^{-1} D_t^* C_t \mathbf{x}(t) + \tilde{\mathbf{u}}(t)$ with $\tilde{\mathbf{u}}(t)$ to be designed. Then (5.92) is changed into

$$\begin{bmatrix} \mathbf{x}(t+1) \\ \mathbf{z}(t) \end{bmatrix} = \begin{bmatrix} \tilde{A}_t & B_{2t} \\ \tilde{C}_t & D_t \end{bmatrix} \begin{bmatrix} \mathbf{x}(t) \\ \tilde{\mathbf{u}}(t) \end{bmatrix} + \begin{bmatrix} B_{1t} \\ \mathbf{0} \end{bmatrix} \mathbf{v}(t).$$

The result in Theorem 5.19 can then be applied to obtain the DRE in (5.94) and the optimal control gain as $\tilde{F}_t = -\left(R_t + B_{2t}^* X_t B_{2t}\right)^{-1} B_{2t}^* X_t \tilde{A}_t$. Hence, $\mathbf{u}(t) = -R_t^{-1} D_t^* C_t \mathbf{x}(t) + \tilde{F}_t \mathbf{x}(t) = F_t \mathbf{x}(t)$ with

$$\begin{aligned} F_t = \tilde{F}_t - R_t^{-1} D_t^* C_t &= -\Theta_t^{-1} \left[B_{2t}^* X_t \tilde{A}_t + (R_t + B_{2t}^* X_t B_{2t}) R_t^{-1} D_t^* C_t\right] \\ &= -\Theta_t^{-1} \left[B_{2t}^* X_t \left(A_t - B_{2t} R_t^{-1} D_t^* C_t\right) + D_t^* C_t + B_{2t}^* X_t B_{2t} R_t^{-1} D_t^* C_t\right] \\ &= -\left(R_t + B_{2t}^* X_t B_{2t}\right)^{-1} \left(B_{2t}^* X_t A_t + D_t^* C_t\right) \end{aligned}$$

which is identical to (5.93) where $\Theta_t = (R_t + B_{2t}^* X_t B_{2t})$ is used. □

Corollary 5.2 shows that it has no loss of generality to study the minimum variance control for the case $D_t^* C_t = \mathbf{0}$ from which the results can be easily carried to the case $D_t^* C_t \neq \mathbf{0}$. Moreover, there is no loss of generality to study the LQR problem in place of the minimum variance control. Both result in the same linear feedback control law. Hence, the rest of the section will focus on the LQR problem under the condition $D_t^* C_t = \mathbf{0}$ for simplicity.

5.2.2 *Stability*

State-feedback control was briefly discussed in Chap. 3 in connection with the notion of stabilizability. The LQR control is an effective way to design state-feedback control laws for the system model in (5.82) and is aimed at minimizing the quadratic performance index (5.83). A more general LQR control problem can be found in Problem 5.24 in Exercises that includes the cross term for the performance index. It is natural to study stability of the closed-loop system for the LQR feedback control which is governed by

$$\mathbf{x}(t+1) = [A_t + B_t F_t]\mathbf{x}(t), \quad \mathbf{x}(0) = \mathbf{x}_0 \tag{5.95}$$

with F_t given in Theorem 5.17. An important question to be answered is under what conditions the closed-loop system (5.95) is asymptotically or exponentially stable as $T \to \infty$ in the performance index J_T. It needs to be pointed out that difficulties exist in computing the LQR control law in the limiting case because of the time-varying realization for the system in (5.82) and time-varying weighting matrices in the performance index. Nevertheless, theoretical analysis can be made to obtain similar stability results to those for the Kalman filter. The next result is dual to Theorem 5.8 in the previous section but with strengthened stability property.

Theorem 5.20. *For the state-space model in (5.82) and the performance index in (5.83), assume that (A_t, B_t) is uniformly stabilizable and (C_t, A_t) is uniformly detectable with $Q_t = C_t^* C_t$. Then the closed-loop system (5.95) for the LQR control as described in Theorem 5.17 is exponentially stable as $T \to \infty$.*

The proof of this theorem is again left as an exercise (Problem 5.26). It is noted that exponential stability can be concluded for the LQR control different from that for the Kalman filter which is only asymptotically stable. Its reason lies in the Lyapunov stability criteria as discussed in Chap. 3. Recall that the filtering DRE in (5.36) can be written as the forward Lyapunov difference equation (Problem 5.10 in Exercises) for which the result of Lemma 3.4 can only ensure the asymptotic stability. On the other hand, the control DRE in (5.86) can be written as the backward Lyapunov difference equation (Problem 5.23 in Exercises) for which the result of Theorem 3.37 can in fact ensure the exponential stability.

While a stronger stability result holds for the LQR control than that for the Kalman filter, optimal state-feedback gain is difficult to compute for the limiting case $T \to \infty$. The exception is the stationary LQR control when the realization and the weighting matrices in the performance index J_T are all time invariant, and the time horizon $T \to \infty$. Consider the ARE

$$\begin{aligned} X &= A^* X A - A^* X B (R + B^* X B)^{-1} B^* X A + C^* C \\ &= A^* \left(I_n + X B R^{-1} B^*\right)^{-1} X A + C^* C, \end{aligned} \tag{5.96}$$

where $R = D^* D$ and $D^* C = \mathbf{0}$. The above is the same as the DRE in (5.86) except that all the time indices are removed. It is often called the control ARE, versus the filtering ARE (5.50) for the stationary Kalman filtering.

Example 5.21. Consider the flight control system introduced in Problem 1.11 in Exercises of Chap. 1. Its state-space realization after discretization with sampling period $T_s = 0.025$ is obtained in Example 3.17 of Chap. 3. Suppose that the LQR control is employed with $Q = C'C$ and $R = I$. The Matlab command "dare" can be used to compute the solution X to the ARE in (5.96) that is given by

$$X = \begin{bmatrix} 112.3476 & 18.2824 & 164.7632 & 25.6960 & -102.8198 \\ 18.2824 & 43.9026 & 38.9169 & 8.1070 & -24.7280 \\ 164.7632 & 38.9169 & 382.4594 & 81.6639 & -213.8244 \\ 25.6960 & 8.1070 & 81.6639 & 23.1625 & -42.9162 \\ -102.8198 & -24.7280 & -213.8244 & -42.9162 & 129.6328 \end{bmatrix}.$$

The corresponding stationary state-feedback gain is obtained as

$$F = -\left(R + B^*XB\right)^{-1} B^*XA = \begin{bmatrix} 0.1230 & 0.0096 & -0.0520 & -0.0853 & -0.0399 \\ -0.4477 & -1.0709 & -0.9530 & -0.1986 & 0.6062 \\ 0.8648 & 0.2892 & 2.9645 & 0.8771 & -1.5290 \end{bmatrix}.$$

It can be easily verified with Matlab that $(A + BF)$ is a stability matrix by examining its eigenvalues, implying that the state vector under this LQR control approaches zero asymptotically.

A similar notion to that for Kalman filtering is defined next.

Definition 5.2. The solution X to the *ARE* (5.96) is said to be stabilizing, if the state-feedback gain $F = -\left(R + B^*XB\right)^{-1} B^*XA$ is stabilizing.

With the feedback gain $F = -\left(R + B^*XB^*\right)^{-1} B^*XA$, the ARE in (5.96) can be written into the Lyapunov equation (refer to Problem 5.23 in Exercises)

$$X = (A + BF)^*X(A + BF) + C^*C + F^*RF. \tag{5.97}$$

The following is the stability result for the stationary LQR control.

Theorem 5.22. *Let $Q = C^*C$ and (C,A) be detectable. If the* ARE *(5.96) admits a stabilizing solution X, then the solution $\{X_t(T)\}$ to the* DRE

$$X_t(T) = A^*X_{t+1}(T)\left[I_n + BR^{-1}B^*X_{t+1}(T)\right]^{-1}A + Q, \quad X_T(T) = \mathbf{0} \tag{5.98}$$

converges to X as $T \to \infty$. In this case, the closed-loop system

$$\mathbf{x}(t+1) = (A + BF)\mathbf{x}(t), \quad F = -\left(R + B^*XB\right)^{-1} B^*XA \tag{5.99}$$

for the stationary LQR control is stable.

Proof. The solution to (5.98) satisfies (refer to Problem 5.31 in Exercises):

$$X_{t+1}(T) \le X_t(T) = X_{t+1}(T+1),$$

where $X_{t+1}(T+1)$ is the solution to the same DRE in (5.98) with T replaced by $T+1$. Hence, $\{X_t(T)\}$ is monotonically increasing with respect to T. Let $J_T(t)$, $t \geq 0$, be the performance index associated with DRE (5.98). Then

$$0 \leq \lim_{T\to\infty} J_T(t) = \lim_{T\to\infty} \mathbf{x}(t)^* X_t(T)\mathbf{x}(t) \leq \mathbf{x}(t)^* X \mathbf{x}(t)$$

implying $0 \leq X_t(T) = X_t(T)^* \leq X \; \forall t < T$ by the fact that the stabilizing solution X is maximal (Problem 5.27 in Exercises). Thus, it has a unique limit $\overline{X} = \overline{X}^*$ satisfying $0 \leq \overline{X} \leq X$. Since the ARE (5.96) is the limit of the DRE (5.98), $\overline{X}$ is a solution to (5.96) which is the same as the Lyapunov equation (5.97) with X replaced by $\overline{X}$ and F replaced by $\overline{F} = -\left(R + B^*\overline{X}B^*\right)^{-1} B^*\overline{X}A$. The detectability of (C,A) and $\overline{X} \geq 0$ imply that $\left(A + B\overline{F}\right)$ is a stability matrix in light of the Lyapunov stability result or $\overline{X}$ is a stabilizing solution to the ARE in (5.96). By the uniqueness of the stabilizing solution to the ARE (Problem 5.13 in Exercises), $\overline{X} = X$. It follows that the closed-loop system in (5.99) is stable. □

Theorem 5.22 offers a numerical algorithm for computing the unique stabilizing solution X to the ARE (5.96) through computing iteratively the solution to (5.98). That is, one may set $X^{(0)} = X_T(T) = \mathbf{0}$ then compute

$$X^{(i+1)} = A^* \left[I_n + BR^{-1}B^*X^{(i)}\right]^{-1} X^{(i)}A + Q$$

for $i = 1, 2, \ldots$ until $\left\|X^{(N+1)} - X^{(N)}\right\| \leq \varepsilon$ with $\varepsilon > 0$ some prespecified error tolerance and then take $X = X^{(N+1)}$. The next result answers under what condition there exists a stabilizing solution to the ARE (5.96). Since the ARE (5.50) for the stationary Kalman filter is dual to the ARE (5.96), it also provides the proof for Theorem 5.11.

Theorem 5.23. *Let $Q = C^*C$ and $R > 0$. There exists a stabilizing solution to the* ARE *in (5.96), if and only if (A,B) is stabilizable and*

$$\text{rank}\left\{\begin{bmatrix} A - e^{j\omega}I_n \\ C \end{bmatrix}\right\} = n \quad \forall\ \omega \in \mathbb{R}. \tag{5.100}$$

Proof. It is obvious that stabilizability of (A,B) is a necessary condition for the ARE (5.96) to have a stabilizing solution. To confirm that (5.100) is also a necessary condition, assume on the contrary that (5.100) does not hold but the ARE (5.96) admits a stabilizing solution X. The Lyapunov form of the ARE in (5.97) implies that $(A+BF)$ is a stability matrix with F as in (5.99), and thus, $X = X^* \geq 0$. Since (5.100) does not hold,

$$A\mathbf{q} = e^{j\theta}\mathbf{q}, \quad C\mathbf{q} = 0 \tag{5.101}$$

for some θ real and $\mathbf{q} \neq \mathbf{0}$. That is, (C,A) has at least one unobservable mode on the unit circle. Multiplying both sides of the ARE in (5.96) by $\mathbf{q}^*$ from left and $\mathbf{q}$ from right, and using the relation in (5.101) yield

$$\mathbf{q}^* XB\left(R+B^* XB\right)^{-1} B^* X \mathbf{q} = 0 \implies B^* X \mathbf{q} = \mathbf{0}.$$

By the expression of F, the above leads to

$$(A+BF)\mathbf{q} = \left[A - BB^* \left(R+B^* XB\right)^{-1} B^* XA\right] \mathbf{q} = \mathrm{e}^{j\theta}\mathbf{q}.$$

So $\mathrm{e}^{j\theta}$ remains an eigenvalue of $A+BF$ contradicting the stabilizing assumption on X. This concludes the necessity of (5.100).

For the sufficiency part of the proof, assume that (A,B) is stabilizable and (5.100) holds. Then some F_0 exists such that $(A+BF_0)$ is a stability matrix. It is claimed that the following recursion

$$X_i = (A+BF_i)^* X_i (A+BF_i) + F_i^* R F_i + Q, \tag{5.102}$$

$$F_{i+1} = -\left(R+B^* X_i B\right)^{-1} B^* X_i A, \quad i = 0, 1, \ldots, \tag{5.103}$$

converges to the stabilizing solution X of the ARE (5.96). The proof of the claim proceeds in three steps. At the first step, it will be shown that stability of $(A+BF_i)$ implies stability of $(A+BF_{i+1})$ for $i \geq 0$. For this purpose, rewrite (5.102) as (refer to Problem 5.27 in Exercises)

$$\begin{aligned} X_i &= A^* \left(I_n + X_i B R^{-1} B^*\right)^{-1} X_i A + Q + \Delta_F(i)^* \left[R + B^* X_i B\right] \Delta_F(i) \\ &= (A+BF_{i+1})^* X_i (A+BF_{i+1}) + F_{i+1}^* R F_{i+1} + Q \\ &\quad + \Delta_F(i)^* \left[R + B^* X_i B\right] \Delta_F(i) \end{aligned} \tag{5.104}$$

with $\Delta_F(i) = F_{i+1} - F_i$ where (5.97) is used with X replaced by X_i and F by F_{i+1} to obtain the second equality. Now suppose that

$$(A+BF_{i+1})\mathbf{v} = \lambda \mathbf{v}, \quad |\lambda| \geq 1. \tag{5.105}$$

Multiplying both sides of (5.104) by $\mathbf{v}^*$ from left and $\mathbf{v}$ from right yields

$$\left(1-|\lambda|^2\right) \mathbf{v}^* X_i \mathbf{v} = \mathbf{v}^* \left[F_{i+1}^* R F_{i+1} + Q + \Delta_F(i)^* \left(R + B^* X_i B\right) \Delta_F(i)\right] \mathbf{v},$$

where (5.105) is used. Because the left-hand side ≤ 0 by $|\lambda| \geq 1$ and positivity of X_i due to stability of $(A+BF_i)$ and the right-hand side ≥ 0 by positivity of R, Q, and X_i, it is concluded that $|\lambda| = 1$ and

$$C\mathbf{v} = \mathbf{0}, \quad F_{i+1}\mathbf{v} = \mathbf{0}, \quad \Delta_F(i)\mathbf{v} = \mathbf{0} \implies F_i \mathbf{v} = \mathbf{0}.$$

The above together with (5.105) imply $A\mathbf{v} = \lambda\mathbf{v}$ which in turn implies

$$(A + BF_i)\mathbf{v} = A\mathbf{v} = \lambda\mathbf{v}.$$

Because $(A + BF_i)$ is a stability matrix, $\mathbf{v} = \mathbf{0}$ concluding that λ is not an eigenvalue of $(A + BF_{i+1})$. As λ with $|\lambda| \geq 1$ is arbitrary, $(A + BF_{i+1})$ is also a stability matrix. The fact that $(A + BF_0)$ is a stability matrix implies that F_{i+1} in (5.103) is stabilizing for each $i \geq 0$. As a second step, it is noted that (5.104) and the definition of X_{i+1} imply

$$\Delta_X(i) = (A + BF_{i+1})^* \Delta_X(i)(A + BF_{i+1}) + \Delta_F(i)^*(R + B^* X_i B)\Delta_F(i)$$

with $\Delta_X(i) = X_i - X_{i+1}$. Stability of $(A + BF_{i+1})$ implies that $\Delta_X(i) \geq \mathbf{0}$ or $\{X_i\}$ is a decreasing matrix sequence. Since $X_i \geq \mathbf{0}$ by stability of $(A + BF_i)$, the recursion in (5.102) and (5.103) converges with limits $X \geq \mathbf{0}$ satisfying the ARE (5.96) and F as given in (5.99). Finally, as $(A + BF_i)$ is stable for all $i \geq 0$, the n eigenvalues of $(A + BF_i)$ converge to the n eigenvalues of $(A + BF)$ on the closed unit disk. The condition (5.100) prohibits any eigenvalues of $(A + BF)$ from being on the unit circle because if it does, then

$$(A + BF)\mathbf{v} = e^{j\theta}\mathbf{v}, \quad \mathbf{v} \neq \mathbf{0}$$

for some θ real. Multiplying both sides of (5.97) by $\mathbf{v}^*$ from left and $\mathbf{v}$ from right leads to $F\mathbf{v} = \mathbf{0}$ and $C\mathbf{v} = \mathbf{0}$, and thus $A\mathbf{v} = e^{j\theta}\mathbf{v}$ with the same argument as before. This contradicts the condition (5.100). The proof is now complete. □

The proof of Theorem 5.23 shows that the condition (5.100) is indispensable. Stabilizability of (A, B) alone does not ensure that the LQR problem is well posed. If the condition (5.100) is violated, then any unobservable mode of (C, A) on the unit circle does not contribute to the LQR performance index. Thus, in this case, even if the ARE (5.96) admits a solution $X = X^* \geq 0$, the optimal performance index (for the stationary LQR control)

$$J_{\text{opt}} = \sum_{t=0}^{\infty} \mathbf{x}(t)^* Q\mathbf{x}(t) + \mathbf{u}(t)^* R\mathbf{u}(t) = \sum_{t=0}^{\infty} \mathbf{x}(t)^* (Q + F^* RF)\mathbf{x}(t) = \mathbf{x}_0^* X\mathbf{x}_0$$

and stability of $(A + BF)$ cannot be achieved simultaneously. The reason lies in the facts that stabilization of any unobservable mode of (C, A) on the unit circle will increase the energy cost of the control input by $R > 0$ and that such unstable modes of (C, A) do not contribute to the LQR performance index anyway. This is illustrated in the following example.

Example 5.24. Consider the stationary LQR control with

$$A = \begin{bmatrix} -1 & 0 \\ 0 & 0 \end{bmatrix}, \quad B = \begin{bmatrix} 1 \\ 0 \end{bmatrix}, \quad C = \begin{bmatrix} 0 & 1 \end{bmatrix}$$

and $R = 1$. Clearly, (A, B) is stabilizable, but the condition (5.100) does not hold. It can be verified that with $F_0 = \begin{bmatrix} \frac{1}{2} & 0 \end{bmatrix}$, $(A + BF_0)$ is a stability matrix. The recursive algorithm as in (5.102) and (5.103) gives

$$X_i = \begin{bmatrix} \frac{1}{2^{i+2}-1} & 0 \\ 0 & 1 \end{bmatrix}, \quad A+BF_i = \begin{bmatrix} \frac{1}{2^{i+1}} - 1 & 0 \\ 0 & 0 \end{bmatrix}$$

for $0 \le i < \infty$. Hence, $(A+BF_i)$ is stable for any finite i and $X_i \ge \mathbf{0}$ is monotonically decreasing. However, the limits

$$\lim_{i\to\infty} X_i = \begin{bmatrix} 0 & 0 \\ 0 & 1 \end{bmatrix}, \quad \lim_{i\to\infty} F_i = \lim_{k\to\infty} \begin{bmatrix} \frac{1}{2^{i+1}} & 0 \end{bmatrix} = \begin{bmatrix} 0 & 0 \end{bmatrix}$$

and thus $(A+BF_i) \to A$ as $i \to \infty$ which is unstable.

Example 5.24 leads to the deduction that if (C,A) has unobservable modes strictly outside the unit circle but the stabilizability of (A,B) and (5.100) hold, then the ARE has more than one nonnegative definite solutions. One is the stabilizing solution X. There is at least one more, denoted by X_u, which is not stabilizing. That is, the unobservable modes of (C,A) strictly outside the unit circle are not stabilized by $F_u = -(R+B^*X_uB)^{-1}B^*X_uA$. Since the unstable modes strictly outside the unit circle do not contribute to the performance index by the hypothesis, $X \ge X_u \ge 0$. In this case, the maximal solution of the ARE is always the stabilizing solution. It is now clear why the detectability of (C,A) is required in Theorem 5.22, without which $X_T(t)$ may converge to X_u as $T \to \infty$ with $t \ge 0$ finite.

The next result states the solution to the general stationary LQR control.

Corollary 5.3. *For* $\mathbf{x}(t+1) = A\mathbf{x}(t) + B\mathbf{u}(t)$ *with* $\mathbf{x}(0) = \mathbf{x}_0 \ne \mathbf{0}$, *let*

$$J_\infty = \sum_{t=0}^{\infty} \|C\mathbf{x}(t) + D\mathbf{u}(t)\|^2, \quad R = D^*D > 0, \quad D^*C \ne \mathbf{0}.$$

Let $\tilde{A} = \left(A - BR^{-1}D^*C\right)$ *and* $\tilde{C} = \left(I - DR^{-1}D^*\right)C$. *Suppose that the* ARE

$$X = \tilde{A}^*\left(I + XBR^{-1}B^*\right)^{-1} X\tilde{A} + \tilde{C}^*\tilde{C} \tag{5.106}$$

has a stabilizing solution X. *Then the optimal control law is given by*

$$\mathbf{u}(t) = F\mathbf{x}(t), \quad F = -(R+B^*XB)^{-1}(B^*XA + D^*C) \tag{5.107}$$

which is stabilizing and minimizes J_∞. *Moreover, the* ARE *(5.106) admits a stabilizing solution, if and only if* (A,B) *is stabilizable and*

$$\text{rank}\left\{\begin{bmatrix} A - e^{j\omega}I_n & B \\ C & D \end{bmatrix}\right\} = n+m \quad \forall\ \omega \in \mathbb{R}, \tag{5.108}$$

where m *is the dimension of the input and* n *the dimension of the state vector.*

Proof. Since the general LQR control is the same as that for Theorems 5.22 and 5.23 with A replaced by $\tilde{A}$ and C replaced by $\tilde{C}$, the proof of the first part of the corollary is simple and skipped. For the second part of the corollary, it is noted that stabilizability of $(\tilde{A}, B)$ is the same as stabilizability of (A, B), and thus, the proof can be completed by showing that the condition

$$\text{rank}\left\{\begin{bmatrix} \tilde{A} - e^{j\omega} I_n \\ \tilde{C} \end{bmatrix}\right\} = n \tag{5.109}$$

is equivalent to (5.108). It is straightforward to compute

$$\begin{bmatrix} I_n & -BR^{-1}D^* \\ \mathbf{0} & I - DR^{-1}D^* \\ \mathbf{0} & R^{-1}D^* \end{bmatrix} \begin{bmatrix} A - e^{j\omega} I_n & B \\ C & D \end{bmatrix} = \begin{bmatrix} \tilde{A} - e^{j\omega} I_n & \mathbf{0} \\ \tilde{C} & \mathbf{0} \\ R^{-1}D^*C & I_m \end{bmatrix}.$$

The first matrix on the left is an elementary matrix that does not alter the rank of the second matrix on the left. It follows that

$$\text{rank}\left\{\begin{bmatrix} A - e^{j\omega} I_n & B \\ C & D \end{bmatrix}\right\} = m + \text{rank}\left\{\begin{bmatrix} \tilde{A} - e^{j\omega} I_n \\ \tilde{C} \end{bmatrix}\right\},$$

and hence, the condition (5.108) is equivalent to the one in (5.109). □

5.2.3 Full Information Control

In minimum variance control, the controlled output $\mathbf{z}(t)$ in (5.80) does not involve the disturbance input. This is the main reason why the optimal feedback control law is a function of only $\mathbf{x}(t)$. Suppose that the state-space model and the controlled output are specified respectively by

$$\begin{aligned} \mathbf{x}(t+1) &= A_t \mathbf{x}(t) + B_{1t} \mathbf{v}(t) + B_{2t} \mathbf{u}(t), \\ \mathbf{z}(t) &= C_t \mathbf{x}(t) + D_{1t} \mathbf{v}(t) + D_{2t} \mathbf{u}(t). \end{aligned} \tag{5.110}$$

It can be expected that the optimal feedback control law will be a function of not only $\mathbf{x}(t)$ but also of $\mathbf{v}(t)$ which is the white noise process satisfying (5.20). Such a control law is termed *full information control*. One needs to keep in mind that often, in the practice of feedback control, both $\mathbf{x}(t)$ and $\mathbf{v}(t)$ are not measurable directly for which output estimators in the previous section can be employed to provide information on $\mathbf{x}(t)$ and $\mathbf{v}(t)$. The next result provides the optimal solution to full information control.

Theorem 5.25. *Consider the state-space system (5.110) where* $\{\mathbf{v}(t)\}$ *is the white noise process satisfying (5.20). Suppose that* $R_t = D_{2t}^* D_{2t} > 0$. *Let*

$$\tilde{A}_t = A_t - B_{2t} R_t^{-1} D_{2t}^* C_t, \quad \tilde{C}_t = \left[I - D_{2t} R_t^{-1} D_{2t}^* \right] C_t.$$

Let X_t *be the solution to the* DRE *(5.94). Then the optimal control law that minimizes* $\mathrm{E}\{\|\mathbf{z}(t)\| \,|\, \mathcal{U}_T\}$ *with* $\mathcal{U}_T = \{\mathbf{u}(t)\}_{t=0}^{T-1}$ *is* $\mathbf{u}(t) = F_{1t}\mathbf{x}(t) + F_{2t}\mathbf{v}(t)$ *with*

$$\begin{aligned} F_{1t} &= -\left(R_t + B_{2t}^* X_{t+1} B_{2t}\right)^{-1} \left(B_{2t}^* X_{t+1} A_t + D_{2t}^* C_t\right), \\ F_{2t} &= -\left(R_t + B_{2t}^* X_{t+1} B_{2t}\right)^{-1} \left(B_{2t}^* X_{t+1} B_{1t} + D_{2t}^* D_{1t}\right). \end{aligned} \tag{5.111}$$

Theorem 5.25 for full information control is dual to Theorem 5.14 for output estimation. Its proof is similar to that of Theorem 5.14 and is thus left as an exercise (Problem 5.33).

It is noted that the closed-loop system for (5.110) under the full information control law (5.111) is given by

$$\begin{aligned} \mathbf{x}(t+1) &= (A_t + B_{2t}F_{1t})\mathbf{x}(t) + (B_{1t} + B_{2t}F_{2t})\mathbf{v}(t), \\ \mathbf{z}(t) &= (C_t + D_{2t}F_{1t})\mathbf{x}(t) + (D_{1t} + D_{2t}F_{2t})\mathbf{v}(t). \end{aligned} \tag{5.112}$$

The above is dual to (5.143) in Exercises for output estimation. The optimality of the full information control shows that the static feedback gains (F_{1t}, F_{2t}) in (5.111) outperform any other controllers such as dynamic or nonlinear ones in minimization of $\mathbf{E}\left\{\|\mathbf{z}(t)\|^2 \,|\, \mathcal{U}_T\right\}$ under the white noise disturbance $\{\mathbf{v}(t)\}$ for all $t \in [0,\ T)$. This observation is important as shown in the next example.

Example 5.26. In wireless data communications, the processing burden at the receiver site is sometimes shifted to the transmitter site which often has more computational power for the downlink channels (from the station to the cellular users). A precoder is designed at the transmitter site to compensate the distorted channel so that the receivers can pick up the data directly without further digital processing. The block diagram below shows the use of such precoders in data communications where the state-space model with realization (A_t, B_t, C_t, D_t) represents the (downlink) wireless channel which is asymptotically stable. For simplicity, the additive noise at the receiver site is taken to be zero, and $\det(D_t^* D_t) \neq 0$ is assumed for each t.

Our objective is to design the linear precoder that minimizes the error variance of $\mathbf{e}_s(t)$ under the assumption that the transmitted signal $\mathbf{s}(t)$ is white with zero mean and identity covariance. It is claimed that any linear, causal, and stable precoder has the form

$$\mathbf{x}_p(t+1) = (A_t + B_t F_t)\mathbf{x}_p(t) + B_t \mathbf{w}(t), \quad \mathbf{u}(t) = F_t \mathbf{x}_p(t) + \mathbf{w}(t) \tag{5.113}$$

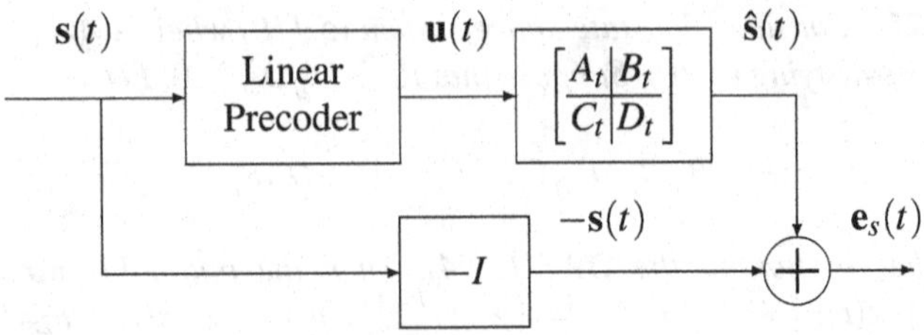

Fig. 5.7 Precoder in data detection

for some asymptotically stabilizing F_t and $\mathbf{w}(t) = Q(t,k) \star \mathbf{s}(t)$ with $\{Q(t,k)\}$ the impulse response of some causal and stable LTV system at time t. Indeed, given a linear, causal, and stable precoder with impulse response $\{G(t,k)\}$, consider the inverse of the system in (5.113):

$$\tilde{\mathbf{x}}_p(t+1) = A_t\tilde{\mathbf{x}}_p(t) + B_t\mathbf{u}(t), \quad \mathbf{w}(t) = -F_t\tilde{\mathbf{x}}_p(t) + \mathbf{u}(t).$$

Denote the impulse response of the above system by $L(t,k)$. Then $Q(t,k) = L(t,k) \star G(t,k)$ is causal and stable. Thus, $G(t,k)$ can be implemented by (5.113) with $\mathbf{w}(t) = Q(t,k) \star \mathbf{s}(t)$. The channel is now described by

$$\begin{aligned} \mathbf{x}(t+1) &= A_t\mathbf{x}(t) + B_t\mathbf{u}(t) = A_t\mathbf{x}(t) + B_tF_t\mathbf{x}_p(t) + B_tQ_t\mathbf{s}(t), \\ \hat{\mathbf{s}}(t) &= C_t\mathbf{x}(t) + D_t\mathbf{u}(t) = C_t\mathbf{x}(t) + D_tF_t\mathbf{x}_p(t) + D_tQ_t\mathbf{s}(t), \end{aligned}$$

by (5.113) and $\mathbf{w}(t) = Q_t\mathbf{s}(t)$ where $Q(t,k) = Q_t$ is taken as a static gain for each t temporarily. The overall system in Fig. 5.7 has a realization

$$\left[\begin{array}{cc|c} A_t & B_tF_t & B_tQ_t \\ \mathbf{0} & A_t + B_tF_t & B_tQ_t \\ \hline C_t & D_tF_t & D_tQ_t - I \end{array}\right] \implies \left[\begin{array}{c|c} A_t + B_tF_t & B_tQ_t \\ \hline C_t + D_tF_t & D_tQ_t - I \end{array}\right]$$

after using the similarity transform

$$T = \begin{bmatrix} I & -I \\ \mathbf{0} & I \end{bmatrix}$$

to eliminate the unreachable modes. Therefore, the overall system in Fig. 5.7 is described by

$$\begin{aligned} \hat{\mathbf{x}}(t+1) &= (A_t + B_tF_t)\hat{\mathbf{x}}(t) + B_tQ_t\mathbf{s}(t), \\ \mathbf{e}_s(t) &= (C_t + D_tF_t)\hat{\mathbf{x}}(t) + (D_tQ_t - I)\mathbf{s}(t), \end{aligned}$$

which has the same form as in (5.112) by taking $B_{1t} = \mathbf{0}$, $B_{2t} = B_t$, $D_{1t} = -I$, and $D_{2t} = D_t$. Hence, the results in Theorem 5.25 for optimal full information control can be applied to compute the optimal precoder gains F_t and Q_t. It is noted that the use of dynamic gains $Q(t,k)$ do not improve its performance any further.

The closed-loop system for full information control as in Theorem 5.25 admits the same stability properties as those for LQR control and minimum variance control in light of the fact that they share the same DRE (5.94). Hence, all the stability results in the previous subsection apply to the case of full information control which will not be repeated here. In the case of stationary full information control, the realization matrices in both the state-space model and the controlled signal are all time invariant and the time horizon $T \to \infty$ for the performance index. It can be expected that the DRE (5.94) converges to the ARE

$$X = \tilde{A}^* \left(I + XB_2R^{-1}B_2^*\right)^{-1} X\tilde{A} + \tilde{C}^*\tilde{C} \tag{5.114}$$

with $R = D_2^*D_2$ which is identical to (5.106) except that B is replaced by B_2 and D by D_2. In this case, the transfer matrix from $\mathbf{v}(t)$ to $\mathbf{z}(t)$ is given by

$$\mathbf{T}(z) = (D_1 + D_2F_2) + (C + D_2F_1)(zI - A - B_2F_1)^{-1}(B_1 + B_2F_2),$$

where F_1 and F_2 are the same as in (5.111) with all the time indices removed. It is interesting to note that with the white noise disturbance $\{\mathbf{v}(t)\}$ WSS satisfying (5.20), there holds $\mathbf{E}\left\{\|\mathbf{z}(t)\|^2\right\} = \|\mathbf{T}\|_2^2$ where

$$\|\mathbf{T}\|_2 = \sqrt{\mathrm{Tr}\left\{(D_1 + D_2F_2)^*(D_1 + D_2F_2) + (B_1 + B_2F_2)^*X(B_1 + B_2F_2)\right\}}$$

with X the stabilizing solution to (5.114). By the optimality of the solution to full information control, $\|\mathbf{T}\|_2$ is minimized by static feedback controllers F_1 and F_2. In fact, dynamic feedback controllers do not outperform static feedback controllers for stationary full information control.

5.3 LTI Systems and Stationary Processes

This section intends to explore further optimal estimation and control for LTI state-space models and stationary white noises. As shown in the previous two sections, both Kalman filters and LQR controllers tend to stationary ones as the time horizon approaches to infinity. Hence, the results for optimal estimation and control can have frequency domain interpretations which will help deepen the understanding of the results in the previous two sections. Several results will be presented which have applications to various problems in design of communication and control systems in later chapters.

5.3.1 Spectral Factorizations

A PSD transfer matrix $\Psi(z)$ has the form

$$\Psi(z) = \sum_{k=-\infty}^{\infty} \Gamma_k z^{-k}, \quad \Gamma_k^* = \Gamma_{-k}, \tag{5.115}$$

and $\Psi(e^{j\omega}) \geq \mathbf{0}$ for all real ω. There exist spectral factorizations

$$\Psi(z) = \mathbf{H}_L(z)\mathbf{H}_L(z)^\sim = \mathbf{H}_R(z)^\sim \mathbf{H}_R(z), \tag{5.116}$$

where $\mathbf{H}_L(z)$ and $\mathbf{H}_R(z)$ are both causal, stable, and minimum phase. The transfer matrices $\mathbf{H}_L(z)$ and $\mathbf{H}_R(z)$ are called left and right spectral factors of $\Psi(z)$, respectively. This section shows how Kalman filtering and LQR control can be used to compute spectral factorizations.

Recall the random process in form of state-space model

$$\mathbf{x}(t+1) = A\mathbf{x}(t) + B\mathbf{v}(t), \quad \mathbf{y}(t) = C\mathbf{x}(t) + D\mathbf{v}(t), \tag{5.117}$$

where $\mathbf{v}(t)$ is the white noise satisfying (5.20) and $\mathbf{y}(t)$ is the observed output. Assume that $\mathbf{x}(0) = \mathbf{x}_0$ is independent of $\{\mathbf{v}(t)\}$, has mean $\bar{\mathbf{x}}_0 = 0$, and covariance P satisfying the Lyapunov equation (5.47). Then the white noise hypothesis on $\mathbf{v}(t)$ implies that the PSD of the observed output is

$$\Psi_\mathbf{y}(\omega) = \mathbf{G}\left(e^{j\omega}\right)\mathbf{G}\left(e^{j\omega}\right)^*, \quad \mathbf{G}(z) = D + C(zI - A)^{-1}B. \tag{5.118}$$

The zero mean initial condition for $\mathbf{x}_0$ yields the ACS of $\mathbf{y}(t)$ given by

$$R_\mathbf{y}(k) = \mathrm{E}\{\mathbf{y}(t)\mathbf{y}(t-k)^*\} = \begin{cases} CA^{k-1}(APC^* + BD^*), & k > 0, \\ R + CPC^*, & k = 0, \\ (CPA^* + D^*B)(A^*)^{k-1}C^*, & k < 0, \end{cases}$$

by Problem 5.7 in Exercises. Hence, $\Psi_\mathbf{y}(\omega)$ is the Fourier transform of $\{R_\mathbf{y}(k)\}$ which exists, if A is a stability matrix. Let $\tilde{A} = A - BD^*R^{-1}C$ and $R = DD^* > 0$. Then the associated filtering ARE is (5.52) which is copied below:

$$\Sigma = \tilde{A}\left(I_n + \Sigma C^* R^{-1} C\right)^{-1} \Sigma \tilde{A}^* + B\left(I - D^* R^{-1} D\right) B^*.$$

Lemma 5.1. *Consider the state-space system in (5.117) with $\mathbf{v}(t)$ of dimension m and $\mathbf{y}(t)$ of dimension p. Assume that $m \geq p$ and $R = DD^* > 0$. Let K be the Kalman gain as in (5.51) with Σ satisfying (5.52). Then the* PSD *$\Psi_\mathbf{y}(\omega)$ as in (5.118) has the expression*

$$\Psi_{\mathbf{y}}(\omega) = \left[I - C(e^{j\omega}I - A)^{-1}K\right](R + C\Sigma C^*)\left[I - C(e^{j\omega}I - A)^{-1}K\right]^*. \tag{5.119}$$

Proof. Direct state-space computations yield

$$\begin{aligned}
\mathbf{W}(z) &= \left[I + C(zI - A - KC)^{-1}K\right]\mathbf{G}(z) \\
&= \left[\begin{array}{c|c} A+KC & K \\ \hline C & I \end{array}\right]\left[\begin{array}{c|c} A & B \\ \hline C & D \end{array}\right] \quad \left\{T = \left[\begin{array}{c|c} I & 0 \\ \hline I & I \end{array}\right]\right\} \\
&= \left[\begin{array}{cc|c} A & \mathbf{0} & B \\ KC & A+KC & KD \\ \hline C & C & D \end{array}\right] = \left[\begin{array}{c|c} A+KC & B+KD \\ \hline C & D \end{array}\right],
\end{aligned} \tag{5.120}$$

where the similarity transform T is used to eliminate the unobservable subsystem. It is claimed that

$$\mathbf{F}(z) = \left[I + C(zI - A - KC)^{-1}K\right] = \left[I - C(zI - A)^{-1}K\right]^{-1} \tag{5.121}$$

is a "whitening" filter in the sense that

$$\Phi_{\mathbf{W}}(z) = W(z)W(z)^{\sim} = R + C\Sigma C^*. \tag{5.122}$$

Indeed, denote $A_K = A + KC$, $B_K = B + KD$, and $\Pi = B_K B_K^*$. Then the ARE (5.52) can be written as

$$\begin{aligned}
\Pi = B_K B_K^* &= (B + KD)(B + KD)^* = \Sigma - A_K \Sigma A_K^* \\
&= (zI - A_K)\Sigma\left(z^{-1}I - A_K^*\right) + (zI - A_K)\Sigma A_K^* + A_K\Sigma\left(z^{-1}I - A_K^*\right).
\end{aligned}$$

Multiplying both sides of the above equation by $C(zI - A_K)^{-1}$ from left and $\left(z^{-1}I - A_K^*\right)^{-1}C^*$ from right gives

$$\begin{aligned}
\Phi_{\Pi}(z) &= C(zI - A_K)^{-1}B_K B_K^*\left(z^{-1}I - A_K^*\right)C^* \\
&= C\Sigma C^* + C(zI - A_K)^{-1}A_K\Sigma C^* + C\Sigma A_K^*\left(z^{-1}I - A_K^*\right)^{-1}C^*.
\end{aligned}$$

It follows from the state-space realization of $\mathbf{W}(z)$ that

$$\begin{aligned}
\Phi_{\mathbf{W}}(z) &= R + C(zI - A_K)^{-1}B_K D^* + D B_K^*\left(z^{-1}I - A_K^*\right)^{-1}C^* + \Phi_{\Pi}(z) \\
&= R + C\Sigma C^* + C(zI - A_K)^{-1}(B_K D^* + A_K\Sigma C^*) \\
&\quad + (B_K D^* + A_K\Sigma C^*)^*\left(z^{-1}I - A_K^*\right)^{-1}C^*.
\end{aligned}$$

By the expression of the Kalman gain,

$$\begin{aligned} B_K D^* + A_K \Sigma C^* &= (B + KD)D^* + (A + KC)\Sigma C^* \\ &= BD^* + A\Sigma C^* + K(DD^* + C\Sigma C^*) = \mathbf{0}. \end{aligned}$$

Therefore, substituting the expression of $\Phi_{\Pi}(z)$ into $\Phi_{\mathbf{W}}(z)$ yields (5.122) concluding the fact that $\mathbf{F}(z)$ is a "whitening" filter. In light of (5.120) or $\mathbf{G}(z) = \left[I + C(zI - A - KC)^{-1}K\right]^{-1}\mathbf{W}(z) = \left[I - C(zI - A)^{-1}K\right]\mathbf{W}(z)$,

$$\Psi_{\mathbf{y}}(\omega) = \left[I - C\left(e^{j\omega}I - A\right)^{-1}K\right]\Phi_{\mathbf{W}}\left(e^{j\omega}\right)\left[I - C\left(e^{j\omega}I - A\right)^{-1}K\right]^*$$

which is the same as (5.119). The proof is thus completed. □

In the case when A is a stability matrix and the stabilizing solution to the ARE (5.52) exits, then $I - C(zI - A)^{-1}K$ is not only causal and stable but also admits a causal and stable inverse. Let $R + C\Sigma C^* = \Omega\Omega^*$ be the Cholesky factorization and $\mathbf{G}_o(z) = \left[I - C(zI - A)^{-1}K\right]\Omega$. Then

$$\Phi(z) = \mathbf{G}(z)\mathbf{G}(z)^{\sim} = \mathbf{G}_o(z)\mathbf{G}_o(z)^{\sim}, \tag{5.123}$$

and thus, $\mathbf{G}_o(z)$ is the left *spectral factor* of $\Phi(z)$. Kalman filtering provides an algorithm to compute spectral factorization of $\Phi(z) = \mathbf{G}(z)\mathbf{G}(z)^{\sim}$. Conversely, (left) spectral factorization can be used to compute the Kalman filtering gain K by the expression of (5.119). It is noted that $\mathbf{F}(z)$ in (5.121) satisfies the following state-space equation

$$\hat{\mathbf{x}}_{t+1|t} = (A + KC)\hat{\mathbf{x}}_{t|t-1} - K\mathbf{y}(t), \quad \delta\mathbf{y}(t) = \mathbf{y}(t) - C\hat{\mathbf{x}}_{t|t-1}, \tag{5.124}$$

where $\hat{\mathbf{x}}_{k|k-1}$ is the stationary MMSE estimate of $\mathbf{x}(k)$ based on the observation up to time $(k-1)$. Hence, the output of $\mathbf{F}(z)$ is the innovation sequence.

Example 5.27. In the traditional Wiener filtering (refer to Example 5.16), a whitening filter is designed first to obtain the innovation sequence, and an estimator is then designed for smoothing, filtering, or prediction. The whitening filter can clearly be obtained using the spectral factorization for

$$\Phi(z) = \mathbf{G}_1(z)\mathbf{G}_1(z)^{\sim} + \mathbf{G}_2(z)\mathbf{G}_2(z)^{\sim} = \mathbf{G}_o(z)\mathbf{G}_o(z)^{\sim}$$

by the fact that $\mathbf{v}_1(t)$ and $\mathbf{v}_2(t)$ are independent of each other and have zero means. Thus, $\Psi_{\mathbf{y}}(\omega) = \mathbf{G}_1\left(e^{j\omega}\right)\mathbf{G}_1\left(e^{j\omega}\right)^* + \mathbf{G}_2\left(e^{j\omega}\right)\mathbf{G}_2\left(e^{j\omega}\right)^*$. To proceed, a realization (A, B, C, D) for $\left[\mathbf{G}_1(z)\ \mathbf{G}_2(z)\right]$ needs to be obtained before applying Lemma 5.1 for computing the whitening filter $\mathbf{F}_o(z) = \mathbf{G}_o^{-1}(z)$. It is noted that Wiener filtering can be approached by Kalman filtering, if $\mathbf{G}_2(z) = D$ and $\mathbf{G}_1(z) = C(zI - A)^{-1}B$ in Example 5.16. In this case,

$$\Phi(z) = R + C(zI - A)^{-1}BB^*\left(z^{-1}I - A^*\right)^{-1}C^*$$

which is identical to $\Psi_y(\omega)$ at $z = e^{j\omega}$ as in (5.118) provided that $BD^* = \mathbf{0}$. Hence, Kalman filtering can be employed to compute the whitening filter for Wiener filtering. If in addition $\mathbf{v}_1(t) = \mathbf{v}_2(t) = \mathbf{v}(t)$, then Wiener filtering for $m = 1$ in Fig. 5.5 coincides with Kalman filtering in Fig. 5.2. Recall that optimal output estimation is the same as the optimal state estimation for one-step prediction or strictly causal filtering.

The next result is dual to Lemma 5.1,and thus, the proof is omitted.

Lemma 5.2. *Let $\tilde{\mathbf{G}}(z) = D + C(zI - A)^{-1}B$. Assume that $R = D^*D > 0$, A is a stability matrix and the* ARE *(5.106) has a unique stabilizing solution $X \geq 0$ so that F in (5.107) is stabilizing. There holds factorization*

$$\tilde{\mathbf{G}}(z)^{\sim}\tilde{\mathbf{G}}(z) = \left[I - F(zI - A)^{-1}B\right]^{\sim}(R + B^*XB)\left[I - F(zI - A)^{-1}B\right]. \tag{5.125}$$

Let $R + B^*XB = \tilde{\Omega}^*\tilde{\Omega}$, and $\tilde{\mathbf{G}}_o(z) = \tilde{\Omega}\left[I - F(zI - A)^{-1}B\right]$. Then

$$\Phi(z) = \tilde{\mathbf{G}}(z)^{\sim}\tilde{\mathbf{G}}(z) = \tilde{\mathbf{G}}_o(z)^{\sim}\tilde{\mathbf{G}}_o(z). \tag{5.126}$$

Hence, $\tilde{\mathbf{G}}_o(z)$ is a right spectral factor of $\Phi(z)$. Spectral factors $\mathbf{G}_o(z)$ in (5.123) and $\tilde{\mathbf{G}}_o(z)$ in (5.126) are also called *outers* because both are stable and their inverses are analytic outside the unit circle. Moreover,

$$\mathbf{G}_i(z) = \mathbf{G}_o^{-1}(z)\mathbf{G}(z), \quad \tilde{\mathbf{G}}_i(z) = \tilde{\mathbf{G}}(z)\tilde{\mathbf{G}}_o^{-1}(z) \tag{5.127}$$

satisfy $\mathbf{G}_i(z)\mathbf{G}_i(z)^{\sim} = I$ and $\tilde{\mathbf{G}}_i(z)^{\sim}\tilde{\mathbf{G}}_i(z) = I$. Hence, all transmission zeros of $\mathbf{G}_i(z)$ and $\tilde{\mathbf{G}}_i(z)$ are unstable, or their inverses are analytic inside the unit circle. For this reason, $\tilde{\mathbf{G}}_i(z)$ is called *inner* and $\mathbf{G}_i(z)$ called *co-inner*. In light of (5.127) and Lemmas 5.1 and 5.2,

$$\mathbf{G}(z) = \mathbf{G}_o\mathbf{G}_i(z), \quad \tilde{\mathbf{G}}(z) = \tilde{\mathbf{G}}_i(z)\tilde{\mathbf{G}}_o(z)$$

which are termed inner-outer factorizations. The next result is thus true.

Theorem 5.28. *Let A be a stability matrix and D have size $p \times m$.* (i) *If $p \leq m, R = DD^* > 0$, and the* ARE *(5.52) admits a unique solution $\Sigma \geq 0$, then $\mathbf{G}(z) = D + C(zI - A)^{-1}B$ admits inner-outer factorization $\mathbf{G}(z) = \mathbf{G}_o\mathbf{G}_i(z)$ with*

$$\mathbf{G}_o = \left[\begin{array}{c|c} A & K\Omega \\ \hline -C & \Omega \end{array}\right], \quad \mathbf{G}_i(z) = \left[\begin{array}{c|c} A + KC & B + KD \\ \hline \Omega^{-1}C & \Omega^{-1}D \end{array}\right], \tag{5.128}$$

where K is the Kalman gain as defined in (5.51) and $\Omega = (R + C\Sigma C^*)^{1/2}$. (ii) *If $p \geq m, R = D^*D > 0$, and the* ARE *(5.106) admits a unique solution $X \geq 0$, then $\tilde{\mathbf{G}}(z) = D + C(zI - A)^{-1}B$ admits inner-outer factorization $\tilde{\mathbf{G}}(z) = \tilde{\mathbf{G}}_i(z)\tilde{\mathbf{G}}_o(z)$ with*

$$\tilde{\mathbf{G}}_{\text{o}} = \left[\begin{array}{c|c} A & -B \\ \hline \tilde{\Omega}F & \tilde{\Omega} \end{array}\right], \quad \tilde{\mathbf{G}}_{\text{i}}(z) = \left[\begin{array}{c|c} A+BF & B\tilde{\Omega}^{-1} \\ \hline C+DF & D\tilde{\Omega}^{-1} \end{array}\right], \tag{5.129}$$

where F is defined as in (5.107) and $\tilde{\Omega} = (R+B^*XB)^{1/2}$.

Remark 5.1. The hypothesis $R > \mathbf{0}$ in Theorem 5.28 can be weakened to

$$\text{(i) rank}\left\{\begin{bmatrix} C & D \end{bmatrix}\right\} = p, \quad \text{(ii) rank}\left\{\begin{bmatrix} B \\ D \end{bmatrix}\right\} = m, \tag{5.130}$$

respectively, even if D may not have the full rank. Indeed, for (i) there holds

$$\mathbf{G}(z)\mathbf{G}(z)^{\sim} = \left[I - C(zI-A)^{-1}K\right](R+C\Sigma C^*)\left[I - C(zI-A)^{-1}K\right]^{\sim} \tag{5.131}$$

in light of (5.119) in Lemma 5.1. Hence, (i) of (5.130) implies that $\mathbf{G}(z)$ has normal rank equal to p that in turn implies that $(R+C\Sigma C^*)$ is nonsingular. Similarly, for (ii) there holds

$$\mathbf{G}(z)^{\sim}\mathbf{G}(z) = \left[I - F(zI-A)^{-1}B\right]^{\sim}(R+B^*XB)\left[I - F(zI-A)^{-1}B\right] \tag{5.132}$$

that is dual to (5.131). Hence, (ii) of (5.130) implies that $(R+B^*XB)$ is nonsingular. Consequently, the formulas in Theorem 5.28 for computing inner-outer factorizations are valid under the weak conditions in (5.130).

5.3.2 Normalized Coprime Factorizations

Coprime factorizations have been studied in Sect. 3.2.3. For a given plant model

$$\mathbf{P}(z) = D + C(zI-A)^{-1}B \tag{5.133}$$

coprime factorizations search for $\{\mathbf{M}(z),\mathbf{N}(z)\}$ and $\{\tilde{\mathbf{M}}(z),\tilde{\mathbf{N}}(z)\}$ which are stable transfer matrices such that

$$\mathbf{P}(z) = \mathbf{M}(z)^{-1}\mathbf{N}(z) = \tilde{\mathbf{N}}(z)\tilde{\mathbf{M}}(z)^{-1}$$

and the augmented transfer matrices

$$\tilde{\mathbf{G}}(z) = \begin{bmatrix} \tilde{\mathbf{M}}(z) \\ \tilde{\mathbf{N}}(z) \end{bmatrix}, \quad \mathbf{G}(z) = \begin{bmatrix} \mathbf{M}(z) & \mathbf{N}(z) \end{bmatrix} \tag{5.134}$$

void zeros on and outside the unit circle. In other words, $\tilde{\mathbf{G}}(z)$ and $\mathbf{G}(z)$ are outers. Normalized coprime factorizations search for coprime factors such that $\tilde{\mathbf{G}}(z)$ and

$\mathbf{G}(z)$ are not only outers but also inner and co-inner respectively:

$$\tilde{\mathbf{G}}(z)^{\sim}\tilde{\mathbf{G}}(z) = \tilde{\mathbf{M}}(z)^{\sim}\tilde{\mathbf{M}}(z) + \tilde{\mathbf{N}}(z)^{\sim}\tilde{\mathbf{N}}(z) = I,$$
$$\mathbf{G}(z)\mathbf{G}(z)^{\sim} = \mathbf{M}(z)\mathbf{M}(z)^{\sim} + \mathbf{N}(z)\mathbf{N}(z)^{\sim} = I.$$

Such $\mathbf{G}(z)$ and $\tilde{\mathbf{G}}(z)$ are termed *power complementary* in the signal processing literature. The following result shows that normalized coprime factorizations can be solved via Kalman filtering and LQR control.

Theorem 5.29. *Denote $R_o = I + DD^*/\tilde{R}_o = I + D^*D$ for $\mathbf{P}(z)$ in (5.133).*

(i) *Assume that (C,A) is detectable and (A,B) has no unreachable modes on the unit circle. Let $A_o = A - BD^*R_o^{-1}C$. Then the following* ARE

$$\Sigma = A_o\Sigma\left(I + C^*R_o^{-1}C\Sigma\right)^{-1}A_o^* + B\tilde{R}_o^{-1}B^* \tag{5.135}$$

admits a unique stabilizing solution $\Sigma = \Sigma^* \geq \mathbf{0}$. A state-space realization of the normalized (right) coprime factors is given by

$$\mathbf{G}(z) = \begin{bmatrix} \mathbf{M}(z) & \mathbf{N}(z) \end{bmatrix} = \left[\begin{array}{c|cc} A+KC & K & B+KD \\ \hline \Omega_o^{-1}C & \Omega_o^{-1} & \Omega_o^{-1}D \end{array}\right], \tag{5.136}$$

where $K = -(A\Sigma C^* + BD^*)(R_o + C\Sigma C^*)^{-1}$ and $\Omega_o = (R_o + C\Sigma C^*)^{1/2}$.

(ii) *Assume that (A,B) is stabilizable and (C,A) has no unobservable modes on the unit circle. Let $\tilde{A}_o = A - B\tilde{R}_o^{-1}D^*C$. Then the following* ARE

$$X = \tilde{A}_o^*X\left(I + B\tilde{R}_o^{-1}B^*X\right)^{-1}\tilde{A}_o + C^*R_o^{-1}C \tag{5.137}$$

admits a unique stabilizing solution $X = X^* \geq \mathbf{0}$. A state-space realization of the normalized (left) coprime factors is given by

$$\tilde{\mathbf{G}}(z) = \begin{bmatrix} \tilde{\mathbf{M}}(z) \\ \tilde{\mathbf{N}}(z) \end{bmatrix} = \left[\begin{array}{c|c} A+BF & B\tilde{\Omega}_o^{-1} \\ \hline F & \tilde{\Omega}_o^{-1} \\ C+DF & D\tilde{\Omega}_o^{-1} \end{array}\right], \tag{5.138}$$

where $F = -\left(\tilde{R}_o + B^*XB\right)^{-1}(B^*XA + D^*C)$ and $\tilde{\Omega}_o = \left(\tilde{R}_o + B^*XB\right)^{1/2}$.

Proof. For (i), the pair $\{\mathbf{M}(z), \mathbf{N}(z)\}$ in (5.136) is a pair of left coprime factors for K is stabilizing. To show that $\{\mathbf{M}(z), \mathbf{N}(z)\}$ is normalized, denote $B_v = \begin{bmatrix} \mathbf{0} & B \end{bmatrix}$ and $D_v = \begin{bmatrix} I & D \end{bmatrix}$. Let

$$\mathbf{T}(z) = D_v + C(zI - A)^{-1}B_v \tag{5.139}$$

and associate $\mathbf{T}(z)$ with the following Kalman filtering problem:

$$\mathbf{x}(t+1) = A\mathbf{x}(t) + B_{\mathrm{v}}\mathbf{v}(t), \quad \mathbf{y}(t) = C\mathbf{x}(t) + D_{\mathrm{v}}\mathbf{v}(t),$$

where $\mathbf{v}(t)$ is an independent white noise process with zero mean and identity covariance. Applying the results of the stationary Kalman filter yields the ARE (5.135) and the required Kalman gain K which is stabilizing by the hypothesis. In light of the proof of Lemma 5.1, the filter

$$\mathbf{W}(z) = [I + C(zI - A - KC)^{-1}K]\mathbf{T}(z) = \left[\begin{array}{c|c} A+KC & B_{\mathrm{v}} + KD_{\mathrm{v}} \\ \hline C & D_{\mathrm{v}} \end{array}\right]$$

$$= \left[\begin{array}{c|cc} A+KC & K & B+KD \\ \hline C & I & D \end{array}\right]$$

has the white PSD. That is, $\mathbf{W}(z)\mathbf{W}(z)^{\sim} = \Omega_{\mathrm{o}}\Omega_{\mathrm{o}}^* = R_{\mathrm{o}} + C\Sigma C^*$, and hence, $\Omega_{\mathrm{o}}^{-1}\mathbf{W}(z)$ is co-inner and has the same realization as in (5.136). It follows that $\{\mathbf{M}(z), \mathbf{N}(z)\}$ is a pair of the normalized left coprime factors. Since (ii) is dual to (i), the proof for (ii) is similar and omitted. □

For the given left and right normalized coprime factors in (5.136) and (5.138), respectively, the following result gives their respective reachability and observability gramians.

Theorem 5.30. *Consider Theorem 5.29. The reachability gramian P and observability gramian Q of $\mathbf{G}(z)$ as in (5.136) are given respectively by:*

$$P = \Sigma, \quad Q = (I + X\Sigma)^{-1}X \tag{5.140}$$

while the reachability gramian $\tilde{P}$ and observability gramian $\tilde{Q}$ of $\tilde{\mathbf{G}}(z)$ as in (5.138) are given respectively by

$$\tilde{P} = (I + \Sigma X)^{-1}\Sigma, \quad \tilde{Q} = X. \tag{5.141}$$

Proof. By definition the controllability gramian of $\mathbf{G}(z)$ in (5.136) satisfies

$$P = (A+KC)^* P (A+KC) + (B_{\mathrm{v}} + KD_{\mathrm{v}})(B_{\mathrm{v}} + KD_{\mathrm{v}})^*$$

with $B_{\mathrm{v}} = \begin{bmatrix} \mathbf{0} & B \end{bmatrix}$ and $D_{\mathrm{v}} = \begin{bmatrix} I & D \end{bmatrix}$. The above is the same as the ARE (5.135) if $P = \Sigma$. Hence, Σ is indeed the controllability gramian of $\mathbf{G}(z)$. Now assume temporarily that $\det(A_{\mathrm{o}}) \neq 0$ and $\det(A+KC) \neq 0$. Since $D^* R_{\mathrm{o}}^{-1} = \tilde{R}_{\mathrm{o}}^{-1} D^*$, $A_{\mathrm{o}} = \tilde{A}_{\mathrm{o}}$. The ARE in (5.137) can then be written as

$$\begin{bmatrix} -X & I \end{bmatrix} S \begin{bmatrix} I \\ X \end{bmatrix} = \mathbf{0}, \quad S = \begin{bmatrix} A_{\mathrm{o}} + \Gamma (A_{\mathrm{o}}^*)^{-1}\Pi & -\Gamma (A_{\mathrm{o}}^*)^{-1} \\ -(A_{\mathrm{o}}^*)^{-1}\Pi & (A_{\mathrm{o}}^*)^{-1} \end{bmatrix},$$

where $\Pi = C^* R_o^{-1} C$ and $\Gamma = B\tilde{R}_o^{-1} B^*$ (refer to Appendix A). Denote

$$T = \begin{bmatrix} I & \Sigma \\ \mathbf{0} & I \end{bmatrix} \implies T^{-1} = \begin{bmatrix} I & -\Sigma \\ \mathbf{0} & I \end{bmatrix}.$$

Let $\tilde{S} = TST^{-1}$. The ARE in (5.137) can be written as

$$\begin{bmatrix} -Z & I \end{bmatrix} \tilde{S} \begin{bmatrix} I \\ Z \end{bmatrix} = \mathbf{0}, \quad Z = (I + X\Sigma)^{-1} X. \tag{5.142}$$

Direction computation yields

$$\tilde{S} = \begin{bmatrix} \tilde{S}_{11} & \tilde{S}_{12} \\ \tilde{S}_{21} & \tilde{S}_{22} \end{bmatrix} = \begin{bmatrix} A_o + (\Gamma - \Sigma)(A_o^*)^{-1}\Pi & \mathbf{0} \\ -(A_o^*)^{-1}\Pi & (A_o^*)^{-1}(\Pi\Sigma + I) \end{bmatrix}$$

due to $\tilde{S}_{12} = -\left[A_o \Sigma (I + \Pi\Sigma)^{-1} A_o^* + \Gamma - \Sigma\right] (A_o^*)^{-1} (\Pi\Sigma + I) = \mathbf{0}$ by the ARE in (5.135). On the other hand, the results on Kalman filtering with the dynamic model in (5.139) show that

$$A + KC = A_o (I + \Pi\Sigma)^{-1} \implies \tilde{S}_{22} = [(A + KC)^*]^{-1}.$$

Since the ARE in (5.135) can be written as $\Gamma - \Sigma = -A_o \Sigma (I + \Pi\Sigma)^{-1} A_o^*$,

$$\tilde{S}_{11} = A_o - A_o \Sigma (I + \Pi\Sigma)^{-1} \Pi = A_o - A_o \Sigma C^* (R_o + C\Sigma C^*)^{-1} C = A + KC.$$

Finally, by the expression of $\tilde{S}_{22}$,

$$\begin{aligned} \tilde{S}_{21} &= -(A_o^*)^{-1}\Pi = -[(A + KC)^*]^{-1} (I + \Pi\Sigma)^{-1} \Pi \\ &= -[(A + KC)^*]^{-1} C^* (R_o + C\Sigma C^*) C = -[(A + KC)^*]^{-1} C_\Omega^* C_\Omega, \end{aligned}$$

where $C_\Omega = \Omega_o^{-1} C$. Substituting the above into (5.142) yields

$$\begin{aligned} \mathbf{0} &= \begin{bmatrix} -Z & I \end{bmatrix} \begin{bmatrix} A + KC & \mathbf{0} \\ -[(A + KC)^*]^{-1} C_\Omega^* C_\Omega & [(A + KC)^*]^{-1} \end{bmatrix} \begin{bmatrix} I \\ Z \end{bmatrix} \\ &= -Z(A + KC) + [(A + KC)^*]^{-1} Z - [(A + KC)^*]^{-1} C_\Omega^* C_\Omega. \end{aligned}$$

Multiplying the above by $(A + KC)^*$ from left leads to

$$Z = (A + KC)^* Z (A + KC) + C_\Omega^* C_\Omega$$

which verifies that $Q = Z = (I + X\Sigma)^{-1} X$ is the observability gramian of $\mathbf{G}(z)$. If A and $(A + KC)$ are singular, then A and $(A + KC)$ can be perturbed to A_ε and $A_{\varepsilon K}$, respectively, by adding εI such that both are nonsingular. Similar proof can thus be adopted to obtain the observability gramian Z_ε. The limit $\varepsilon \to 0$ can be taken to conclude the proof for the case when A and $(A+KC)$ are singular. As (5.141) is dual to (5.140), its proof is skipped. □

Notes and References

There are many papers and books on optimal control for continuous-time systems. See [5, 39, 57, 58, 74, 122, 126] for a sample of references. For linear discrete-time systems, readers are referred to [1, 7, 11, 16, 25, 68, 69] for a glimpse of work on optimal control. For optimal estimation or filtering, most of work has been focused on discrete-time systems, except the Kalman–Bucy filter [60]. Many books are available with [8, 54] as the representative.

Exercises

5.1. Let X be a random vector of dimension $n > 1$ that is Gaussian distributed with mean zero and covariance $\Sigma_{\mathbf{xx}}$. Suppose that $\Sigma_{\mathbf{xx}}$ has rank $m < n$. Show that its PDF has the form

$$p_X(X=\mathbf{x}) = \frac{1}{\sqrt{(2\pi)^m \prod_{i=1}^m \sigma_i^2}} \exp\left\{-\frac{1}{2}\mathbf{x}^* \Sigma_{\mathbf{xx}}^+ \mathbf{x}\right\},$$

where $\Sigma_{\mathbf{xx}}^+$ is the pseudoinverse of $\Sigma_{\mathbf{xx}}$ and $\{\sigma_i^2\}_{i=1}^m$ are the m nonzero singular values of $\Sigma_{\mathbf{xx}}$. (*Hint:* Consider first $\Sigma_{\mathbf{xx}} = \mathrm{diag}(\sigma_1^2, \ldots, \sigma_m^2, 0, \ldots, 0)$ and then extend it to the general case.)

5.2. Suppose that the system is described by state-space model

$$\mathbf{x}(t+1) = A\mathbf{x}(t) + B\mathbf{v}(t), \quad \mathbf{y}(t) = C\mathbf{x}(t) + D\mathbf{v}(t),$$

where $\{\mathbf{v}(t)\}$ is a WSS white noise with mean zero and covariance Q_v. Let $\mathbf{H}(z) = D + C(zI-A)^{-1}B$ be the transfer matrix. Show that

$$\begin{aligned}\|\mathbf{y}\|_{\mathscr{P}} = \left\|\mathbf{H}Q_v^{1/2}\right\|_2 &:= \sqrt{\mathrm{Tr}\left\{\frac{1}{2\pi}\int_{-\pi}^{\pi} \mathbf{H}\left(e^{j\omega}\right) Q_v \mathbf{H}\left(e^{j\omega}\right)^* \, d\omega\right\}} \\ &= \sqrt{\mathrm{Tr}\{CPC^* + DQ_vD^*\}},\end{aligned}$$

where $P = APA^* + BQ_dB^*$ is the covariance of the state vector $\mathbf{x}(t)$. Recall that $\|\cdot\|_{\mathscr{P}}$ is the power norm as defined by (2.49) in Chap. 2.

5.3. Consider the nth order state-space system

$$\mathbf{x}(t+1) = A\mathbf{x}(t), \quad \mathbf{y}(t) = C\mathbf{x}(t) + \mathbf{v}(t)$$

with $\mathbf{x}(0) = \mathbf{x}_0 \neq \mathbf{0}$ and $\mathbf{v}(t)$ the measurement noise. Assume that (C,A) is observable. Let $\mathscr{O}_\ell$ be the observability matrix of size $\ell \geq n$ and

$$\mathscr{Y}_\ell = \text{vec}\left\{\left[\mathbf{y}(0)\ \mathbf{y}(1)\ \cdots\ \mathbf{y}(\ell-1)\right]\right\}.$$

Show that the estimate $\hat{\mathbf{x}}_0$ which minimizes the estimation error $\|\mathscr{Y}_\ell - \mathscr{O}_\ell \hat{\mathbf{x}}_0\|$ is given by $\hat{\mathbf{x}}_0 = \left(\mathscr{O}_\ell^* \mathscr{O}_\ell\right)^{-1} \mathscr{O}_\ell^* \mathscr{Y}_\ell$.

5.4. Prove the expression for the conditional PDF in (5.6). What modifications are needed for the PDFs of X and Y and for the conditional PDF in (5.6), if the dimensions of X and Y are different from each other?

5.5. Let X and Y be two jointly distributed random variables. Let $\hat{x}$ be the optimal estimate, given observation $Y = y$, such that

$$\text{E}\{|X - \hat{x}|\ |Y = y\} \leq \text{E}\{|X - z|\ |Y = y\} \quad \forall z.$$

That is, $\hat{x}$ minimizes the absolute error of the estimation. Show that $\hat{x}$ is the median of the conditional density $p_{X|Y}(x|y)$; i.e.,

$$P_{X|Y}\left[X \leq \hat{x}|y\right] = P_{X|Y}\left[X \geq \hat{x}|y\right] = 0.5.$$

5.6. Let X and Y be jointly distributed. If $\text{E}\{XY^*\} = \mathbf{0}$, then X and Y are termed orthogonal. Show that the linear MMSE estimate $\hat{X}$ in (5.15) as in Theorem 5.5 satisfies the orthogonality condition

$$\text{E}\left\{\left(X - \hat{X}\right) Y^*\right\} = \mathbf{0}.$$

Give a geometric interpretation for the above orthogonality condition.

5.7. Suppose that $B_t D_t^* \neq \mathbf{0}$ for the random process in (5.19). Show that for $t \geq k$,

$$\begin{aligned} Q_{t,k} &= \text{E}\left\{\left[\mathbf{y}(t) - \bar{\mathbf{y}}_t\right]\left[\mathbf{y}(k) - \bar{\mathbf{y}}_k\right]^*\right\} \\ &= C_t \Phi_{t,k} P_k C_k^* + C_t \Phi_{t,k+1} B_k D_k^* + D_t D_t^* \delta(t-k), \\ \Gamma_{t,k} &= \text{E}\left\{\left[\mathbf{x}(t) - \bar{\mathbf{x}}_t\right]\left[\mathbf{y}(k) - \bar{\mathbf{y}}_k\right]^*\right\} = \Phi_{t,k} P_k C_k^* + \Phi_{t,k+1} B_k D_k^*. \end{aligned}$$

5.8. Suppose that Ψ and Θ are both square and hermitian, which may not necessarily have the same dimensions. Assume $\Psi > 0$. Show that

$$Z^{-1} = \begin{bmatrix} \Psi & \Omega^* \\ \Omega & \Theta \end{bmatrix}^{-1} = \begin{bmatrix} \Psi^{-1} & 0 \\ 0 & 0 \end{bmatrix} + \begin{bmatrix} \Psi^{-1}\Omega^* \\ -I \end{bmatrix} \nabla^{-1} \begin{bmatrix} \Omega\Psi^{-1} & -I \end{bmatrix}$$

whenever Z is also square and hermitian positive definite, where

$$\nabla = \Theta - \Omega\Psi^{-1}\Omega^*$$

is called *Schur complement*. (*Hint:* Use factorization

$$Z = \begin{bmatrix} \Psi & \Omega^* \\ \Omega & \Theta \end{bmatrix} = \begin{bmatrix} I & 0 \\ \Omega\Psi^{-1} & I \end{bmatrix} \begin{bmatrix} \Psi & 0 \\ 0 & \nabla \end{bmatrix} \begin{bmatrix} I & \Psi^{-1}\Omega^* \\ 0 & I \end{bmatrix}$$

to compute the inverse of Z). What if $\Theta > 0$ but Ψ is singular?

5.9. Prove Theorem 5.7.

5.10. (i) Use the matrix inversion formula (refer to Appendix A)

$$\left(F + HJ^{-1}G\right)^{-1} = F^{-1} - F^{-1}H\left(J + GF^{-1}H\right)^{-1} GF^{-1}$$

to show that with the Kalman gain in (5.35),

$$A_t + K_tC_t = A_t\left(I_n + \Sigma_tC_t^*R_t^{-1}C_t\right)^{-1}.$$

(ii) Show that the DRE (5.36) can be equivalently written as

$$\begin{aligned}\Sigma_{t+1} &= A_t\left(I_n + \Sigma_tC_t^*R_t^{-1}C_t\right)^{-1}\Sigma_tA_t^* + B_tB_t^* \\ &= (A_t + K_tC_t)\,\Sigma_t\,(A_t + K_tC_t)^* + B_tB_t^* + K_tR_tK_t^*.\end{aligned}$$

(iii) Show that for $B_tD_t^* \neq \mathbf{0}$, the DRE in (5.43) can be written as

$$\begin{aligned}\Sigma_{t+1} &= \tilde{A}_t\left(I_n + \Sigma_tC_t^*R_t^{-1}C_t\right)^{-1}\Sigma_t\tilde{A}_t^* + B_t\left(I_m - D_t^*R_t^{-1}D_t\right)B_t^* \\ &= (A_t + K_tC_t)\,\Sigma_t\,(A_t + K_tC_t)^* + (B_t + K_tD_t)\,(B_t + K_tD_t)^*,\end{aligned}$$

where $\tilde{A}_t = A_t - B_tD_t^*R_t^{-1}C_t$ and $K_t = -(A_t\Sigma_tC_t^* + B_tD_t^*)(R_t + C_t\Sigma_tC_t^*)^{-1}$.

5.11. Consider linear estimator

$$\tilde{\mathbf{x}}_{t+1} = (A + L_tC_t)\tilde{\mathbf{x}}_t - L_t\mathbf{y}(t), \quad \tilde{\mathbf{x}}_0 = \overline{\mathbf{x}}_0$$

for the process in (5.19). Let $Q_t = \mathrm{E}\{[\mathbf{x}(t) - \tilde{\mathbf{x}}_t][\mathbf{x}(t) - \tilde{\mathbf{x}}_t]^*\}$ be its error covariance. Show that $Q_t \geq \Sigma_t$ for all $t \geq 0$ with Σ_t the error covariance for the Kalman filter.

5.12. (Kalman filter as a whitening filter) For the random process described in (5.19), consider the linear estimator of the form

$$\hat{\mathbf{x}}(t+1) = (A_t + L_t C_t)\hat{\mathbf{x}}(t) - L_t \mathbf{y}(t), \quad \delta\mathbf{y}(t) = \mathbf{y}(t) - \hat{\mathbf{y}}(t)$$

with $\hat{\mathbf{y}}(t) = C_t \hat{\mathbf{x}}(t)$ and $\hat{\mathbf{x}}(0) = \overline{\mathbf{x}}_0$. Note that $\{\delta\mathbf{y}(t)\}$ is the innovation sequence. Show that the output process $\{\delta\mathbf{y}(t)\}$ is white (i.e.,

$$\mathrm{E}\{\delta\mathbf{y}(t)\delta\mathbf{y}(t-k)^*\} = \mathrm{E}\{\delta\mathbf{y}(t)\delta\mathbf{y}(t)^*\}\delta(k)$$

for all t and k), if and only if

$$(B_t + L_t D_t)D_t^* + (A_t + L_t C_t)X_t C_t^* = \mathbf{0},$$

where $X_t = \mathrm{E}\{[\mathbf{x}(t) - \hat{\mathbf{x}}(t)][\mathbf{x}(t) - \hat{\mathbf{x}}(t)]^*\}$ is the error covariance. Show also in this case that L_t is necessarily the Kalman gain K_t as in Corollary 5.1 and $\hat{\mathbf{x}}(t+1) = \hat{\mathbf{x}}_{t+1|t}$ is the linear MMSE estimate of $\mathbf{x}(t+1)$ based on $\mathscr{Y}_t$.

5.13. Show that the stabilizing solution to the ARE (5.50), if it exists, is unique. (*Hint:* Assume Σ_1 and Σ_2 are both stabilizing solutions to (5.50). Show that:

$$\Delta_\Sigma = (A + K_1 C)\Delta_\Sigma (A + K_2 C)^*, \quad \Delta_\Sigma = \Sigma_1 - \Sigma_2,$$

where $K_i = -A\Sigma_i C^*(R + C\Sigma_i C^*)^{-1}$ for $i = 1, 2$.)

5.14. For Example 5.12, find the optimal linear receiver in the case $H_0(t) \neq \mathbf{0}$ and discuss its performance in comparison with that of the optimal linear receivers designed in Example 5.15 assuming that H_k are the same for $1 \leq k \leq \ell$.

5.15. Prove Theorem 5.14 for the case $B_t D_{2t}^* \neq \mathbf{0}$.

5.16. Suppose that the random process has the state-space form:

$$\mathbf{x}(t+1) = A_t \mathbf{x}(t) + B_t \mathbf{v}(t), \quad \mathbf{y}(t) = C_t \mathbf{x}(t) + D_t \mathbf{v}(t),$$

where $\mathbf{v}(t)$ is a white process satisfying (5.20) and $\mathbf{x}(0) = \mathbf{x}_0$ is random and independent of $\{\mathbf{v}(t)\}$ with mean $\overline{\mathbf{x}}_0$ and covariance P_0. (i) Find the linear MMSE estimator for $\mathbf{x}(t)$ based on observation $\{\mathbf{y}(k)\}_{k=0}^{t}$. (ii) Find the linear MMSE estimator for $\mathbf{v}(t)$ based on observation $\{\mathbf{y}(k)\}_{k=0}^{t}$. (*Hint:* Use Theorem 5.14).

5.17. Use Simulink toolbox to program and simulate data detection for a SISO channel with gains $h_k = 1/\sqrt{5}$ for $0 \leq k \leq \ell = 4$. The symbol detector is the linear receiver (based on Kalman filter) followed by a quantizer $Q_n(\cdot) = \mathrm{sign}(\cdot)$. The observation noise $\{v(t)\}$ can be generated by normal distributed uncorrelated or white random variables with variance 0.1. The data block of the same size can generated in a similar way followed by $Q_n(\cdot) = \mathrm{sign}(\cdot)$ to produce ± 1 sequence. It is emphasized that the data and noise sequences are uncorrelated. In the context of Example 5.15, do the following:

(i) Design an MMSE estimator to estimate $s(t-m)$ with $m = 2\ell$ followed by a quantizer based on observation of the channel output up to time t.
(ii) Simulate and access the average performance of the detector by counting the number of detection errors in each block of 10^4 data assuming that the receiver knows the first ℓ transmitted data.

5.18. For the output estimator in Theorem 5.14, show that the output error variance $\mathrm{E}\{\|\mathbf{e}_z(t)\|^2\}$ is given by

$$\mathrm{Tr}\{(D_{1t}+L_tD_{2t})(D_{1t}+L_tD_{2t})^* + (C_{1t}+L_tC_{2t})\Sigma_t(C_{1t}+L_tC_{2t})^*\}$$

(*Hint:* Let $\mathbf{e}(k) = \mathbf{x}(k) - \hat{\mathbf{x}}_{k|k-1}$ for $k = t, t+1$. Show first that

$$\begin{aligned} \mathbf{e}(t+1) &= (A_t + K_tC_{2t})\mathbf{e}(t) + (B_t + K_tD_{2t})\mathbf{v}(t) \\ \mathbf{e}_z(t) &= (C_{1t}+L_tC_{2t})\mathbf{e}(t) + (D_{1t}+L_tD_{2t})\mathbf{v}(t) \end{aligned} \tag{5.143}$$

and then compute the variance of $\mathbf{e}_z(t)$.)

5.19. Consider the equivalent Wiener filtering as in Fig. 5.6 where $\mathbf{G}_1(z)$ and $\mathbf{G}_2(z)$ are both stable and causal. Suppose that $z^{-m}\mathbf{G}_1(z)$ is noncausal in the case $m < 0$. Decompose

$$z^{-m}\mathbf{G}_1(z) = \mathbf{G}_C(z) + \mathbf{G}_A(z),$$

where $\mathbf{G}_C(z)$ is causal and $\mathbf{G}_A(z)$ is anticausal. Show that the optimal estimation for the output of $\mathbf{G}_A(z)$ with white noise input is zero and thus conclude that the optimal estimate $\hat{\mathbf{z}}_{t|t-m}$ is independent of $\mathbf{G}_A(z)$. Provide a design procedure for the optimal output estimation. (*Hint:* Use the result from the solution to Problem 2.19.)

5.20. Consider the nth order state-space system

$$\mathbf{x}(t+1) = A\mathbf{x}(t) + B\mathbf{u}(t), \quad \mathbf{x}(0) = \mathbf{x}_0 \neq \mathbf{0}.$$

Assume that (A,B) is controllable with $\mathscr{C}_\ell$ the controllability matrix of size $\ell > n$. Let

$$\tilde{\mathscr{U}}_\ell = \mathrm{vec}\left\{\left[\mathbf{u}(\ell-1)\ \mathbf{u}(\ell-2)\ \cdots\ \mathbf{u}(0)\right]\right\}.$$

Show that the control input $\tilde{\mathscr{U}}_\ell = -\mathscr{C}_\ell^*(\mathscr{C}_\ell\mathscr{C}_\ell^*)^{-1}A^\ell\mathbf{x}_0$ has the minimum energy $\|\tilde{\mathscr{U}}_\ell\|^2$ among all possible control inputs which brings $\mathbf{x}(\ell)$ to the origin.

5.21. Use direct computation to show that the solution X_t to the DRE (5.86) with boundary condition $X_T = \mathbf{0}$ satisfies

$$X_0 \geq X_1 \geq \cdots X_{T-1} \geq X_T = \mathbf{0}.$$

5.22. Show that the closed-loop system under the LQR control (5.87) is $\mathbf{x}(t+1) = \left(I + B_t R_t^{-1} B_t^* X_{t+1}\right)^{-1} A_t \mathbf{x}(t)$. Show also that the DRE (5.86) can be written as

$$X_t = A_t^* X_{t+1} \left(I + B_t R_t^{-1} B_t^* X_{t+1}\right)^{-1} A_t + Q_t, \quad X_T = Q_T.$$

5.23. Show that the control DRE in (5.86) can be written into:

$$X_t = (A_t + B_t F_t)^* X_{t+1} (A_t + B_t F_t) + C_t^* C_t + F_t^* R_t F_t$$

with F_t in (5.87) being the optimal state-feedback gain for the LQR control. If the control law $\mathbf{u}(t) = G_t \mathbf{x}(t)$ is used with $G_t \neq F_t$, show that the solution to

$$Y_t = (A_t + B_t G_t)^* Y_{t+1} (A_t + B_t G_t) + C_t^* C_t + G_t^* R_t G_t$$

satisfies $Y_t \geq X_t$ for $0 \leq t \leq T$ where $X_T = Y_T = \mathbf{0}$ is assumed.

5.24. For the system model in (5.82), let the controlled output be $\mathbf{z}(t) = C_t \mathbf{x}(t) + D_t \mathbf{u}(t)$ and the performance index be

$$J_T = \mathbf{x}(T)^* Q_T \mathbf{x}(T) + \sum_{t=0}^{T-1} \|\mathbf{z}(t)\|^2. \tag{5.144}$$

Denote $\tilde{A}_t = A_t - B_{2t} R_t^{-1} D_t^* C_t$ and $\tilde{C}_t = \left(I - D_t \left(D_t^* D_t\right)^{-1} D_t^*\right) C_t$. Show that the control law which minimizes J_T is given by $\mathbf{u}(t) = F_t \mathbf{x}(t)$ where

$$\begin{aligned} F_t &= -\left(R_t + B_t^* X_{t+1} B_t\right)^{-1} \left(B_{2t}^* X_{t+1} A_t + D_t^* C_t\right), \quad R_t = D_t^* D_t > 0 \\ X_t &= \tilde{A}_t^* X_{t+1} \left(I + B_t R_t^{-1} B_t^* X_{t+1}\right)^{-1} \tilde{A}_t + \tilde{C}_t^* \tilde{C}_t, \quad X_T = Q_T. \end{aligned}$$

5.25. Let the state-space model be as in (5.82). Show that if the open-loop system $\mathbf{x}(t+1) = A_t \mathbf{x}(t)$ is asymptotically (exponentially) stable, then the closed-loop system (5.95) for the LQR control as described in Theorem 5.17 is asymptotically (exponentially) stable as $T \to \infty$.

5.26. Prove Theorem 5.20.

5.27. For any stabilizing state-feedback gain F, show that

$$\begin{aligned} X &= (A + BF)^* X (A + BF) + F^* R F + Q \\ &= A^* \left(I_n + X B R^{-1} B^*\right)^{-1} X A + Q + \Delta_F^* (R + B^* X B) \Delta_F, \end{aligned}$$

where $\Delta_F = F + (R + B^* X B)^{-1} B^* X A$. Establish that (a) the LQR control law minimizes $\mathrm{Tr}\{X\}$ and (b) the stabilizing solution to the ARE (5.96) is maximal among all possible nonnegative solutions to the ARE (5.96).

5.28. Consider $\mathbf{x}(t+1) = A\mathbf{x}(t) + B\mathbf{u}(t)$ and assume that $(A+BF)$ is a stability matrix for some state-feedback gain F. (i) Let $\mathbf{u}(t) = F\mathbf{x}(t)$ be the state-feedback control law. Show that

$$J(F) = \sum_{t=0}^{\infty} \|\mathbf{u}(t)\|^2 + \|C\mathbf{x}(t)\|^2 = \mathbf{x}_0' X \mathbf{x}_0,$$

where $X = (A+BF)^* X (A+BF) + F^* F + C^* C$ and $\mathbf{x}(0) = \mathbf{x}_0$ is the initial condition. (ii) Let $X_{\mathrm{m}} \geq \mathbf{0}$ be the stabilizing solution to

$$X_{\mathbf{m}} = A^* X_{\mathbf{m}} (I + BB^* X_{\mathbf{m}})^{-1} A + C^* C.$$

Show that $X_{\mathrm{m}} \leq X$.

5.29. Suppose that the ARE (5.96) admits a stabilizing solution. Show that X is positive definite, if and only if all stable modes of (C,A) are observable where $Q = C^* C$.

5.30. (i) Let $\tilde{A} = A - BR^{-1}D^*C$ and $\tilde{C} = \left[I - DR^{-1}D^*\right] C$ with $R = D^*D > \mathbf{0}$. Show that the ARE $X = \tilde{A}^* X \left[I + BR^{-1}B^*X\right]^{-1} \tilde{A} + \tilde{C}^* \tilde{C}$ can be equivalently written as the following ARE:

$$X = A^*XA - (A^*XB + C^*D)(R + B^*XB)^{-1}(B^*XA + D^*C) + C^*C.$$

(ii) What are the equivalent AREs in the case of optimal estimation?

5.31. Consider the DRE

$$X_t(T) = A^* X_{t+1}(T) \left[I_n + BR^{-1}B^* X_{t+1}(T)\right]^{-1} A + Q, \quad X_T(T) = \mathbf{0}.$$

Show that $\{X_t(T)\}$ satisfy $X_t(T) \geq X_{t+1}(T)$ for $0 \leq t < T$. (*Hint:* Use the same idea as in the proof of Proposition 5.1).

5.32. Suppose that (A,B) is stabilizable and the condition (5.100) holds. Construct a numerical example for which the ARE (5.96) has a solution $X_u \geq 0$ that is not stabilizing.

5.33. Prove Theorem 5.25. (*Hint:* Consider the augmented state vector $\breve{\mathbf{x}}(t) = \left[\mathbf{x}(t)^* \ \mathbf{v}(t)^*\right]^*$ and then convert the full information control to the state-feedback control problem as in Theorem 5.19.)

5.34. Let (A,B,C,D) be a minimal realization of $\mathbf{G}(z)$ with $A \in \mathbb{C}^{n\times n}$, $B \in \mathbb{C}^{n\times m}$, and $C \in \mathbb{C}^{p\times n}$. Assume that D has full rank. (i) If $p \leq m$, show that $z = z_0$ is a transmission zero of $\mathbf{G}(z)$, if and only if z_0 is an unreachable mode of $(\tilde{A}, \tilde{B})$ with

$$\tilde{A} = A - BD^*(DD^*)^{-1}C, \quad \tilde{B} = B\left(I - D^*(DD^*)^{-1}D\right).$$

(ii) If $p \geq m$, show that $z = z_0$ is a transmission zero of $\mathbf{G}(z)$, if and only if z_0 is an unobservable mode of $(\tilde{C}, \tilde{A})$ with

$$\tilde{A} = A - B(D^*D)^{-1}D^*C, \quad \tilde{C} = \left(I - D(D^*D)^{-1}D^*\right)C.$$

5.35. Let $\mathbf{G}(z)$ in (5.118) be stable. (i) Show that

$$\Phi(z) = \mathbf{G}(z)\mathbf{G}(z)^{\sim} = DD^* + CPC^* + C(zI - A)^{-1}L + L^*\left(z^{-1}I - A^*\right)^{-1}C^*,$$

where $L = APC^* + BD^*$ and $P = APA^* + BB^*$. (ii) Show that

$$\Phi(z) = \mathbf{G}(z)^{\sim}\mathbf{G}(z) = D^*D + B^*QB + H(zI - A)^{-1}B + B^*\left(z^{-1}I - A^*\right)^{-1}H^*,$$

where $H = B^*QA + D^*C$ and $Q = A^*QA + C^*C$.

5.36. Let $\mathbf{G}(z) = D + C(zI - A)^{-1}B$ with A a stability matrix, $R = DD^*$ nonsingular and $BD^* = \mathbf{0}$. (i) Show that

$$\mathbf{G}\left(e^{j\omega}\right)\mathbf{G}\left(e^{j\omega}\right)^* = R + C\left(e^{j\omega}I - A\right)^{-1}BB^*\left(e^{-j\omega}I - A^*\right)^{-1}C^* \geq R.$$

(ii) Use Lemma 5.1 and (i) to show that for each real ω,

$$\left[I - C(e^{j\omega}I - A)^{-1}K\right](R + C\Sigma C^*)\left[I - C(e^{j\omega}I - A)^{-1}K\right]^* \geq R.$$

Chapter 6
Design of Feedback Control Systems

Feedback control has triumphed for more than half a century, but its earliest invention was dated some 2000 years ago by ancient Greeks in float regulators of water clocks. More active use of feedback control was around the time of *Industrial Revolution* in Europe when many control devices and regulators were invented. The flourish of such control devices and regulators attracted mathematicians' attention from Maxwell, Routh, Lyapunov, and Hurwitz, who contributed to the stability theory for systems governed by differential equations. However, it was until after Black reinvented negative feedback into design of amplifiers that initiated mathematical analysis of feedback control systems in frequency domain. The role of the Bell Lab and the market need of mass communications are essential in the birth of classical control theory symbolized by the work of Bode and Nyquist in the late 1930s. The World War II stimulated further the development of the classical control technology with wide applications, which proliferated to academia postwar.

While being highly successful, the classical control theory is limited in its applications to primarily SISO systems due to the graphical nature of its design tools such as methods of Bode and root locus. For multi-input/multi-output (MIMO) systems, it is the modern control theory that grows into a mature design methodology, termed as *linear quadratic Gaussian* (LQG) optimal control. It assumes a linear state-space model for the underlying multivariable system with stochastic descriptions for exogenous noises and disturbances impinged on the feedback system. The performance index is a combination of the mean powers of the control input and error signal. Minimization of the performance index is the design objective that admits a closed-form solution for the optimal feedback controller.

LQG control was motivated largely by the space program massively funded in the 1950s. It had a huge success in space applications because accurate models can be developed for space vehicles, white noise descriptions are appropriate for external disturbances, and the fuel consumption is crucial in space missions. Owing to the work of Kalman and other scientists, LQG control has since become an effective design methodology for multivariable feedback control systems. This chapter considers $\mathscr{H}_2$ optimal control that is modified from the LQG control

G. Gu, *Discrete-Time Linear Systems: Theory and Design with Applications*,
DOI 10.1007/978-1-4614-2281-5_6, © Springer Science+Business Media, LLC 2012

focusing on LTI systems. The optimal solution to linear quadratic control will be derived for the case of output feedback rather than state feedback. Design of multivariable feedback control systems will then be studied with the objective to satisfying performance specifications in frequency and time domains. Design tools such as frequency loop shaping and eigenvalue assignment will be developed which can be regarded as extension of Bode and root locus in classic control, respectively. Various other design issues such as controller reduction and stability margins will be investigated as well.

6.1 Output Feedback Control

A familiar feedback system is shown in Fig. 6.1 with $\mathbf{P}(z)$, the physical system to be controlled, and $\mathbf{K}(z)$, the feedback controller to be designed. Exogenous signals are $\mathbf{d}_1(t)$, the disturbance at the input of the system; $\mathbf{d}_2(t)$, the noise corrupted at the output of the system; and $-\mathbf{r}(t)$, the command signal for the system to track. Recall that the negative sign of the feedback path has been absorbed into the feedback controller $\mathbf{K}(z)$ as discussed in Sect. 3.2.3.

In design of feedback control systems, stability is the first priority. On the other hand, performance measure or index is dependent on applications. Possible performance index includes variance of the tracking error signal and the mean power of the control signal, assuming white noises for $\mathbf{d}_1(t)$ and $\mathbf{d}_2(t)$, and WSS process for $\mathbf{r}(t)$.

Example 6.1. Suppose that $\mathbf{d}_1(t) \equiv \mathbf{0}$, $\mathbf{d}_2(t) \equiv \mathbf{0}$, and $\mathbf{r}(t) = \mathbf{H}(q)\mathbf{d}(t)$ for some proper and stable transfer matrix $\mathbf{H}(z)$ with $\mathbf{d}(t)$ white noise of zero mean and identity covariance. The frequency shape of $\mathbf{H}(e^{j\omega})$ represents the frequency contents of the reference signal $\mathbf{r}(t)$. Let α and β be positive satisfying $\alpha + \beta = 1$. The signal to be controlled is chosen as convex combination of the tracking error variance and mean power of the control signal specified by

$$\mathbf{w}(t) := \begin{bmatrix} \alpha\mathbf{u}(t) \\ \beta\mathbf{e}(t) \end{bmatrix} = \begin{bmatrix} \alpha\mathbf{K} \\ \beta I \end{bmatrix} (I - \mathbf{PK})^{-1}\mathbf{Hd}(t) =: \mathbf{T}_{dw}(q)\mathbf{d}(t). \tag{6.1}$$

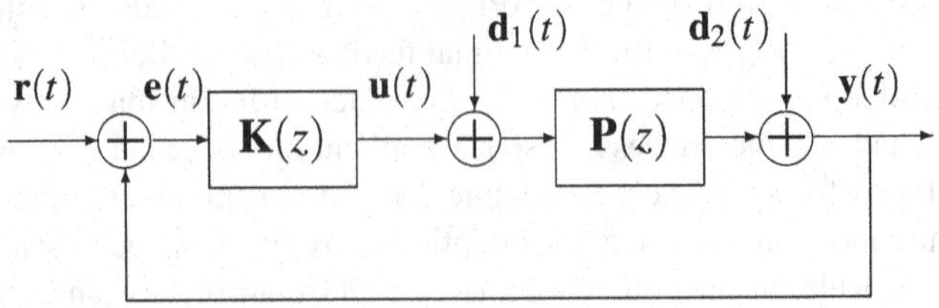

Fig. 6.1 Feedback control system in presence of disturbances

Indeed, the variance of $\mathbf{w}(t)$ is the same as square of its power norm given by

$$\begin{aligned}\|\mathbf{w}\|_{\mathscr{P}}^2 &= \alpha^2\|\mathbf{e}\|_{\mathscr{P}}^2+\beta^2\|\mathbf{u}\|_{\mathscr{P}}^2 = \mathrm{E}\{\|\mathbf{T}_{dw}(q)\mathbf{d}(t)\|^2\} \\ &= \mathrm{Tr}\left\{\tfrac{1}{2\pi}\int_{-\pi}^{\pi}\mathbf{T}_{dw}(e^{j\omega})\mathbf{T}_{dw}(e^{j\omega})^*\,d\omega\right\}.\end{aligned} \tag{6.2}$$

Recall definition of the power norm and its relation to variance in Chap. 2. Note that (6.2) holds only if $\mathbf{T}_{dw}(z)$ is a stable transfer matrix in which case

$$\|\mathbf{w}\|_{\mathscr{P}} = \|\mathbf{T}_{dw}\|_2 := \sqrt{\mathrm{Tr}\left\{\frac{1}{2\pi}\int_{-\pi}^{\pi}\mathbf{T}_{dw}(e^{j\omega})\mathbf{T}_{dw}(e^{j\omega})^*\,d\omega\right\}}.$$

Hence, feedback control design amounts to computing the stabilizing controller $\mathbf{K}(z)$ such that $\|\mathbf{w}\|_{\mathscr{P}} = \|\mathbf{T}_{dw}\|_2$ is minimized. Such a problem is referred to as $\mathscr{H}_2$ optimal control owing to minimization of the $\mathscr{H}_2$ norm of $\mathbf{T}_{dw}(z)$. The scalar weights (α,β) provide trade-offs between minimization of the tracking error and control power. Diagonal and frequency dependent matrices can be employed in place of scalar weights (α,β) to offer more freedom of trade-offs among different output channels and different frequency bands.

Example 6.1 formulates the tracking performance mixed with control power in absence of disturbances. The following presents an example of disturbance rejection without taking tracking performance into consideration explicitly.

Example 6.2. Suppose that $\mathbf{r}(t) \equiv \mathbf{0}$. Define

$$\mathbf{w}(t) := \begin{bmatrix}\mathbf{u}(t)\\ \mathbf{d}(t)\end{bmatrix}, \quad \mathbf{d}(t) := \begin{bmatrix}\mathbf{d}_1(t)\\ \mathbf{d}_2(t)\end{bmatrix}.$$

It can be verified that $\mathbf{w}(t) = \mathbf{T}_{dw}(q)\mathbf{d}(t)$ where

$$\mathbf{T}_{dw} = \begin{bmatrix}\mathbf{K}\\ I\end{bmatrix}(I-\mathbf{PK})^{-1}\begin{bmatrix}\mathbf{P} & I\end{bmatrix}. \tag{6.3}$$

If $\mathbf{d}(t)$ is white with mean zero and covariance Q_d, then $\|\mathbf{w}\|_{\mathscr{P}} = \|\mathbf{T}_{dw}Q_d^{1/2}\|_2$. Disturbance rejection now amounts to design of the stabilizing controller $\mathbf{K}(z)$ such that the following performance index

$$J = \sqrt{\mathrm{Tr}\left\{\frac{1}{2\pi}\int_{-\pi}^{\pi}\mathbf{T}_{dw}(e^{j\omega})Q_d\mathbf{T}_{dw}(e^{j\omega})^*\,d\omega\right\}} \tag{6.4}$$

is minimized. By noting that $\mathbf{T}_{dw}(z)$ contains $(I-\mathbf{PK})^{-1}$ as a submatrix that is the transfer matrix from $\mathbf{r}(t)$ to $\mathbf{e}(t)$, minimization of J in (6.4) ensures small tracking error even though it is not minimized directly.

Two-degree-of-freedom controllers are often used in engineering practice which are illustrated in the next figure.

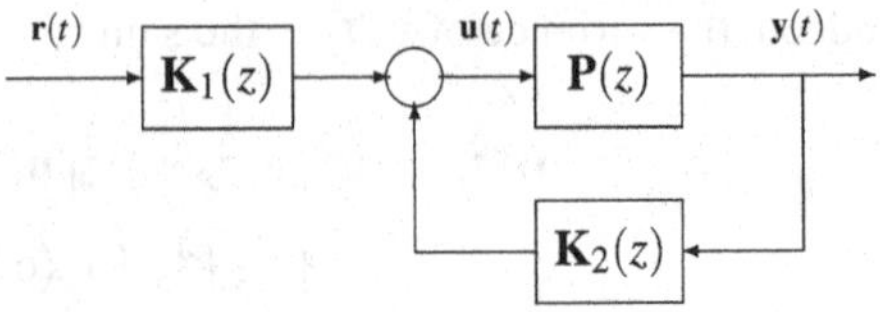

Fig. 6.2 Feedback system with two-degree-of-freedom controller

In comparison with Example 6.1, the extra precompensator $\mathbf{K}_1$ shapes the command signal $\mathbf{r}(t)$ and helps to achieve better tracking performance.

Example 6.3. Let us reexamine the tracking performance by employing the feedback structure in Fig. 6.2. It is easy to see that

$$\begin{aligned}\mathbf{y}(t) &= \mathbf{P}(I-\mathbf{K}_2\mathbf{P})^{-1}\mathbf{K}_1\mathbf{r}(t),\\ \mathbf{u}(t) &= (I-\mathbf{K}_2\mathbf{P})^{-1}\mathbf{K}_1\mathbf{r}(t).\end{aligned}$$

Since the tracking error is $\mathbf{e}(t) = \mathbf{y}(t) - \mathbf{r}(t)$, its expression is given by

$$\mathbf{e}(t) = \left[I - \mathbf{P}(I-\mathbf{K}_2\mathbf{P})^{-1}\mathbf{K}_1\right]\mathbf{r}(t).$$

Consider the following signal to be controlled:

$$\mathbf{w}(t) := \begin{bmatrix}\mathbf{e}(t)\\ \varepsilon\mathbf{u}(t)\end{bmatrix} = \left(\begin{bmatrix} I\\ \mathbf{0}\end{bmatrix} + \begin{bmatrix}-\mathbf{P}\\ \varepsilon I\end{bmatrix}(I-\mathbf{K}_2\mathbf{P})^{-1}\mathbf{K}_1\right)\mathbf{r}(t).$$

The parameter $\varepsilon > 0$ provides the trade-offs between minimization of the tracking error and of the control power. Specifically,

$$\|\mathbf{w}\|_{\mathscr{P}}^2 = \|\mathbf{e}\|_{\mathscr{P}}^2 + \varepsilon^2\|\mathbf{u}\|_{\mathscr{P}}^2.$$

Hence, the above is equivalent to the performance index in Example 6.1 by taking $\alpha = 1/\sqrt{1+\varepsilon^2}$ and $\beta = \varepsilon/\sqrt{1+\varepsilon^2}$. However, there is an extra freedom for control or precompensator $\mathbf{K}_1(z)$ that may help to minimize $\|\mathbf{w}\|_{\mathscr{P}}^2$. Consequently, a similar $\mathscr{H}_2$ optimal control problem is resulted in.

The preceding examples consider minimization of linear combination of the power norms of the tracking error and control signal. Although other signals in the feedback system of Fig. 6.1 or Fig. 6.2 can also be considered for optimization, most exhibits the same structure for $\mathscr{H}_2$ optimal control. This structure is characterized by *linear fractional transformation* (LFT) that is defined by

$$\mathscr{F}(\mathbf{G},\mathbf{P}) = \mathbf{G}_{11} + \mathbf{G}_{12}\mathbf{K}(I-\mathbf{G}_{22}\mathbf{K})^{-1}\mathbf{G}_{21}$$

where $\mathbf{G}$ is partitioned as

$$\mathbf{G} = \left[\begin{array}{c|c}\mathbf{G}_{11} & \mathbf{G}_{12}\\ \hline \mathbf{G}_{21} & \mathbf{G}_{22}\end{array}\right].$$

It can be shown that the closed-loop transfer matrix $\mathbf{T}_{dw}(z)$ in the aforementioned three examples can all be written as LFT of the generalized plant $\mathbf{G}(z)$ and the feedback controller $\mathbf{K}(z)$, i.e.,

$$\mathbf{T}_{dw}(z) = \mathscr{F}[\mathbf{G}(z), \mathbf{K}(z)]$$

for some $\mathbf{G}(z)$ and $\mathbf{K}(z)$. Indeed, in Example 6.1, the generalized plant is given by

$$\mathbf{G}(z) = \left[\begin{array}{c|c} \mathbf{0} & \alpha I \\ \beta \mathbf{H}(z) & \beta \mathbf{P}(z) \\ \hline \mathbf{H}(z) & \mathbf{P}(z) \end{array}\right]. \tag{6.5}$$

In Example 6.2, the generalized plant is given by

$$\mathbf{G}(z) = \left[\begin{array}{cc|c} \mathbf{0} & \mathbf{0} & I \\ \mathbf{P}(z) & I & \mathbf{P}(z) \\ \hline \mathbf{P}(z) & I & \mathbf{P}(z) \end{array}\right]. \tag{6.6}$$

Example 6.3 is a little more subtle. Denote

$$\mathbf{K}(z) = \begin{bmatrix} \mathbf{K}_1(z) & \mathbf{K}_2(z) \end{bmatrix}, \quad \mathbf{P}_a(z) = \begin{bmatrix} \mathbf{0} \\ \mathbf{P}(z) \end{bmatrix}.$$

Recall $\mathbf{r}(t) = \mathbf{H}(q)\mathbf{d}(t)$ where $\mathbf{H}(z)$ is some proper and stable transfer matrix and $\mathbf{d}(t)$ some white process of zero mean and identity covariance. Then $\mathbf{w}(t) = \mathbf{T}_{dw}(q)\mathbf{d}(t)$ with

$$\begin{aligned} \mathbf{T}_{dw} &= \begin{bmatrix} \mathbf{H} \\ \mathbf{0} \end{bmatrix} + \begin{bmatrix} -\mathbf{P} \\ \varepsilon I \end{bmatrix} (I - \mathbf{K}\mathbf{P}_a)^{-1}\mathbf{K} \begin{bmatrix} \mathbf{H} \\ \mathbf{0} \end{bmatrix} \\ &= \begin{bmatrix} \mathbf{H} \\ \mathbf{0} \end{bmatrix} + \begin{bmatrix} -\mathbf{P} \\ \varepsilon I \end{bmatrix} \mathbf{K}(I - \mathbf{P}_a\mathbf{K})^{-1} \begin{bmatrix} \mathbf{H} \\ \mathbf{0} \end{bmatrix}. \end{aligned} \tag{6.7}$$

The generalized plant is thus given by

$$\mathbf{G}(z) = \left[\begin{array}{c|c} \mathbf{H}(z) & -\mathbf{P}(z) \\ \mathbf{0} & \varepsilon I \\ \hline \mathbf{H}(z) & \mathbf{0} \\ \mathbf{0} & \mathbf{P}(z) \end{array}\right]. \tag{6.8}$$

All the three preceding examples show that the related $\mathscr{H}_2$ optimal control problems have the same LFT structure. The difference lies in the forms of the

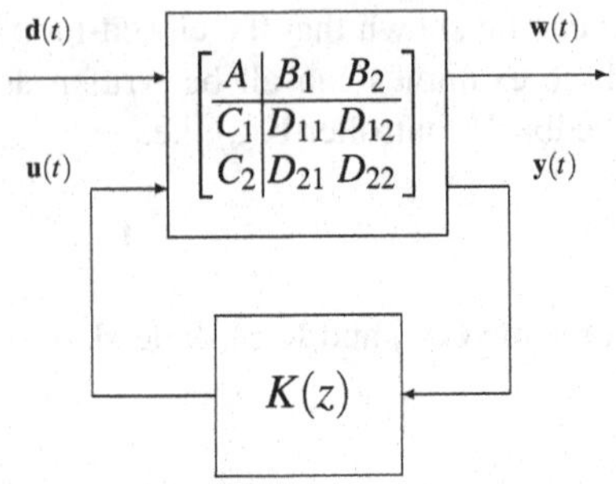

Fig. 6.3 LFT feedback control system

generalized plant and feedback controller which are problem dependent. An optimal solution will be presented in this section to synthesize the feedback controller $\mathbf{K}(z)$ such that

$$\mathbf{T}_{dw}(z) = \mathscr{F}[\mathbf{G}(z), \mathbf{K}(z)].$$

is internally stable and the performance index $\|\mathbf{T}_{dw}\|_2 = \|\mathscr{F}(\mathbf{G}, \mathbf{K})\|_2$ is minimized. In fact, the optimal controller has the same order as that of the generalized plant that can be high. Hence, controller reduction will also be studied in a later part of the section.

6.1.1 Optimal Controllers

As demonstrated previously, many optimal control problems admit the LFT structure. A state-space realization is assumed for the generalized plant giving rise to

$$\mathbf{G}(z) = \left[\begin{array}{c|cc} A & B_1 & B_2 \\ \hline C_1 & D_{11} & D_{12} \\ C_2 & D_{21} & D_{22} \end{array}\right] \tag{6.9}$$

where $\mathbf{G}_{ij}(z) = D_{ij} + C_i(zI - A)^{-1}B_j$ for $i, j = 1, 2$. For this reason, $\mathscr{H}_2$ optimal control for the LFT feedback system in Fig. 6.3 is considered next.

The corresponding state-space system is described by

$$\begin{aligned} \mathbf{x}(t+1) &= A\mathbf{x}(t) + B_1\mathbf{d}(t) + B_2\mathbf{u}(t), \quad \mathbf{x}(t_0) = \mathbf{0}, \\ \mathbf{w}(t) &= C_1\mathbf{x}(t) + D_{11}\mathbf{d}(t) + D_{12}\mathbf{u}(t), \\ \mathbf{y}(t) &= C_2\mathbf{x}(t) + D_{21}\mathbf{d}(t) + D_{22}\mathbf{u}(t), \end{aligned} \tag{6.10}$$

where $\mathbf{d}(t) \in \mathbb{R}^{m_1}$, $\mathbf{u}(t) \in \mathbb{R}^{m_2}$, $\mathbf{w}(t) \in \mathbb{R}^{p_1}$, $\mathbf{y}(t) \in \mathbb{R}^{p_2}$, and $\mathbf{x}(t) \in \mathbb{R}^n$. The zero initial condition is owing to the stationary nature of the $\mathscr{H}_2$ control.

Generically, both m_2 and p_2 are strictly smaller than n that is the state dimension. It is thus difficult to design the optimal feedback controller $\mathbf{K}(z)$ directly based on output measurement $\mathbf{y}(t)$. Our strategy is to tackle the simple problem first by considering the special case of full information when state $\mathbf{x}(t)$ and disturbance $\mathbf{d}(t)$ are available for feedback that corresponds to

$$C_2 = \begin{bmatrix} I_n \\ \mathbf{0} \end{bmatrix}, \quad D_{21} = \begin{bmatrix} \mathbf{0} \\ I_{m_1} \end{bmatrix}, \quad D_{22} = \begin{bmatrix} \mathbf{0} \\ \mathbf{0} \end{bmatrix}. \tag{6.11}$$

This strategy has the advantage in that the solution to full information control in Chap. 5 can be utilized. More importantly, when the conditions in (6.11) fail to hold (that is generic and corresponds to output feedback), full information control signal can be estimated based on output measurements which will be employed in place of the true state and disturbance. The surprising fact is that such a strategy leads to $\mathscr{H}_2$ optimal control based on output feedback provided that both full information control and output estimation are optimal. Such a nice property is referred to as *separation principle* that will be made more precise.

Under full information, the feedback control law has the form

$$\mathbf{u}(t) = \mathbf{F}(q)\mathbf{x}(t) + \mathbf{F}_0(q)\mathbf{d}(t). \tag{6.12}$$

The feedback controller $\mathbf{K}(z) = \begin{bmatrix} \mathbf{F}(z) & \mathbf{F}_0(z) \end{bmatrix}$ is restricted to linear shift-invariant systems so as to preserve the WSS property that is also the reason why zero initial condition is taken in (6.10). It follows that

$$\begin{aligned} \mathbf{x}(t+1) &= (A + B_2\mathbf{F})\mathbf{x}(t) + (B_1 + B_2\mathbf{F}_0)\mathbf{d}(t), \quad \mathbf{x}(t_0) = \mathbf{0}, \\ \mathbf{w}(t) &= (C_1 + D_{12}\mathbf{F}_0)\mathbf{x}(t) + (D_{11} + D_{12}\mathbf{F}_0)\mathbf{d}(t). \end{aligned}$$

The closed-loop transfer matrix under full information is given by

$$\mathbf{T}_{\mathrm{FI}}(z) = \left[\begin{array}{c|c} A + B_2\mathbf{F} & B_1 + B_2\mathbf{F}_0 \\ \hline C_1 + D_{12}\mathbf{F} & D_{11} + D_{12}\mathbf{F}_0 \end{array}\right] \tag{6.13}$$

by a slight abuse of notation. The objective is to design the full information control law in (6.12) that ensures internal stability of $\mathbf{T}_{\mathrm{FI}}(z)$ and minimizes $\|\mathbf{T}_{\mathrm{FI}}\|_2$. The optimal solution is restated next.

Theorem 6.4. *Consider state-space system (6.10) satisfying the conditions in (6.11) and $R = D_{12}^* D_{12}$ being nonsingular. Denote $A_R = A - B_2 R^{-1} D_{12}^* C_1$. If (A, B_2) is stabilizable and*

$$\operatorname{rank}\left\{\begin{bmatrix} zI - A & B_2 \\ C_1 & D_{12} \end{bmatrix}\right\} = n + m_2 \;\forall\; |z| = 1, \tag{6.14}$$

then the following (control) ARE

$$X = A_R^* X (I + B_2 R^{-1} B_2^* X)^{-1} A_R + C_1^* (I - D_{12} R^{-1} D_{12}^*) C_1 \tag{6.15}$$

admits a stabilizing solution $X \geq 0$. In this case, the optimal full information control gains are static and specified by

$$\begin{aligned}\mathbf{F} &= F = -(R + B_2^* X B_2)^{-1}(B_2^* X A + D_{12}^* C_1), \\ \mathbf{F}_0 &= F_0 = -(R + B_2^* X B_2)^{-1}(B_2^* X B_1 + D_{12}^* D_{11}).\end{aligned} \tag{6.16}$$

The minimum performance index $J_{\text{FI}} = \inf \|\mathbf{T}_{dw}\|_2^2$ over all stabilizing feedback controllers has the following expression:

$$\text{Tr}\{(D_{11} + D_{12}F_0)^*(D_{11} + D_{12}F_0) + (B_1 + B_2F_0)^* X (B_1 + B_2F_0)\}. \tag{6.17}$$

Theorem 6.4 is the stationary version of Theorem 5.25 in Chap. 5. Its proof is left as an exercise (Problem 6.1). When full information is unavailable or conditions in (6.11) fail, it is customary to introduce the variable substitution

$$\mathbf{u}(t) = \mathbf{s}(t) + F\mathbf{x}(t) + F_0\mathbf{d}(t) \tag{6.18}$$

where F and F_0 are the optimal feedback gains for full information control in (6.16). Substituting $\mathbf{u}(t)$ in (6.18) to the state-space system in (6.10) leads to

$$\begin{aligned}\mathbf{x}(t+1) &= (A + B_2F)\mathbf{x}(t) + (B_1 + B_2F_0)\mathbf{d}(t) + B_2\mathbf{s}(t), \\ \mathbf{w}(t) &= (C_1 + D_{12}F)\mathbf{x}(t) + (D_{11} + D_{12}F_0)\mathbf{d}(t) + D_{12}\mathbf{s}(t).\end{aligned}$$

It follows that $\mathbf{w}(t) = \mathbf{T}_{\text{FI}}(q)d(t) + \mathbf{T}_{\text{tmp}}(q)\mathbf{s}(t)$ where $\mathbf{T}_{\text{FI}}(z)$ is the same as in (6.13) with $\mathbf{F}$ and $\mathbf{F}_0$ replaced by F and F_0 in (6.16), respectively, and

$$\mathbf{T}_{\text{tmp}}(z) = \left[\begin{array}{c|c} A + B_2F & B_2 \\ \hline C_1 + D_{12}F & D_{12} \end{array}\right]. \tag{6.19}$$

The following lemma is important and shows that $\mathbf{T}_{\text{FI}}(z)$ and $\mathbf{T}_{\text{tmp}}(z)$ are orthogonal in the sense that

$$\frac{1}{2\pi}\int_{-\pi}^{\pi} \mathbf{T}_{\text{FI}}(e^{j\omega})^* \mathbf{T}_{\text{tmp}}(e^{j\omega})\, d\omega = \mathbf{0}. \tag{6.20}$$

Lemma 6.1. *Let $\mathbf{F} = F$ and $\mathbf{F}_0 = F_0$ be the optimal feedback gains for full information control as in (6.16). Then $A_F = A + B_2F$ is a stability matrix and $\mathbf{H}(z) = \mathbf{T}_{\text{FI}}(z)^\sim \mathbf{T}_{\text{tmp}}(z)$ is a strictly proper transfer matrix given by*

$$\mathbf{H}(z) = (B_F^* X A_F + D_F^* C_F)(zI - A_F)^{-1} B_2 \tag{6.21}$$

where $D_F = D_{11} + D_{12}F_0$, $B_F = B_1 + B_2F_0$, and $C_F = C_1 + D_{12}F$. Moreover,

$$\mathbf{T}_{\text{tmp}}(z)^\sim \mathbf{T}_{\text{tmp}}(z) = D_{12}^* D_{12} + B_2^* X B_2. \tag{6.22}$$

Proof. The optimality hypothesis implies that $A_F = A + B_2 F$ is a stability matrix. The expression of the feedback gains F and F_0 in (6.16) yields

$$M_{F_0} := D_{12}^* D_F + B_2^* X B_F = (D_{12}^* D_{11} + B_2^* X B_1) + (R + B_2^* X B_1) F_0 = \mathbf{0},$$
$$M_F := D_{12}^* C_F + B_2^* X A_F = (D_{12}^* C_1 + B_2^* X A) + (R + B_2^* X B_2) F = \mathbf{0}.$$

In addition, the stabilizing solution $X \geq 0$ to the ARE (6.15) is the observability gramian of both $\mathbf{T}_{\mathrm{FI}}(z)$ and $\mathbf{T}_{\mathrm{tmp}}(z)$ or

$$X = A_F^* X A_F + C_F^* C_F. \tag{6.23}$$

Denote $\mathbf{T}_F(z) = C_F(zI - A_F)^{-1}$. Then

$$\mathbf{T}_F(z)^{\sim} \mathbf{T}_F(z) = X + X A_F (zI - A_F)^{-1} + (z^{-1} I - A_F^*)^{-1} A_F^* X. \tag{6.24}$$

Similar equations to (6.23) and (6.24) are used in proof of Lemma 5.1 of the previous chapter. Verifications of (6.23) and (6.24) are left in Exercises (Problem 6.3). By the expression of $\mathbf{T}_F(z)$,

$$\mathbf{T}_{\mathrm{FI}}(z) = D_F + \mathbf{T}_F(z) B_F, \quad \mathbf{T}_{\mathrm{tmp}}(z) = D_{12} + \mathbf{T}_F(z) B_2.$$

It follows from (6.24) that

$$\mathbf{H}(z) = (B_F^* X A_F + D_F^* C_F)(zI - A_F)^{-1} B_2 + M_{F_0}^* + B_J^* (z^{-1} I - A_F^*)^{-1} M_F^*$$

after lengthy calculations. Because $M_F = \mathbf{0}$ and $M_{F_0} = \mathbf{0}$, $\mathbf{H}(z)$ is indeed strictly proper as given in (6.21). On the other hand, (6.24) yields

$$\mathbf{T}_{\mathrm{tmp}}^{\sim} \mathbf{T}_{\mathrm{tmp}} = D_{12}^* D_{12} + B_2^* X B_2 + B_2^* (z^{-1} I - A_F^*)^{-1} M_F^* + M_F (zI - A_F)^{-1} B_2$$

that verifies (6.22) due to $M_F = \mathbf{0}$. In fact, $\mathbf{T}_{\mathrm{tmp}}(z)(R + B_2^* X B_2)^{-1/2}$ is inner as already shown in (ii) of Theorem 5.28 from the previous chapter. □

In light of stability of $\mathbf{H}(z)$ or A_F,

$$\frac{1}{2\pi} \int_{-\pi}^{\pi} \|\mathbf{H}(e^{j\omega})\|_{\mathscr{F}} \, d\omega < \infty$$

where $\| \cdot \|_{\mathscr{F}}$ is the Frobenius norm. The orthogonality in (6.20) thus follows from the absolute integrability of $\mathbf{H}(z)$ on the unit circle and the fact that $\mathbf{H}(z)$ is strictly proper by

$$\mathbf{H}(e^{j\omega}) = \mathbf{T}_{\mathrm{FI}}(e^{j\omega})^* \mathbf{T}_{\mathrm{tmp}}(e^{j\omega}) = \sum_{k=1}^{\infty} H_k e^{-j\omega}$$

for some $\{H_k\}_{k=1}^{\infty}$. Because the LFT system in Fig. 6.3 assumes linear and time-invariant feedback controller $\mathbf{K}(z)$ that is stabilizing,

$$\mathbf{s}(t) = \mathbf{u}(t) - [F\mathbf{x}(t) + F_0\mathbf{d}(t)] \tag{6.25}$$

is causal and has bounded variance. As a result, $\mathbf{s}(t) = \mathbf{Q}(q)\mathbf{d}(t)$ for some proper and stable transfer matrix $\mathbf{Q}(z)$, and hence,

$$\mathbf{w}(t) = [\mathbf{T}_{\rm FI}(q) + \mathbf{T}_{\rm tmp}(q)\mathbf{Q}_{\rm tmp}(q)]\mathbf{d}(t).$$

Recall that $\mathbf{d}(t)$ is the only exogenous signal that is a white process with zero mean and identity covariance. The result in Lemma 6.1 or orthogonality in (6.20) implies that the variance of the output is given by

$$\mathrm{E}\{\|\mathbf{w}(t)\|^2\} = \|\mathbf{T}_{\rm FI}\|_2^2 + \|\mathbf{T}_{\rm tmp}\mathbf{Q}\|_2^2 \tag{6.26}$$

where the minimum $J_{\rm FI} = \|\mathbf{T}_{\rm FI}\|_2^2$ is specified in (6.17). The remaining problem is minimization of the second term in (6.26) or

$$\|\mathbf{T}_{\rm tmp}\mathbf{Q}\|_2^2 = \mathrm{E}\{\|\mathbf{T}_{\rm tmp}(q)\mathbf{s}(t)\|^2\} = \mathrm{E}\{\|(D_{12}^*D_{12} + B_2^*XB_2)^{1/2}\mathbf{s}(t)\|^2\}$$

in light of (6.22) in Lemma 6.1. Therefore, minimization of $\|\mathbf{T}_{\rm tmp}\mathbf{Q}\|_2^2$ amounts to minimization of variance of the augmented error signal

$$\tilde{\mathbf{w}}(t) = (D_{12}^*D_{12} + B_2^*XB_2)^{1/2}[\mathbf{u}(t) - \mathbf{u}_{\rm FI}(t)]$$

by designing $\mathbf{u}(t)$ as the optimal estimate to the full information control law $\mathbf{u}_{\rm FI}(t) = F\mathbf{x}(t) + F_0\mathbf{d}(t)$. This is recognized as (stationary) output estimation. The nonsingular matrix $(D_{12}^*D_{12} + B_2^*XB_2)^{1/2}$ does not alter the solution to the problem of output estimation but changes its error variance.

Specifically, consider first the case

$$D_{22} = \mathbf{0}, \quad \det(D_{21}D_{21}^*) \neq 0. \tag{6.27}$$

The corresponding problem of output estimation is described by

$$\begin{aligned} \mathbf{x}(t+1) &= A\mathbf{x}(t) + B_1\mathbf{d}(t) + B_2\mathbf{u}(t), \\ \mathbf{u}_{\rm FI}(t) &= F\mathbf{x}(t) + F_0\mathbf{d}(t), \\ \mathbf{y}(t) &= C_2\mathbf{x}(t) + D_{21}\mathbf{d}(t). \end{aligned} \tag{6.28}$$

In accordance with the results in Sect. 5.1.4 of the previous chapter, the output estimator has the form

$$\begin{aligned} \hat{\mathbf{x}}(t+1) &= (A + LC_2)\hat{\mathbf{x}}(t) - L\mathbf{y}(t) + B_2\mathbf{u}(t), \\ \hat{\mathbf{u}}_{\rm FI}(t) &= (F + L_0C_2)\hat{\mathbf{x}}(t) - L_0\mathbf{y}(t) \end{aligned} \tag{6.29}$$

for some (L, L_0) with L stabilizing. The next result is obtained from extending Theorem 5.14 in Sect. 5.1.4 to the stationary case.

Theorem 6.5. *Consider output estimation for the system described in (6.28) with* $\tilde{R} = D_{21}D_{21}^*$ *being nonsingular. Denote* $A_{\tilde{R}} = A - B_2D_{21}^*\tilde{R}^{-1}D_{21}C_2$. *If* (C_2, A) *is stabilizable and*

$$\text{rank}\left\{\begin{bmatrix} zI - A & B_1 \\ C_2 & D_{21} \end{bmatrix}\right\} = n + p_2 \ \forall\ |z| = 1, \tag{6.30}$$

then the following (filtering) ARE

$$Y = A_{\tilde{R}}Y(I + C_2^*\tilde{R}^{-1}C_2Y)^{-1}A_{\tilde{R}}^* + B_1(I - D_{21}^*\tilde{R}^{-1}D_{21})B_1^* \tag{6.31}$$

admits a stabilizing solution $Y \geq 0$. *In this case, the optimal output estimation gains are static and specified by*

$$\begin{aligned} L &= -(AYC_2^* + B_1D_{21}^*)(\tilde{R} + C_2YC_2^*)^{-1}, \\ L_0 &= -(FYC_2^* + F_0D_{21}^*)(\tilde{R} + C_2YC_2^*)^{-1}. \end{aligned} \tag{6.32}$$

The minimum performance index $J_{\text{OE}} = \inf \text{E}\{\|\tilde{\mathbf{w}}(t)\|^2\}$ *over all stabilizing feedback estimators is given by*

$$\begin{aligned} J_{\text{OE}} = \text{Tr}\{&(D_{12}^*D_{12} + B_2^*XB_2)(F_0 + L_0D_{21})(F_0 + L_0D_{21})^* \\ &+ (D_{12}^*D_{12} + B_2^*XB_2)(F + L_0C_2)Y(F + L_0C_2)^*\}. \end{aligned} \tag{6.33}$$

Proof. By the assumption $\mathbf{x}_0 = \mathbf{0}$, $\hat{\mathbf{x}}(0) = \mathbf{0}$ with covariance $P_0 = Y$ can be taken for the output estimator in (6.29) where $Y \geq \mathbf{0}$ is the stabilizing solution to ARE (6.31). Hence, an application of Theorem 5.14 concludes that the corresponding DRE solutions agree with Y at each time index t leading to the stationary output estimator in (6.29) with L and L_0 as specified in (6.32) that is optimal over all L and L_0 subject to stability of $(A + LC_2)$. Denote

$$\mathbf{e}_x(t) = \mathbf{x}(t) - \hat{\mathbf{x}}(t), \quad \mathbf{e}_u(t) = \mathbf{u}_{\text{FI}}(t) - \hat{\mathbf{u}}(t).$$

It is left as an exercise (Problem 6.5) to show that

$$\begin{aligned} \mathbf{e}_x(t+1) &= (A + LC_2)\mathbf{e}_x(t) + (B_1 + LD_{21})\mathbf{d}(t), \\ \mathbf{e}_u(t) &= (F + L_0C_2)\mathbf{e}_x(t) + (F_0 + L_0D_{21})\mathbf{d}(t), \end{aligned} \tag{6.34}$$

as $D_{22} = \mathbf{0}$ is assumed. The associated reachability gramian is precisely the solution $Y \geq 0$ to ARE in (6.31). It follows that the minimum variance is given by $\text{E}\{\|\mathbf{e}_u(t)\|^2\} = \text{Tr}(\text{E}\{\mathbf{e}_u(t)\mathbf{e}_u(t)^*\})$ or

$$\text{E}\{\|\mathbf{e}_u(t)\|^2\} = \text{Tr}\{(F_0 + L_0D_{21})(F_0 + L_0D_{21})^* + (F + L_0C_2)Y(F + L_0C_2)^*\}.$$

See also Problem 5.18 in Exercises of the previous chapter. Consequently, the minimum of $J_{\rm OE}$ is given by

$$J_{\rm OE} = {\rm Tr}\left([D_{12}^* D_{12} + B_2^* X B_2] {\rm E}\{\mathbf{e}_u(t)\mathbf{e}_u(t)^*\}\right)$$

that is the same as in (6.33). □

Our results in Theorems 6.4 and 6.5 show that the optimal $\mathscr{H}_2$ control system in Fig. 6.3 can be designed with full information control and output estimation separately that is referred to as *separation principle*. That is, if full information is available or the conditions in (6.11) hold, then full information control in Sect. 5.2.3 is the optimal $\mathscr{H}_2$ control that completes the design of optimal $\mathscr{H}_2$ feedback control system in Fig. 6.3. If full information is unavailable or the conditions in (6.11) fail to hold, the optimal output estimator needs to be synthesized to estimate the full information control law according to the result in Sect. 5.1.4. Together, they produce the optimal $\mathscr{H}_2$ controller $\mathbf{K}(z)$ by taking $\mathbf{u}(t) = \hat{\mathbf{u}}_{\rm FI}(t)$, i.e., by taking the estimated full information control law as the control signal. Indeed, by substituting $\mathbf{u}(t) = \hat{\mathbf{u}}_{\rm FI}(t)$ in the second equation into the first equation of (6.29) gives the state-space description of the feedback controller

$$\begin{aligned} \hat{\mathbf{x}}(t+1) &= (A + B_2F + LC_2 + B_2L_0C_2)\hat{\mathbf{x}}(t) - (L + B_2L_0)\mathbf{y}(t), \\ \mathbf{u}(t) &= (F + L_0C_2)\hat{\mathbf{x}}(t) - L_0\mathbf{y}(t). \end{aligned}$$

For convenience, denote $\mathbf{K}_0(z) = -[\hat{D} + \hat{C}(zI - \hat{A})^{-1}\hat{B}]$ in the case $D_{22} = \mathbf{0}$. That is,

$$\begin{aligned} \hat{A} &= (A + B_2F + LC_2 + B_2L_0C_2), \\ \hat{B} &= (B_2L_0 + L), \quad \hat{C} = F + L_0C_2, \end{aligned} \tag{6.35}$$

and $\hat{D} = L_0$. The optimal feedback controller $\mathbf{K}(z)$ for $D_{22} \neq \mathbf{0}$ can be easily obtained based on $\mathbf{K}_0(z)$. Indeed, $\mathbf{K}_0(z)$ is recognized as the optimal feedback controller for the case when the output measurement is

$$\mathbf{y}_{\rm d}(t) := \mathbf{y}(t) - D_{22}\mathbf{u}(t) = C_2\mathbf{x}(t) + D_{21}\mathbf{d}(t).$$

In other words, $\mathbf{u}(t) = \mathbf{K}_0(q)\mathbf{y}_{\rm d}(t)$ is the optimal control law or

$$\mathbf{u}(t) = \mathbf{K}_0(q)[\mathbf{y}(t) - D_{22}\mathbf{u}(t)]. \tag{6.36}$$

The equivalent state-space description is given by

$$\begin{aligned} \hat{\mathbf{x}}(t+1) &= \hat{A}\hat{\mathbf{x}}(t) + \hat{B}[\mathbf{y}(t) - D_{22}\mathbf{u}(t)], \\ -\mathbf{u}(t) &= \hat{C}\hat{\mathbf{x}}(t) + \hat{D}[\mathbf{y}(t) - D_{22}\mathbf{u}(t)]. \end{aligned} \tag{6.37}$$

Recall that the feedback controller $\mathbf{K}(z)$ has $\mathbf{y}(t)$ as input and $\mathbf{u}(t)$ as output. Solving $\mathbf{u}(t)$ from the second equation of (6.37) yields

$$\mathbf{u}(t) = -(I - \hat{D}D_{22})^{-1}[\hat{C}\hat{\mathbf{x}}(t) + \hat{D}\mathbf{y}(t)],$$

assuming $\det(I+\hat{D}D_{22}) \neq 0$. Thus, the state-space equations in (6.37) can be rewritten as

$$\begin{aligned} \hat{\mathbf{x}}(t+1) &= [\hat{A}+\hat{B}D_{22}(I-\hat{D}D_{22})^{-1}\hat{C}]\hat{\mathbf{x}}(t)+\hat{B}(I-D_{22}\hat{D})^{-1}\mathbf{y}(t), \\ -\mathbf{u}(t) &= (I-\hat{D}D_{22})^{-1}\hat{C}\hat{\mathbf{x}}(t)-(I-\hat{D}D_{22})^{-1}\hat{D}\mathbf{y}(t). \end{aligned}$$

The above gives the optimal feedback controller as

$$\mathbf{K}(z) = -\left[\begin{array}{c|c} \hat{A}+\hat{B}D_{22}(I-\hat{D}D_{22})^{-1}\hat{C} & \hat{B}(I-D_{22}\hat{D})^{-1} \\ \hline (I-\hat{D}D_{22})^{-1}\hat{C} & (I-\hat{D}D_{22})^{-1}\hat{D} \end{array}\right]. \tag{6.38}$$

Remark 6.1. The conventional LQG control aims at state feedback and state estimation without considering disturbances. It results in a strictly proper controller in the form of observer as follows:

$$K(z) = -F(zI-A-B_2F-LC_2-LD_{22}F)^{-1}L. \tag{6.39}$$

Such an LQG controller can be obtained by setting $L_0 = \mathbf{0}$ in (6.35) and by setting $\hat{D} = 0$ in (6.38), if $D_{22} \neq 0$. The LQG controller in (6.39) is optimal among all strictly proper and stabilizing feedback controllers.

Example 6.6. Consider Example 6.2. The corresponding generalized plant model $\mathbf{G}(z)$ is given in (6.6). Assume that $\mathbf{P}(z) = C(zI-A)^{-1}B$. Then

$$\begin{aligned} \mathbf{G}(z) &= \left[\begin{array}{c|c} \mathbf{G}_{11}(z) & \mathbf{G}_{12}(z) \\ \hline \mathbf{G}_{21}(z) & \mathbf{G}_{22}(z) \end{array}\right] = \left[\begin{array}{cc|c} \mathbf{0} & \mathbf{0} & I \\ \mathbf{P}(z) & I & \mathbf{P}(z) \\ \hline \mathbf{P}(z) & I & \mathbf{P}(z) \end{array}\right] \\ &= \left[\begin{array}{cc|c} \mathbf{0} & \mathbf{0} & I \\ \mathbf{0} & I & \mathbf{0} \\ \hline \mathbf{0} & I & \mathbf{0} \end{array}\right] + \left[\begin{array}{c} \mathbf{0} \\ C \\ \hline C \end{array}\right] (zI-A)^{-1} \left[\begin{array}{cc|c} B & \mathbf{0} & B \end{array}\right]. \end{aligned}$$

It can be verified that the conditions in Theorems 6.4 and 6.5 hold, if (A,B) is stabilizable and (C,A) is detectable. These conditions imply the existence of the stabilizing solutions to the following two AREs:

$$X = A^*X(I+B_2B_2^*X)^{-1}A+C_2^*C_2, \tag{6.40}$$

$$Y = AY(I+C_2^*C_2Y)^{-1}A^*+B_2B_2^*, \tag{6.41}$$

with $B_2 = B$ and $C_2 = C$. The controller and estimator gains are given by

$$F = -(I+B^*XB)^{-1}B^*XA, \quad \left[\begin{array}{c} L \\ \hline L_0 \end{array}\right] = -\left[\begin{array}{c} A \\ \hline F \end{array}\right] YC^*(I+CYC^*)^{-1},$$

respectively. The expression of the gain F_0 is skipped, because it is not used in the construction of the following optimal $\mathscr{H}_2$ controller:

$$\mathbf{K}(z) = -\left[\begin{array}{c|c} A+BF+LC+BL_0C & L+BL_0 \\ \hline F+L_0C & L_0 \end{array}\right].$$

Let $\mathbf{P}(z)$ be the same as in Example 4.2 from Chap. 4. Straightforward calculation using Matlab yields

$$F = \begin{bmatrix} -0.5706 & 0.0852 & -0.1319 & -0.2414 & 0.2625 \\ -0.8030 & 0.1610 & -0.1633 & 0.3420 & -0.3745 \end{bmatrix},$$

$$L = \begin{bmatrix} -0.3926 & 0.2038 \\ -0.4369 & 0.1754 \\ -0.0825 & -0.0542 \\ -0.0265 & 0.6595 \\ -0.0949 & 0.2663 \end{bmatrix}, \quad L_0 = \begin{bmatrix} 0.2570 & 0.0552 \\ 0.3469 & -0.3538 \end{bmatrix}.$$

It is interesting to observe that $(A+BF)$ and $(A+LC)$ have the same eigenvalues, which are eigenvalues of the closed-loop system. In addition, D_{11} is not involved in design of the optimal feedback controller, but contributes to the performance index.

The $\mathscr{H}_2$ control in the previous example is very close to the normalized $\mathscr{H}_2$ control, referred to the case when the generalized plant is given by

$$\mathbf{G}(z) = \left[\begin{array}{cc|c} \mathbf{0} & \mathbf{0} & I \\ \mathbf{0} & \mathbf{P}(z) & \mathbf{P}(z) \\ \hline I & \mathbf{P}(z) & \mathbf{P}(z) \end{array}\right]$$

$$= \left[\begin{array}{cc|c} \mathbf{0} & \mathbf{0} & I \\ \mathbf{0} & \mathbf{0} & \mathbf{0} \\ \hline \mathbf{0} & I & \mathbf{0} \end{array}\right] + \left[\begin{array}{c} \mathbf{0} \\ C \\ \hline C \end{array}\right] (zI-A)^{-1} \left[\begin{array}{cc|c} \mathbf{0} & B & B \end{array}\right].$$

It is left as an exercise (Problem 6.7) to show that the $\mathscr{H}_2$ optimal controller is the same as the one in Example 6.6. If L_0 in the optimal controller is set to zero, then the normalized $\mathscr{H}_2$ control reduces to *normalized* LQG *control* that is commented in Remark 6.1.

6.1.2 Controller Reduction

The optimal feedback controller $\mathbf{K}(z)$ is basically an output estimator that is built upon the state estimation. Hence, the order of the optimal controller can be high leading to high complexity in its implementation. For this reason, model reduction is often a necessary step in design of feedback control systems.

There are two considerations in controller reduction. The first is stability. That is, the reduced order controller is required to stabilize the full order plant. The second is performance. The error variance or the $\mathscr{H}_2$ performance needs to be close to the optimal one achieved by the full order controller. This section focuses on stability and derives the a priori bounds that ensure closed-loop stability. The common methods in controller reduction include reduction of the plant order first and design of reduced order controller later or design of the full controller first and reduction of the controller later. Both will be studied in this text, but the focus will be on simultaneous reduction of the plant and controller.

Generically, $\mathbf{G}_{22}(z) = \mathbf{P}(z)$ that holds for the three examples discussed at the beginning of the chapter. Thus, stability requirement is translated to stabilization of $\mathbf{G}_{22}(z)$ by the feedback controller that is why stabilizability of (A, B_2) and detectability of (C_2, A) are assumed in design of the optimal $\mathscr{H}_2$ controller $\mathbf{K}(z)$. Since the discretized system is strictly causal, $D_{22} = \mathbf{0}$ often holds. Even if $D_{22} \neq \mathbf{0}$, reduction of $\mathbf{K}_0(z)$ with realization in (6.35) can be considered that is synthesized for the case $D_{22} = \mathbf{0}$. For these reasons, $D_{22} = \mathbf{0}$ has no loss of generality and will be assumed throughout this subsection and next. It follows that coprime factorization results from Sect. 3.2.3 can be used to compute left and right coprime factors of $\mathbf{G}_{22}(z)$ and $\mathbf{K}(z)$. Specifically, let F and L be the optimal state-feedback and state estimation gains, respectively, which are computed as in Theorems 6.4 and 6.5, respectively. Then both $(A + B_2F)$ and $(A + LC_2)$ are stability matrices. Accordingly, $\mathbf{G}_{22}(z) = \mathbf{P}(z)$ and $\mathbf{K}(z)$ admit coprime factorizations

$$\mathbf{P}(z) = \mathbf{M}(z)^{-1}\mathbf{N}(z) = \tilde{\mathbf{N}}(z)\tilde{\mathbf{M}}(z)^{-1},$$
$$-\mathbf{K}(z) = \mathbf{V}(z)^{-1}\mathbf{U}(z) = \tilde{\mathbf{U}}(z)\tilde{\mathbf{V}}(z)^{-1},$$

where $-\mathbf{K}(z)$ admits a state-space representation

$$-\mathbf{K}(z) = \left[\begin{array}{c|c} A + B_2F + LC_2 + B_2L_0C_2 & L + B_2L_0 \\ \hline F + L_0C_2 & L_0 \end{array}\right]. \tag{6.42}$$

It is left as an exercise (Problem 6.8) to show that

$$\tilde{\mathbf{H}}(z) = \begin{bmatrix} \tilde{\mathbf{H}}_P(z) & \tilde{\mathbf{H}}_K(z) \end{bmatrix} := \begin{bmatrix} \tilde{\mathbf{M}}(z) & -\tilde{\mathbf{U}}(z) \\ \tilde{\mathbf{N}}(z) & \tilde{\mathbf{V}}(z) \end{bmatrix}$$
$$= \left[\begin{array}{c|cc} A + B_2F & B_2 & -(L + B_2L_0) \\ \hline F & I & -L_0 \\ C_2 & 0 & I \end{array}\right], \tag{6.43}$$

$$\mathbf{H}(z) = \begin{bmatrix} \mathbf{H}_K(z) \\ \mathbf{H}_P(z) \end{bmatrix} := \begin{bmatrix} \mathbf{V}(z) & \mathbf{U}(z) \\ -\mathbf{N}(z) & \mathbf{M}(z) \end{bmatrix}$$
$$= \left[\begin{array}{c|cc} A + LC_2 & -B_2 & L \\ \hline F + L_0C_2 & I & L_0 \\ C_2 & 0 & I \end{array}\right], \tag{6.44}$$

satisfy the double Bezout identity

$$\mathbf{H}(z)\tilde{\mathbf{H}}(z) = \begin{bmatrix} \mathbf{H}_K(z) \\ \mathbf{H}_P(z) \end{bmatrix} \begin{bmatrix} \tilde{\mathbf{H}}_P(z) \; \tilde{\mathbf{H}}_K(z) \end{bmatrix} = I. \tag{6.45}$$

It follows that $\tilde{\mathbf{H}}_K(z)$ is a stable right inverse of $\mathbf{H}_P(z)$ or $\tilde{\mathbf{H}}_K(z) = \mathbf{H}_P(z)^+$ and $\mathbf{H}_K(z)$ is a stable left inverse of $\tilde{\mathbf{H}}_P(z)$ or $\mathbf{H}_K(z) = \tilde{\mathbf{H}}_P(z)^+$.

Our model reduction method focuses on simultaneous reduction of $\tilde{\mathbf{H}}(z)$ and $\mathbf{H}(z)$ through inverse balanced truncation (IBT). Since both $\tilde{\mathbf{H}}(z)$ and $\mathbf{H}(z)$ consist of the coprime factors of the plant and controller, it results in simultaneous reduction of the plant and controller models. While the performance is quantified by $\mathscr{H}_2$ norm, stability and stability margin are difficult to be ensured and to be studied using $\mathscr{H}_2$ norm. Instead, $\mathscr{H}_\infty$ norm is often employed to study robust stability or stability in presence of model error due to either modeling error of the plant or reduction error of the controller. Recall that for a stable rational transfer matrix $\mathbf{T}(z)$, its $\mathscr{H}_\infty$ norm is defined by

$$\|\mathbf{T}\|_\infty := \sup_{|z|\geq 1} \overline{\sigma}[\mathbf{T}(z)] = \sup_{\omega} \overline{\sigma}[\mathbf{T}(e^{j\omega})].$$

The following is the well-known small gain theorem.

Lemma 6.2. *Let* $\mathbf{T}(z)$ *be a square stable rational transfer matrix. Its inverse* $[I + \mathbf{T}(z)]^{-1}$ *exists and is a stable transfer matrix, if* $\|\mathbf{T}\|_\infty < 1$.

Proof. The condition $\|\mathbf{T}\|_\infty < 1$ implies that $[I + \mathbf{T}(z)]^{-1}$ exists and is bounded for each $|z| \geq 1$. In fact, its $\mathscr{H}_\infty$ norm satisfies

$$\|[I + \mathbf{T}(z)]^{-1}\|_\infty \leq \frac{1}{1 - \|\mathbf{T}\|_\infty} \tag{6.46}$$

that is bounded. Thus, $[I + \mathbf{T}(z)]^{-1}$ is rational and stable. □

Let $\mathbf{S}(z) = D_s + C_s(zI_n - A_s)^{-1}B_s$ be a square matrix with $\det(D_s) \neq 0$. Its inverse is given by

$$\mathbf{S}(z)^{-1} = \left[\begin{array}{c|c} A_s - B_s D_s^{-1} C_s & B_s D_s^{-1} \\ \hline -D_s^{-1} C_s & D_s^{-1} \end{array}\right]. \tag{6.47}$$

Assume that both A_s and $(A_s - B_s D_s^{-1} C_s)$ are stability matrices, i.e., $\mathbf{S}(z)$ is both stable and strictly minimum phase. The IBT algorithm balances the reachability gramian of $\mathbf{S}(z)$ against the observability gramian of $\mathbf{S}(z)^{-1}$, or vice versa, prior to truncation (refer to Chap. 4). Let Q_s and P_s be respective solutions to the following Lyapunov equations:

$$Q_s = A_s Q_s A_s^* + B_s B_s^*, \quad R_s = D_s D_s^*, \tag{6.48}$$

$$P_s = (A_s - B_s D_s^{-1} C_s)^* P_s (A_s - B_s D_s^{-1} C_s) + C_s^* R_s^{-1} C_s. \tag{6.49}$$

Then $Q_s \geq \mathbf{0}$ is the reachability gramian of $\mathbf{S}(z)$ and $P_s \geq \mathbf{0}$ is the observability gramian of $\mathbf{S}(z)^{-1}$. The realization matrices $\{A_s, B_s, C_s, D_s\}$ are called inverse balanced, if

$$P_s = Q_s = \operatorname{diag}(s_1 I_{i_1}, s_2 I_{i_2}, \cdots, s_\eta I_\eta) \tag{6.50}$$

where $s_1 > s_2 > \cdots > s_\eta > 0$ and $i_1 + i_2 + \cdots + i_\eta = n$. Suppose that $r = i_1 + \cdots + i_\rho$ for $\rho < \eta$, and the reduced model

$$\widehat{\mathbf{S}}(z) = D_s + \widehat{C}_s (zI_r - \widehat{A}_s)^{-1} \widehat{B}_s$$

of order r is obtained via direction truncation:

$$\widehat{A}_s = \begin{bmatrix} I_r & \mathbf{0} \end{bmatrix} A_s \begin{bmatrix} I_r \\ \mathbf{0} \end{bmatrix}, \quad \widehat{B}_s = \begin{bmatrix} I_r & \mathbf{0} \end{bmatrix} B_s, \quad \widehat{C}_s = C_s \begin{bmatrix} I_r \\ \mathbf{0} \end{bmatrix}.$$

The above is the IBT algorithm studied in Chap. 4. The model reduction error associated with IBT has the relative and multiplicative forms and admits an $\mathscr{H}_\infty$-norm bound. Specifically, the following relations

$$\mathbf{S}(z) = \widehat{\mathbf{S}}(z)\left[I + \Delta_{\text{mul}}(z)\right], \quad \widehat{\mathbf{S}}(z) = \mathbf{S}(z)\left[I + \Delta_{\text{rel}}(z)\right] \tag{6.51}$$

hold where $\Delta_{\text{mul}}(z)$ and $\Delta_{\text{rel}}(z)$ represent the multiplicative and relative error, respectively, and satisfy

$$\begin{aligned} \|\Delta_{\text{mul}}\|_\infty &\leq \prod_{i=\rho+1}^{\eta} \left(1 + 2s_i\left(\sqrt{1+s_i^2} + s_i\right)\right) - 1, \\ \|\Delta_{\text{rel}}\|_\infty &\leq \prod_{i=\rho+1}^{\eta} \left(1 + 2s_i\left(\sqrt{1+s_i^2} + s_i\right)\right) - 1, \end{aligned} \tag{6.52}$$

in light of Theorem 4.6. Recall that the multiplicative and relative errors are explicitly given by

$$\Delta_{\text{mul}}(z) = \widehat{\mathbf{S}}(z)^{-1}\left(\mathbf{S}(z) - \widehat{\mathbf{S}}(z)\right), \tag{6.53}$$

$$\Delta_{\text{rel}}(z) = \mathbf{S}(z)^{-1}\left(\widehat{\mathbf{S}}(z) - \mathbf{S}(z)\right). \tag{6.54}$$

Model reduction with multiplicative/relative error bounds takes a crucial role in controller reduction. Indeed, applying the IBT algorithm to $\mathbf{H}(z)$ and $\mathbf{H}(z)^{-1} = \tilde{\mathbf{H}}(z)$ yields the following observability and reachability gramians:

$$P = (A + B_2 F)^* P (A + B_2 F) + F^* F + C_2^* C_2, \tag{6.55}$$

$$Q = (A + LC_2) Q (A + LC_2)^* + LL^* + B_2 B_2^*. \tag{6.56}$$

It can be shown (Problem 6.10 in Exercises) that under the assumption

$$D_{12}^*\left[D_{12}\ C_1\right] = \left[I\ \mathbf{0}\right], \quad \begin{bmatrix} B_1 \\ D_{21} \end{bmatrix} D_{21}^* = \begin{bmatrix} I \\ \mathbf{0} \end{bmatrix}, \tag{6.57}$$

and the matching condition

$$B_1B_1^* = B_2B_2^*, \quad C_1^*C_1 = C_2^*C_2, \tag{6.58}$$

$P = X$ is the stabilizing solution to ARE (6.40), and $Q = Y$ is the stabilizing solution to ARE (6.41). Feedback control for this special case corresponds to either normalized $\mathscr{H}_2$ control or normalized LQG control. Recall the discussion in or after Example 6.6 in the previous subsection.

Suppose that the realization of $\mathbf{H}(z)$ is inverse balanced, i.e.,

$$P = Q = \mathrm{diag}(s_1 I_{i_1}, s_2 I_{i_2}, \cdots, s_\eta I_\eta) \tag{6.59}$$

where $s_1 > s_2 > \cdots > s_\eta \geq 0$ and $i_1 + i_2 + \cdots + i_\eta = n$. Since $D_{22} = \mathbf{0}$ and L_0 is a direct transmission matrix. Partition

$$\left[\begin{array}{c|c} A & L\ B_2 \\ \hline F & \\ C_2 & \end{array}\right] = \left[\begin{array}{cc|cc} A_{11} & A_{12} & L_1 & B_{21} \\ A_{21} & A_{22} & L_2 & B_{22} \\ \hline F_1 & F_2 & & \\ C_{21} & C_{22} & & \end{array}\right] \tag{6.60}$$

where A_{11} has size $r \times r$ and $r = i_1 + \cdots + i_\rho < n$. Denote

$$\begin{aligned} \widehat{A}_K &= A_{11} + B_{21}F_1 + L_1C_{21} + B_{21}L_0C_{21}, \\ \widehat{B}_K &= B_{21}L_0 + L_1, \quad \widehat{C}_K = F_2 + L_0C_{21}. \end{aligned} \tag{6.61}$$

The reduced plant and controller of order r are given, respectively, by

$$\widehat{\mathbf{G}}_{22}(z) = \left[\begin{array}{c|c} A_{11} & B_{21} \\ \hline C_{21} & \mathbf{0} \end{array}\right], \quad \widehat{\mathbf{K}}(z) = \left[\begin{array}{c|c} \widehat{A}_K & \widehat{B}_K \\ \hline \widehat{C}_K & L_0 \end{array}\right]. \tag{6.62}$$

The next result states under what condition the feedback system in Fig. 6.3 is stable when the optimal feedback controller $\mathbf{K}(z)$ is replaced by its reduced order model $\widehat{\mathbf{K}}(z)$.

Theorem 6.7. *Consider $\tilde{\mathbf{H}}(z)$ in (6.43) and $\mathbf{H}(z)$ in (6.44) for the optimal $\mathscr{H}_2$ feedback control system in* Fig. 6.3. *Let P and Q be the observability and reachability gramians in (6.55) and (6.56), respectively, which are balanced in the sense of (6.59). Let $\widehat{\mathbf{G}}_{22}(z)$ and $\widehat{\mathbf{K}}(z)$ in (6.62) be reduced plant and controller of order $r = i_1 + \cdots + i_\rho < n$, respectively, obtained through applying the* IBT *algorithm to*

$\mathbf{H}(z)$ *and* $\tilde{\mathbf{H}}(z)$. *Then the feedback control system in* Fig. 6.3 *remains stable when* $\mathbf{K}(z)$ *is replaced by* $\widehat{K}(z)$, *provided that*

$$\prod_{i=\rho+1}^{\eta}\left(1+2s_i\left(\sqrt{1+s_i^2}+s_i\right)\right)<1+\gamma^{-1} \tag{6.63}$$

where $\gamma := \|\tilde{\mathbf{H}}_P\mathbf{H}_K\|_\infty = \|\tilde{\mathbf{H}}_K\mathbf{H}_P\|_\infty$.

Proof. If $\mathbf{H}(z)$ and $\tilde{\mathbf{H}}(z)$ are inverse balanced as in (6.59), direct truncation yields $\widehat{\mathbf{H}}(z)$ and $\widehat{\tilde{\mathbf{H}}}(z)$ leading to the expressions of the reduced order plant and controller in (6.62). Moreover, Theorem 4.6 in Chap. 4 asserts that

$$\widehat{\mathbf{H}}(z) = \mathbf{H}(z)\left[I+\Delta_{\text{rel}}(z)\right].$$

Substituting the partition of $\mathbf{H}(z)$ in (6.44) into the above equality leads to

$$\begin{bmatrix}\widehat{\mathbf{H}}_K(z)\\ \widehat{\mathbf{H}}_P(z)\end{bmatrix} = \begin{bmatrix}\mathbf{H}_K(z)\\ \mathbf{H}_P(z)\end{bmatrix}(I+\Delta_{\text{rel}}(z)). \tag{6.64}$$

With $\mathbf{K}(z)$ in Fig. 6.3 replaced by $\widehat{\mathbf{K}}(z)$ in (6.62), the closed-loop stability is equivalent to

$$\det[\widehat{\mathbf{H}}_K(z)\tilde{\mathbf{H}}_P(z)] \neq 0 \;\; \forall\, |z| \geq 1 \tag{6.65}$$

in light of the feedback stability in Sect. 3.2.3. It follows from (6.64) that

$$\begin{aligned}\det[\widehat{\mathbf{H}}_K(z)\tilde{\mathbf{H}}_P(z)] &= \det[\mathbf{H}_K(z)\,(I+\Delta_{\text{rel}}(z))\,\tilde{\mathbf{H}}_P(z)]\\ &= \det[I+\mathbf{H}_K(z)\Delta_{\text{rel}}(z)\tilde{\mathbf{H}}_P(z)]\\ &= \det[I+\tilde{\mathbf{H}}_P(z)\mathbf{H}_K(z)\Delta_{\text{rel}}(z)]\end{aligned}$$

where $\mathbf{H}_K(z)\tilde{\mathbf{H}}_P(z) = I$ in (6.45) is used. Hence, (6.65) holds if

$$\|\tilde{\mathbf{H}}_P\mathbf{H}_K\Delta_{\text{rel}}\|_\infty \leq \gamma\|\Delta_{\text{rel}}\|_\infty < 1, \tag{6.66}$$

in light of the small gain theorem in Lemma 6.2 and $\gamma = \|\tilde{\mathbf{H}}_P\mathbf{H}_K\|_\infty$. The multiplicative property in (i) of Problem 6.11 is used in arriving at the inequality (6.66). The stability condition in (6.63) can be obtained rather easily by using the bound for $\|\Delta_{\text{rel}}\|_\infty$ from Theorem 4.6. For each z on the unit circle, $\tilde{\mathbf{H}}_P(z)\mathbf{H}_K(z)$ and $\tilde{\mathbf{H}}_K(z)\mathbf{H}_P(z)$ are projection matrices and

$$\tilde{\mathbf{H}}_P(z)\mathbf{H}_K(z)+\tilde{\mathbf{H}}_K(z)\mathbf{H}_P(z) = I$$

by the double Bezout identity (6.45). The equality $\|\tilde{\mathbf{H}}_P\mathbf{H}_K\|_\infty = \|\tilde{\mathbf{H}}_K\mathbf{H}_P\|_\infty$ holds by using the result in Problem 6.12 for each z on the unit circle and by the definition on $\mathscr{H}_\infty$ norm. □

Theorem 6.7 shows that $\gamma = \|\tilde{\mathbf{H}}_P\mathbf{H}_K\|_\infty = \|\tilde{\mathbf{H}}_K\mathbf{H}_P\|_\infty$ affects adversely the stability condition in (6.63). By direct calculation,

$$\tilde{\mathbf{H}}_P\mathbf{H}_K = \begin{bmatrix} I \\ \mathbf{P} \end{bmatrix} (I - \mathbf{KP})^{-1} \begin{bmatrix} I \; \mathbf{K} \end{bmatrix},$$

$$\tilde{\mathbf{H}}_K\mathbf{H}_P = \begin{bmatrix} -\mathbf{K} \\ I \end{bmatrix} (I - \mathbf{PK})^{-1} \begin{bmatrix} -\mathbf{P} \; I \end{bmatrix}.$$

So it is desirable to minimize γ in feedback control system design, but this is beyond the scope of this book. Note that $\|\tilde{\mathbf{H}}_K\mathbf{H}_P\|_\infty = \|\mathbf{T}_{dw}\|_\infty$ with $\mathbf{T}_{dw}(z)$ given in Example 6.2.

The next result derives an equivalent stability condition to that in Theorem 6.7. The proof is left as an exercise (Problem 6.13).

Corollary 6.1. *Under the same hypotheses/conditions in* Theorem 6.7, *the feedback control system in* Fig. 6.3 *remains stable when the* $\mathscr{H}_2$ *optimal controller* $\mathbf{K}(z)$ *is replaced by* $\widehat{\mathbf{K}}(z)$, *provided that*

$$\|\tilde{\mathbf{H}}_P(\mathbf{H}_K - \widehat{\mathbf{H}}_K)\|_\infty = \|\mathbf{H}_K^+(\mathbf{H}_K - \widehat{\mathbf{H}}_K)\|_\infty < 1.$$

The above result is especially attractive for normalized $\mathscr{H}_2$ control. Indeed, for the plant model $\mathbf{P}(z) = C(zI_n - A)^{-1}B$, the feedback controller

$$\mathbf{K}(z) = -F(zI - A - BF - LC)^{-1}L$$

is obtained by setting $F = -(I + B^*XB)^{-1}B^*XA$ and $L = -AYC^*(I + CYC^*)^{-1}$ where $X \geq \mathbf{0}$ and $Y \geq \mathbf{0}$ are stabilizing solutions to

$$X = A^*X(I + BB^*X)^{-1}B + C^*C, \quad Y = AY(I + C^*CY)^{-1}A^* + BB^*,$$

respectively. Let $\Omega = (I + B^*XB)^{1/2}$. Then $\tilde{\mathbf{H}}_P(z)\Omega^{-1}$ is an orthogonal matrix for each z on the unit circle that is referred to as normalized coprime factorization (refer to Sect. 5.3.2 in the previous chapter). Hence,

$$\|\tilde{\mathbf{H}}_P(\mathbf{H}_K - \widehat{\mathbf{H}}_K)\|_\infty = \|\Omega(\mathbf{H}_K - \widehat{\mathbf{H}}_K)\|_\infty.$$

Direct balanced truncation can be applied to $\Omega\mathbf{H}_K(z)$ to obtain the reduced order controller represented by $\Omega\widehat{\mathbf{H}}_K(z)$, and thus, the inequality

$$\|\tilde{\mathbf{H}}_P(\mathbf{H}_K - \widehat{\mathbf{H}}_K)\|_\infty \leq 2 \sum_{k=\rho+1}^{\eta} \sigma_k$$

holds true where σ_k is the kth distinct Hankel singular value of $\Omega \mathbf{H}_K(z)$. Recall the notation in Chap. 4. Consequently,

$$2 \sum_{k=\rho+1}^{\eta} \sigma_k < 1$$

ensures the feedback stability. If $\mathbf{K}(z)$ is not a normalized $\mathscr{H}_2$ controller, balanced truncation can still be applied to $\mathbf{H}_K(z)$ to obtain the reduced order controller but the above stability condition is weakened to

$$2 \sum_{k=\rho+1}^{\eta} \sigma_k < \|\tilde{\mathbf{H}}_P\|_\infty^{-1}.$$

Optimal Hankel-norm approximation is not employed here because it may not preserve the structure of the feedback controller that is in contrast to balaned truncation.

Example 6.8. Consider the plant model $\mathbf{P}(z)$ and the controller $\mathbf{K}(z)$ in Example 6.6. It can be verified by Matlab that $\gamma = 4.8376$ in the case of normalized $\mathscr{H}_2$ control and $\gamma = 5.3$ in the case of normalized LQG control ($L_0 = \mathbf{0}$). Under the inverse balanced realization,

$$X = Y = \text{diag}(2.3703, 2.3061, 0.4234, 0.3090, 0.0089).$$

Hence, the left-hand side of (6.63) in Theorem 6.7 is equal to 0.0180, if the order of the controller is reduced by only 1 which satisfies the stability condition in Theorem 6.7. However, the left-hand side of (6.63) is equal to 1.871, if the order of the controller is reduced by 2, which violates the stability condition in (6.63). It is important to notice that the stability condition established in Theorem 6.7 is only sufficient. In fact, even if the order of the controller is reduced to all the way to 1, stability of the closed-loop system remains intact, owing probably to stability of the plant.

The results in Theorem 6.7 and Corollary 6.1 assume that the full order controller is designed first and reduced later, even though the plant and controller are reduced simultaneously. A dual scenario is the case when the plant is reduced first, and the reduced order controller $\widehat{\mathbf{K}}(z)$ is then designed based on the reduced order plant $\widehat{\mathbf{P}}(z) = \widehat{\mathbf{G}}_{22}(z)$ in which the multiplicative error, rather than the relative error, helps to derive the stability condition.

Theorem 6.9. *Let* $\widehat{\mathbf{G}}_{22}(z) = C_{21}(zI_r - A_{11})^{-1}B_{21}$ *be the reduced order model obtained through applying the* IBT *algorithm to* $\mathbf{G}_{22}(z)$ *and*

$$\widehat{\mathbf{G}}(z) = \left[\begin{array}{c|cc} A_{11} & B_{11} & B_{21} \\ \hline C_{11} & D_{11} & D_{12} \\ C_{21} & D_{21} & D_{22} \end{array}\right]$$

where B_{21} and C_{21} are obtained via direct truncation in (6.60) and $D_{22} = \mathbf{0}$. Let $\widehat{\mathbf{K}}(z)$ be the optimal $\mathscr{H}_2$ controller for $\widehat{\mathbf{G}}(z)$ with $\widehat{F}$, $\widehat{L}$ the state-feedback and estimation gains, respectively, and $\widehat{L}_0$ the output estimation gain. Then

$$\widehat{\mathbf{G}}_{22} = \widehat{\mathbf{M}}(z)^{-1}\widehat{\mathbf{N}}(z), \quad \widehat{\mathbf{K}}(z) = -\widehat{\tilde{\mathbf{U}}}(z)\widehat{\tilde{\mathbf{V}}}(z)^{-1}$$

are coprime factorizations as specified by

$$\widehat{\mathbf{H}}_P(z) = \begin{bmatrix} -\widehat{\mathbf{N}}(z) & \widehat{\mathbf{M}}(z) \end{bmatrix} = \left[\begin{array}{c|cc} A_{11} + \widehat{L}C_{21} & -B_{21} & \widehat{L} \\ \hline C_{21} & \mathbf{0} & I \end{array}\right],$$

$$\tilde{\mathbf{H}}_{\widehat{K}}(z) = \begin{bmatrix} -\widehat{\tilde{\mathbf{U}}}(z) \\ \widehat{\tilde{\mathbf{V}}}(z) \end{bmatrix} = \left[\begin{array}{c|c} A_{11} + B_{21}\widehat{F} & -(\widehat{L} + B_{21}\widehat{L}_0) \\ \hline \widehat{F} & -\widehat{L}_0 \\ C_{21} & I \end{array}\right],$$

satisfying $\widehat{\mathbf{H}}_P(z)\tilde{\mathbf{H}}_{\widehat{K}}(z) \equiv I$. The $\mathscr{H}_2$ controller $\widehat{\mathbf{K}}(z)$ stabilizes $\mathbf{G}_{22}(z)$, if

$$\prod_{i=r+1}^{n}\left(1 + 2s_i\left(\sqrt{1 + s_i^2} + s_i\right)\right) < 1 + \widehat{\gamma}^{-1} \tag{6.67}$$

where $\widehat{\gamma} := \|\tilde{\mathbf{H}}_{\widehat{K}}\widehat{\mathbf{H}}_P\|_\infty = \|\widehat{\mathbf{H}}_P\tilde{\mathbf{H}}_{\widehat{K}}\|_\infty$ and $\{s_i\}$ are the same as in (6.59).

Proof. The hypotheses and Theorem 4.6 imply

$$\mathbf{H}_P(z) = \widehat{\mathbf{H}}_P(z)\left[I + \Delta_{\text{mul}}(z)\right] \tag{6.68}$$

with $\mathbf{H}_P(z)$ as in (6.43) and $\|\Delta_{\text{mul}}\|_\infty$ bounded as in (6.52). Similar to the proof for Theorem 6.7, $\widehat{\mathbf{K}}(z)$ stabilizes $\mathbf{G}_{22}(z)$, if and only if

$$\det\left(\mathbf{H}_P(z)\tilde{\mathbf{H}}_{\widehat{K}}(z)\right) \neq 0 \;\forall\, |z| \geq 1. \tag{6.69}$$

It follows from (6.68) and $\widehat{\mathbf{H}}_P(z)\tilde{\mathbf{H}}_{\widehat{K}}(z) \equiv I$ that

$$\begin{aligned} \det\left(\mathbf{H}_P(z)\tilde{\mathbf{H}}_{\widehat{K}}(z)\right) &= \det\left(\widehat{\mathbf{H}}_P(z)\left[I + \Delta_{\text{mul}}(z)\right]\tilde{\mathbf{H}}_{\widehat{K}}(z)\right) \\ &= \det\left(I + \widehat{\mathbf{H}}_P(z)\Delta_{\text{mul}}(z)\tilde{\mathbf{H}}_{\widehat{K}}(z)\right) \\ &= \det\left(I + \tilde{\mathbf{H}}_{\widehat{K}}(z)\widehat{\mathbf{H}}_P(z)\Delta_{\text{mul}}(z)\right). \end{aligned}$$

Hence, the inequality (6.69) holds if (6.67) is true in light of the small gain theorem in Lemma 6.2 that concludes the proof. □

The following is a similar stability condition to Corollary 6.1, and its proof is again left as an exercise (Problem 6.14).

Corollary 6.2. *Under the same hypotheses/conditions in* Theorem 6.9, *the feedback control system in* Fig. 6.3 *remains stable when the $\mathscr{H}_2$ optimal controller* $\mathbf{K}(z)$ *is replaced by* $\widehat{\mathbf{K}}(z)$, *provided*

$$\left\|\tilde{\mathbf{H}}_{\widehat{K}}\left(\mathbf{H}_P-\widehat{\mathbf{H}}_P\right)\right\|_\infty=\left\|\widehat{\mathbf{H}}_P^+\left(\mathbf{H}_P-\widehat{\mathbf{H}}_P\right)\right\|_\infty<1. \tag{6.70}$$

The above result suggests that balanced truncation can be applied to $\mathbf{H}_P(z)$ that results in an additive model error or

$$\mathbf{H}_P(z)=\widehat{\mathbf{H}}_P(z)+\Delta_{\text{add}}(z),\quad \|\Delta_{\text{add}}\|_\infty\leq 2\sum_{k=\rho+1}^{\eta}\sigma_k,$$

where σ_k is the kth distinct Hankel singular value of $\mathbf{H}_P(z)$. Recall again the notation in Chap. 4. The reduced order controller $\widehat{\mathbf{K}}(z)$ can be synthesized by computing $\tilde{\mathbf{H}}_{\widehat{K}}(z)=\widehat{\mathbf{H}}_P(z)^+$ or stable left inverse of $\widehat{\mathbf{H}}_P(z)$ that has as small $\mathscr{H}_\infty$ norm as possible in order to maximize the stability margin. Hence, the stability condition in (6.70) is ensured by

$$2\sum_{k=\rho+1}^{\eta}\sigma_k<\left\|\widehat{\mathbf{H}}_P^+\right\|_\infty^{-1}.$$

Even though the stability condition is less conservative than the previous one due to the additive form of the error bound, it is still a sufficient condition for the closed-loop stability. The error bound on left can again be considerably greater than the true error.

6.2 Control System Design

Design of the feedback control system in Fig. 6.1 begins with analysis of the plant model, performance objective, and exogenous disturbances/noises. Because of the difference in engineering practice, design objectives differ from one control system to another. More importantly, there is no single magic index that covers all design tasks which, if it is true, would actually trivialize control system design. Hence, the optimal solution to $\mathscr{H}_2$ control alone does not fulfill the design objective. In fact, analysis plays a major role in control system design. The goal of this chapter is to provide tools and guidelines for designing MIMO feedback control systems. Two methods which will be studied in this section are (frequency) loop shaping and eigenvalue assignment, akin to Bode plot and root locus in classic control, respectively. Special cases such as state feedback or state estimation will be treated first. Loop transfer recovery (LTR) will then be employed to deal with the case of output feedback.

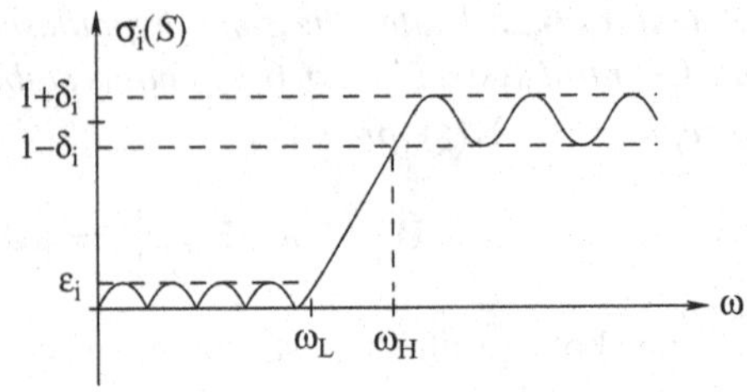

Fig. 6.4 Magnitude response of the desired sensitivity

6.2.1 Loop Shaping via Frequency Weighting

The synthesis method of loop shaping has its origin in the classic Bode design. The difference lies in its applicability to MIMO systems. It assumes that the performance requirement of the feedback system is specified in frequency domain with magnitude response replaced by singular value plots. For instance, the design objective can be the ith singular value of the desired sensitivity in frequency domain as shown in the next figure where the maximum ripples in low and high frequency bands specify the design specification.

Alternatively, the desired frequency shape can be imposed on the loop transfer matrix to ensure the desired sensitivity. For instance, low gain of the sensitivity can be ensured by high gain or high singular values of the loop transfer matrix while gain close to 1 for the sensitivity can be ensured by low gain or low singular values of the loop transfer matrix. Let $\mathbf{P}(z) = C(zI - A)^{-1}B$ be the plant. The problem is design of $\mathbf{K}(z)$ such that the MIMO feedback control system in Fig. 6.1 is not only internally stable but also meets the frequency domain specification such as the one specified in Fig. 6.4. The state-space approach suggests the estimator structure for $\mathbf{K}(z)$ and to begin with the synthesis of the state-feedback or state estimation gain. The output feedback controller $\mathbf{K}(z)$ can then be obtained by designing the other gain.

A problem arises in multivariable systems because the sensitivity at the plant input is different from that at the plant output given by

$$\mathbf{S}_{\text{in}}(z) = [I - \mathbf{K}(z)\mathbf{P}(z)]^{-1}, \quad \mathbf{S}_{\text{out}}(z) = [I - \mathbf{P}(z)\mathbf{K}(z)]^{-1},$$

respectively. Let $p \times m$ be the size of $\mathbf{P}(z)$. In the case of $p < m$, the sensitivity magnitude responses $\sigma_i[\mathbf{S}_{\text{in}}(e^{j\omega})]$ cannot be made smaller than 1 for the first $(m - p)$ singular values due to $\text{rank}\{\mathbf{K}(z)\mathbf{P}(z)\} \leq p$ for all $|z| = 1$. Hence, if $p < m$, sensitivity optimization needs to be carried out at the plant output rather than at the plant input. Similarly, in the case of $p > m$, the sensitivity magnitude responses $\sigma_i[\mathbf{S}_{\text{out}}(e^{j\omega})]$ cannot be made smaller than 1 for the first $(p - m)$ singular values due to $\text{rank}\{\mathbf{P}(z)\mathbf{K}(z)\} \leq m$ for all $|z| = 1$. Hence, if $p > m$, sensitivity optimization needs to be carried out at the plant input rather than at the plant output. Fortunately, it is often unnecessary to optimize the frequency shape of the sensitivity at both the plant input and output in practice.

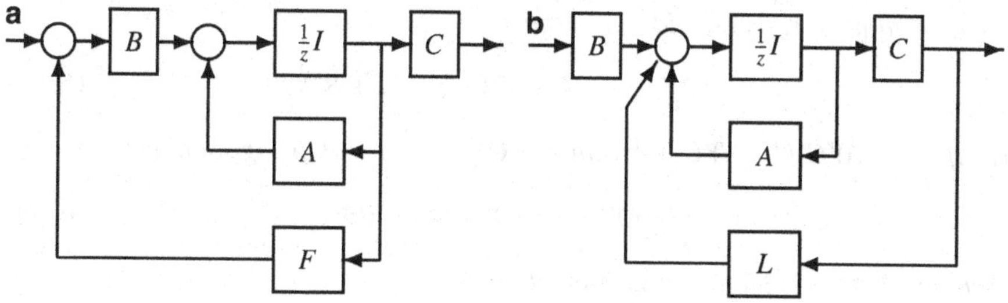

Fig. 6.5 State feedback and output injection

Consider first when the state is accessible by the controller as depicted in block diagram (a) of Fig. 6.5. Then with F as the state-feedback gain, the sensitivity at the plant input ($p \geq m$ is normally assumed) and overall transfer matrix are given by

$$\begin{aligned} \mathbf{S}_F(z) &= I + F(zI - A - BF)^{-1}B, \\ \mathbf{T}_F(z) &= C(zI - A - BF)^{-1}B, \end{aligned} \tag{6.71}$$

respectively. On the other hand, block diagram (b) of Fig. 6.5 corresponds to state estimation in which sensitivity at the plant output ($p \leq m$ is normally assumed) and overall transfer matrix are given by

$$\begin{aligned} \mathbf{S}_L(z) &= I + C(zI - A - LC)^{-1}L, \\ \mathbf{T}_L(z) &= C(zI - A - LC)^{-1}B, \end{aligned} \tag{6.72}$$

respectively, with L the state estimation gain. However, strictly speaking, (b) of Fig. 6.5 does not represent a state estimator which is often referred to as output injection. By convention, the overall transfer matrix is referred to as complementary sensitivity inherited from the case of output feedback.

It is interesting to observe that (refer to Theorem 3.28 in Chap. 3)

$$\mathbf{P}(z) = \mathbf{T}_F(z)\mathbf{S}_F(z)^{-1} = \mathbf{S}_L(z)^{-1}\mathbf{T}_L(z) = C(zI - A)^{-1}B,$$

and thus, sensitivity and complementary sensitivity constitute the coprime factors of the plant model, if F and L are stabilizing. Moreover, state feedback corresponds to right coprime factorization and output injection corresponds to left coprime factorization of the plant. A particular set of coprime factors that interest us is the set of normalized ones. Recall Sect. 5.3.2 in the previous chapter. The following result can be stated.

Corollary 6.3. *Consider the plant model* $\mathbf{P}(z) = C(zI - A)^{-1}B$ *where* (A, B) *is stabilizable and* (C, A) *is detectable.* (i) *If* $F = -(I + B^*XB)^{-1}B^*XA$ *with* $X \geq \mathbf{0}$ *being the stabilizing solution to*

$$X = A^*X(I + BB^*X)^{-1}A + C^*C, \tag{6.73}$$

then for all $|z| = 1$, *there holds the identity*

$$\mathbf{S}_F(z)^*\mathbf{S}_F(z) + \mathbf{T}_F(z)^*\mathbf{T}_F(z) = I + B^*XB; \tag{6.74}$$

(ii) *If* $L = -AYC^*(I + CYC^*)^{-1}$ *with* $Y \geq \mathbf{0}$ *being the stabilizing solution to*

$$Y = AY(I + C^*CY)^{-1}A^* + BB^*, \tag{6.75}$$

then for all $|z| = 1$, *there holds the identity*

$$\mathbf{S}_L(z)\mathbf{S}_L(z)^* + \mathbf{T}_L(z)\mathbf{T}_L(z)^* = I + CYC^*. \tag{6.76}$$

Let $I + B^*XB = USU^*$ be its SVD. Then $I + B^*XB = \Omega_F^2$ by taking $\Omega_F = U\sqrt{S}U^*$. In light of (6.74),

$$\{\tilde{\mathbf{M}}(z), \tilde{\mathbf{N}}(z)\} = \{\mathbf{S}_F(z)\Omega_F^{-1}, \mathbf{T}_F(z)\Omega_F^{-1}\}$$

constitutes the pair of normalized right coprime factorization of $\mathbf{P}(z)$. Let $\Omega_L^2 = I + CYC^*$. The equality (6.74) implies that

$$\{\mathbf{M}(z), \mathbf{N}(z)\} = \{\Omega_L^{-1}\mathbf{S}_L(z), \Omega_L^{-1}\mathbf{T}_L(z)\}$$

constitutes the pair of normalized left coprime factorization of $\mathbf{P}(z)$. More importantly, the normalization property yields

$$\begin{aligned} \tilde{\mathbf{M}}(e^{j\omega})\tilde{\mathbf{M}}(e^{j\omega})^* &= \left[I + \mathbf{P}(e^{j\omega})^*\mathbf{P}(e^{j\omega})\right]^{-1}, \\ \mathbf{M}(e^{j\omega})^*\mathbf{M}(e^{j\omega}) &= \left[I + \mathbf{P}(e^{j\omega})\mathbf{P}(e^{j\omega})^*\right]^{-1}. \end{aligned} \tag{6.77}$$

Hence, if $\sigma_i[\mathbf{P}(e^{j\omega})]$ admits the same shape as that of the desired loop transfer matrix, then for each integer i,

$$\sigma_i[\tilde{\mathbf{M}}(e^{j\omega})] = \sigma_i[\mathbf{S}_F(e^{j\omega})\Omega_F^{-1}]$$

represents the ideal sensitivity shape at the plant input and

$$\sigma_i[\mathbf{M}(e^{j\omega})] = \sigma_i[\Omega_L^{-1}\mathbf{S}_L(e^{j\omega})]$$

represents the ideal sensitivity shape at the plant output. In addition, by the normalization property there hold

$$\begin{aligned} \tilde{\mathbf{N}}(e^{j\omega})\tilde{\mathbf{N}}(e^{j\omega})^* &= \mathbf{P}(e^{j\omega})\left[I + \mathbf{P}(e^{j\omega})^*\mathbf{P}(e^{j\omega})\right]^{-1}\mathbf{P}(e^{j\omega})^*, \\ \mathbf{N}(e^{j\omega})^*\mathbf{N}(e^{j\omega}) &= \mathbf{P}(e^{j\omega})^*\left[I + \mathbf{P}(e^{j\omega})\mathbf{P}(e^{j\omega})^*\right]^{-1}\mathbf{P}(e^{j\omega}). \end{aligned} \tag{6.78}$$

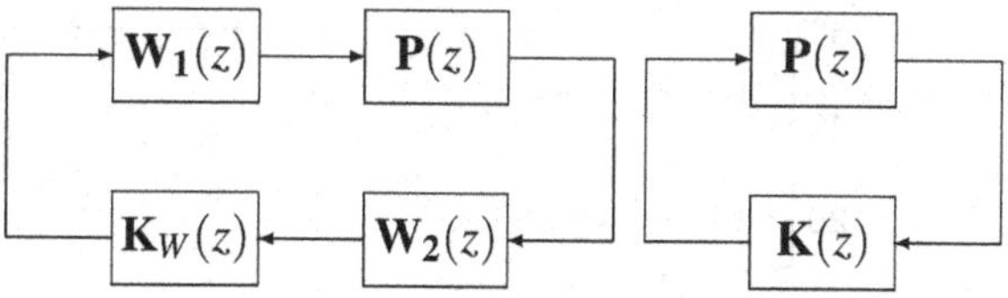

Fig. 6.6 Feedback controller $\mathbf{K}(z) = \mathbf{W}_1(z)\mathbf{K}_W(z)\mathbf{W}_2(z)$

Hence, under the same hypothesis on $\sigma_i[\mathbf{P}(e^{j\omega})]$ for each i, both

$$\sigma_i[\tilde{\mathbf{N}}(e^{j\omega})] = \sigma_i[\mathbf{T}_F(e^{j\omega})\Omega_F^{-1}], \quad \sigma_i[\mathbf{N}(e^{j\omega})] = \sigma_i[\Omega_L^{-1}\mathbf{T}_L(e^{j\omega})]$$

represent the ideal shape of the complementary sensitivity.

The desired loop shape for the plant model is rarely true in practice. A procedure to modify the frequency shape of $\mathbf{P}(z)$ is by means of frequency weighting functions. Simple compensators such as static gains, accumulators, and lead/lag transfer functions can be introduced to each row or column of $\mathbf{P}(z)$ so as to shape its singular value plots. That is, stable and minimum phase transfer matrices $\mathbf{W}_1(z)$ and $\mathbf{W}_2(z)$ are searched for such that

$$\mathbf{P}_W(z) = \mathbf{W}_2(z)\mathbf{P}(z)\mathbf{W}_1(z) \tag{6.79}$$

admits the desired loop shape. A feedback controller $\mathbf{K}_W(z)$ is then designed for the weighted plant $\mathbf{P}_W(z)$. The feedback controller $\mathbf{K}(z)$ as in Fig. 6.1 can then be recovered via

$$\mathbf{K}(z) = \mathbf{W}_1(z)\mathbf{K}_W(z)\mathbf{W}_2(z). \tag{6.80}$$

This is illustrated in Fig. 6.6.

The following example illustrates the loop shaping procedure.

Example 6.10. Consider the flight control example in Problem 1.11 from Chap. 1. Under the sampling frequency of 40 Hz, the discretized model has the following singular value plots in frequency domain (Fig. 6.7).

In order to have zero steady-state error for step inputs, weight function with pole at 1 is employed. After a few trials, $\mathbf{W}(z) = \text{diag}(-2, 1, -2)W(z)$ is taken where $W(z) = 24.12 + 0.24/(z-1)$. The singular value plots for the weighted plant $\mathbf{P}_W(z) = \mathbf{P}(z)\mathbf{W}(z)$ are shown in Fig. 6.8.

It can be seen that the magnitudes of the shaped plant are raised greatly at low frequencies. In addition, it maintains good roll-off at high frequencies. In fact, all the three singular values are below 0 dB after $\omega \geq 0.2$ or 8π rad/s in terms of the physical frequency, since $\omega = 1$ corresponds to 40π rad/s. With $T_s = 0.025$ s, the discretized realization matrices are found to be

$$A_d = \begin{bmatrix} 1 & 0.0001 & 0.0283 & 0.0000 & -0.0248 \\ 0 & 0.9986 & -0.0043 & -0.0000 & 0.0017 \\ 0 & 0.0000 & 1.0000 & 0.0247 & -0.0003 \\ 0 & 0.0013 & -0.0000 & 0.9785 & -0.0248 \\ 0 & -0.0072 & 0.0000 & 0.0258 & 0.9827 \end{bmatrix},$$

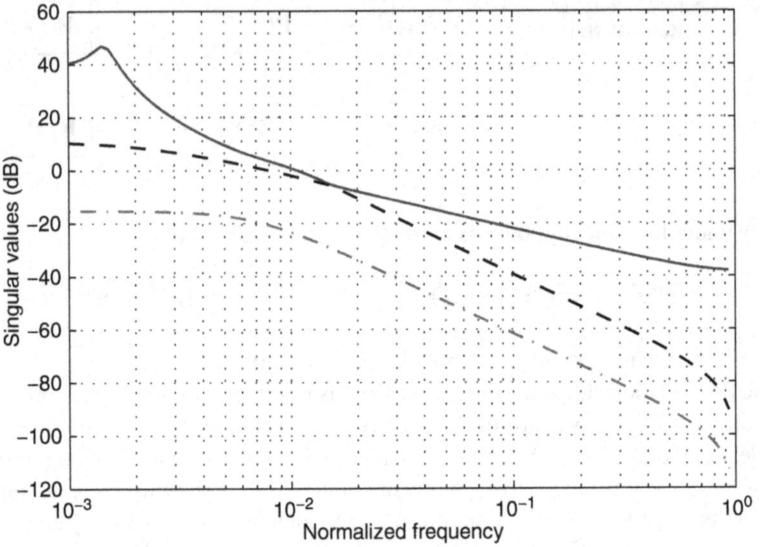

Fig. 6.7 Plant singular value plots versus normalized frequency

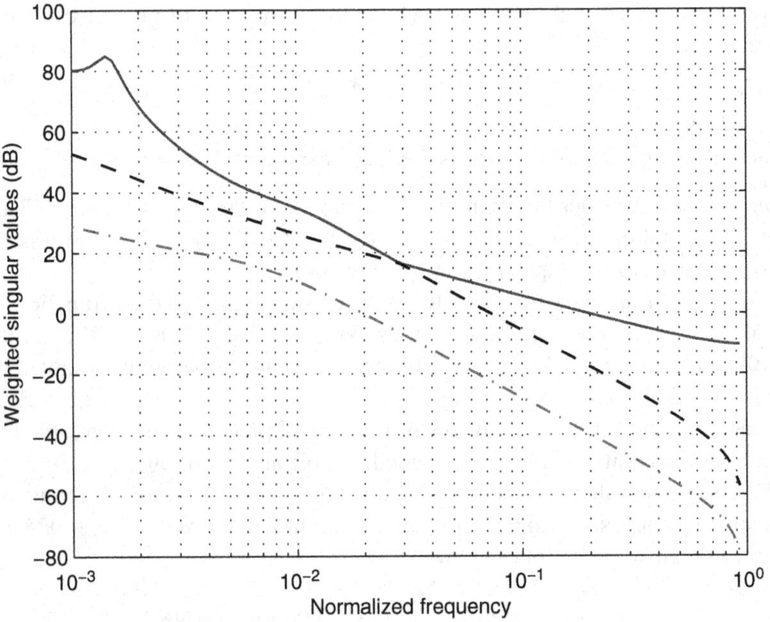

Fig. 6.8 Plant weighted singular value plots vs. normalized frequency

$$B_d = \begin{bmatrix} -0.0000 & 0.0000 & 0.0000 \\ -0.0000 & 0.0250 & -0.0000 \\ 0.0001 & 0.0000 & -0.0005 \\ 0.0109 & 0.0000 & -0.0411 \\ 0.0040 & -0.0001 & -0.0024 \end{bmatrix},$$

and C_d remains the same as that of the continuous-time one.

6.2.2 LQR Design via Root Locus

Root locus in classic control aims at placing the closed-loop poles to the desired locations dictated by performance specifications. The LQR method, on the other hand, reassigns the eigenvalues for multi-input systems to which the graphical method of root locus is inapplicable. A significant problem in LQR design is how to select the weighting matrices Q and R in order to achieve the desired locations for closed-loop eigenvalues under state feedback. The answer turns out to be the root locus.

Let the plant model be represented by $\mathbf{P}(z) = C(zI - A)^{-1}B$ that has size $p \times m$ and is of order n. In general, eigenvalues of A are not in the right locations to ensure the performance. Feedback controllers need to be designed to reassign eigenvalues for the closed-loop system in order to achieve the desirable output responses to exogenous inputs such as step or disturbances. Control system design in the case of $p \geq m$ begins with the LQR performance index

$$J = \sum_{t=0}^{\infty} \|\mathbf{u}(t)\|^2 + \rho^2 \|C_0 \mathbf{x}(t)\|^2 = \sum_{t=0}^{\infty} \|\mathbf{w}(t)\|^2,$$

assuming that the state vector of $\mathbf{P}(z)$ is available for feedback. Thus, the LQR problem admits the following state-space description:

$$\mathbf{x}(t+1) = A\mathbf{x}(t) + B\mathbf{u}(t), \quad \mathbf{x}(0) = \mathbf{x}_0,$$
$$\mathbf{w}(t) = \begin{bmatrix} \rho C_0 \\ \mathbf{0} \end{bmatrix} \mathbf{x}(t) + \begin{bmatrix} \mathbf{0} \\ I \end{bmatrix} \mathbf{u}(t). \qquad (6.81)$$

Selection of C_0 impacts eigenvalue assignment directly.

Specifically, the corresponding LQR state-feedback control law is $\mathbf{u}(t) = F_\rho \mathbf{x}(t)$ that is parameterized by ρ and specified by

$$F_\rho = -(I + B^* X_\rho B)^{-1} B^* X_\rho A,$$
$$X_\rho = A^* X_\rho (I + BB^* X_\rho)^{-1} A + \rho^2 C_0^* C_0, \qquad (6.82)$$

with $X_\rho \geq \mathbf{0}$ the stabilizing solution. Eigenvalues associated with the closed-loop system are poles of

$$[I - F_\rho(zI - A)^{-1}B]^{-1} = I + F_\rho(zI - A - BF_\rho)^{-1}B.$$

Denote $\mathbf{G}_O(z) = C_O(zI - A)^{-1}B$ and

$$\mathbf{G}(z) = \begin{bmatrix} \mathbf{0} \\ I \end{bmatrix} + \rho \begin{bmatrix} C_O \\ \mathbf{0} \end{bmatrix} (zI - A)^{-1}B.$$

Recall that $\mathbf{L}(z) := F_\rho(zI - A)^{-1}B$ is the loop transfer matrix under the state feedback in (a) of Fig. 6.5, which is often termed as the return difference. It is left as an exercise (Problem 6.15) to show that for all $|z| = 1$, there holds

$$\begin{aligned} \mathbf{G}(z)^*\mathbf{G}(z) &= I + \rho^2 \mathbf{G}_O(z)^*\mathbf{G}_O(z) \\ &= [I - \mathbf{L}(z)]^* (I + B^*X_\rho B) [I - \mathbf{L}(z)]. \end{aligned} \tag{6.83}$$

In the case of $m = 1$, the above equality implies

$$|1 - \mathbf{L}(e^{j\omega})| \geq r := (1 + B^*X_\rho B)^{-1/2} \ \forall \ \omega. \tag{6.84}$$

Hence, the Nyquist plot of $-\mathbf{L}(e^{j\omega})$ for $\omega \in [-\pi, \pi]$ does not intersect the disk of radius r centered at -1 as shown next.

Consequently, the system admits phase margin of $2\sin^{-1}(r/2)$ and gain margin of either $(1 + r)$ or $(1 - r)^{-1}$. See Problem 6.16 in Exercises. If $m > 1$, the inequality (6.84) is replaced by

$$\underline{\sigma}[I - \mathbf{L}(e^{j\omega})] \geq r := \underline{\sigma}[(I + B^*X_\rho B)^{-1/2}] \ \forall \ \omega \tag{6.85}$$

where $\underline{\sigma}(\cdot)$ denotes the minimum singular value. Similar interpretation of the stability margins can also be obtained.

Denote $\mathbf{a}(z) = \det(zI - A)$ whose roots are eigenvalues of A and $\mathbf{a}_F(z) = \det(zI - A - BF_\rho)$ whose roots are eigenvalues of $(A + BF_\rho)$. If $m = 1$,

$$\mathbf{G}_O(z) = C_O(zI - A)^{-1}B = \frac{\mathbf{b}_O(z)}{\mathbf{a}(z)}.$$

For real rational $\mathbf{G}(z)$, the equality (6.83) leads to

$$1 + \rho^2 \frac{\mathbf{b}_O(z^{-1})\mathbf{b}_O(z)}{\mathbf{a}(z^{-1})\mathbf{a}(z)} = \frac{\mathbf{a}_F(z^{-1})\mathbf{a}_F(z)}{r^2\mathbf{a}(z^{-1})\mathbf{a}(z)}.$$

It follows from the above equation that

$$r^{-2}\mathbf{a}_F(z^{-1})\mathbf{a}_F(z) = \mathbf{a}(z^{-1})\mathbf{a}(z) + \rho^2 \mathbf{b}_O(z^{-1})\mathbf{b}_O(z). \tag{6.86}$$

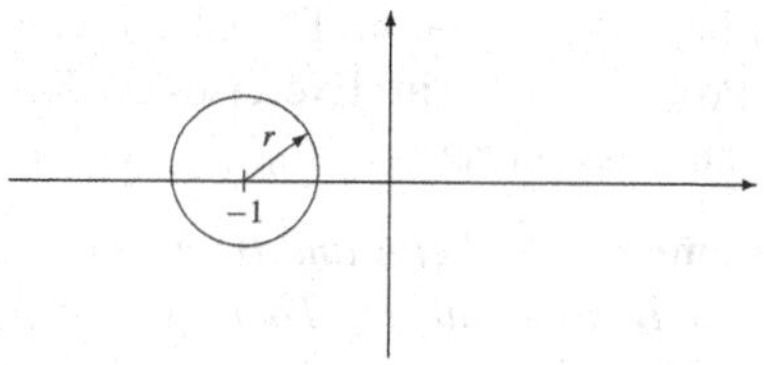

Fig. 6.9 Stability margins represented by the disk on $\mathscr{Z}$-plane

Root locus of $z^n \mathbf{a}_F(z^{-1})\mathbf{a}_F(z) = 0$ can be sketched for $0 \le \rho^2 \le \infty$ where n is the dimension of the state vector $\mathbf{x}(t)$. The root locus of $\mathbf{a}_F(z) = 0$ are those inside the unit circle.

The case of $\rho \to 0$ corresponds to low state weighting that is equivalent to expensive control. In this case, roots of $\mathbf{a}_F(z) = 0$ are stable roots of $\mathbf{a}(z) = 0$ plus the mirror images of unstable roots of $\mathbf{a}(z) = 0$ about the unit circle that is less inspiring. The case that enlightens the design is when $\rho \to \infty$. Let d be the degree difference or $(n-d)$ be the degree of $\mathbf{b}_0(z)$. For $\rho \to \infty$, roots of $\mathbf{a}_F(z) = 0$ approach stable roots of $\mathbf{b}_0(z) = 0$, the mirror images of unstable roots of $\mathbf{b}_0(z) = 0$ about the unit circle, and plus d roots approaching the origin. Because roots of $\mathbf{b}_0(z) = 0$ are transmission zeros of $\mathbf{G}_0(z) = C_0(zI - A)^{-1}B$, eigenvalue assignment can be conveniently achieved through synthesis of C_0 such that $\mathbf{G}_0(z)$ has transmission zeros in some special locations of the complex plane plus those at the origin. These zeros are used to attract the n stable root loci of the right-hand side in (6.86) with parameter ρ^2 so that the desired eigenvalues of $(A + BF_\rho)$ can be achieved at some $\rho > 0$.

An important issue worth discussion is that large ρ yields small r, the radius of the circle in Fig. 6.9. Indeed, by replacing X_ρ with $\rho^2\overline{X}$ and taking limit $\rho \to \infty$, the ARE in (6.82) can be written as

$$\overline{X} = A^*\overline{X}A - A^*\overline{X}B(B^*\overline{X}B)^+B^*\overline{X}A + C_0^*C_0$$

in which $B^*\overline{X}B$ may be singular, and thus, the generalized inverse is used. The stabilizing solution $\overline{X}$ exists and is finite under the stabilizability of (A,B) and observability of (C_0,A) on the unit circle. If in addition $B^*\overline{X}B$ is invertible, then $X_\rho \to \rho^2\overline{X} \to \infty$ as $\rho \to \infty$. As a result, $r = [\underline{\sigma}(I + B^*XB)]^{-1/2} \to 0$. That is, the stability margin shrinks to zero as $\rho \to \infty$. For this reason, the value of ρ is prohibited from being large. But if zeros of $C_0(zI - A)^{-1}B$ can be arbitrarily assigned with C_0, a reasonable value of $\rho > 0$ can be used to place roots of $\mathbf{a}_F(z) = 0$ to the desired locations including those near the origin, provided that zeros of $C_0(zI - A)^{-1}B$ are assigned to some suitable locations obtained through trials and errors. It is important to observe that there is an inherent conflict between stability margins and the performance. In fact, desired eigenvalue locations are achieved at the expense of stability margins. In the case of $m > 1$, the root locus needs to be carried out for

$$\det\left[I + \rho^2 B^*(z^{-1}I - A^*)^{-1}C_0^*C_0(zI - A)^{-1}B\right] = 0 \tag{6.87}$$

where $\rho = 0 \to \infty$. The above can be converted to computation of eigenvalues (Problem 6.17 in Exercises). The next result answers the question when the transmission zeros of $C_0(zI - A)^{-1}B$ can be arbitrarily assigned with C_0.

Lemma 6.3. *Let A and B be matrices of size $n \times n$ and $n \times m$, respectively. Suppose that B has rank m. Then* (i) *the transmission zeros of $\mathbf{G}_0(z) = C_0(zI - A)^{-1}B$ can be arbitrarily assigned with C_0 of size $m \times n$, if and only if (A,B) is reachable;* (ii) *the transmission zeros of $\mathbf{G}_0(z) = C_0(zI - A)^{-1}B$ are all strictly inside the unit circle for some C_0 of size $m \times n$, if and only if (A,B) is stabilizable.*

Proof. Without loss of generality, (A,B) can be assumed to be of the following form:

$$A = \begin{bmatrix} A_{11} & A_{12} \\ A_{21} & A_{22} \end{bmatrix}, \quad B = \begin{bmatrix} R_B \\ \mathbf{0} \end{bmatrix} \tag{6.88}$$

where A_{11} and R_B are square and of the same size $m \times m$. If it is not, QR factorization can be applied to B to yield $B = Q_B R_B$ where Q_B is orthogonal and R_B is square and upper triangular matrix. Hence, an orthogonal $Q_\perp$ exists such that $Q_a = \begin{bmatrix} Q_B & Q_\perp \end{bmatrix}$ is square and unitary. A similarity transform $T = Q_a^*$ can be applied to obtain (TAT^*, TB) which have the same form as (A,B) in (6.88).

Let $C_0 = \begin{bmatrix} C_{01} & C_{02} \end{bmatrix}$ with $C_{01} = I$. Then the transmission zeros of $\mathbf{G}_0(z)$ are the roots of

$$\lambda_B(z) := \det\left(\begin{bmatrix} A - zI & B \\ C_0 & \mathbf{0} \end{bmatrix}\right) = 0.$$

Substituting the expressions of C_0 and (A,B) in (6.88) gives

$$\begin{aligned} \lambda_B(z) &= \det\left(\begin{bmatrix} A_{11} - zI & A_{12} & R_B \\ A_{21} & A_{22} - zI & \mathbf{0} \\ I & C_{02} & \mathbf{0} \end{bmatrix}\begin{bmatrix} I & -C_{02} & \mathbf{0} \\ \mathbf{0} & I & \mathbf{0} \\ \mathbf{0} & \mathbf{0} & I \end{bmatrix}\right) \\ &= \det\left(\begin{bmatrix} * & * & R_B \\ * & A_{22} - A_{21}C_{02} - zI & \mathbf{0} \\ I & \mathbf{0} & \mathbf{0} \end{bmatrix}\right) \\ &= \pm\det(R_B)\det(A_{22} - A_{21}C_{02} - zI). \end{aligned}$$

Hence, the transmission zero assignment with C_0 is the same as eigenvalue assignment of $(A_{22} - A_{21}C_{02})$ with $-C_{02}$ as the "state-feedback gain" and A_{21} as the equivalent "B" matrix. Since

$$\begin{aligned} \operatorname{rank}\left\{\begin{bmatrix} A - zI & B \end{bmatrix}\right\} &= \operatorname{rank}\left\{\begin{bmatrix} A_{11} - zI & A_{12} & R_B \\ A_{21} & A_{22} - zI & \mathbf{0} \end{bmatrix}\right\} \\ &= m + \operatorname{rank}\left\{\begin{bmatrix} A_{22} - zI & A_{21} \end{bmatrix}\right\}, \end{aligned}$$

the reachability and stabilizability of (A,B) are equivalent to those of (A_{22},A_{21}), respectively. Therefore, Theorem 3.25 can be applied to conclude (i), and Theorem 3.26 can be applied to conclude (ii). □

Example 6.11. Consider the flight control example in Chap. 1. Its discretized plant model is computed in Example 6.10 under the sampling frequency of 40 Hz. Let $A = A_d$ and $B = B_d$. The QR factorization yields

$$T = Q_a^* = \begin{bmatrix} -0.0042 & -0.0023 & 0.0118 & 0.9372 & 0.3486 \\ -0.0000 & -1.0000 & -0.0000 & -0.0034 & 0.0026 \\ -0.0118 & 0.0036 & -0.0043 & -0.3486 & 0.9372 \\ 0.9287 & 0.0000 & -0.3705 & 0.0045 & 0.0117 \\ 0.3705 & 0.0000 & 0.9287 & -0.0117 & 0.0046 \end{bmatrix}.$$

The above similarity transform results in

$$A_{22} = \begin{bmatrix} 0.9900 & 0.0244 \\ -0.0039 & 1.0094 \end{bmatrix}, \quad A_{21} = \begin{bmatrix} -0.0162 & -0.0000 & -0.0188 \\ 0.0188 & -0.0001 & -0.0169 \end{bmatrix}.$$

Suppose that the desired damping ratio is $\zeta = 0.5901$, corresponding to 10% overshot, and $\omega_n = 10$, which yield the settling time of 0.6779 s (within 2% of the final value). The desired closed-loop poles are thus $-5.9010 \pm j8.0733$ that are mapped to $0.8453 \pm j0.1730$ in discrete time under the sampling frequency of 40 Hz. By taking

$$C_{02} = -A_{21}^*(A_{21}A_{21}^*)^{-1}S_0 \tag{6.89}$$

for some square matrix S_0, it leads to

$$A_{22} - A_{21}C_{02} = A_{22} + S_0 = \begin{bmatrix} 0.8453 & 0.1730 \\ -0.1730 & 0.8453 \end{bmatrix}$$

that admits the desired eigenvalues at $0.8453 \pm j0.1730$. The above yields

$$S_0 = \begin{bmatrix} 0.8453 & 0.1730 \\ -0.1730 & 0.8453 \end{bmatrix} - A_{22} = \begin{bmatrix} -0.1446 & 0.1486 \\ -0.1691 & -0.1641 \end{bmatrix}$$

and C_{02} in (6.89). It follows from $C_0 = \begin{bmatrix} I & C_{02} \end{bmatrix} Q_a^*$ that

$$C_0 = \begin{bmatrix} 4.3962 & -0.0022 & 7.8782 & 0.8377 & 0.4034 \\ -0.0390 & -1.0000 & -0.0057 & -0.0033 & 0.0021 \\ -8.0280 & 0.0035 & 3.4316 & -0.3903 & 0.8365 \end{bmatrix}.$$

It can be verified that $C_0(zI - A)^{-1}B$ has finite transmission zeros only at $0.8453 \pm j0.1730$ and $\det(C_0B) \neq 0$.

As discussed in the previous subsection, state feedback as depicted in (a) of Fig. 6.4 focuses on the sensitivity at the plant input that is given by

$$\mathbf{S}_F(z) = I + F_\rho(zI - A - BF_\rho)^{-1}B.$$

But if the design objective is the sensitivity at the plant output, then output injection in (b) of Fig. 6.4 will be under consideration that is dual to LQR. In this case, the state estimation gain L_ρ is computed first according to

$$\begin{aligned} Y_\rho &= AY_\rho(I + C^*CY_\rho)^{-1}A^* + \rho^2 B_0 B_0^*, \\ L_\rho &= -AY_\rho C^*(I + CY_\rho C^*)^{-1}, \end{aligned} \tag{6.90}$$

where B_0 is chosen such that transmission zeros of $C(zI - A)^{-1}B_0$ are in the right place to help assign eigenvalues of $(A + L_\rho C)$ to the desirable locations with the aid of the parameter $\rho \geq 0$. The computation of L_ρ is dual to that of F_ρ in (6.82) and previous discussions on selections of C_0 and ρ are applicable to selections of B_0 and ρ here. As a result, the sensitivity at the plant output is given by

$$\mathbf{S}_L(z) = I + C(zI - A - L_\rho C)^{-1}L_\rho.$$

In the next subsection, the design procedure of LTR will be introduced to recover the desired sensitivity in the case of output feedback.

6.2.3 Loop Transfer Recovery

Consider the feedback system in Fig. 6.1. LTR is a design procedure for the output feedback controller $\mathbf{K}(z)$ to recover the desired transfer matrix achieved under state feedback or output injection. As such, the desired sensitivity and stability margins are also recovered under LTR. The key lies in the estimator structure of $\mathbf{K}(z)$. In fact, the control signal is based on the estimated state from a Kalman filter or its dual, to be precise.

Suppose that the desired sensitivity is at the plant input. Let the state-feedback gain F be synthesized with either the frequency loop shaping or LQR-based eigenvalue assignment. LTR aims at synthesis of the state estimation gain for the output feedback control system to recover the same performance achieved under state feedback. Perfect recovery is possible only under some restrictive conditions. This problem is dual to state feedback by considering the associated filtering ARE:

$$Y = AYA^* - AYC^*(q^{-2}I + CYC^*)^{-1}CYA^* + BB^*. \tag{6.91}$$

By denoting $\overline{Y} = q^2 Y$, the above ARE can be rewritten as

$$\begin{aligned} \overline{Y} &= A\overline{Y}A^* - A\overline{Y}C^*(I + C\overline{Y}C^*)^{-1}C\overline{Y}A^* + q^2BB^* \\ &= A\overline{Y}(I + CC^*\overline{Y})^{-1}A^* + q^2BB^* \end{aligned}$$

that is similar to (6.75) or (6.90).

Lemma 6.4. *Assume that the plant* $\mathbf{P}(z) = C(zI-A)^{-1}B$ *has an equal number of inputs and outputs and* $\det(CB) \neq 0$. *If, in addition, the realization of* $\mathbf{P}(z)$ *is strictly minimum phase, then* $Y = BB^*$ *is the stabilizing solution to the filtering* ARE (6.91) *in the limiting case of* $q \to \infty$.

Proof. In the limiting case of $q \to \infty$, ARE (6.91) is reduced to

$$Y = AYA^* - AYC^*(CYC^*)^{-1}CYA^* + BB^*. \tag{6.92}$$

Since $\det(CB) \neq 0$, $Y = BB^*$ clearly satisfies the above ARE. The corresponding state estimation gain is given by

$$L = -AYC^*(CYC^*)^{-1} = -AB(CB)^{-1}. \tag{6.93}$$

It follows that $A + LC = A[I_n - B(CB)^{-1}C]$. Since $\det(CB) \neq 0$, the plant model $\mathbf{P}(z) = C(zI-A)^{-1}B$ has precisely $(n-m)$ finite zeros that are strictly inside the unit circle in light of the hypotheses. By noting that

$$\begin{aligned} zC(zI-A)^{-1}B &= CB + C(zI-A)^{-1}AB \\ &= \left[I + C(zI-A)^{-1}AB(CB)^{-1}\right]CB \\ &= \left[I - C(zI-A)^{-1}L\right]CB, \end{aligned}$$

there holds the equality

$$z^{-1}\left[C(zI-A)^{-1}B\right]^{-1} = (CB)^{-1}\left[I + C(zI-A-LC)^{-1}L\right].$$

Hence, eigenvalues of $(A+LC)$ are $(n-m)$ zeros of $\mathbf{P}(z)$ plus the remaining m at the origin which are all stable. It is thus concluded that $Y = BB^*$ is indeed the stabilizing solution to the filtering ARE (6.91). □

It is commented that the hypothesis of strictly minimum phase on realization of square $\mathbf{P}(z)$ in Lemma 6.4 implies stabilizability of (A,B) and detectability of (C,A). Hence, the existence of the stabilizing solution to ARE (6.91) is ensured and is given by $Y = BB^*$.

Let L be the estimation gain obtained in accordance with (6.93). The LTR design method sets the output feedback controller to be

$$\mathbf{K}(z) = -zF_1(zI-A-LC)^{-1}L, \quad F_1 = -(I+B^*XB)^{-1}B^*X, \tag{6.94}$$

where $X \geq \mathbf{0}$ is the stabilizing solution to the control ARE in either (6.73) or in (6.82) dependent on which design method is used. There holds $F = F_1A$. The controller in (6.94) is of the dual form to the state estimator that will be clarified at a later stage. The following result demonstrates that

$$\mathbf{K}(z)\mathbf{P}(z) = F(zI-A)^{-1}B, \tag{6.95}$$

if $\mathbf{P}(z)$ is square, strictly minimum phase, and $\det(CB) \neq 0$.

Theorem 6.12. *Consider the square plant* $\mathbf{P}(z) = C(zI - A)^{-1}B$. *If* (i) $\det(CB) \neq 0$ *and* (ii) *the realization of* $\mathbf{P}(z)$ *is strictly minimum phase, then the equality* (6.95) *holds where* $\mathbf{K}(z)$ *is the output feedback controller in* (6.94).

Proof. It is recognized that $\Pi = B(CB)^{-1}C$ is a projection matrix, and there hold $LC = -A\Pi$ and $(I - \Pi)B = \mathbf{0}$. Denote

$$\Delta(z) = F(zI - A)^{-1}B - \mathbf{K}(z)\mathbf{P}(z).$$

The proof is hinged to proving $\Delta(z) \equiv \mathbf{0}$ that is equivalent to the equality (6.95). By direct calculation,

$$\begin{aligned}
\Delta(z) &= F_1A(zI - A)^{-1}B + zF_1(zI - A - LC)^{-1}LC(zI - A)^{-1}B \\
&= F(zI - A)^{-1}B - zF_1(zI - A + A\Pi)^{-1}A\Pi(zI - A)^{-1}B \\
&= F(zI - A)^{-1}B - zF[zI - (I - \Pi)A]^{-1}\Pi(zI - A)^{-1}B \\
&= F[zI - (I - \Pi)A]^{-1}[zI - (I - \Pi)A - z\Pi](zI - A)^{-1}B \\
&= F[zI - (I - \Pi)A]^{-1}[(I - \Pi)(zI - A)](zI - A)^{-1}B \equiv \mathbf{0},
\end{aligned}$$

in light of the fact $(I - \Pi)B = \mathbf{0}$. □

Theorem 6.12 shows that the loop transfer properties in the case of state feedback can be perfectly recovered by the output feedback controller provided that the two assumptions in the theorem hold. Roughly speaking, $\mathbf{K}(z)$ has m hidden modes at the origin which are canceled by m zeros at the origin due to z in the front. The remaining $(n - m)$ poles of $\mathbf{K}(z)$ are canceled by the $(n - m)$ finite zeros of $\mathbf{P}(z)$ which are all stable under the given assumptions. Because the loop transfer matrix is the same as in (6.95), the closed-loop poles are eigenvalues of $(A + BF)$ which lie in the desired locations.

The LTR procedure requires that the open-loop plant $\mathbf{P}(z)$ be square with full normal rank. The root locus argument in the previous subsection can also be used to fulfill this requirement in the case of $p \neq m$ that is called square down. Cautions need to be taken without increasing the number of unstable zeros of the plant in the process of square down.

Example 6.13. Consider the inverted pendulum discussed in Chap. 1. Under the sampling frequency of 4 Hz, the discretized plant is obtained as

$$\mathbf{P}_d(z) = \begin{bmatrix} P_1(z) \\ P_2(z) \end{bmatrix} := \frac{\begin{bmatrix} 0.0383z^3 - 0.1430z^2 - 0.0284z + 0.0160 \\ 0.1408z^3 - 0.2283z^2 + 0.0344z + 0.0532 \end{bmatrix}}{z^4 - 5.5556z^3 + 6.0481z^2 - 1.5526z + 0.0601}.$$

The transfer function $P_2(z)$ is discretization of $G_{u\theta}(s)$ and has a zero at 1 due to the zero of $G_{u\theta}(s)$ at the origin. For this reason, $P_1(z)$ and $P_2(z)$ are low pass and

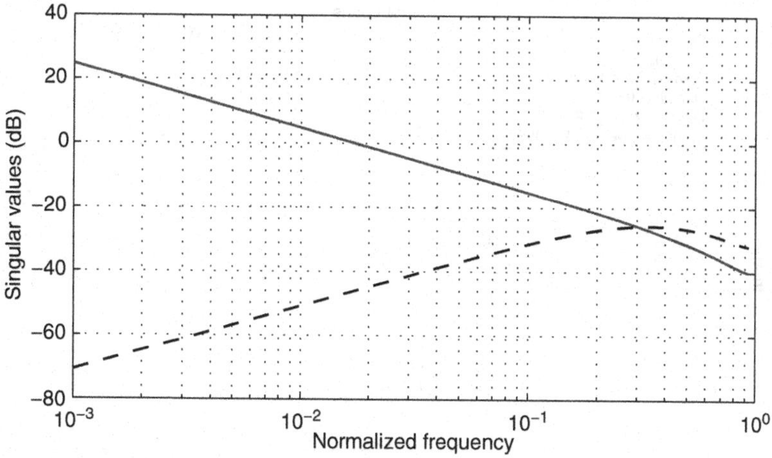

Fig. 6.10 Magnitude responses versus normalized frequency

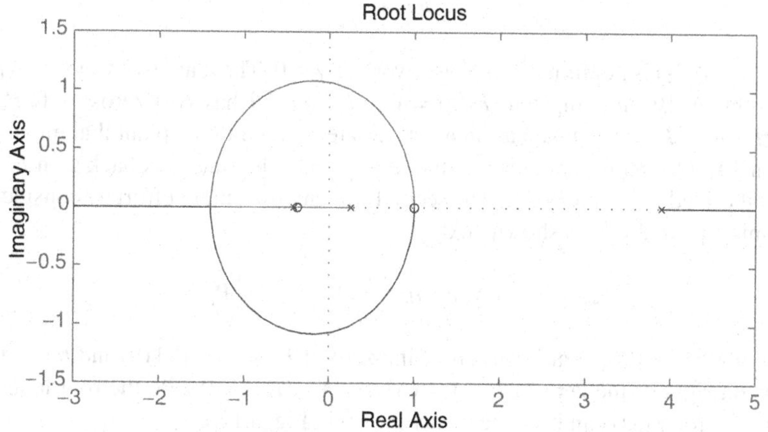

Fig. 6.11 Root locus showing zeros of the squared-down plant

high pass, respectively. To better illustrate its frequency response, magnitude responses of the two transfer functions in $\mathbf{P}_d(z)$ are plotted separately below with the solid line for the first and the dashed line for the second (Fig. 6.10).

Hence, the sampling frequency of 4 Hz is warranted by well below -20 dB for both of the magnitude responses at half of the sampling frequency. To square down the plant, let $\mathbf{P}_d(z) = C(zI - A)^{-1}B$ with C consisting of C_1 and C_2 as the first and second row of C, respectively. By taking $C_\gamma = C_2 - \gamma C_1$, the root locus for $\mathbf{P}_\gamma(z) = C_\gamma(zI - A)^{-1}B = 0$ with respect to $\gamma > 0$ can be sketched. From Fig. 6.11, it can be seen that the squared-down plant model $\mathbf{P}_\gamma(z)$ is not minimum phase for all $\gamma > 0$.

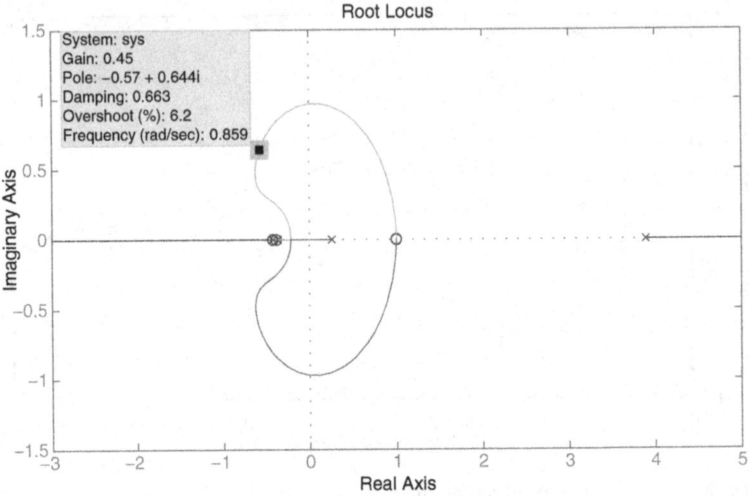

Fig. 6.12 Root locus showing zeros of the squared-down plant

In fact, $\mathbf{P}_\gamma(z)$ is not minimum phase even for $\gamma < 0$. There are two ways to bypass the problem. By noticing that $P_2(z) = C_2(zI - A)^{-1}B$ has two zeros at 1, $P_{2\varepsilon} = C_{2\varepsilon}(zI - A)^{-1}B$ can be taken as an approximate squared-down plant that moves the zero at 1 to $(1-\varepsilon)$ for some very small $\varepsilon \in (0,\ 1)$. The state-feedback gain F_p can be synthesized based on $P_{2\varepsilon}(z)$. The second way involves more effort by considering dynamic square down as shown next:

$$P_{\rm sd}(z) = \left[-\gamma(1 - az^{-1})\ (1 - bz^{-1})\right]\mathbf{P}_d(z).$$

By taking $a = -0.42$ that is on left of the zero of $P_2(z)$ at -0.419, and $b = -0.37$ that is on right of the zero of $P_1(z) = C_1(zI - A)^{-1}B$ at -0.378, the root locus for $P_{\rm sd}(z) = 0$ for $\gamma > 0$ can be sketched as shown in Fig. 6.12.

The dynamically squared-down plant $P_{\rm sd}(z) = C_{\rm sd}(zI - A_{\rm sd})^{-1}B_{\rm sd}$ is indeed strictly minimum phase and satisfies $C_{\rm sd}B_{\rm sd} \neq 0$ at $\gamma \approx 0.45$. □

Under the equality (6.95), the transfer matrix

$$\mathbf{S}_{\rm in}(z) = [I - \mathbf{K}(z)\mathbf{P}(z)]^{-1} = I + F(zI - A - BF)^{-1}B$$

is the sensitivity at the plant input for the feedback system in Fig. 6.1 that is identical to $\mathbf{S}_F(z)$ in (6.71) under state feedback. It is important to note the difference of $\mathbf{S}_{\rm in}(z)$ from the sensitivity at the plant output. That is, the desired performance and stability margins are achieved at the plant input rather than at the plant output. If, instead, the stability margin at the plant output are the design goal due to the tracking performance or disturbance rejection at the plant output, then state estimation gain needs to be synthesized first.

Suppose that L is synthesized via either frequency loop shaping or eigenvalue assignment with the dual LQR. That is,

$$L = -AYC^*(I + CYC^*)^{-1}$$

where $Y \geq 0$ is the stabilizing solution to the filtering ARE either in (6.75) for the case of loop shaping or in (6.90) with $Y = Y_p$ for the case of eigenvalue assignment. Let $\mathbf{y}(t)$ be the plant output at time t. The corresponding Kalman filter is given by

$$\begin{aligned} \hat{\mathbf{x}}_{t+1|t} &= (A + LC)\hat{\mathbf{x}}_{t|t-1} - L\mathbf{y}(t) + B\mathbf{u}(t) \\ \hat{\mathbf{x}}_{t|t} &= (I + L_1 C)\hat{\mathbf{x}}_{t|t-1} - L_1\mathbf{y}(t), \end{aligned} \tag{6.96}$$

where $L_1 = -YC^*(I + CYC^*)^{-1}$ and thus $L = AL_1$.

Under the same two assumptions as in Theorem 6.12, the dual LTR method computes the state-feedback gain $F = -(CB)^{-1}CA$ with $X = C^*C$ the stabilizing solution to the control ARE

$$X = A^*XA - A^*XB(B^*XB)^{-1}B^*XA + C^*C.$$

The control input signal is set as $\mathbf{u}(t) = F\hat{\mathbf{x}}_{t|t}$ that results in a feedback controller based on Kalman filter. Indeed, by time update,

$$\hat{\mathbf{x}}_{t+1|t} = A\hat{\mathbf{x}}_{t|t} + B\mathbf{u}(t) = (A + BF)\hat{\mathbf{x}}_{t|t}.$$

In conjunction with the second equation in Kalman filter (6.96), it yields

$$\begin{aligned} \hat{\mathbf{x}}_{t+1|t} &= (A + BF)(I + L_1C)\hat{\mathbf{x}}_{t|t-1} - (A + BF)L_1\mathbf{y}(t), \\ \mathbf{u}(t) &= F\hat{\mathbf{x}}_{t|t} = F(I + L_1C)\hat{\mathbf{x}}_{t|t-1} - FL_1\mathbf{y}(t), \end{aligned} \tag{6.97}$$

that is the state-space description for the output feedback controller $\mathbf{K}(z)$.

Remark 6.2. The output feedback controller described in (6.97) admits the transfer matrix $\mathbf{K}(z) = -[\hat{D} + \hat{C}(zI - \hat{A})^{-1}\hat{B}]$ and has the same form as the $\mathscr{H}_2$ controller specified in (6.35) for the case $D_{22} = \mathbf{0}$. Indeed, by noting $AL_1 = L$ and setting $L_0 = FL_1$, there hold

$$\begin{aligned} &\hat{A} = (A + BF)(I + L_1C) = A + BF + LC + BL_0C, \\ &\hat{B} = L + BL_0, \quad \hat{C} = F + L_0C, \quad \hat{D} = L_0, \end{aligned}$$

that has the identical from to the realization of the $\mathscr{H}_2$ controller in (6.35).

Now recall the expression of $F = -(CB)^{-1}CA$ leading to $BF = -\Pi A$ with $\Pi = B(CB)^{-1}C$ a projector. It follows that $A + BF = (I - \Pi)A$ and

$$(I + L_1C)(A + BF) = (A + BF) + L_1C(I - \Pi)A = A + BF \tag{6.98}$$

by $C(I-\Pi)=\mathbf{0}$. Multiplying the first equation of (6.97) by $(I+L_1C)$ from left and setting $\hat{\mathbf{x}}(k)=(I+L_1C)\hat{\mathbf{x}}_{k|k-1}$ for $k=t+1$ and $k=t$ lead to a new state-space description for $\mathbf{K}(z)$:

$$\begin{aligned}\hat{\mathbf{x}}(t+1) &= (A+BF)\hat{\mathbf{x}}(t)-(A+BF)L_1\mathbf{y}(t)\\ \mathbf{u}(t) &= F\hat{\mathbf{x}}(t)-FL_1\mathbf{y}(t)\end{aligned} \tag{6.99}$$

in light of (6.98). The transfer matrix of $\mathbf{K}(z)$ is obtained as

$$\begin{aligned}\mathbf{K}(z) &= -[FL_1+F(zI-A-BF)^{-1}(A+BF)L_1]\\ &= -zF(zI-A-BF)^{-1}L_1\end{aligned} \tag{6.100}$$

that is dual to the feedback controller in (6.94) and provides the estimator interpretation for the output feedback controller. The next result follows from Theorem 6.12, and thus, the proof is skipped.

Corollary 6.4. *Consider the square plant* $\mathbf{P}(z)=C(zI-A)^{-1}B$. *If* (i) $\det(CB)\neq 0$ *and* (ii) *the realization of* $\mathbf{P}(z)$ *is strictly minimum phase, then the feedback controller* $\mathbf{K}(z)$ *in (6.100), with* $F=-(CB)^{-1}CA$ *and* $L_\rho=AL_1$ *computed from (6.90), achieves perfect* LTR, *that is,*

$$\mathbf{P}(z)\mathbf{K}(z)=C(zI-A)^{-1}L. \tag{6.101}$$

Under the equality (6.101), the transfer matrix

$$\mathbf{S}_{\text{out}}(z)=[I-\mathbf{K}(z)\mathbf{P}(z)]^{-1}=I+C(zI-A-LC)^{-1}L$$

is the sensitivity at the plant output for the feedback system in Fig. 6.5, identical to $\mathbf{S}_L(z)$ in (6.72) under output injection. Hence, performance and stability margins are perfectly recovered at the plant output, which is dual to the preceding LTR method.

It is noted that perfect LTR is achieved under the two assumptions in Theorem 6.12 and Corollary 6.4 which are questionable. Clearly, the minimum phase assumption is violated under high sampling rate that is a known fact. On the other hand, the assumption $\det(CB)\neq 0$ can also be violated if the underlying continuous-time system has pure time delay exceeding the sampling period. It is intuitive that perfect LTR is not possible when both assumptions fail to hold. Two questions will be answered next. The first one is how the LTR design method should be modified so that it is applicable to feedback design for more general systems. The second one is how the recovery error can be characterized in terms of the desirable loop transfer matrix. The following result in instrumental.

Lemma 6.5. *Consider the square plant model* $\mathbf{P}(z)=C(zI-A)^{-1}B$ *with stabilizable and detectable realization. If* $\mathbf{P}(z)$ *has full normal rank void zeros on the unit circle, then there holds following factorization:*

$$\mathbf{P}(z)=\mathbf{C}_{\text{a}}(z)\mathbf{P}_{\text{m}}(z),\quad \mathbf{P}_{\text{m}}(z)=C_{\text{m}}(zI-A)^{-1}B, \tag{6.102}$$

where $\mathbf{C}_{\mathrm{a}}(z)$ *is a square inner (all pass and stable), the realization of* $\mathbf{P}_{\mathrm{m}}(z)$ *is strictly minimum phase, and* $\det(C_{\mathrm{m}}B) \neq 0$.

Proof. By the hypothesis, there exists an integer $\kappa \geq 0$ such that

$$z^{\kappa+1}\mathbf{P}(z) = D_\kappa + C_\kappa(zI - A)^{-1}B, \quad D_\kappa = CA^\kappa B \neq \mathbf{0}.$$

In light of Remark 5.1 and (ii) of Theorem 5.28 in the previous chapter,

$$z^{\kappa+1}\mathbf{P}(z) = \mathbf{P}_{\mathrm{i}}(z)\mathbf{P}_{\mathrm{o}}(z)$$

with $\mathbf{P}_{\mathrm{i}}(z)$ an inner and $\mathbf{P}_{\mathrm{o}}(z)$ an outer if A is a stability matrix. If A is not a stability matrix, then $\mathbf{P}_{\mathrm{o}}(z)$ is strictly minimum phase. Specifically, let $R_\kappa = D_\kappa^* D_\kappa$ and $Z \geq \mathbf{0}$ be the stabilizing solution to the ARE

$$Z = A^*ZA + C_\kappa^* C_\kappa - (A^*ZB + C_\kappa^* D_\kappa)(R_\kappa + B^*ZB)^{-1}(B^*ZA + D_\kappa^* C_\kappa).$$

The full normal rank condition for $\mathbf{P}(z)$ implies that $\Omega^2 = R_\kappa + B^*ZB$ is nonsingular. The state-feedback gain $F_\kappa = -(R_\kappa + B^*ZB)^{-1}(D_\kappa^* C_\kappa + B^*ZA)$ is stabilizing. Moreover, $\mathbf{P}_{\mathrm{i}}(z)$ and $\mathbf{P}_{\mathrm{o}}(z)$ have the following expressions:

$$\mathbf{P}_{\mathrm{i}}(z) = \left[\begin{array}{c|c} A + BF_\kappa & B \\ \hline C_\kappa + D_\kappa F_\kappa & D_\kappa \end{array}\right]\Omega^{-1}, \quad \mathbf{P}_{\mathrm{o}}(z) = \Omega\left[\begin{array}{c|c} A & B \\ \hline -F_\kappa & I \end{array}\right].$$

It is noted that $F_\kappa = \widehat{F}_\kappa A$ where

$$\widehat{F}_\kappa = -(R_\kappa + B^*ZB)^{-1}(D_\kappa^* CA^\kappa + B^*Z).$$

The expressions of R_κ and D_κ lead to

$$\begin{aligned} R_\kappa + B^*ZB &= B^*(A^{*\kappa}C^*CA^\kappa + Z)B, \\ D_\kappa^* CA^\kappa + B^*Z &= B^*(A^{*\kappa}C^*CA^\kappa + Z). \end{aligned}$$

It follows that $I = -\widehat{F}_\kappa B$, and thus:

$$\mathbf{P}_{\mathrm{o}}(z) = -\Omega\left[\begin{array}{c|c} A & B \\ \hline \widehat{F}_\kappa A & \widehat{F}_\kappa B \end{array}\right] = -z\Omega\widehat{F}_\kappa(zI - A)^{-1}B.$$

Setting $\mathbf{C}_{\mathrm{a}}(z) = z^{-\kappa}\mathbf{P}_{\mathrm{i}}(z)$ and $C_{\mathrm{m}} = -\Omega\widehat{F}_\kappa$ leads to the factorization in (6.102). Since $\det(C_{\mathrm{m}}B) = \det(\Omega) \neq 0$, and zeros of $\mathbf{P}_{\mathrm{o}}(z)$ are eigenvalues of $(A + BF_\kappa)$ which are all stable, $\mathbf{P}_{\mathrm{m}}(z) = C_{\mathrm{m}}(zI - A)B$ is indeed strictly minimum phase that completes the proof. □

Example 6.14. Consider the plant model

$$P(z) = \frac{z-2}{z(z+1.5)(z-0.5)}.$$

A simple realization is obtained as

$$A = \begin{bmatrix} -1 & 1 & 0 \\ 0.75 & 0 & 1 \\ 0 & 0 & 0 \end{bmatrix}, \quad B = \begin{bmatrix} 0 \\ 1 \\ -2 \end{bmatrix}, \quad C = \begin{bmatrix} 1 & 0 & 0 \end{bmatrix}.$$

Taking $\kappa = 1$ and carrying out the calculations in the proof of Lemma 6.5 yield $Z = \mathbf{z}\mathbf{z}^*$ with $\mathbf{z}^* = \begin{bmatrix} -0.9073 & 0.4124 & -0.6598 \end{bmatrix}$. As a result,

$$\widehat{F}_\kappa = \begin{bmatrix} 0.6429 & -0.4286 & 0.2857 \end{bmatrix},$$

$$F_\kappa = \begin{bmatrix} -0.9643 & 0.6429 & -0.4286 \end{bmatrix}.$$

Hence, $\mathbf{P}_{\rm i}(z)$ and $\mathbf{P}_{\rm o}(z)$ can be easily computed accordingly. Both do not admit minimum realizations. Specifically, $\mathbf{P}_{\rm i}(z)$ has two unobservable modes at the origin, while $\mathbf{P}_{\rm o}(z)$ has two unobservable modes with one at the origin and the other at 0.5. After eliminating these unobservable modes, it leads to

$$\mathbf{C}_{\rm a}(z) = \frac{0.5z-1}{z(z-0.5)}, \quad \mathbf{P}_{\rm m}(z) = \frac{2z-1}{z+1.5}.$$

Recall $\mathbf{C}_{\rm a}(z) = z^{-\kappa}\mathbf{P}_{\rm i}(z)$ and $\mathbf{P}_{\rm m}(z) = z^{-1}\mathbf{P}_{\rm o}(z)$.

The dual LTR design method synthesizes the estimation gain $L = AL_1$ in accordance with either (6.75) or (6.90) regardless of the assumptions on minimum phase and $\det(CB) \neq 0$. However, the state-feedback gain is replaced by $F = -(C_{\rm m}B)^{-1}C_{\rm m}A$ based on the plant model $\mathbf{P}_{\rm m}(z)$ rather than $\mathbf{P}(z)$, because $\mathbf{P}_{\rm m}(z)$ satisfies the two assumptions in Theorem 6.12 and Corollary 6.4. As discussed earlier, perfect LTR is not possible. The next result characterizes the recovery error.

Theorem 6.15. *Suppose that* $\mathbf{P}(z)$ *is factorized as in* Lemma 6.5 *and the dual* LTR *is applied to* $\mathbf{P}_m(z)$. *Under the stabilizability and detectability condition,*

$$\mathbf{P}(z)\mathbf{K}(z) = [\mathbf{H}(z) - \mathbf{E}(z)][I - \mathbf{E}(z)]^{-1} \tag{6.103}$$

where $\mathbf{H}(z) = C(zI - A)^{-1}L$ *is the target loop transfer matrix and* $\mathbf{E}(z) = [C - \mathbf{C}_a(z)C_{\rm m}](zI - A)^{-1}L$ *is the error function.*

Proof. In light of (6.97), the dual LTR feedback controller is given by

$$\mathbf{K}(z) = -zF[zI - (I + L_1C)(A + BF)]^{-1}L_1$$

by taking $\hat{\mathbf{x}}(t) = (I + L_1C)\hat{\mathbf{x}}_{t|t-1}$ as the state vector where $L = L_1A$, $L_1 = -YC^*(I + CYC^*)^{-1}$ with $Y \geq \mathbf{0}$, the stabilizing solution to some filtering ARE, and $F = -(C_{\rm m}B)^{-1}C_{\rm m}A$. Denote $\Phi = (zI - A)^{-1}$ for convenience. The feedback controller can be written as

$$\begin{aligned}\mathbf{K}(z) &= -zF(zI - A - BF - L_1CA - L_1CBF)^{-1}L_1 \\ &= -zF(I - \Phi BF - \Phi L_1CA - \Phi L_1CBF)^{-1}\Phi L_1.\end{aligned}$$

Since $\Phi L_1CA\Phi BF - z\Phi L_1C\Phi BF = -\Phi L_1CBF$, there holds

$$\begin{aligned}\Psi &:= (I - \Phi L_1CA)(I - \Phi BF) - z\Phi L_1C\Phi BF \\ &= I - \Phi BF - \Phi L_1CA - \Phi L_1CBF.\end{aligned}$$

Consequently, $\mathbf{K}(z) = -zF\Psi^{-1}\Phi L_1$ is given by

$$\mathbf{K}(z) = -zF[(I - \Phi L_1CA)(I - \Phi BF) - z\Phi L_1C\Phi BF]^{-1}\Phi L_1.$$

Using identity $(I - \Phi L_1CA)^{-1}\Phi L_1 = \Phi L_1(I - CA\Phi L_1)^{-1}$ yields

$$(I - \Phi L_1CA)^{-1}\Phi L_1 = \Phi L_1(I - C\Phi L)^{-1} = \Phi L_1(I - \mathbf{H})^{-1}$$

by $AL_1 = L$ where $\mathbf{H}(z)$ is the desired loop transfer matrix. It follows that

$$\mathbf{K}(z) = -z[I - F\Phi B - F\Phi L_1(I - \mathbf{H})^{-1}(zC\Phi B)F]^{-1}F\Phi L_1(I - \mathbf{H})^{-1}.$$

A similar notation is $\mathbf{H}_m(z) = C_{\rm m}(zI - A)^{-1}L = C_{\rm m}A\Phi L_1$. Upon substitution of $F = -(C_{\rm m}B)^{-1}C_{\rm m}A$ into the above $\mathbf{K}(z)$ with suitable rearrangement yields

$$\mathbf{K}(z) = z[C_{\rm m}B + C_{\rm m}A\Phi B + \mathbf{H}_{\rm m}(I - \mathbf{H})^{-1}(zC\Phi B)]^{-1}\mathbf{H}_{\rm m}(I - \mathbf{H})^{-1}.$$

It is recognized that $C_{\rm m}B + C_{\rm m}A\Phi B = zC_{\rm m}\Phi B$ and recall that $C\Phi B = \mathbf{C}_{\rm a}C_{\rm m}\Phi B$ where $\mathbf{C}_{\rm a}(z)$ is inner (stable and allpass). The following expression of $\mathbf{K}(z)$ is obtained as

$$\mathbf{K}(z) = [C_{\rm m}\Phi B + \mathbf{H}_{\rm m}(I - \mathbf{H})^{-1}\mathbf{C}_a C_{\rm m}\Phi B]^{-1}\mathbf{H}_{\rm m}(I - \mathbf{H})^{-1}.$$

The loop transfer matrix under output feedback is

$$\mathbf{P}(z)\mathbf{K}(z) = \mathbf{C}_{\rm a}(z)C_{\rm m}\Phi B\mathbf{K}(z).$$

Upon substituting the previous expression of $\mathbf{K}(z)$ gives

$$\begin{aligned}\mathbf{P}(z)\mathbf{K}(z) &= \mathbf{C}_{\rm a}[I - \mathbf{H}_{\rm m}(z)(I - \mathbf{H})^{-1}\mathbf{C}_{\rm a}]^{-1}\mathbf{H}_{\rm m}(I - \mathbf{H})^{-1} \\ &= \mathbf{C}_{\rm a}\mathbf{H}_{\rm m}(I - \mathbf{H} + \mathbf{C}_{\rm a}\mathbf{H}_{\rm m})^{-1} \\ &= [\mathbf{H} - (\mathbf{H} - \mathbf{C}_{\rm a}\mathbf{H}_{\rm m})][I - (\mathbf{H} - \mathbf{C}_{\rm a}\mathbf{H}_{\rm m})]^{-1}.\end{aligned}$$

By recognizing $\mathbf{H} - \mathbf{C}_a\mathbf{H}_m = (C - \mathbf{C}_a C_m)(zI - A)^{-1}L = \mathbf{E}(z)$ to be the error function, the proof is concluded. □

The equality in (6.103) can be rewritten as

$$\mathbf{E}(z) = [I - \mathbf{P}(z)\mathbf{K}(z)]^{-1}[\mathbf{H}(z) - \mathbf{P}(z)\mathbf{K}(z)]$$

providing an alternative expression for the error function. For the dual LTR design method, the overall transfer matrix or the complementary sensitivity for the feedback system in Fig. 6.1 is obtained as

$$\mathbf{T}_{ry}(z) = \mathbf{PK}(I - \mathbf{PK})^{-1} = \mathbf{C}_a C_m (zI - A - LC)^{-1}L \tag{6.104}$$

that is simpler than the corresponding one obtained with the original LTR design method. In addition, the sensitivity at the plant output is given by

$$\mathbf{S}_{\text{out}}(z) = [I - \mathbf{E}(z)][I + L(zI - A - LC)^{-1}B]. \tag{6.105}$$

Its proof and its dual are left as an exercise (Problem 6.18 in Exercises). While

$$(C - \mathbf{C}_a C_m)(zI - A)^{-1}B \equiv \mathbf{0},$$

the error function $\mathbf{E}(z) = (C - \mathbf{C}_a C_m)(zI - A)^{-1}L$ is not identically zero. In fact, magnitude response of $\mathbf{E}(z)$ determines the recovery performance.

It is worth to mentioning that good tracking performance also depends on the frequency shape of the loop transfer matrix $\mathbf{P}(z)\mathbf{K}(z)$. Accumulators and lead/lag compensators can be employed to shape the plant model prior to carrying out the LTR design that will be clarified with design examples. Since Theorem 6.15 is concerned with the dual LTR design procedure, there is a need to provide the LTR design procedure that is summarized in the following without proof.

Given a square plant $\mathbf{P}(z) = C(zI - A)^{-1}B$ and the state-feedback gain $F = F_1 A$ obtained via either the loop shaping method in (6.73) or the LQR method in (6.82) where $F_1 = -(I + B^*XB)^{-1}B^*X$, the LTR controller is synthesized as follows:

- Compute factorization $\mathbf{P}(z) = \mathbf{P}_m(z)\mathbf{B}_a(z)$ where $\mathbf{B}_a(z)$ is square inner and the realization of $\mathbf{P}_m(z) = C(zI - A)^{-1}B_m$ is strictly minimum phase and satisfies $\det(CB_m) \neq 0$:
  - Find integer $\kappa \geq 0$ such that $CA^iB = \mathbf{0}$ for $1 \leq i < \kappa$ and $CA^\kappa B \neq \mathbf{0}$; Set $D_\kappa = CA^\kappa B$ and $B_\kappa = A^{\kappa+1}B$. Thus,

    $$z^{\kappa+1}\mathbf{P}(z) = D_\kappa + C(zI - A)^{-1}B_\kappa.$$

  - Compute the stabilizing solution to the ARE

    $$Z = AZA^* + B_\kappa B_\kappa^* - L_\kappa(D_\kappa D_\kappa^* + CZC^*)L_\kappa^*$$

    where $L_\kappa = -(B_\kappa D_\kappa^* + AZC^*)(D_\kappa D_\kappa^* + CZC^*)^{-1}$.

– Set $\Omega = (D_K D_K^* + CZC^*)^{1/2}$ that is nonsingular and

$$\mathbf{B}_\mathrm{a}(z) = z^{-\kappa}\Omega^{-1}\left[\begin{array}{c|c} A + L_K C & B_K + L_K D_K \\ \hline C & D_K \end{array}\right].$$

– Set $B_\mathrm{m} = -\widehat{L}_K \Omega$ where

$$\widehat{L}_K = -(A^\kappa B D_K^* + ZC^*)(D_K D_K^* + CZC^*)^{-1}.$$

- Set $\mathbf{K}(z) = -zF_1[zI - (A + LC)(I + BF_1)]^{-1}L$ with $L = -AB_\mathrm{m}(CB_\mathrm{m})^{-1}$.

It is commented that if the two assumptions in Theorem 6.12 hold, then $\mathbf{B}_\mathrm{a}(z) \equiv I$ for all z, and thus, the factorization step can be skipped.

Remark 6.3. A similar $\mathscr{H}_2$ interpretation to that in Remark 6.2 holds for the feedback controller. Specifically, by taking $L_0 = F_1 L$ and noting $F = F_1 A$, realization of the LTR controller $\mathbf{K}(z)$ is given by

$$\begin{aligned} -\mathbf{K}(z) &= \left[\begin{array}{c|c} (A + LC)(I + BF_1) & L \\ \hline F_1(A + LC)(I + BF_1) & BF_1 \end{array}\right] \\ &= \left[\begin{array}{c|c} (I + BF_1)(A + LC) & (I + BF_1)L \\ \hline F_1(A + LC) & L_0 \end{array}\right] \\ &= \left[\begin{array}{c|c} A + BF + LC + BL_0C & L + BL_0 \\ \hline F + L_0C & L_0 \end{array}\right] \end{aligned}$$

that indeed has the form of $\mathscr{H}_2$ controller in (6.35) for the case $D_{22} = \mathbf{0}$. However, the closed-loop transfer matrix $\mathbf{T}_{ry}(z)$ for the feedback system in Fig. 6.1 is not as nice as that of the dual case in (6.104). For this reason, the following two-degree-of-freedom control system is suggested:

where $\mathbf{V}(z)$ is the left coprime denominator of $\mathbf{K}(z)$ given by

$$\mathbf{V}(z) = I - (F + L_0C)[zI - (A + LC)]^{-1}B,$$

copied from (6.44) and $\Omega_F = (I + B^*XB)^{1/2}$. It follows that

$$\begin{aligned} \mathbf{T}_{ry}(z) &= \mathbf{P}(I - \mathbf{KP})^{-1}\mathbf{V}^{-1}\Omega_F^{-1} \\ &= C(zI - A - BF)^{-1}B\Omega_F^{-1} \end{aligned}$$

by the double Bezout identity and the expression of the coprime factors in (6.43). Hence, good tracking performance achieved by state-feedback control is carried to the case of output feedback control.

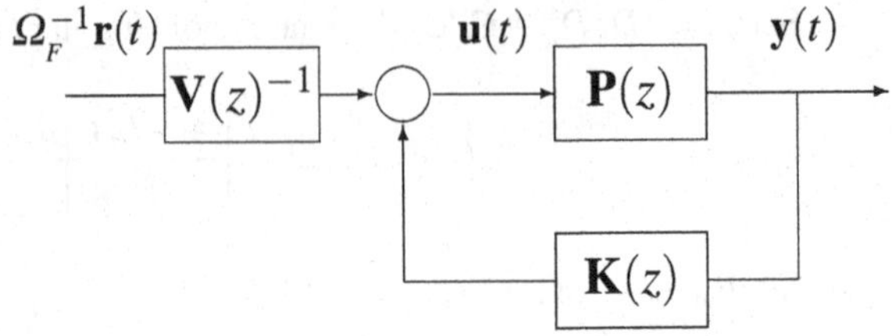

Fig. 6.13 Two-degree-of-freedom control system

Often the feedback controller $\mathbf{K}_W(z)$ is synthesized based on the weighted plant $\mathbf{P}_W(z)$. The weighted form of $\mathbf{P}_W(z) = \mathbf{P}(z)\mathbf{W}(z)$ is more preferred, because it does not alter the output to be controlled. Hence, the feedback system in Fig. 6.13 remains the same by setting

$$\mathbf{K}(z) = \mathbf{W}(z)\mathbf{K}_W(z), \quad \mathbf{V}(z)^{-1} = \mathbf{W}(z)\mathbf{V}_W(z)^{-1}.$$

If the controller reduction is carried out, then the simple expression of the overall transfer matrix $\mathbf{T}_{ry}(z)$ in Remark 6.3 does not hold anymore. It is left as an exercise (Problem 6.22) to derive the new expression of $\mathbf{T}_{ry}(z)$.

6.3 Design Examples

Two design examples are presented in this section with one for a flight control system and the other for the inverted pendulum. Both loop shape and eigenvalue assignment methods will be used to synthesize the MIMO feedback controller. Detailed steps are worked out to illustrate the synthesis procedure. These steps include reduction of the feedback controller.

6.3.1 Design of Flight Control System

The first design example is the flight control system considered in Example 6.10. The state variables and control inputs are

x_1	−relative altitude (m),	u_1	−spoilder angle ($^o \times 10^{-1}$),
x_2	−forward speed (m/s),	u_2	−forward acceleration (m/s^2),
x_3	−pitch angle (^{o}),	u_3	−elevator angle (^{o}),
x_4	−pitch rate (^{o}/s),	x_5	−vertical speed (m/s).

The three outputs are simply the measurements of the first three states, and thus, $C = \begin{bmatrix} I_3 & \mathbf{0}_{2\times 2} \end{bmatrix}$. The discretized plant under the sampling frequency of 40 Hz has one real pole at 1 and four complex poles at

$$0.9804 \pm j0.0252, \quad 0.9995 \pm j0.0046.$$

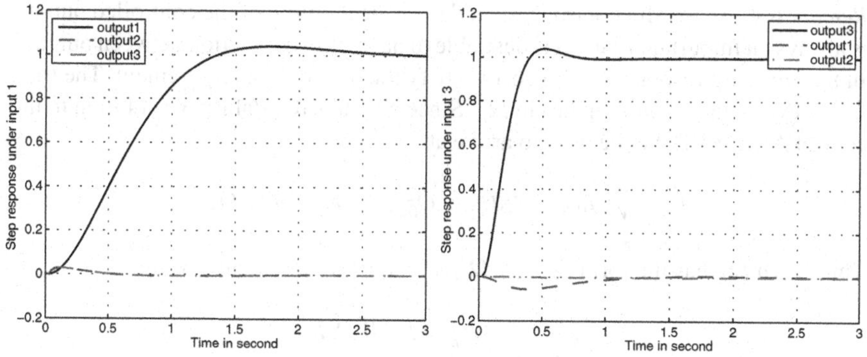

Fig. 6.14 Step responses under input 1 (left) and input 3 (right)

The continuous-time plant has no finite zeros, but the discretized model introduces two transmission zeros at -0.9960 and -0.9912.

The design specifications require that the percentage overshot is no more than 5%, the settling time for the first output be no more than 2.5 s, or 100 sampling periods, and be as small as possible for the other two outputs.

Using the same weighting as in Example 6.10 obtained after several trials, singular values of the frequency response for the shaped plant is plotted in Fig. 6.8. The resulting state-feedback gain F for the shaped plant is given by

$$\begin{bmatrix} -0.958 & -0.0004 & 0.0055 & -0.922 & -0.006 & -0.348 & 0.0028 & 0.43 \\ -0.015 & -0.9881 & -0.0043 & -0.027 & -0.739 & -0.010 & -0.0009 & 0.01 \\ -0.036 & 0.0015 & -0.9972 & -0.261 & -0.002 & -1.017 & -0.1396 & 0.16 \end{bmatrix}$$

The step responses under the state feedback are plotted next which satisfy the design requirements.

The three step responses under input 2 are not shown because they are much better than the two in Fig. 6.14 and have no overshot with settling time no more than 0.5 s. In light of the Remark 6.3, the overall transfer matrix is the same as

$$\mathbf{T}_{ry}(z) = C_W(zI - A_W - B_W F)^{-1} B_W \Omega_F^{-1}$$

with (A_W, B_W, C_W) the realization for $\mathbf{P}_W(z) = \mathbf{P}(z)\mathbf{W}(z)$ due to the use of weighting to shape the plant. Hence, the step responses under output feedback control are identical to those under state-feedback control.

For the LTR step, it is noted that the plant $\mathbf{P}(z) = C(zI - A_d)^{-1}B_d$ is strictly minimum phase and $\det(CB_d) \neq 0$. In addition, the weighting functions are also strictly minimum phase. Hence, the ideal loop transfer matrix can be perfectly recovered. The output feedback controller can be computed according to (6.94), but the expressions of its realization are omitted.

Remark 6.4. In MIMO control system design, decoupling of the controlled outputs is always an important issue. It is desirable to have little interaction between outputs. In fact, the step responses in Fig. 6.14 satisfy the decoupling requirement. The trick in achieving the decoupling lies in Ω_F. Since $X \geq \mathbf{0}$ is the stabilizing solution to the control ARE, SVD can be used to arrive at

$$I + B_W^* X B_W = USU^* = \Omega_{0_F}^2, \quad \Omega_{0_F} = U\sqrt{S}U^*.$$

Thus, each Ω_F has the form $\Omega_F = V\Omega_{0_F}$ for some unitary matrix V by

$$\Omega_F^* \Omega_F = \Omega_{0_F}^2 = I + B_W^* X B_W.$$

The above results in the overall transfer matrix:

$$\mathbf{T}_{ry}(z) = C_W(zI - A_W - B_W F)^{-1} B_W \Omega_{0_F}^{-1} V^*.$$

Recall the normalized coprime factors in Sect. 6.2.1 and in particular (6.78). Since $\mathbf{T}_{ry}(z) = \tilde{\mathbf{N}}_W(z)$ due to the use of the weighted plant, there holds ω,

$$\mathbf{T}_{ry}(\mathrm{e}^{\mathrm{j}\omega})\mathbf{T}_{ry}(\mathrm{e}^{\mathrm{j}\omega})^* = \mathbf{P}_W(\mathrm{e}^{\mathrm{j}\omega})\left[I + \mathbf{P}_W(\mathrm{e}^{\mathrm{j}\omega})^*\mathbf{P}_W(\mathrm{e}^{\mathrm{j}\omega})\right]^{-1}\mathbf{P}_W(\mathrm{e}^{\mathrm{j}\omega})^*$$

at each frequency. Since both the plant and weighting do not have transmission zero at 1, and $\mathbf{P}_W(z)$ has pole at 1 for each of its entry due to the use of weighting,

$$\lim_{\omega \to 0} \mathbf{T}_{ry}(\mathrm{e}^{\mathrm{j}\omega})\mathbf{T}_{ry}(\mathrm{e}^{\mathrm{j}\omega})^* = \lim_{z \to 1} \mathbf{T}_{ry}(z)\mathbf{T}_{ry}(z)^{\sim} = I.$$

That is, $\mathbf{T}_{ry}(1)$ is a unitary matrix. By taking V appropriately, $\mathbf{T}_{ry}(1) = I$ can be made true. Hence, the output responses of the closed-loop system are asymptotically decoupled. It is left as an exercise to show that the procedure for asymptotic decoupling works for other types of the weighted plant models as well. □

The IBT algorithm from Sect. 6.2 is employed to reduce the order of the feedback controller $\mathbf{K}_W(z)$, designed based on the shaped plant $\mathbf{P}_W(z)$ using LQR/LTR. The inverse balanced singular values are

$$3156, 472.1, 3.1516, 1.7677, 1.5887, 0.0465, 0.0076, 0.0017.$$

Truncation of three modes based on IBT yields very small loss of performance. However, the controller $\mathbf{K}_W(z)$ cannot be reduced to lower than 5th order, since it results in unstable feedback system. Singular values of the frequency response of the loop transfer matrix based on the reduced 5th order controller $\widehat{\mathbf{K}}_W(z)$ is plotted in Fig. 6.15, together with step responses. It can be seen that the reduced controller performs very well. In fact, both frequency and time responses are very

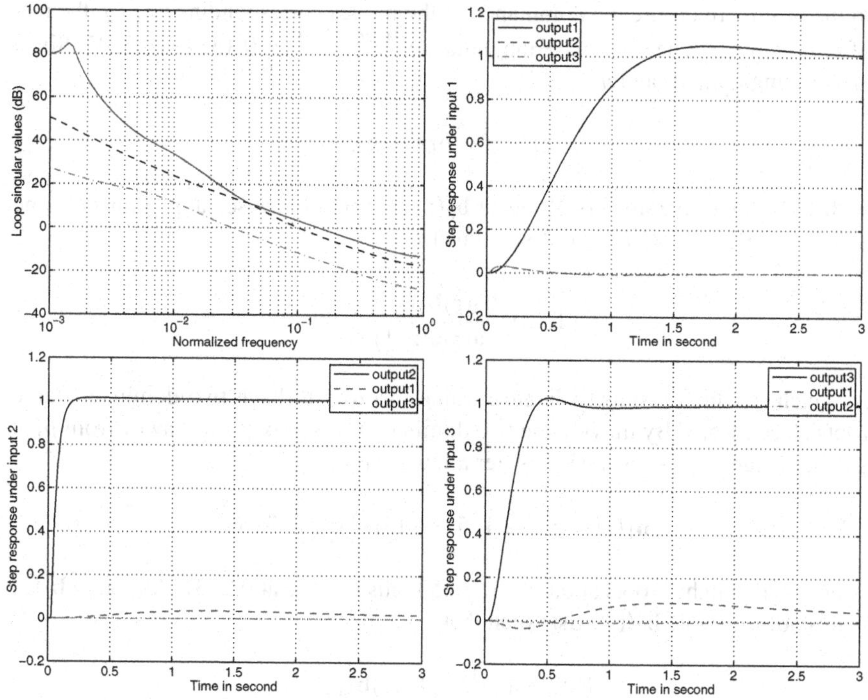

Fig. 6.15 Loop shape and step responses based on the reduced controller

close to those obtained with the full order controller. The actual feedback controller $\widehat{\mathbf{K}}(z) = \mathbf{W}(z)\widehat{\mathbf{K}}_W(z)$ is of 8th order after including the weighting function which is still greater than the order of the plant.

6.3.2 Design of Inverted Pendulum Control System

The discretized plant model under the sampling frequency of 4 Hz is given in Example 6.13. The objective is to stabilize the inverted pendulum with the dominant poles at $-1.5 \pm j$ that correspond to damping ratio of 0.832 and natural frequency of 1.803 for the continuous-time system. Under the sampling period of $T_s = 0.25$s, the desired dominant poles for the discretized closed-loop system are at $p_{1,2} = 0.6659 \pm j0.1700$.

Different from the previous example, this control system design employs eigenvalue assignment to compute the state-feedback gain F_ρ. Moreover, the plant has two inputs and one output. Thus, square down of the plant model is necessary and is discussed in Example 6.13. Let us consider first $\mathbf{P}_{2\varepsilon}(z)$ as the approximate squared-down plant. The goal is to assign eigenvalues of $(A + BF_\rho)$ at $p_{1,2}$ with the rest close to the origin. To ensure the dominance of the modes corresponding to $p_{1,2}$, the rest of the eigenvalues needs to have magnitude $e^{-1.5T_s\times(3\sim5)} = 0.1534 \sim 0.3247$. Since $m = 1$ (single input) and $n = 4$,

$$\mathbf{b}_0(z) = z^3 + b_1 z^2 + b_2 z + b_3$$

with $\{b_i\}_{i=1}^3$ free parameters. Roots of $\mathbf{b}_0(z)$ are initially chosen at $p_{1,2}$ plus the one at 0.2. Let $\mathbf{a}(z) = \det(zI - A)$. Root locus for

$$1 + \rho^2 \frac{\mathbf{b}_0(z)\mathbf{b}_0(z^{-1})}{\mathbf{a}(z)\mathbf{a}(z^{-1})} = 0$$

is sketched which results in large ρ value in order to have two dominant stable roots close to $p_{1,2}$. By moving the two dominant roots of $\mathbf{b}_0(z) = 0$ away from $p_{1,2}$, smaller value of ρ is resulted in. After a few trials,

$$\mathbf{b}_1(z) = z^3 - 2.125z^2 + 1.6441z - 0.3109$$

is obtained, and has roots at $0.9259 \pm j0.53$ plus the one at 0.2731. Although $\mathbf{b}_1(z)$ is unstable, a stable $\mathbf{b}_0(z)$ exists such that

$$\mathbf{b}_0(z)\mathbf{b}_0(z^{-1}) = \mathbf{b}_1(z)\mathbf{b}_1(z^{-1}).$$

The corresponding root locus is shown next.

It is seen from Fig. 6.16 that at $\rho^2 = 9.65$, the dominant stable roots are at $0.656 \pm j0.17$. Although these roots are not exactly the same as $p_{1,2}$, they are very close to each other, respectively. The other two stable roots of the loci are at 0.238 and 0.2731 that do not affect the dominant poles.

Next, C_0 is synthesized such that

$$\mathbf{G}_0(z) = C_0(zI - A)^{-1}B = \frac{\mathbf{b}_0(z)}{\mathbf{a}(z)}.$$

The procedure demonstrated in the proof of Lemma 6.3 and Example 6.11 can be followed. By using the canonical controller realization,

$$A_{21} = \begin{bmatrix} 1 \\ 0 \\ 0 \end{bmatrix}, \quad A_{22} = \begin{bmatrix} 0 & 0 & 0 \\ 1 & 0 & 0 \\ 0 & 1 & 0 \end{bmatrix},$$

and $R_B = 1$. Thus, (A_{22}, A_{21}) is in canonical controller form as well. Hence, for $C_{01} = 1$ and $C_{02} = \begin{bmatrix} b_1 & b_2 & b_3 \end{bmatrix}$,

$$\det(zI - A_{22} + A_{21}C_{02}) = z^3 + b_1 z^2 + b_2 z + b_3 = \mathbf{b}_0(z).$$

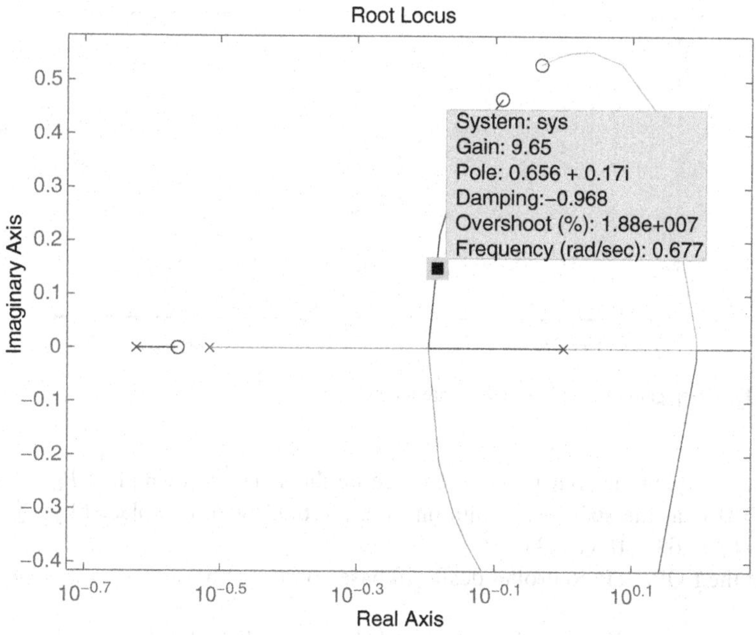

Fig. 6.16 Root locus for inverted pendulum design

After C_0 is available, the stabilizing solution to the ARE in (6.82) can be computed yielding the state-feedback gain

$$F_\rho = \begin{bmatrix} -3.9561 \ 5.1981 \ -1.4140 \ 0.0585 \end{bmatrix}.$$

The eigenvalues of $(A+BF_\rho)$ are at $0.6580 \pm j0.2016$ and $0.2715, 0.0121$. The discrepancy from the stable roots of the root locus is probably caused by the numerical problem in computing root locus.

It needs to be pointed out that $r = 1/\sqrt{1+B'X_\rho B} = 0.1607$ that is rather small. The gain and phase of $-F_\rho(zI-A)^{-1}B$ are plotted in Fig. 6.17. The phase margin at ω_c is about 11.65^o that is close to $2\sin^{-1}(r/2) = 9.1975^o$.

The root locus method is very different from the loop-shaping method. In fact, for the flight control system example, eigenvalues of $(I+B'XB)^{-1/2}$ are between 0.74084 and 0.9588 that are considerably greater than 0.1607, indicating large gain and phase margins for the loop-shaping method. These attributes do not exist for the root locus method. On the other hand, the mathematical model for the inverted pendulum is very accurate, implying that the small gain and phase margins do not destroy the performance under modeling error. It will be seen next that the step response of the inverted pendulum control system is actually very decent.

It is noticed that $\mathbf{P}_d(z) = C(zI-A)^{-1}B$ and $P_{2\varepsilon}(z) = C_{2\varepsilon}(zI-A)^{-1}B$ share the same A and B. Hence, the value of ε is not used in the LQR part but is needed for the LTR part of the design. By taking $\varepsilon = 0.001$, $P_{2\varepsilon}(z)$ is strictly minimum

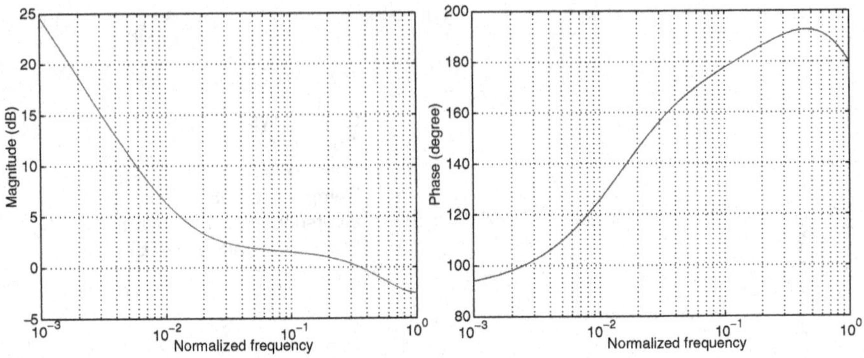

Fig. 6.17 Loop gain/phase plots under state feedback

phase. Applying the LTR procedure based on the squared-down plant $P_{2\varepsilon}(z)$ leads to $Y = BB^*$ as the stabilizing solution to ARE (6.92) with C replaced by $C_{2\varepsilon}$, and $L = -\begin{bmatrix} 39.463 \ 7.1034 \ 0 \ 0 \end{bmatrix}^*$.

Let the LQR/LTR controller designed based on $P_{2\varepsilon}(z)$ be

$$K_\varepsilon(z) = zF_{1\rho}(zI - A - LC_{2\varepsilon})^{-1}L, \quad F_\rho = F_{1\rho}A.$$

Then the two-degree-of-freedom control system in Fig. 6.13 can be used to implement $\mathbf{K}(z) = K_\varepsilon(z)W$ with $W = \begin{bmatrix} 0 \ 1 \end{bmatrix}$, yielding

$$\mathbf{T}_{ry}(z) = \mathbf{P}(I + K_\varepsilon W\mathbf{P})^{-1}V_\varepsilon^{-1}\Omega^{-1} \approx C(zI - A - BF_\rho)^{-1}B\Omega^{-1}$$

for the closed-loop system, in light of the perfect LTR. By taking

$$\Omega = [C_1(I - A - BF_\rho)^{-1}B]^{-1}$$

with C_1 the first row of C corresponding to the position of the cart, near perfect tracking for the cart position can be achieved. The step responses of the closed-loop system with transfer matrix $\mathbf{T}_{ry}(z)$ are given in Fig. 6.18.

The solid line shows the step response of the position output, while the dashed line shows the response of the angle position. Small overshot and fast settling time are due to relatively large damping ratio and natural frequency used for the dominant pole location. Unfortunately, the order of the feedback controller cannot be reduced. In fact, the reduced order controller based on the IBT algorithm does not stabilize the inverted pendulum. The reason probably lies in the poor stability margin and near unit circle zero for the squared-down plant $P_{2\varepsilon}(z)$.

A second design is based on the dynamically squared-down plant $P_{\text{sd}}(z)$ studied in Example 6.13. Let

$$\mathbf{W}(z) = \begin{bmatrix} -\gamma(1 + 0.42z^{-1}) & (1 + 0.37z^{-1}) \end{bmatrix} = D_w + C_w(zI - A_w)^{-1}B_w$$

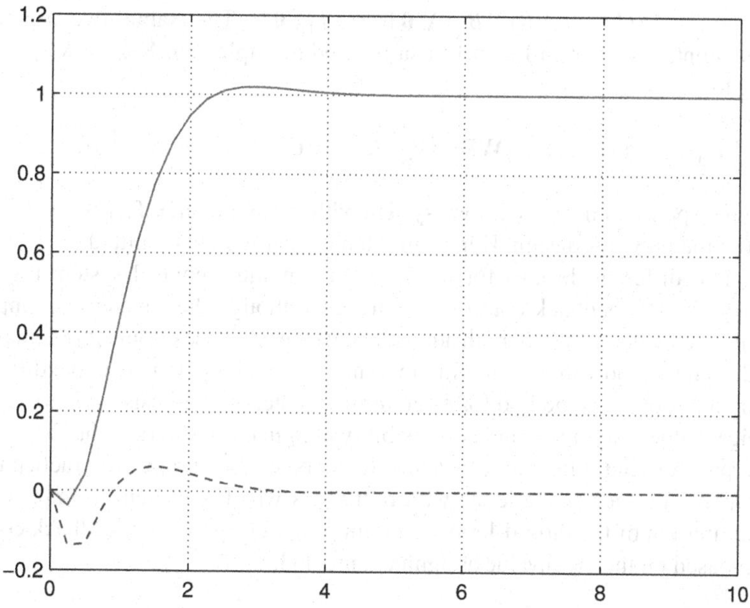

Fig. 6.18 Step responses for the closed-loop inverted pendulum system

with $\gamma = 0.45$. The squared-down plant admits the following expression:

$$P_{\rm sd}(z) = \mathbf{W}(z)\mathbf{P}_d(z) = \left[\begin{array}{cc|c} A & \mathbf{0} & B \\ B_wC & A_w & \mathbf{0} \\ \hline D_wC & C_w & \mathbf{0} \end{array}\right].$$

Denote the realization of $P_{\rm sd}(z)$ by $(A_{\rm sd}, B_{\rm sd}, C_{\rm sd})$. By taking $\tilde{C}_0 = \begin{bmatrix} C_0 & \mathbf{0} \end{bmatrix}$ as the new "C_0" matrix,

$$G_0(z) = \tilde{C}_0(zI - A_{\rm sd})^{-1}B_{\rm sd} = \frac{\mathbf{b}_0(z)}{\mathbf{a}_0(z)}$$

remains the same. Hence, the same root locus procedure as earlier can be used. However, the state-feedback gain F_ρ is now replaced by $\tilde{F}_\rho = \begin{bmatrix} F_\rho & \mathbf{0} \end{bmatrix}$ with F_ρ the same as that in the first design. See Problem 6.23 in Exercises.

The LTR design part is more different from that of the first design. Its main reason lies in the higher order of the squared-down plant $P_{\rm sd}(z)$. Nevertheless, the model $P_{\rm sd}(z)$ is strictly minimum phase, and $C_{\rm sd}B_{\rm sd} = D_wCB$ is nonzero. Hence, perfect LTR can be achieved. The required state estimation gain $L_{\rm sd}$ can also be easily computed based on $\tilde{Y} = B_{\rm sd}B_{\rm sd}^*$ that is the stabilizing solution to the corresponding ARE in the LTR step. The LQR/LTR controller for the second design is obtained as

$$K_{\rm sd}(z) = z\tilde{F}_{1\rho}(zI - A_{\rm sd} - L_{\rm sd}C_{\rm sd})^{-1}L_{\rm sd},$$

where $\tilde{F}_{1\rho} = -(I + B^*XB)^{-1}B^*\begin{bmatrix} X & \mathbf{0} \end{bmatrix} = \begin{bmatrix} F_{1\rho} & \mathbf{0} \end{bmatrix}$. The same two-degree-of-freedom control system in Fig. 6.13 can be used to implement $\mathbf{K}(z) = K_{\rm sd}(z)\mathbf{W}(z)$ leading to

$$\mathbf{T}_{ry}(z) = \mathbf{P}(I + K_{\rm sd}(z)\mathbf{WP})^{-1}V_{\rm sd}^{-1}\Omega^{-1} \approx C(zI - A - BF_\rho)^{-1}B\Omega^{-1}.$$

The step response of the closed-loop system with transfer matrix $\mathbf{T}_{ry}(z)$ is very close to that of the previous design. Hence, the step response plot is omitted.

The two different designs for the inverted pendulum control system result in two very similar feedback control systems, even though the second one employs dynamic square down. While both admit decent step responses, some problems exist in its design method with the stability margin as the major one. Due to the difficulty in using the root locus method for discrete-time systems, synthesis of F_ρ to achieve both eigenvalue assignment and good stability margin is considerably harder than its counterpart for continuous-time systems. Readers are encouraged to practice more in order to gain more experience with root locus which will help to improve the stability margin of the closed-loop system in future design of the feedback control system based on the eigenvalue assignment method.

Notes and References

LQG control has been studied in many papers and books. The presentation in this chapter is translated from $\mathscr{H}_2$ control [27, 126]. The design method of loop shaping is based on [81, 115] which is developed for continuous-time systems. The LQR/LTR design has been studied in [20, 26, 123], and is generalized to discrete-time systems in [79, 124]. See also [9, 44, 99, 105, 109]. This chapter also provides the state-space formula for factorizing the unstable zeros and $\mathscr{H}_2$ controller interpretation.

Exercises

6.1. Verify the expressions in (6.5) and (6.6).

6.2. Prove Theorem 6.4.

6.3. Verify (6.23), (6.24), and (6.17).

6.4. Show that the stabilizing solution $X \geq 0$ to ARE (6.15) is the reachability gramian of $\mathbf{T}_{\rm FI}(z)$ in (6.13) or of $\mathbf{T}_{\rm tmp}(z)$ in (6.19).

6.5. Prove the error equations in (6.34) and show that $Y \geq 0$ to ARE in (6.31) is the associated reachability gramian.

6.6. Write Matlab function or m-file to compute the optimal feedback controller $\mathbf{K}(z)$ for the LFT feedback system in Fig. 6.3 under the assumptions of Theorems 6.4 and 6.5. The m-file needs to take realization of the generalized plant $\mathbf{G}(z)$ as input variables and realization of the controller $\mathbf{K}(z)$ as output variables.

6.7. Show that the normalized $\mathscr{H}_2$ control after Example 6.6 has the identical optimal controller to the one in Example 6.6.

6.8. Verify the realizations in (6.43) and (6.44) and the double Bezout identity

$$\begin{bmatrix} \mathbf{V}(z) & \mathbf{U}(z) \\ -\mathbf{N}(z) & \mathbf{M}(z) \end{bmatrix} \begin{bmatrix} \tilde{\mathbf{M}}(z) & -\tilde{\mathbf{U}}(z) \\ \tilde{\mathbf{N}}(z) & \tilde{\mathbf{V}}(z) \end{bmatrix} = I$$

(*Hint:* Consider $\hat{\mathbf{P}}(z) = -\mathbf{K}(z) = \hat{D} + \hat{C}(zI - \hat{A})^{-1}\hat{B}$ with $(\hat{A}, \hat{B}, \hat{C}, \hat{D})$ the same as in (6.35) and $\hat{\mathbf{K}}(z) = -\mathbf{G}_{22}(z)$ with $D_{22} = \mathbf{0}$. Show that

$$\hat{\mathbf{K}}(z) = -\left[\begin{array}{c|c} \hat{A} + \hat{B}\hat{F} + \hat{L}\hat{C} + \hat{L}\hat{D}\hat{F} & \hat{L} \\ \hline \hat{F} & \mathbf{0} \end{array}\right]$$

where $\hat{F} = -C_2$ and $\hat{L} = -B_2$, and then use the results in Sect. 3.2.3 to derive the expressions in (6.43) and (6.44) for the feedback system consisting of $\hat{\mathbf{P}}(z)$ and $\hat{\mathbf{K}}(z)$.)

6.9. Prove inequality in (6.46) by first proving

$$\overline{\sigma}[(I+S)^{-1}] \leq \frac{1}{1 - \overline{\sigma}(S)}$$

for each square matrix S satisfying $\overline{\sigma}(S) < 1$.

6.10. Show that under the assumption in (6.57) plus (6.58), the observability gramian P in (6.55) is the same as X to the ARE in (6.15) and the reachability gramian Q in (6.56)is the same as Y to the ARE in and (6.31).

6.11. For $\mathbf{T}(z) = \mathbf{T}_1(z)\mathbf{T}_2(z)$ with $\mathbf{T}(z)$, $\mathbf{T}_1(z)$, and $\mathbf{T}_2(z)$, all stable proper transfer matrices, show that

$$\text{(i) } \|\mathbf{T}\|_\infty \leq \|\mathbf{T}_1\|_\infty \|\mathbf{T}_2\|_\infty, \quad \text{(ii) } \|\mathbf{T}\|_2 \leq \|\mathbf{T}_1\|_\infty \|\mathbf{T}_2\|_2$$

6.12. Let H and W be square complex matrices of the dimension. Partition H and W compatibly as

$$H = \begin{bmatrix} H_1 & H_2 \end{bmatrix}, \quad W = \begin{bmatrix} W_1 \\ W_2 \end{bmatrix}$$

Suppose that $WH = I$, and thus, $HW = H_1W_1 + H_2W_2 = I$. Show that $\sigma_{\max}(H_1W_1) = \sigma_{\max}(H_2W_2)$ with $\sigma_{\max}(\cdot)$ the maximum singular value.

6.13. Prove Corollary 6.1.

6.14. Prove the stability condition (6.70) in Corollary 6.2.

6.15. Let $\mathbf{G}_{12}(z) = D_{12} + C_1(zI - A)^{-1}B_2$ satisfying the condition

$$D_{12}^*\begin{bmatrix} D_{12} & C_1 \end{bmatrix} = \begin{bmatrix} R & \mathbf{0} \end{bmatrix}$$

Let $X \geq 0$ be the stabilizing solution to the ARE (6.15) and F be as in (6.16). Show that for each real ω, $\mathbf{G}_{12}(e^{j\omega})^*\mathbf{G}_{12}(e^{j\omega}) \geq R$ and

$$[I - F(e^{j\omega}I - A)^{-1}B_2]^*(R + B_2^*XB_2)[I - F(e^{j\omega}I - A)^{-1}B_2] \geq R$$

(*Hint:* This is dual to Problem 5.36 and thus one may begin by showing that the left-hand side is the same as $\mathbf{G}_{12}(e^{j\omega})^*\mathbf{G}_{12}(e^{j\omega})$ for each real ω.)

6.16. Show that if the Nyquist plot does not intersect the disk of radius $r > 0$ centered at -1 as shown in Fig. 6.7, then the feedback system admits phase margin of $2\sin^{-1}(r/2)$ and gain margins of $(1+r)$ or $((1-r)^{-1}$.

6.17. Consider root locus for (6.87). Suppose that $\det(A) \neq 0$ and $\lambda_i(A)\bar{\lambda}_j(A) \neq 1$ for all $i \neq j$. Let Q_0 be the unique solution to

$$Q_0 = A^*Q_0A + C_0^*C_0.$$

Show that the roots of (6.87) are eigenvalues of the following matrix:

$$\begin{bmatrix} A & \mathbf{0} \\ \mathbf{0} & (A^*)^{-1} \end{bmatrix} - \rho^2 \begin{bmatrix} B \\ Q_0B \end{bmatrix} \begin{bmatrix} B^*Q_0A & -(A^{-1}B)^* \end{bmatrix}.$$

(*Hint*: Use the result in Problem 5.35 of Chap. 5 to show that

$$\mathbf{G}_0(z)^\sim\mathbf{G}_0(z) = B^*Q_0B + B^*Q_0A(zI - A)^{-1}B + B^*(z^{-1}I - A)^{-1}A^*Q_0B.$$

The zero eigenvalues of A correspond to z^{-1} which can removed, but eigenvalues on the unit circle can be difficult to deal with.)

6.18. (i) Verify the expression of the complementary sensitivity and sensitivity at the plant output in (6.104) and (6.105), respectively. (ii) Find the results dual to (6.104) and (6.105).

6.19. (i) Given $(A, \mathbf{b})$ that is reachable with $\mathbf{b}$ a column vector, show that there exists a nonsingular similarity transform T such that

$$TAT^{-1} = \begin{bmatrix} -a_1 & -a_2 & \cdots & \cdots & -a_n \\ 1 & 0 & \cdots & \cdots & 0 \\ 0 & \ddots & \ddots & & \vdots \\ \vdots & \ddots & \ddots & \ddots & \vdots \\ 0 & \cdots & 0 & 1 & 0 \end{bmatrix}, \quad T\mathbf{b} = \begin{bmatrix} 1 \\ 0 \\ \vdots \\ \vdots \\ 0 \end{bmatrix}$$

6.20. Let C and A be matrices of size $p \times n$ and $n \times n$, respectively. Suppose that C has rank p. Show that (i) the transmission zeros of $\mathbf{G}_0(z) = C(zI - A)^{-1}B_0$ can be arbitrarily assigned with B_0 of size $n \times p$, if and only if (C, A) is observable; (ii) the transmission zeros of $\mathbf{G}_0(z) = C(zI - A)^{-1}B_0$ are all strictly inside the unit circle for some B_0 of size $n \times p$, if and only if (A, B) is stabilizable. (The proofs are dual to those for (i) and (ii) in Lemma 6.3.)

6.21. Suppose that $\mathbf{P}_W(z)$ is square and has neither poles nor zeros at 1. Generalize the result in Remark 6.4 to the case when all singular values of $\mathbf{P}_W(1)$ are considerably greater than 1.

6.22. Let $\mathbf{P}(z) = C(zI - A)^{-1}B$ and $\mathbf{K}(z) = \mathbf{V}(z)^{-1}\mathbf{U}(z)$ with

$$\begin{bmatrix} \mathbf{V}(z) & \mathbf{U}(z) \end{bmatrix} = \left[\begin{array}{c|cc} A_K & B_V & B_U \\ \hline C_K & D_V & D_U \end{array}\right]$$

for the feedback system in Fig. 6.13. Show that

$$\mathbf{T}_{ry}(z) = \left[\begin{array}{cc|c} A - BD_V^{-1}D_UC & BD_V^{-1}C_K & BD_V^{-1} \\ B_VD_V^{-1}D_UC - B_UC & A_K - B_VD_V^{-1}C_K & -B_VD_V^{-1} \\ \hline C & \mathbf{0} & \mathbf{0} \end{array}\right] \Omega_F^{-1}.$$

(*Hint*: Consider $\mathbf{P}_\varepsilon(z) = C(zI - A)^{-1}B + D$ and find realization for

$$\mathbf{T}_{ry}(z) = [\mathbf{V}(z)\mathbf{P}_\varepsilon(z)^{-1} + \mathbf{U}(z)]^{-1}\Omega_F^{-1}$$

and then take $D = \varepsilon I \to \mathbf{0}$.)

6.23. For the second design of the inverted pendulum control system, the squared-down plant is given by $P_{\rm sd}(z)$. Assume that A_w is a stability matrix. Show that the stabilizing solution to

$$X_{\rm sd} = A_{\rm sd}^* X_{\rm sd}(I + B_{\rm sd}B_{\rm sd}^* X_{\rm sd})^{-1}A_{\rm sd} + \rho^2 \tilde{C}_0^* \tilde{C}_0$$

has the form

$$X_{\rm sd} = \begin{bmatrix} X & \mathbf{0} \\ \mathbf{0} & \mathbf{0} \end{bmatrix}$$

where $X \geq \mathbf{0}$ is the stabilizing solution to

$$X = A^*X(I + BB^*X)^{-1}A + \rho^2 CC^*.$$

Use the above to obtain the expression of the state-feedback gain

$$\tilde{F}_\rho = \begin{bmatrix} F_\rho & \mathbf{0} \end{bmatrix}, \quad F_\rho = -(I + B^*XB)^{-1}B^*XA.$$

Chapter 7
Design of Wireless Transceivers

Wireless communications have undergone a remarkable development since, early 1980s when the first cellular and coreless systems were introduced. Indeed, users have experienced three different generations of cellular and coreless systems, which are now being evolved into the fourth generation (4G). From initial narrowband circuit-switched voice services based on FM (frequency modulation) technology of 1G to broadband wireless access with asymmetric bit rate approaching 1 gigabits per second (Gb/s) of 4G, the wireless systems nowadays provide not only voice, but also data and multimedia services seamlessly and ubiquitously. The wireless Internet and online videos/games are increasingly part of our daily life, and smart phones and other mobile devices are becoming essential components of our professional life.

The rapid development of wireless communications is not possible without advancement of the wireless transceiver technology. Regardless the wide area network (WAN) or local area network (LAN), wireless transceivers constitute the physical layer of any wireless network. Their implementations usually involve a combination of different technologies from CMOS (complementary metal oxide semiconductor) to silicon bipolar or BiCMOS. Notwithstanding the hardware progress, algorithms for digital signal processing in wireless transceivers have also experienced major developments in the past two decades. Notables include those for MIMO wireless channels in order to increase the data rate and to satisfy the more stringent quality of service (QoS) requirements. This text focuses on signal processing side of the wireless transceiver technology, including channel estimation, equalization, precoding, and data or symbol detection which are fundamental to design of wireless transceivers.

In this chapter, several wireless systems will be introduced, with emphasis on DS (direct sequence) spread sequence CDMA (code division multiple access) systems and OFDM (orthogonal frequency-division multiplex) modulated systems over fading channels. Optimal algorithms will be presented for zero-forcing channel equalization, precoding, and linear receivers. These algorithms are developed for MIMO channels due to the multiuser nature of the wireless communication system and multiple antenna technology. Finally, optimal multiuser symbol detection will

G. Gu, *Discrete-Time Linear Systems: Theory and Design with Applications*,
DOI 10.1007/978-1-4614-2281-5_7, © Springer Science+Business Media, LLC 2012

be considered jointly with decision feedback. It will be seen that Kalman filtering and LQR control play pivotal roles in developing optimal design algorithms for wireless transceivers.

7.1 MIMO Wireless Communication Systems

There are quite a few different systems available for wireless communications. This chapter is aimed at illustrating the application of optimal estimation and control, and thus, the focus will be placed on CDMA and OFDM which dominate 3G and 4G wireless systems. Since multirate signal processing underlies the wireless transceiver technology, our presentation will begin with multirate devices and filterbanks.

7.1.1 *Multirate Transmultiplexer*

Multirate digital signal processing has been widely used in telecommunications. The following depicts the two frequently used sampling converters where L and M are integers.

The L-fold expander on the left is the upsampling converter that increases the sampling rate by a factor of L. Mathematically, the input and output are related according to

$$y_E(t) = \begin{cases} x(t/L), & t = 0, \pm L, \pm 2L, \cdots, \\ 0, & \text{otherwise.} \end{cases}$$

In Matlab when $\{x(t)\}$ is given, its L-fold upsampled version can be easily created with

$$y = \text{zeros}(1, L * \text{length}(x)); \quad y_E(1:L:\text{length}(y)) = x.$$

That is, with the sampling frequency ω_s for $\{x(t)\}$, the output signal $\{y_E(t)\}$ of the expander admits a new sampling frequency of $\omega_{\text{new}} = L\omega_s$.

For the M-fold decimator on the right of Fig. 7.1, there holds $y_D(t) = x(tM)$. The downsampled version of $\{x(t)\}$ can be easily created with the Matlab command:

$$y_D = x(1:M:\text{length}(x)).$$

Clearly, the sampling frequency of $\{y(t)\}$ is only the Mth sampling frequency of $\{x(t)\}$. That is, $100 \times \left(1 - \frac{1}{M}\right)\%$ of total samples are discarded during the downsampling conversion with a factor of M.

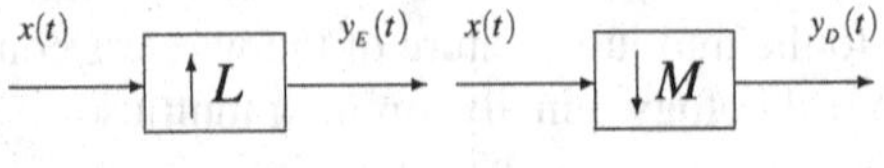

Fig. 7.1 L-fold expander (*left*) and M-fold decimator (*right*)

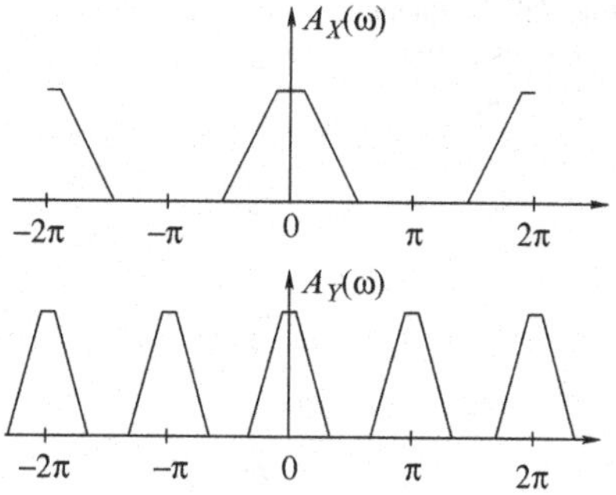

Fig. 7.2 Frequency responses showing imaging for $L = 2$

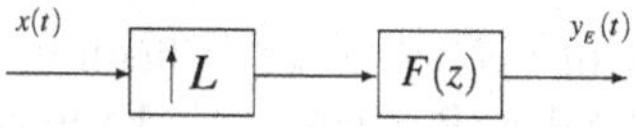

Fig. 7.3 L-fold interpolator with $\left[-\frac{\pi}{L}, \frac{\pi}{L}\right]$ the passband of $F(z)$

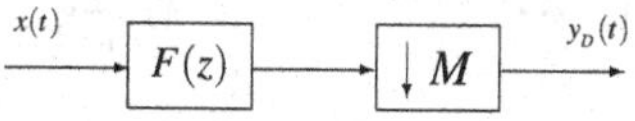

Fig. 7.4 M-fold decimator precedent by the low-pass filter $F(z)$

Suppose that the input signal has frequency response $X(e^{j\omega})$. The frequency response after the upsampling is found to be

$$Y_E(e^{j\omega}) = \sum_{t=-\infty}^{\infty} y(t)e^{-jt\omega} = \sum_{k=-\infty}^{\infty} x(k)e^{-jkL\omega} = X(e^{jL\omega}).$$

Hence, the frequency response images over the interval of length $2L\pi$ are compressed to fit in the interval of length 2π as shown in the following Fig. 7.2 for the case $L = 2$ in which $A_X(\omega) = |X(e^{j\omega})|$ and $A_Y(\omega) = |Y_E(e^{j\omega})|$. This phenomenon is called "imaging effect."

In order to preserve the original signal, a low-pass filter is often employed at the output of the expander to remove the duplicated images in $Y_E(\omega)$.

The expander followed by low-pass filter in Fig. 7.3 is now called *interpolator*, because it replaces the added zeros in $\{y_E(t)\}$ by the "interpolated" values. Specifically, when the discrete-time signal $\{x(t)\}$ satisfies the condition of the sampling theorem and the filter is ideal, the output signal $\{y_E(t)\}$ will be no different from the sampled signal of the original continuous-time signal under the new sampling frequency ω_{new}. See Problem 7.1 in Exercises.

For decimator, the downsampled signal may violate, even if the input signal satisfies, the condition of the sampling theorem. For this reason, the low-pass filter is often employed to filter the input signal first prior to downsampling as shown in Fig. 7.4.

Fig. 7.5 Frequency responses showing aliasing for $M = 2$

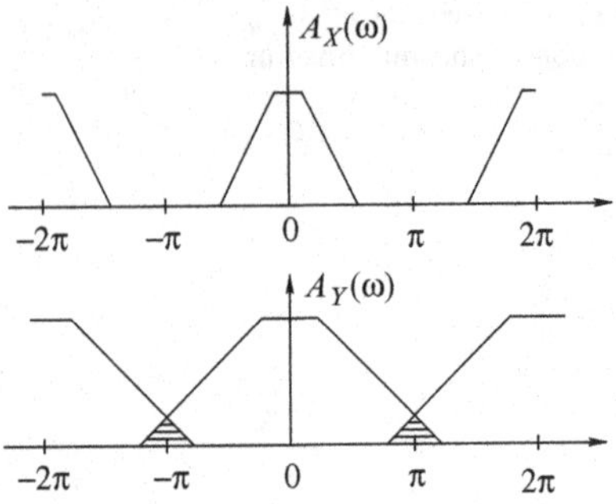

The frequency response of $y_D(t)$ is more difficult to derive than that of $y_E(t)$. Nonetheless, it can be shown that (Problem 7.2 in Exercises)

$$Y_D(e^{j\omega}) = \frac{1}{M}\sum_{i=0}^{M-1} X\left(e^{j\frac{\omega-2i\pi}{M}}\right) = \frac{1}{M}\sum_{i=0}^{M-1} X\left(W_M^i e^{j\frac{\omega}{M}}\right) \tag{7.1}$$

in the case $F(z) \equiv 1$ where $W_M = e^{-j2\pi/M}$. The above hints that a decimator in absence of $F(z)$ may result in aliasing as seen in Fig. 7.5.

The above figure shows that the low-pass filter $F(z)$ needs to have passband $[-\frac{\pi}{M}, \frac{\pi}{M}]$ with magnitude M, instead of 1, in order to avoid aliasing and restore the amplitude of the signal.

A discrete-time signal $\{s(t)\}$ can be put into block form to obtain a vector-valued signal, that is, a vector-valued signal

$$\mathbf{s}(t) = \begin{bmatrix} s_0(t) \\ s_1(t) \\ \vdots \\ s_{M-1}(t) \end{bmatrix} := \begin{bmatrix} s(tM) \\ s(tM+1) \\ \vdots \\ s(tM+M-1) \end{bmatrix} \tag{7.2}$$

for $t = 0, \pm 1, \cdots$, with M the block size. The following figure illustrates the operation of blocking with z standing for the unit advance operation:

Clearly, the blocking does not lose information. The kth subsignal $\{s_k(t)\}$ is called kth polyphase component of $\{s(t)\}$. Applying DTFT to $\{s_k(t)\}$ yields

$$S_k(e^{j\omega}) = \sum_{t=-\infty}^{\infty} s_k(t)e^{-jt\omega}. \tag{7.3}$$

It follows that $S_k(e^{j\omega})$ is the kth entry of $\mathbf{S}(e^{j\omega})$ which is the DTFT of $\{\mathbf{s}(t)\}$. Moreover, there holds polyphase decomposition

$$S(e^{j\omega}) = \sum_{t=-\infty}^{\infty} s(t)e^{-jt\omega} = \sum_{k=0}^{M-1} e^{-jk\omega} S_k(e^{jM\omega}). \tag{7.4}$$

Fig. 7.6 Blocking a signal with size M

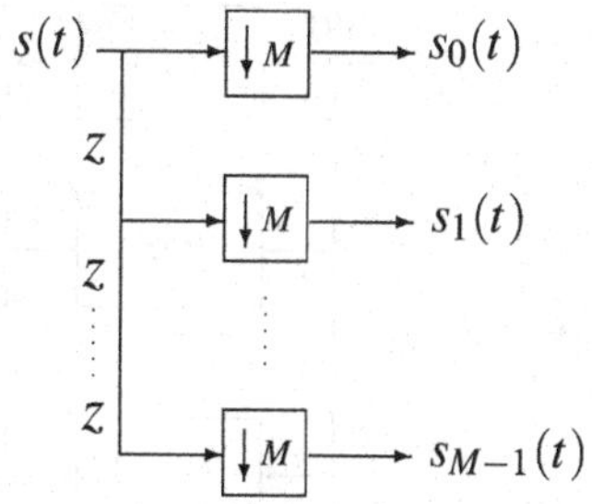

The interpolator and decimator are the elementary devices in filter banks, widely used in telecommunications. In Fig. 7.6, $\{H_i(z)\}_{i=0}^{M-1}$ on left are called analysis filters, and $\{G_i(z)\}_{i=0}^{M-1}$ on right are synthesis filters. The signal $s(t)$ is first filtered by $\{H_i(z)\}$ and split into M subband signals $\{v_i(t)\}$ which are then coded and individually transmitted. The bits are judiciously allocated to each subband signal $v_i(t)$ according to its energy which is termed *subband coding*. On the receiver side, the M signals are first upsampled, then decoded and filtered which are finally summed together to form $\hat{s}(t)$. If

$$\hat{s}(t) = \alpha s(t-\beta)$$

for some nonzero α and integer β, then the filterbank is said to have achieved perfect reconstruction (PR).

For downsampled signals, there holds

$$V_k(z) = \frac{1}{M}\sum_{i=0}^{M-1} H_k\left(z^{\frac{1}{M}}W_M^{-i}\right)S\left(z^{\frac{1}{M}}W_M^{-i}\right)$$

for $0 \le k < M$ in light of (7.1). The received signal at the output is

$$\begin{aligned}\hat{S}(z) &= \sum_{k=0}^{M-1} F_k(z)V_k(z^M)\\ &= \frac{1}{M}\sum_{i=0}^{M-1}\sum_{k=0}^{M-1} F_k(z)H_k(zW_M^i)S(zW_M^i).\end{aligned}$$

The terms $\{S(zW_M^i)\}$ for $i \neq 0$ represent aliasing, and need to be eliminated, leading to the condition

$$\sum_{k=0}^{M-1} F_k(z)H_k(zW_M^i) = \underline{H}(zW_M^i)\underline{F}(z) = 0 \tag{7.5}$$

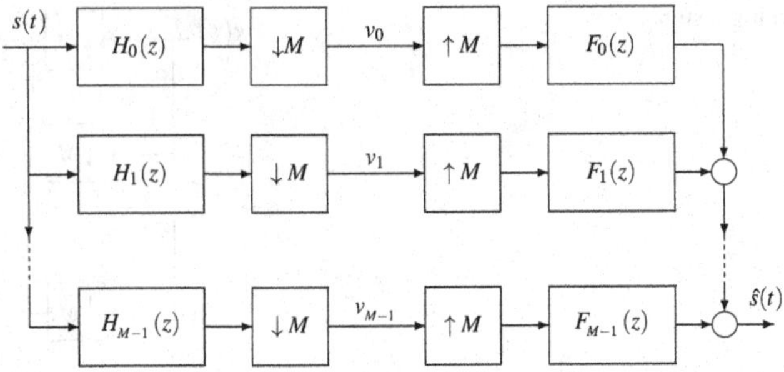

Fig. 7.7 Multirate quadrature mirror filter (QMF) bank

for $i \neq 0$ where

$$\underline{H}(z) = \begin{bmatrix} H_0(z) & \cdots & H_{M-1}(z) \end{bmatrix}, \quad \underline{F}(z) = \begin{bmatrix} F_0(z) \\ \vdots \\ F_{M-1}(z) \end{bmatrix}.$$

For $i = 0$, the PR condition requires $\underline{H}(z)\underline{F}(z) = \alpha z^{-\beta}$ for some $\beta \geq 0$ and $\alpha \neq 0$. It follows that the quadrature mirror filter (QMF) bank in Fig. 7.7 is PR, if and only if

$$\mathbf{H}_{\text{AC}}(z) \begin{bmatrix} F_0(z) \\ F_1(z) \\ \vdots \\ F_{M-1}(z) \end{bmatrix} = \begin{bmatrix} cz^{-d} \\ 0 \\ \vdots \\ 0 \end{bmatrix} \tag{7.6}$$

where $\mathbf{H}_{\text{AC}}(z)$ is called aliasing component matrix given by

$$\mathbf{H}_{\text{AC}}(z) = \begin{bmatrix} H_0(z) & H_1(z) & \cdots & H_{M-1}(z) \\ H_0(zW_M) & H_1(zW_M) & \cdots & H_{M-1}(zW_M) \\ \vdots & \vdots & & \vdots \\ H_0(zW_M^{M-1}) & H_1(zW_M^{M-1}) & \cdots & H_{M-1}(zW_M^{M-1}) \end{bmatrix}.$$

Alternatively, the PR is possible, if and only if the first column of $z^{-\beta}\mathbf{H}_{\text{AC}}(z)^{-1}$ is stable and causal for some $\beta \geq 0$. Normally, design of QMF bank begins with design of the analysis filters $\{H_k(z)\}$, focusing on the filtering specification in order to split the signals into M subbands as perfectly as possible. The synthesis filters $\{F_k(z)\}$ are then designed based on whether or not $\mathbf{H}_{\text{AC}}(z)$ admits a causal and stable inverse. In general, analysis filters designed to satisfy the filtering specification does not yield PR. Synthesis filters will then be designed to minimize the reconstruction error. It is also possible to minimize the filtering error and reconstruction error simultaneously. See Problems 7.4 and 7.5 of Exercises in the case of $M = 2$.

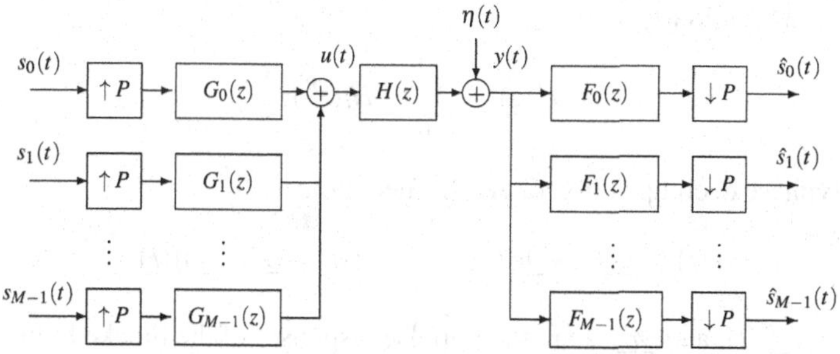

Fig. 7.8 Block diagram for transmultilexer

In high-speed transmission over noisy channels, DMT (digital multitone modulation) is a widely used technique. See the illustrative block diagram of DMT next. The M input signals are M symbol sequences to be transmitted. In the DMT scheme, the channel, rather than signal, is divided into subbands, each with a different frequency band. The transmission power and bits are judiciously allocated according to the SNR in each band. This is similar to the water pouring scheme for discrete transmission channels: more bits are sent over the strong SNR subband channels, while few bits are transmitted over the weak SNR subband channels. The realization of the DMT scheme relies on the design of a transceiver that effectively divides the channel into subbands. Band separation is of particular importance when the SNRs of different bands exhibit large differences. This can happen when the channel or the channel noise is highly frequency selective or non-flat.

For the model in Fig. 7.8, denote the input and output vector signals as

$$\mathbf{s}(t) = \begin{bmatrix} s_0(t) \\ s_1(t) \\ \cdots \\ s_{M-1}(t) \end{bmatrix}, \quad \hat{\mathbf{s}}(t) = \begin{bmatrix} \hat{s}_0(t) \\ \hat{s}_1(t) \\ \cdots \\ \hat{s}_{M-1}(t) \end{bmatrix}. \tag{7.7}$$

Block the input and output of the channel with size P according to

$$\mathbf{u}(t) = \begin{bmatrix} u(tP) \\ \vdots \\ u(tP+P-1) \end{bmatrix}, \quad \mathbf{y}(t) = \begin{bmatrix} y(tP) \\ \vdots \\ y(tP+P-1) \end{bmatrix}. \tag{7.8}$$

Then $\mathbf{u}(t)$, and $\mathbf{y}(t)$ have the same sampling rate as the symbol rate for $\{s_i(t)\}_{i=0}^{M-1}$.

Let $\eta(t)$ be column vector of size P with $\eta(tP+i-1)$ as the ith element. Each $G_k(z)$ and $F_k(z)$ admit polyphase decomposition as

$$G_k(z) = \sum_{i=0}^{P-1} z^{-i} G_{k,i}(z^P), \quad F_k(z) = \sum_{i=0}^{P-1} z^{-i} F_{k,i}(z^P), \tag{7.9}$$

for $0 \le k < M$. Similarly,

$$H(z) = \sum_{i=0}^{P-1} z^{-i} H_i(z^P) \tag{7.10}$$

is the polyphase decomposition of $H(z)$. There holds

$$\hat{\mathbf{s}}(t) = \underline{\mathscr{F}}(t) \star \underline{\mathscr{H}}(t) \star \underline{\mathscr{G}}(t) \star \mathbf{s}(t) + \underline{\mathscr{F}}(t) \star \underline{\eta}(t), \tag{7.11}$$

where $\underline{\mathscr{G}}(t), \underline{\mathscr{F}}(t)$, and $\underline{\mathscr{H}}(t)$ are the impulse responses of the blocked filters whose $\mathscr{Z}$ transform are given, respectively, by

$$\underline{\mathbf{G}}(z) = \begin{bmatrix} G_{0,0}(z) & G_{1,0}(z) & \cdots & G_{M-1,0}(z) \\ G_{0,1}(z) & G_{1,1}(z) & \cdots & G_{M-1,1}(z) \\ \vdots & \vdots & \cdots & \vdots \\ G_{0,P-1}(z) & G_{1,P-1}(z) & \cdots & G_{M-1,P-1}(z) \end{bmatrix}, \tag{7.12}$$

$$\underline{\mathbf{F}}(z) = \begin{bmatrix} F_{0,0}(z) & z^{-1}F_{0,1}(z) & \cdots & z^{-1}F_{0,P-1}(z) \\ F_{1,0}(z) & z^{-1}F_{1,1}(z) & \cdots & z^{-1}F_{1,P-1}(z) \\ \vdots & \vdots & \cdots & \vdots \\ F_{M-1,0}(z) & z^{-1}F_{M-1,1}(z) & \cdots & z^{-1}F_{M-1,P-1}(z) \end{bmatrix},$$

$$\underline{\mathbf{H}}(z) = \begin{bmatrix} H_0(z) & z^{-1}H_{P-1}(z) & \cdots & z^{-1}H_1(z) \\ H_1(z) & H_0(z) & \ddots & \vdots \\ \vdots & \ddots & \ddots & z^{-1}H_{P-1}(z) \\ H_{P-1}(z) & \cdots & H_1(z) & H_0(z) \end{bmatrix}. \tag{7.13}$$

Clearly, the PR condition in absence of the noise amounts to

$$\underline{F}(z)\underline{H}(z)\underline{G}(z) = \text{diag}\left(z^{-d_0}, z^{-d_1}, \cdots, z^{-d_{M-1}}\right) \tag{7.14}$$

for some nonnegative integers d_i with $0 \le i < M$, leading to

$$\hat{s}_i(t) = s_i(t - d_i), \quad 0 \le i < M.$$

7.1.2 Direct Sequence CDMA Systems

The wireless communication system based on code division multiple access (CDMA) allows M multiple asynchronous users to share the same frequency

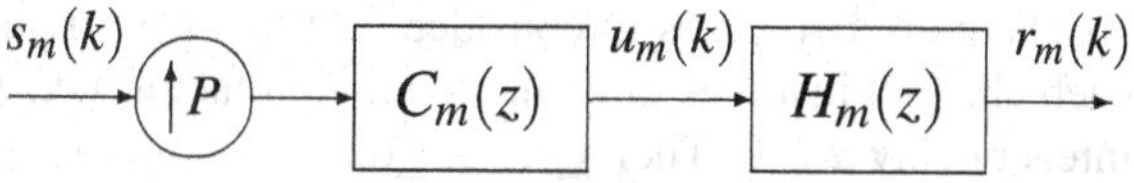

Fig. 7.9 Schematic diagram of CDMA modulation

channel, rather than dividing it into $P \geq M$ subbands. The advantages of such a system mainly lie in (a) spreading the transmission power over a wide bandwidth lowers spectral density, including less interference to another narrowband signals, and (b) By giving each user proper spreading codes, near orthogonality of waveforms can be achieved, allowing multiple users to coexist without mutual interference. This subsection presents the mathematical description of the CDMA modulation scheme.

Let $\{s_m(k)\}$ be the symbol stream of the mth user at the kth symbol sampling instant. Each user is given a unique spread chip sequence $\{c_m(i)\}_{i=0}^{P-1}$ where $P \geq M$ with M the total number of users. The use of spread chip sequence is equivalent to the upsampling with filter

$$C_m(z) = \sum_{i=0}^{P-1} c_m(i) z^{-i},$$

as shown in the following Fig. 7.9 with $H_m(z)$ the channel model.

By adopting the convention that $c_m(i) = 0$ for $i < 0$, or $i \geq P$, there holds

$$u_m(k) = \sum_{i=-\infty}^{\infty} s_m(i) c_m(k - iP).$$

Indeed, for $k = \ell P + j$ with $0 \leq j < P$,

$$u_m(k) = u_m(\ell P + j) = \sum_{i=-\infty}^{\infty} s_m(i) c_m\left[(\ell - i)P + j)\right] = s_m(\ell) c_m(j)$$

for $j = 0, 1, \cdots, P-1$. Let $\psi(\cdot)$ be the chip waveform with support $[0,\ T_c]$ where $T_c = T_s/P$ is the chip period, and T_s is the symbol period. The transmitted signal waveform from the mth user is given by

$$w_m(t) = \sum_{k=-\infty}^{\infty} u_m(k) \psi(t - kT_c)$$

for $m \in \{0, 1, \cdots, M-1\}$. In light of Sect. 1.2, the transmitted waveform will be (match) filtered and then sampled, leading to the equivalent discretized channel model $H_m(z)$, assuming time invariance of the channel. For this reason, the CDMA model is identical to the transmultiplexer in Fig. 7.8 with $G_i(z) = C_i(z)H_i(z)$ and $H(z) \equiv 1$ for $0 \leq i < M$. However, the subband interpretation is lost in the context of CDMA. In fact, narrowband symbol sequences are spread into wideband signals. On the receiver site, the received signals are filtered and downsampled in hope of recovering the desired symbol for each user. It is clear that $\underline{\mathbf{H}}(z) \equiv I_P$.

There are two cases to consider. The first case assumes synchronous users, and each channel involves only an attenuation factor, i.e., $H_m(z) \equiv a_m z^{-d}$ for the same integer delay $d \geq 0$. Then $\underline{\mathbf{G}}(z) \equiv \underline{C}\mathrm{diag}(a_0, \cdots, a_{M-1})$ with

$$\underline{C} := \begin{bmatrix} c_0(0) & c_1(0) & \cdots & c_{M-1}(0) \\ c_0(1) & c_1(1) & \cdots & c_{M-1}(1) \\ \vdots & \cdots & \vdots & \vdots \\ c_0(P-1) & c_1(P-1) & \cdots & c_{M-1}(P-1) \end{bmatrix},$$

and $\underline{\mathbf{F}}(z) \equiv \underline{C}^T \mathrm{diag}(a_0^{-1}, \cdots, a_{M-1}^{-1})$ yields the optimal receiver under the assumption that the spread codes are orthogonal, and thus, $\underline{C}$ becomes an orthogonal matrix. As a result, the mth user can retrieve only his/her symbol sequence $\{s_m(k)\}$.

The second case assumes that

$$H_m(z) = h_{m,0} z^{-n_m} + h_{m,1} z^{-(n_m-1)} + \cdots + h_{m,n_m}, \quad h_{m,0} \neq 0.$$

Even if the continuous-time channel is flat and involves only an attenuation factor and a time delay, asynchronous users may result in the above frequency fading channel (Problem 7.8 in Exercises). Hence, this case is generic. The discretized channel admits polyphase decomposition according to

$$H_m(z) = \sum_{i=0}^{M-1} z^{-i} H_{m,i}(z^P)$$

for $0 \leq m < M$. It follows that

$$\begin{aligned} G_m(z) = H_m(z) C_m(z) &= \left(\sum_{i=0}^{P-1} z^{-i} H_{m,i}(z^P) \right) \left(\sum_{k=0}^{P-1} c_m(k) z^{-k} \right) \\ &= \sum_{i=0}^{P-1} \sum_{k=0}^{P-1} z^{-(i+k)} c_m(k) \Phi_{m,i}(z^P) \\ &= \sum_{\ell=0}^{2P-2} z^{-\ell} \left(\sum_{k=0}^{\ell} c_m(k) H_{m,\ell-k}(z^P) \right) \\ &= \sum_{\ell=0}^{P-1} z^{-\ell} \left(\sum_{k=0}^{\ell} c_m(k) H_{m,\ell-k}(z^P) \right) \\ &\quad + \sum_{\ell=0}^{P-2} z^{-\ell} \left(z^{-P} \sum_{k=\ell+1}^{P-1} c_m(k) H_{m,P+\ell-k}(z^P) \right) \\ &= \sum_{\ell=0}^{P-1} z^{-\ell} G_{m,\ell}(z), \quad G_{m,\ell}(z) = \sum_{k=0}^{P-1} \tilde{G}_{m,\ell}^{(k)}(z) c_m(k) \end{aligned}$$

where

$$\tilde{G}_{m,\ell}^{(k)}(z) = \begin{cases} H_{m,\ell-k}(z), & 0 \le k \le \ell, \\ z^{-1}H_{m,P+\ell-k}(z), & \ell < k \le P-1. \end{cases}$$

By slight abuse of notation, denote

$$\underline{c}_m = \begin{bmatrix} c_m(0) \\ c_m(1) \\ \vdots \\ c_m(P-1) \end{bmatrix}, \quad \underline{\mathbf{G}}_m(z) = \begin{bmatrix} G_{m,0}(z) \\ G_{m,1}(z) \\ \vdots \\ G_{m,P-1}(z) \end{bmatrix},$$

$$\underline{\mathbf{G}_\mathbf{m}}(z) = \begin{bmatrix} H_{m,0}(z) & z^{-1}H_{m,P-1}(z) & \cdots & z^{-1}H_{m,1}(z) \\ H_{m,1}(z) & H_{m,0}(z) & \ddots & \vdots \\ \vdots & \ddots & \ddots & z^{-1}H_{m,P-1}(z) \\ H_{m,P-1}(z) & \cdots & H_{m,1}(z) & H_{m,0}(z) \end{bmatrix}.$$

There holds $\underline{\mathbf{G}}_m(z) = \underline{\mathbf{G}_\mathbf{m}}(z)\underline{c}_m$ for $0 \le m < M$, and thus, $\underline{\mathbf{G}}(z)$ in (7.12) admits the following expression:

$$\underline{\mathbf{G}}(z) = \left[\underline{\mathbf{G}}_0(z)\ \underline{\mathbf{G}}_1(z)\ \cdots\ \underline{\mathbf{G}}_{M-1}(z)\right]. \tag{7.15}$$

In the special case of the identical channel transfer functions, $\underline{\mathbf{G}_\mathbf{m}}(z) = \mathbf{T}_G(z)$ for some $\mathbf{T}_G(z)$ at each m, in which case $\underline{\mathbf{G}}(z) = \mathbf{T}_G(z)\underline{C}$.

7.1.3 *OFDM Model*

The OFDM model is developed to cope with the multipath phenomenon. Let the sequence of symbols be $\{\widehat{b}_k\}_{k=0}^{N-1}$ to be transmitted over the wireless channel represented by $H(z)$. These symbols are complex, assumed to be in the frequency domain, and transformed into the time domain via inverse DFT prior to transmission. Specifically, denote $W_N = \mathrm{e}^{-\mathrm{j}2\pi/N}$,

$$T_F = \frac{1}{\sqrt{N}} \begin{bmatrix} 1 & 1 & \cdots & 1 \\ 1 & W_N & \cdots & W_N^{N-1} \\ \vdots & \vdots & \vdots & \vdots \\ 1 & W_N^{N-1} & \cdots & W_N^{(N-1)^2} \end{bmatrix}$$

as the DFT matrix and $\widehat{\mathbf{b}}$ as a vector of size N with $\widehat{b}_i$ as the ith entry. Then $\mathbf{s} = T_F^*\widehat{\mathbf{s}}$ is the signal vector to be transmitted sequentially.

Let s_i be the ith entry of $\mathbf{s}$ and

$$H(z) = h_0 + h_1 z^{-1} + \cdots + h_{n-1} z^{n-1}$$

be the channel transfer function. Prior to transmitting $\{s_i\}_{i=0}^{N-1}$ sequentially over the channel, the last $(n-1)$ data are added to the beginning of the sequence, which is termed prefix. That is, the data transmitted are arranged in order according to

$$s_{N-n+1}, \cdots, s_{N-2}, s_{N-1}, s_0, s_1, \cdots, s_{N-2}, s_{N-1}.$$

Normally, $N \gg n$, and thus, the overhead due to the redundant $(n-1)$ data is negligible. At the receiver site, the first $(n-1)$ received signals are discarded, and the remaining N data are packed into a column vector $\mathbf{y}$. By the linear convolution relation and prefix, there holds $\mathbf{y} = C_h \mathbf{s} + \mathbf{v}$ where $\mathbf{v}$ is the additive noise vector, and

$$C_h = \begin{bmatrix} h_0 & 0 & \cdots & 0 & h_{n-1} & \cdots & h_1 \\ h_1 & \ddots & \ddots & \ddots & \ddots & \ddots & \vdots \\ \vdots & \ddots & \ddots & \ddots & \ddots & \ddots & h_{n-1} \\ h_{n-1} & \ddots & \ddots & \ddots & \ddots & \ddots & 0 \\ 0 & \ddots & \ddots & \ddots & \ddots & \ddots & \vdots \\ \vdots & \ddots & \ddots & \ddots & \ddots & \ddots & 0 \\ 0 & \cdots & 0 & h_{n-1} & \cdots & h_1 & h_0 \end{bmatrix}$$

is the circulant matrix, consisting of the CIR.

There are several interesting features about the circulant matrix. The first one is its Toeplitz structure. The second one is the shift property in that each row after the first row is the right circular shift of the previous row. The final one is the diagonalizability by the DFT matrix:

$$T_F C_h T_F^* = D_h := \text{diag}(H_0, H_1, \cdots, H_{N-1}) \tag{7.16}$$

where $\{H_k\}_{k=0}^{N-1}$ are the N-point DFT of $\{h_i\}_{i=0}^{n-1}$. See Problem 7.10 in Exercises. It follows from the unitary property of the DFT matrix that

$$T_F \mathbf{y} = (T_F C_h T_F^*)(T_F \mathbf{s}) + T_F \mathbf{v}$$

by multiplying T_F to $\mathbf{y} = C_h \mathbf{s} + \mathbf{v}$ from left. Recall that $\widehat{\mathbf{s}} = T_F \mathbf{s}$ is the symbol vector from the transmitter. Thus, $\widehat{\mathbf{y}} = T_F \mathbf{y}$ and $\widehat{\mathbf{v}} = T_F \mathbf{v}$ can be regarded as the received symbol and noise vectors, respectively, in the frequency domain. The

OFDM scheme removes the intersymbol interference (ISI) induced by multipath, and results in

$$\widehat{y}_k = \widehat{H}_k \widehat{s}_k + \widehat{v}_k, \quad k = 0, 1, \cdots, N-1. \tag{7.17}$$

The number of bits can be judiciously assigned to $\widehat{s}_k$ according to the ratio of $\mathrm{E}\{|\widehat{H}_k|^2\}/\mathrm{E}\{|\widehat{v}_k|^2\}$. Large value of this ratio implies high SNR, and more number of bits can be assigned to meet the requirement of the same bit error probability, and vice versa.

Recovery of the symbol $\widehat{s}_k$ involves detection based on noisy observation $\widehat{y}_k$ and requires the channel information. It is customary to employ n pilot symbols inserted into $\{\widehat{s}_k\}$ for channel estimation. For simplicity, assume that $N = nm$ for some integer $m > 1$. The pilot symbols are those given by $\{\widehat{s}_{im+\ell}\}_{i=0}^{n-1}$ where $\ell \in \{0, 1, \cdots, m-1\}$ is an integer. In light of (7.17),

$$\widehat{y}_{im+\ell} = \widehat{H}_{im+\ell} \widehat{s}_{im+\ell} + \widehat{v}_{im+\ell}$$

for $0 \le i < n$. Hence, n uniform samples of the channel frequency response data $\{\widehat{H}_{im+\ell}\}_{i=0}^{n-1}$ can be estimated based on the pilot symbols $\{\widehat{s}_{im+\ell}\}_{i=0}^{n-1}$ and noisy measurements $\{\widehat{y}_{im+\ell}\}_{i=0}^{n-1}$. Since for each i,

$$\widehat{H}_{im+\ell} = \frac{1}{\sqrt{N}} \sum_{k=0}^{n-1} h_k W_N^{k(im+\ell)} = \frac{1}{\sqrt{N}} \sum_{k=0}^{n-1} (h_k W_N^{k\ell}) W_n^{ik},$$

the CIR $\{h_k\}$ can be obtained from the inverse DFT:

$$W_N^{k\ell} h_k = \frac{1}{\sqrt{N}} \sum_{i=0}^{n-1} \widehat{H}_{im+\ell} W_n^{-ik}, \quad 0 \le k < n,$$

from which the rest of $\{\widehat{H}_k\}$ can be obtained. Channel estimation and its optimality are important research topics in wireless communications, and will studied in more detail in the next chapter.

Orthogonal frequency-division multiplexing can also be described by the block diagram for the transmultiplexer in Fig. 7.8 by using $M = N$. However, there are some variations. The first is that $\{s_i(t)\}$ for $0 \le i < M = N$ originate from the same signal $\{s(t)\}$ which are the N polyphase components of $\{s(t)\}$. That is, for $0 \le i < N$,

$$s_i(t) = s(tN + i), \quad t = 0, \pm 1, \cdots.$$

In other words, from $\{s(t)\}$ to $\{\mathbf{s}(t)\}$ involves serial to parallel conversion where $\mathbf{s}(t)$ is the same as defined in (7.7). The second regards the analysis filters that are implemented with inverse DFT, specified by T_F^*. Thus, $\mathbf{s}(t)$ can be regarded as a complex vector of size N in the frequency domain for each t. The most crucial one is the third that adds a cyclic prefix to the sequence of data $\{u(t)\}$ prior to its transmission.

7.2 Linear Receivers

Multiple antennas have been employed in the past decade to increase the channel capacity. These antennas are placed well apart from each other so that the channel gains are approximately uncorrelated, giving rise to receiver design for MIMO channels. This section considers design of linear receivers for wireless transceivers over MIMO channels. A schematic block diagram is shown in Fig. 7.10 where $\mathbf{H}(z)$ of size $P \times M$ represents the channel transfer matrix, and $\mathbf{F}(z)$ of size $M \times P$ represents the linear receiver. As a result, $\mathbf{s}(k)$ having dimension M is the symbol vector to be transmitted at time index k, $\mathbf{v}(k)$ of dimension P is the additive white noise, and $\hat{\mathbf{s}}(k-d)$ is the detected symbol vector after d samples of delay. The DMT and CDMA represented by the transmultiplexer in the previous section can be regarded as virtual MIMO systems, in which case $\mathbf{H}(z)$ contains both channel and subband analysis filters (DMT), or spreading codes (CDMA). Hence, the design algorithms presented in this section also apply to these virtual MIMO systems.

The focus of this section is design of linear receivers and study of how and why optimal estimation and control theory in Chap. 5 provide design tools in synthesizing optimal linear transceivers. To simplify the presentation, complete channel information is assumed.

7.2.1 Zero-Forcing Equalizers

Zero-forcing (ZF) equalizers aim at synthesizing the receiver $\mathbf{F}(z)$ that achieves

$$\mathbf{F}(z)\mathbf{H}(z) = \text{diag}(z^{-d_1}, z^{-d_2}, \cdots, z^{-d_M}) \tag{7.18}$$

for some $d_i \geq 0$ and $1 \leq i \leq M$. The above is the same as the PR condition. Recall that ISI and interchannel interference (ICI) are major impediments for symbol detection. The PR condition effectively removes both ISI and ICI and thus has been an important approach to design of MIMO receivers.

However, ZF alone is not adequate for optimal symbol detection. The noise impact on the output of the ZF equalizer has to be minimized. By the convention, the additive white noise $\{\mathbf{v}(k)\}$ are assumed to be Gauss and have mean zero and covariance $\Sigma_v = I$. The case when $\Sigma_v \neq I$ is left as an exercise (Problem 7.12). Under ZF equalization in (7.18), there holds

$$\hat{\mathbf{s}}(k) = \mathbf{s}(k-\mathbf{d}) + \hat{\mathbf{v}}(k), \quad \hat{\mathbf{v}}(k) = F(q)\mathbf{v}(k),$$

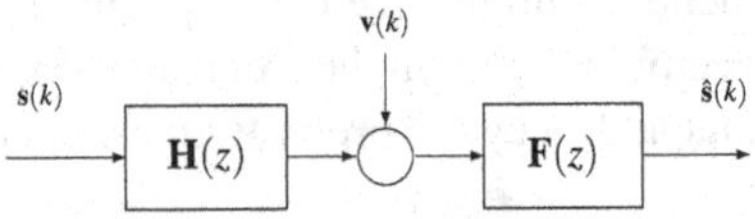

Fig. 7.10 Schematic diagram for MIMO linear receivers

where by an abuse of notation, $\mathbf{s}(k-\mathbf{d})$ is the same as $\mathbf{s}(k)$, except that its ith entry is delayed by d_i samples with d_i the ith entry of $\mathbf{d}$. Hence, $\hat{\mathbf{v}}(k)$ is again Gauss, although it is not white anymore. Optimal detection of $\mathbf{s}(k-\mathbf{d})$ requires to minimize the variance of $\hat{\mathbf{v}}(k)$, given by

$$J_F = \|\mathbf{F}\|_2^2 := \mathrm{Tr}\left\{\frac{1}{2\pi}\int_{-\pi}^{\pi} \mathbf{F}(e^{j\omega})\mathbf{F}(e^{j\omega})^* \, d\omega\right\}.$$

The synthesis of the optimal ZF equalizer seeks a causal and stable $\mathbf{F}(z)$ to satisfy the ZF condition (7.18) and to minimize J_F. Next, lemma provides an equivalent condition for the ZF condition.

Lemma 7.1. *Consider* $\mathbf{H}(z)$ *in* Fig. 7.10 *with size* $P \times M$ *and* $P > M$. *There exists a causal and stable* $\mathbf{F}(z)$ *such that (7.18) holds for some* $d_i \geq 0$ *and* $1 \leq i \leq M$, *if and only if*

$$\mathrm{rank}\left\{\mathbf{H}(z)\mathrm{diag}(z^{d_1}, z^{d_2}, \cdots, z^{d_M})\right\} = M \;\; \forall \; |z| \geq 1 \cup z = \infty.$$

Proof. Since multiplication by another matrix does not increase rank, the failure of the above rank condition implies the failure of the ZF condition (7.18). Conversely, if the above rank condition holds, then there exist more than one causal and stable $\mathbf{F}(z)$ such that (7.18) holds. In fact, all causal and stable ZF equalizers will be parameterized in the next lemma, and thus, the detailed proof for the sufficiency of the rank condition is skipped. □

The nonnegative integers $\{d_i\}$ can be obtained from the inner-outer factorization (Problem 7.14 in Exercises). For convenience, a state-space realization is assumed for the $P \times M$ transfer matrix:

$$\mathbf{T}(z) := \mathbf{H}(z)\mathrm{diag}(z^{d_1}, z^{d_2}, \cdots, z^{d_M}) = D + C(zI - A)^{-1}B. \tag{7.19}$$

The rank condition in Lemma 7.1 is translated to $\mathrm{rank}\{D\} = M$ and strictly minimum phase of (A, B, C, D). There exists an orthogonal $D_\perp$ of size $P \times (P - M)$ such that $D_a = \begin{bmatrix} D & D_\perp \end{bmatrix}$ is square and nonsingular. Let

$$D^+ = (D'D)^{-1}D^*, \quad D_\perp D_\perp^* = I - D(D^*D)^{-1}D^*.$$

It is easy to verify that

$$\begin{bmatrix} D^+ \\ D_\perp^* \end{bmatrix} \begin{bmatrix} D & D_\perp \end{bmatrix} = \begin{bmatrix} I_M & \mathbf{0} \\ \mathbf{0} & I_{P-M} \end{bmatrix}. \tag{7.20}$$

The following presents parameterization of all causal and stable ZF equalizers.

Lemma 7.2. *Suppose that* $\mathbf{T}(z)$ *of size* $P \times M$ *in (7.19) has a stabilizable and detectable realization* (A, B, C, D) *and* $P > M$. *Let* $A_0 = A - BD^+C$ *and* D^+ *and* $D_\perp$ *be specified in (7.20). If* $\mathbf{T}(z)$ *is strictly minimum phase, i.e.,* $\mathrm{rank}\{\mathbf{T}(z)\} = M$

for all $|z| \geq 1$, including $z = \infty$, then $(D_\perp^* C, A_0)$ is detectable, and thus, a stabilizing estimation gain L exists such that $(A_0 + LD_\perp^* C)$ is a stability matrix. Define

$$\begin{bmatrix} \mathbf{T}(z)^+ \\ \mathbf{T}_\perp(z)^+ \end{bmatrix} := \begin{bmatrix} D^+ \\ D_\perp^* \end{bmatrix} \left[\begin{array}{c|c} A_0 + LD_\perp^* C & LD_\perp^* - BD^+ \\ \hline C & I \end{array} \right]. \tag{7.21}$$

All causal and stable channel equalizers are parameterized by

$$\begin{aligned} \mathbf{F}(z) &= \mathbf{T}_{\text{inv}}(z) := \mathbf{T}(z)^+ + \mathbf{Q}(z)\mathbf{T}_\perp(z)^+ \\ &= \left[\begin{array}{c|c} A_0 + LD_\perp^* C & LD_\perp^* - BD^+ \\ \hline (D^+ + \mathbf{Q}D_\perp^*)C & (D^+ + \mathbf{Q}D_\perp^*) \end{array} \right] \end{aligned} \tag{7.22}$$

where $\mathbf{Q}(z)$ is an arbitrary proper and stable transfer matrix.

Proof. Stabilizability of $(D_\perp^* C, A_0)$ is left as an exercise (Problem 7.13). Since $D^+ D = I$ and $D_\perp^* D = \mathbf{0}$,

$$\mathbf{T}(z) = \left[I + C(zI - A)^{-1}(BD^+ + LD_\perp^*) \right] D$$

where $(A_0 + LD_\perp^* C)$ is a stability matrix by the hypothesis on L. A proper and stable left inverse of $\mathbf{T}(z)$ is thus given by

$$\mathbf{T}(z)^+ = D^+ \left[I + C(zI - A)^{-1}(BD^+ + LD_\perp^*) \right]^{-1}$$

yielding the upper part expression in (7.21). With

$$\mathbf{T}_\perp(z) = \left[I + C(zI - A)^{-1}(BD^+ + LD_\perp^*) \right] D_\perp,$$

its proper and stable left inverse $\mathbf{T}_\perp(z)^+$ can be obtained as in the lower part expression in (7.21). There thus holds

$$\begin{bmatrix} \mathbf{T}(z)^+ \\ \mathbf{T}_\perp(z)^+ \end{bmatrix} \begin{bmatrix} \mathbf{T}(z) & \mathbf{T}_\perp(z) \end{bmatrix} = \begin{bmatrix} I_M & \mathbf{0} \\ \mathbf{0} & I_{P-M} \end{bmatrix},$$

identical to the constant case in (7.20). It follows that $\mathbf{T}_{\text{inv}}(z)$ in (7.22) is indeed a proper and stable left inverse, provided that $\mathbf{Q}(z)$ is proper and stable. Conversely, for a given left proper and stable inverse $\mathbf{T}(z)^\dagger$, it satisfies

$$\mathbf{T}(z)^\dagger \begin{bmatrix} \mathbf{T}(z) & \mathbf{T}_\perp(z) \end{bmatrix} = \begin{bmatrix} I & \mathbf{T}(z)^\dagger \mathbf{T}_\perp(z) \end{bmatrix}.$$

On the other hand, $\mathbf{F}(z) = \mathbf{T}_{\text{inv}}(z)$ in (7.22) yields

$$\mathbf{F}(z) \begin{bmatrix} \mathbf{T}(z) & \mathbf{T}_\perp(z) \end{bmatrix} = \begin{bmatrix} I & \mathbf{Q}(z) \end{bmatrix}.$$

Since $\begin{bmatrix} \mathbf{T}(z) & \mathbf{T}_\perp(z) \end{bmatrix}$ is square and admits a stable and proper inverse, setting $\mathbf{Q}(z) = \mathbf{T}(z)^\dagger \mathbf{T}_\perp(z)$ yields $\mathbf{F}(z) = \mathbf{T}(z)^\dagger$ with proper and stable $\mathbf{Q}(z)$, thereby proving the parameterization in (7.22). □

Lemma 7.2 indicates that each ZF channel equalizer indeed achieves

$$\mathbf{F}(z)\mathbf{H}(z) = \text{diag}(z^{-d_1}, z^{-d_2}, \cdots, z^{-d_M}).$$

In addition, the optimal channel equalizer can be synthesized via minimization of $\|\mathbf{F}\|_2$ over all proper and stable $\mathbf{Q}(z)$. While this is feasible, a more efficient method employs the optimal solution to output estimation from Chap. 5, in which case L can be chosen as the optimal state estimation gain and $\mathbf{Q}(z)$ reduces to the (constant) optimal output estimation gain Q. Specifically, consider the random process

$$\begin{aligned} \mathbf{x}(t+1) &= A_0\mathbf{x}(t) - BD^+\mathbf{v}(t), \\ \mathbf{z}(t) &= D^+C\mathbf{x}(t) + D^+\mathbf{v}(t), \\ \mathbf{y}(t) &= D_\perp^* C\mathbf{x}(t) + D_\perp^*\mathbf{v}(t), \end{aligned} \tag{7.23}$$

where $\mathbf{v}(t)$ is white and Gauss, independent of $\mathbf{x}(t)$, with mean zero and covariance I. The problem of stationary output estimation aims to estimate $\mathbf{z}(t)$ based on observations $\{\mathbf{y}(k)\}_{k=-\infty}^{t}$. Compared with the problem setup in Chap. 5, $B_1 = -BD^+$ and

$$D_1 = D^+, \quad D_2 = D_\perp^*, \quad C_1 = D^+C, \quad C_2 = D_\perp^* C.$$

Hence, $D_1 D_2^* = \mathbf{0}$, $B_1 D_2^* = \mathbf{0}$, and $D_2 D_2^* = I$ in light of (7.20). Define $Y \geq 0$ as the stabilizing solution to

$$Y = A(I + YC^*D_\perp D_\perp^* C)^{-1} YA^* + B(D^*D)^{-1}B^*. \tag{7.24}$$

According to Chap. 5, the optimal output estimator is described by

$$\begin{aligned} \hat{\mathbf{x}}_{t+1|t} &= [A_0 + LD_\perp^* C]\hat{\mathbf{x}}_{t|t-1} - L\mathbf{y}(t), \\ \hat{\mathbf{z}}_{t|t} &= [D^+C + QD_\perp^* C]\hat{\mathbf{x}}_{t|t-1} - Q\mathbf{y}(t), \end{aligned} \tag{7.25}$$

where the optimal state and output estimation gains L and Q are given by

$$\begin{bmatrix} L \\ Q \end{bmatrix} = -\begin{bmatrix} A_0 YC^*D_\perp \\ D^+CYC^*D_\perp \end{bmatrix} (I + D_\perp^* CC^* D_\perp)^{-1}, \tag{7.26}$$

respectively. The next theorem shows that the above estimation gains minimizes $J_F = \|\mathbf{F}\|_2^2$.

Theorem 7.1. *Consider the causal and stable* ZF *channel equalizers represented by* $\mathbf{F}(z)$ *in (7.21). Let* $Y \geq \mathbf{0}$ *be the stabilizing solution to the* ARE *in (7.24). Then* $\|\mathbf{F}\|_2$ *is minimized by* L, Q *specified in (7.26).*

Proof. As discussed earlier, the estimator in (7.25) is optimal with L, Q in (7.26) in the sense that it minimizes the variance of $\mathbf{e}_z(t) = \mathbf{z}(t) - \hat{\mathbf{z}}_{t|t}$ among all estimators, including the dynamic ones. Denote $\mathbf{e}_x(t) = \mathbf{x}(t) - \hat{\mathbf{x}}_{t+1|t}$. Taking difference of the equations in (7.23) and (7.25) leads to

$$\begin{aligned}\mathbf{e}_x(t+1) &= [A_0 + LD_\perp^* C]\hat{\mathbf{e}}_x(t) + [LD_\perp^* - BD^+]\mathbf{v}(t),\\ \mathbf{e}_z(t) &= [D^+C + QD_\perp^* C]\hat{\mathbf{e}}_x(t) + [D^+ + QD_\perp^*]\mathbf{v}(t).\end{aligned}$$

The above admits transfer matrix $\mathbf{F}(z) = \mathbf{T}_{\text{inv}}(z)$ in form of (7.21). The optimality of the output estimator concludes the minimum of $J_F = \|\mathbf{F}\|_2^2$ among all causal and stable ZF channel equalizers. □

Example 7.2. The ZF channel equalizer presented in this chapter will be illustrated by a toy example in the sense that both M and P are small. The transmultiplexer-based wireless transceiver is used. The channel model is described by

$$H(z) = 1 - 0.3z^{-1} + 0.5z^{-2} - 0.4z^{-3} + 0.1z^{-4} - 0.02z^{-5} + 0.3z^{-6} - 0.1z^{-7},$$

and $P = 4 > M = 3$ are chosen. The analysis filters $\{G_m(z)\}_{m=0}^{M-1}$ are designed such that $\mathbf{T}(z) = \underline{\mathbf{H}}(z)\underline{\mathbf{G}}(z)$ is both inner and outer. Recall the expressions of $\underline{\mathbf{G}}(z)$ and $\underline{\mathbf{H}}(z)$ in (7.12) and (7.13), respectively. The design of $\{G_m(z)\}$ is carried out by computing spectral factorization of

$$\begin{bmatrix} I_M & \mathbf{0} \end{bmatrix} \underline{\mathbf{H}}(z)^{\sim}\underline{\mathbf{H}}(z) \begin{bmatrix} I_M \\ \mathbf{0} \end{bmatrix} = \mathbf{R}(z)^{\sim}\mathbf{R}(z),$$

where both $\mathbf{R}(z)$ and $\mathbf{R}(z)^{-1}$ are proper and stable, and by setting

$$\underline{\mathbf{G}}(z) = \begin{bmatrix} I_M & \mathbf{0} \end{bmatrix}^* \mathbf{R}(z)^{-1}.$$

Since each entry corresponds to some polyphase of some $G_m(z)$, $\{G_m(z)\}_{m=0}^{M-1}$ can be readily obtained so long as $\underline{\mathbf{G}}(z)$ is available. It can be verified that $\mathbf{T}(z) = \underline{\mathbf{H}}(z)\underline{\mathbf{G}}(z)$ is strictly minimum including at $z = \infty$.

In design of the ZF-based synthesis filters $\{F_m(z)\}$, a state-space realization is set up first for $\mathbf{T}(z)$. Theorem 7.1 is then used to compute the optimal ZF channel equalizer. The optimal receiver filters $\{F_m(z)\}_{m=0}^{3}$ are obtained subsequently. The mean-squared-error J_F to the optimal receivers is $J_F = 1.7856$ that is minimum among all causal and stable ZF channel equalizers, assuming $\mathbf{v}(t)$ is white both spatially and temporally. The average frequency responses of the transmitter and receiver filters are plotted using

$$|\mathscr{G}(f)| := \frac{1}{M}\sum_{m=0}^{M-1} |G_m(e^{j2f\pi})|, \quad |\mathscr{F}(f)| := \frac{1}{M}\sum_{m=0}^{M-1} |F_m(e^{j2f\pi})|. \tag{7.27}$$

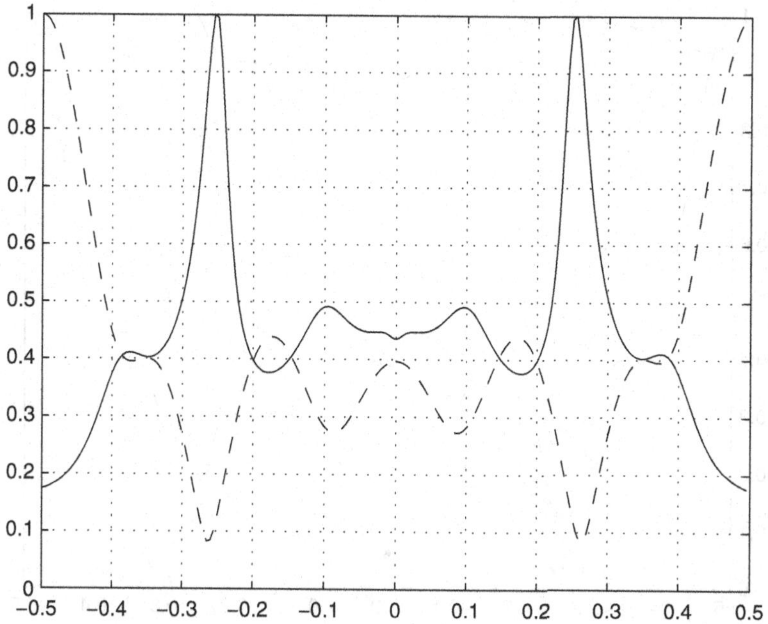

Fig. 7.11 $|\mathscr{G}(f)|$ versus $|H(e^{j2f\pi})|$

See Fig. 7.11 next for $|\mathscr{G}(f)|$ (solid line) versus the channel frequency response (dashed line).

Figure 7.12 plots $|\mathscr{F}(f)|$ (solid line) versus the channel frequency response. All curves are normalized with respect to their respective maximum.

The strictly minimum phase condition in Theorem 7.1 usually holds due to $P > M$ and often $P \gg M$. The drawback of the ZF channel equalizer lies in the possible near singularity of $[\mathbf{T}(e^{j\omega})^*\mathbf{T}(e^{j\omega})]$ at some frequency ω. For this reason, Example 7.2 composes $\mathbf{T}(z)$ in a way that $[\mathbf{T}(e^{j\omega})^*\mathbf{T}(e^{j\omega})] \equiv I$ to remove near frequency nulls.

7.2.2 Zero-Forcing Precoders

ZF channel equalization is effective in removing both ISI and ICI. However, the equalizer can be too complex to implement in the downlink mobile receiver because of the desire to keep the receiver units as simple as possible. The block precoding is a different transmission technique that shifts the signal processing burden from the mobile receiver to the transmitter in the more resourceful base station. In this case, a precoding filter (precoder) is designed at the transmitter side and is applied prior to transmission to equalize the signal at the output of the receive filter. Precoding is used in transmitters to compensate for distortion introduced by the channel response,

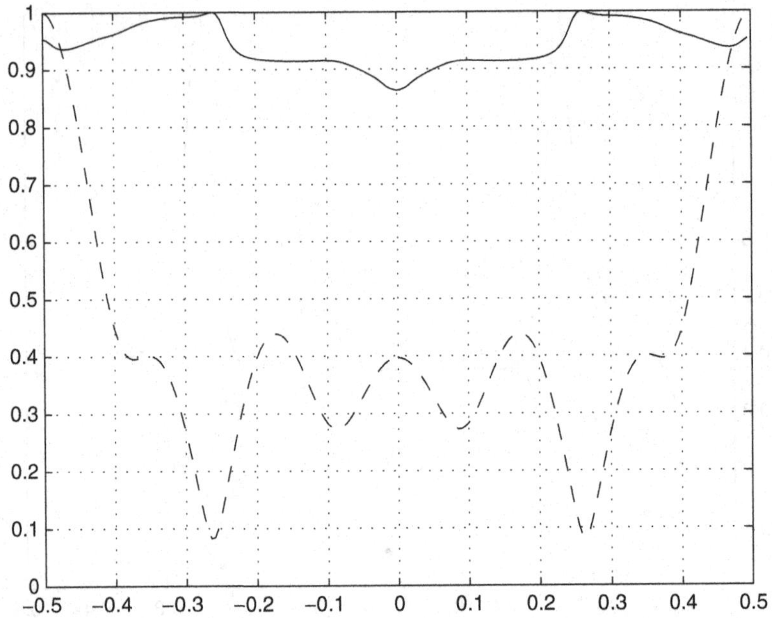

Fig. 7.12 $|\mathscr{F}(f)|$ versus $|H(e^{j2f\pi})|$

Fig. 7.13 Wireless transceiver with precoder

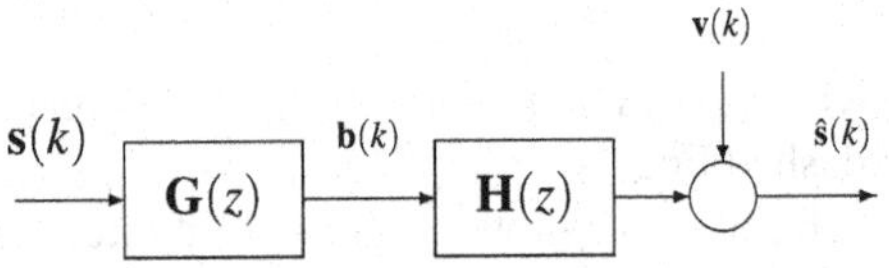

and to reduce the effects of ISI and ICI, thereby allowing for more reliable data transmission at high data rates. Precoder design is expected to allow the receiver to be considerably simplified, which in turn reduces computational complexity and power consumption in the mobile receiver.

The schematic block diagram for precoder receiver is illustrated in Fig. 7.13 that is rather different from the equalizer in Fig. 7.10. First, the channel transfer matrix $\mathbf{H}(z)$ has size $M \times P$ with $P > M$. Second, the precoder $\mathbf{G}(z)$ maps M dimensional input vector symbols $\{\mathbf{s}(k)\}$ to P dimensional vector signals $\{\mathbf{b}(k)\}$ at the transmitter site. The precoded P signal streams $\{\mathbf{b}(k)\}$ are transmitted through the MIMO channel $\mathbf{H}(z)$. The received signal consists of M dimensional data vectors $\{\hat{\mathbf{s}}(k)\}$ corrupted by additive white noise.

Assume that $\{\mathbf{s}(k)\}$ are uncorrelated and white. If not, a prewhitening operation can be performed over the symbol blocks prior to precoding. Hence, both input symbols and additive noise are assumed to be zero mean with covariance matrices given by

$$\Sigma_s = \sigma_s^2 I, \quad \Sigma_v = \sigma_v^2 I, \tag{7.28}$$

respectively. It follows that the SNR equals to σ_s^2/σ_v^2 for each subchannel, which is a constant. The goal is to design precoder $\mathbf{G}(z)$ to eliminate the ISI and ICI and to minimize the bit error rate (BER) of a downlink MIMO communication system by preprocessing the transmitted signal. An effective method is the use of ZF precoders that achieve the PR condition

$$\mathbf{H}(z)\mathbf{G}(z) = \mathrm{diag}(z^{-d_1}, z^{-2}, \cdots, z^{-d_M}) \tag{7.29}$$

for some positive integers $\{d_i\}$. Since $\mathbf{H}(z)$ is a wide matrix at each z, inner-outer factorization $\mathbf{H}(z) = \mathbf{H}_{\mathrm{i}}(z)\mathbf{H}_{\mathrm{o}}(z)$ needs to be carried out as in Problem 7.14. The square inner matrix $\mathbf{H}_{\mathrm{i}}(z)$ contains z^{-d_i} or 0 as entries. With

$$\mathrm{T}(z) = \mathrm{diag}(z^{d_1}, z^{2}, \cdots, z^{d_M})\mathbf{H}(z), \tag{7.30}$$

there holds the ZF condition $\mathbf{T}(z)\mathbf{G}(z) \equiv I$, in which case $\mathbf{G}(z)$ is termed as ZF precoder.

Suppose that (7.29) holds for some causal and stable precoder $\mathbf{G}(z)$. Then

$$\hat{\mathbf{s}}(k) = \mathbf{s}(k-\mathbf{d}) + \mathbf{v}(k).$$

with d_i as the ith entry of $\mathbf{d}$, by an abuse of notation. In other words,

$$\hat{s}_i(k) = s_i(k-d_i) + v_i(k), \quad 1 \le i \le M.$$

Assume constant signal power and white noise, the SNR for subchannel i at the receiver site is given by

$$\mathrm{SNR} = \frac{\sigma_s^2(i)}{\sigma_v^2(i)} = \frac{\sigma_s^2}{\sigma_v^2} = \frac{1}{M}\sum_{i=1}^{M}\frac{\sigma_s^2(i)}{\sigma_v^2(i)} = \text{Average SNR}. \tag{7.31}$$

As a result, the SNR for each subchannel is the same at the receiver site. Under additive Gauss noise, BER is given by

$$\mathrm{BER} = \frac{1}{\sqrt{2\pi}}\int_{\frac{\sigma_s}{\sigma_v}}^{\infty} \exp(-\tau^2/2)\,\mathrm{d}\tau,$$

by assuming the binary phase shift keying (BPSK) modulation. Hence, minimization of BER is equivalent to maximization of σ_s^2. However, the total transmission power at the transmitter site is limited and given by

$$\begin{aligned}\sigma_{\mathbf{b}}^2 &= \mathrm{E}\{\mathbf{b}(k)^*\mathbf{b}(k)\} = \mathrm{Tr}\{\mathrm{E}[\mathbf{b}(k)\mathbf{b}(k)^*]\}\\ &= \mathrm{Tr}\left\{\frac{1}{2\pi}\int_{-\pi}^{\pi}\mathbf{G}(e^{j\omega})\mathrm{E}\{\mathbf{s}(k)\mathbf{s}(k)^*\}\mathbf{G}(e^{j\omega})^*\,\mathrm{d}\omega\right\}\end{aligned}$$

$$= \text{Tr}\left\{\frac{1}{2\pi}\int_{-\pi}^{\pi} \mathbf{G}(e^{j\omega})\mathbf{G}(e^{j\omega})^* \, d\omega\right\}\sigma_s^2 = \|\mathbf{G}\|_2^2\sigma_s^2.$$

Let $\sigma_\mathbf{b}^2 = ME_\mathrm{b}$ with E_b the average bit energy at each transmitter. It follows that the SNR at the receiver site is bounded as

$$\text{SNR} = \frac{\sigma_s^2}{\sigma_v^2} = \frac{\sigma_\mathbf{b}^2}{\sigma_v^2\|\mathbf{G}\|_2^2} = \frac{ME_\mathrm{b}}{\sigma_v^2\|\mathbf{G}\|_2^2} \tag{7.32}$$

with the right-hand side the maximum SNR achievable, yielding the smallest BER. Hence, minimization of BER is equivalent to maximization of SNR, i.e., the right-hand side of (7.32), which is in turn equivalent to minimization of $J_G = \|\mathbf{G}\|_2^2$. Consequently, design of the optimal ZF precoder requires not only the ZF condition in (7.29) but also minimization of $\|\mathbf{G}\|_2$. This conclusion carries through to the case when $\mathbf{s}(k)$ is not composed of binary symbols, due to the constraint of the transmission power.

Associate $\mathrm{T}(z)$ with realization (A,B,C,D). Then D has dimension $M \times P$ with $P > M$. Hence, $D^+ = D^*(DD^*)^{-1}$ is a right inverse of D under the full rank condition for D. There exists $D_\perp$ of dimension $(P-M)\times P$ such that $D_\perp^* D_\perp = I - D^*(DD^*)^{-1}D$ and

$$\begin{bmatrix} D \\ D_\perp \end{bmatrix} \begin{bmatrix} D^+ & D_\perp^* \end{bmatrix} = \begin{bmatrix} I_M & \mathbf{0} \\ \mathbf{0} & I_{P-M} \end{bmatrix}. \tag{7.33}$$

Lemma 7.3. *Suppose that the transfer matrix $\mathbf{T}(z) = C(zI-A)^{-1}B + D$ has size $M \times P$ with $M < P$ and A is a stability matrix. There exists a proper and stable transfer matrix $\mathbf{G}(z)$ of size $P \times M$ such that the ZF condition $\mathbf{T}(z)\mathbf{G}(z) \equiv I$ holds, if and only if* rank$\{D\} = M$ *and $(A - BD^+C, BD_\perp^*)$ is stabilizable.*

Since this result is dual to the one in the previous subsection, the proof is left as an exercise (Problem 7.15). Denote $A_0 = A - BD^+C$. Stabilizability of $(A_0, BD_\perp^*)$ implies the existence of state-feedback gain F such that $(A_0 + BD_\perp^* F)$ is a stability matrix. Hence,

$$\mathrm{T}(z) = D\left[I + (D^+C - D_\perp^* F)(zI-A)^{-1}B\right]$$

Taking right inverse of $\mathbf{T}(z)$ yields

$$\mathbf{T}(z)^+ = \left[\begin{array}{c|c} A_0 + BD_\perp^* F & B \\ \hline -D^+C + D_\perp^* F & I \end{array}\right] D^+ \tag{7.34}$$

that is proper and stable. It follows that

$$\mathbf{T}_\perp(z)^+ = \left[\begin{array}{c|c} A_0 + BD_\perp^* F & B \\ \hline -D^+C + D_\perp^* F & I \end{array}\right] D_\perp^* \tag{7.35}$$

is a proper and stable right inverse of

$$\mathbf{T}_\perp(z) = D_\perp \left[I + (D^+C - D_\perp^* F)(zI - A)^{-1}B\right].$$

There thus holds the identity

$$\begin{bmatrix} \mathbf{T}(z) \\ \mathbf{T}_\perp(z) \end{bmatrix} \begin{bmatrix} \mathbf{T}(z)^+ & \mathbf{T}_\perp(z)^+ \end{bmatrix} = \begin{bmatrix} I_M & \mathbf{0} \\ \mathbf{0} & I_{P-M} \end{bmatrix}. \tag{7.36}$$

The following result is dual to that of Lemma 7.2, and thus, the proof is skipped.

Lemma 7.4. *Suppose that $\mathbf{T}(z)$ of size $M \times P$ in (7.19) has a stabilizable and detectable realization (A,B,C,D) with D full rank and $P > M$. There exist D^+ and $D_\perp$ such that (7.33) holds. If $(A_0, BD_\perp^*)$ is stabilizable where $A_0 = A - BD^+C$, there exists an F such that $(A_0 + BD_\perp^* F)$ is stability matrix. Let $\mathbf{T}(z)^+$ and $\mathbf{T}_\perp(z)^+$ be defined in (7.34) and (7.35), respectively. All causal and stable precoders satisfying the ZF condition are parameterized by $\mathbf{T}_{\text{inv}}(z) = \mathbf{T}(z)^+ + \mathbf{T}_\perp^+(z)\mathbf{Q}(z)$ or by state-space form*

$$\mathbf{T}_{\text{inv}}(z) = \left[\begin{array}{c|c} A_0 + BD_\perp^* F & B(D^+ + D_\perp^* \mathbf{Q}) \\ \hline -D^+C + D_\perp^* F & (D^+ + D_\perp^* \mathbf{Q}) \end{array}\right] \tag{7.37}$$

where $\mathbf{Q}(z)$ is an arbitrary stable transfer function matrix.

Now consider the state-space system

$$\begin{aligned} \mathbf{x}(t+1) &= A_0\mathbf{x}(t) + BD^+\mathbf{v}(t) + BD_\perp^*\mathbf{u}(t), \\ \mathbf{w}(t) &= -D^+C\mathbf{x}(t) + D^+\mathbf{v}(t) + D_\perp^*\mathbf{u}(t), \end{aligned}$$

where $\mathbf{u}(t) = F\mathbf{x}(t) + \mathbf{Q}(q)\mathbf{v}(t)$ is the full information control law. Then the closed-loop transfer matrix is the same as $\mathbf{T}_{\text{inv}}(z)$ in (7.37). It follows that the optimal ZF precoder can be designed using the result from Chap. 5 by taking $A = A_0$, $B_1 = BD^+$, $B_2 = BD_\perp^*$, $D_1 = D^+$, $D_2 = D_\perp^*$, and $C = -D^+C$. The next theorem follows and is dual to Theorem 7.1. The proof is left as an exercise (Problem 7.16).

Theorem 7.3. *Under the hypotheses of* Lemma *7.4, the optimal precoder has the same form as in (7.37) with*

$$\begin{aligned} F &= -(I + D_\perp B^* X B D_\perp^*)^{-1} D_\perp B^* X A_0, \\ Q &= -(I + D_\perp B^* X B D_\perp^*)^{-1} D_\perp B^* X B D^+, \end{aligned}$$

where $X \geq \mathbf{0}$ is the stabilizing solution to the ARE

$$X = A_0^* X (I + BD_\perp^* D_\perp B^* X)^{-1} A_0 + C^*(DD^*)^{-1}C.$$

The next example illustrates the advantage of the optimal precoder based on full-information control.

Example 7.4. The channel is assumed to be specified by

$$H(z) = \frac{1}{9}(1 + 2z^{-1} + 2.5z^{-2} + 2z^{-3} + z^{-4}).$$

The blocked channel matrix with $P = 3$ defined in (7.13) is obtained as

$$\underline{\mathbf{H}}(z) = \frac{1}{9}\begin{bmatrix} 1+2z^{-1} & 2.5z^{-1} & 2z^{-1}+z^{-2} \\ 2+z^{-1} & 1+2z^{-1} & 2.5z^{-1} \\ 2.5 & 2+z^{-1} & 1+2z^{-1} \end{bmatrix}$$

In order to illustrate the advantage of the optimal precoder, it is assumed that the linear receiver admits the transfer matrix

$$\mathbf{R}(z) = 9\begin{bmatrix} -\frac{16}{35} - \frac{56}{35}z^{-1} - \frac{24}{35}z^{-2} & \frac{4}{7} + \frac{12}{7}z^{-1} \\ \frac{68}{35} + \frac{130}{35}z^{-1} + \frac{48}{35}z^{-2} & -\frac{17}{7} - \frac{24}{7}z^{-1} \\ -\frac{34}{35} - \frac{96}{35}z^{-1} - \frac{36}{35}z^{-2} & \frac{12}{7} + \frac{18}{7}z^{-1} \end{bmatrix}^T.$$

It can be verified that the virtual P-input/M-output channel has the combined transfer matrix $\mathbf{H}(z) = \mathbf{R}(z)\underline{\mathbf{H}}(z)$ of size 2×3 given by

$$\mathbf{H}(z) = \begin{bmatrix} 1\ 0 & -0.971 - 0.742z^{-1} - 0.896z^{-2} - 1.z^{-3} - 0.686z^{-4} \\ 0\ 1 & 1.714 + 1.071z^{-1} + 0.571z^{-2} + 1.714z^{-3} \end{bmatrix}.$$

A naive precoder is clearly $\mathbf{G}(z) = G = \begin{bmatrix} I & 0 \end{bmatrix}^*$ of size $P \times M = 3 \times 2$ which results in $\|\mathbf{G}\|_2 = \sqrt{2}$. However, if the optimal ZF precoder $\mathbf{G}_{\text{opt}}(z) = \mathbf{H}_{\text{inv}}(z)$ is employed that has the smallest $\mathscr{H}_2$ norm among all right inverses of $\mathbf{H}(z)$, then $\|\mathbf{G}_{\text{opt}}\|_2 = 1.2292$. The BER performance for these two designs is evaluated, and the BER curves are plotted in Fig. 7.14 with solid line for the optimal precoder and dashed line for the naive precoder. It can be easily seen that the optimal design outperforms the naive one regarding the average BER.

If $P = 4$ and $M = 2$ are chosen, then the blocked channel transfer matrix defined in (7.13) is given by

$$\underline{\mathbf{H}}(z) = \frac{1}{9}\begin{bmatrix} 1+z^{-1} & 2z^{-1} & 2.5z^{-1} & 2z^{-1} \\ 2 & 1+z^{-1} & 2z^{-1} & 2.5z^{-1} \\ 2.5 & 2 & 1+z^{-1} & 2z^{-1} \\ 2 & 2.5 & 2 & 1+z^{-1} \end{bmatrix}.$$

Assume a simple receiver given by the following constant matrix:

$$\mathbf{R}(z) = R = \begin{bmatrix} 0\ 0 & 10 & -8 \\ 0\ 0 & -8 & 10 \end{bmatrix}. \tag{7.38}$$

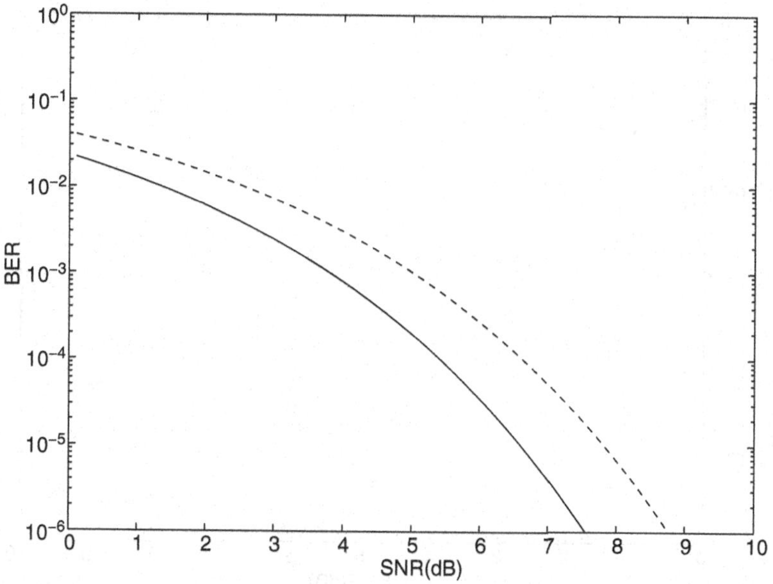

Fig. 7.14 BER comparison for the case $P = 3$ and $M = 2$

In this case, the virtual P-input/M-output channel has the combined transfer matrix $\mathbf{H}(z)$ given by

$$\mathbf{H}(z) = \mathbf{R}(z)\underline{\mathbf{H}}(z) = \frac{1}{9}\begin{bmatrix} 9 & 0 & -6+10z^{-1} & -8+12z^{-1} \\ 0 & 9 & 12-8z^{-1} & 10-6z^{-1} \end{bmatrix}.$$

For the naive design, $\|\mathbf{G}\|_2 = \sqrt{2}$ remains the same. However, for the optimal precoder, $\|\mathbf{G}_{\text{opt}}\|_2 = 1.179$ that becomes smaller. It can be expected that the corresponding BER curves as plotted in Fig. 7.15 demonstrate more clearly the superiority of the optimal ZF precoder.

7.2.3 *Linear MMSE Transceivers*

In the previous two subsections, some ZF condition needs to be assumed. While the ZF condition often holds especially for virtual MIMO channels, it may results in ZF equalizers or ZF precoders with excessive large gains in frequency domain, due to near loss of rank for $\mathbf{H}(e^{j\omega})$ at some frequency bands. In fact, it is possible for $\mathbf{H}(z)$ to have frequency nulls so that exact ZF is not possible, especially when $P \approx M$. This section considers design of linear MMSE transceivers without imposing the ZF condition.

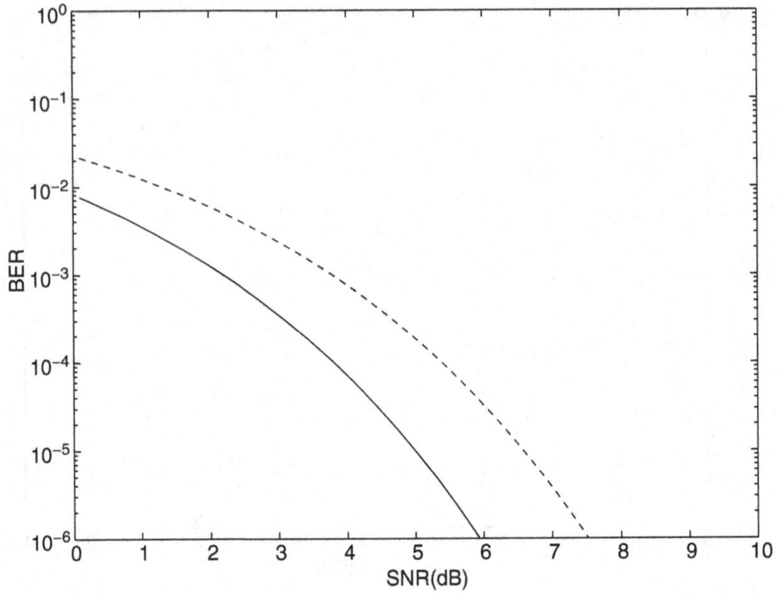

Fig. 7.15 BER comparison for $P = 4$ and $M = 2$

First, consider the case when the channel transfer matrix $\mathbf{H}(z)$ has size $P \times M$ with $P \geq M$. In the time domain, the received signal has the form:

$$\mathbf{y}(t) = \mathbf{H}(q)\mathbf{s}(t) + \mathbf{v}(t),$$

where $\mathbf{s}(t)$ is the symbol vector of dimension M, $\mathbf{v}(t)$ is the additive Gauss noise of dimension P, and q is the unit advance operator. Both $\mathbf{s}(t)$ and $\mathbf{v}(t)$ are temporally white, but may not be spatially white with Σ_s and Σ_v as their respective covariance. Since a causal and stable left inverse of $\mathbf{H}(z)$ is not possible, or undesirable, an alternative is to synthesize the linear receiver model $\mathbf{F}(z)$ that minimizes

$$J_F = \|(\mathbf{FH} - \mathbf{D})\Sigma_s^{1/2}\|_2^2 + \|\mathbf{F}\Sigma_v^{1/2}\|_2^2,$$

where $\mathbf{D}(z) = \text{diag}(z^{-d_1}, z^{-d_2}, \cdots, z^{-d_M})$ for some $d_i \geq 0$. Note that

$$\mathbf{e}_s(t) = [\mathbf{F}(q)\mathbf{H}(q) - \mathbf{D}(q)]\mathbf{s}(t - \mathbf{d}) + \mathbf{F}(q)\mathbf{v}(t) \tag{7.39}$$

is the error signal at the receiver site, and thus,

$$J_F = \text{E}\{\|\mathbf{e}_s(t)\|^2\} = \text{Tr}\{\text{E}[\mathbf{e}_s(t)\mathbf{e}_s(t)^*]\}$$

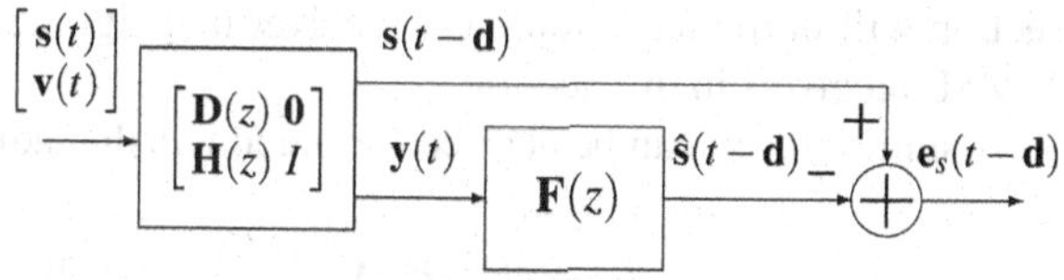

Fig. 7.16 Schematic diagram for linear MMSE receiver

by the temporal white assumption for both signal and noise, and by the independence of the signal and noise. If $\mathbf{F}(z)$ is taken to be causal and stable left inverse of $\mathbf{H}(z)$, then J_F reduces to the performance index for ZF channel equalizer. An MMSE receiver minimizes the mean-squared-error or J_F. It is recognized that minimization of J_F is an output estimation problem studied in Chap. 5. Indeed, a schematic block diagram is shown next:

The above is the same as the one in Fig. 5.4 by assuming

$$\mathbf{T}(z) = \begin{bmatrix}\mathbf{D}(z) & \mathbf{0}\\ \mathbf{H}(z) & I\end{bmatrix} = \begin{bmatrix}D_1 & \mathbf{0}\\ D_2 & I\end{bmatrix} + \begin{bmatrix}C_1\\ C_2\end{bmatrix}(zI-A)^{-1}\begin{bmatrix}B_1 & \mathbf{0}\end{bmatrix}. \tag{7.40}$$

There holds $\mathbf{D}(z) = D_1 + C_1(zI-A)^{-1}B_1$ and $\mathbf{H}(z) = D_2 + C_2(zI-A)^{-1}B_1$. The order of the state-space model for $\mathbf{T}(z)$ clearly depends on $\{d_i\}$. If the integer $d_i > 0$ for each i, then the output estimator in Fig. 7.16 is in fact a smoother. The larger the integers $\{d_i\}$, the smaller the J_F. However, large $\{d_i\}$ increase complexity of the receiver. Hence, there is a trade-off between the complexity and optimality of the linear MMSE receiver. For given $\{d_i\}$, the linear optimal receiver can be easily obtained as follows.

Theorem 7.5. *Consider* $\mathbf{T}(z)$ *in (7.40) with* A *a stability matrix. Let* $R = (I + D_2D_2^*)$, $\tilde{A} = A - B_1D_2^*R^{-1}C_2$, *and* $Y \geq \mathbf{0}$ *be the stabilizing solution to*

$$Y = \tilde{A}(I + YC_2^*R^{-1}C_2)^{-1}Y\tilde{A}^* + B_1(I - D_2^*R^{-1}D_2)B_1^*.$$

Then the optimal receiver in the sense of MMSE *is given by*

$$\mathbf{F}(z) = \left[\begin{array}{c|c} A + LC_2 & L\\ \hline C_1 + QC_2 & Q\end{array}\right]$$

where the stationary estimation gains are given by

$$\begin{bmatrix}L\\ Q\end{bmatrix} = \begin{bmatrix}AYC_2^* + B_1D_2^*\\ C_1YC_2^* + D_1D_2^*\end{bmatrix}(R + C_2YC_2^*)^{-1}.$$

The proof is left as an exercise (Problem 7.17). It is important to emphasize that the receiver $\mathbf{F}(z)$ in Theorem 7.5 is optimal among all linear receivers. For this reason, the non-ZF receiver in Theorem 7.5 is referred to as LE equalizer. The next section will introduce a nonlinear receiver that significantly outperforms the linear MMSE receivers in this section.

A similar result can be derived for precoders that admit performance index

$$J_G = \|(\mathbf{H}\mathbf{G}-\mathbf{D})\Sigma_s^{1/2}\|_2^2 + \rho^2\|\mathbf{G}\Sigma_s^{1/2}\|_2^2 \tag{7.41}$$

for some $\rho > 0$ where $\mathbf{H}(z)$ now has dimension $M \times P$ and $M \leq P$. If the ZF condition $\mathbf{H}(z)\mathbf{G}(z) = \mathbf{D}(z)$ holds, then ZF precoders will be resulted in. Due to possible frequency nulls, ZF is often not preferred. Since the error signal at the receiver is given by

$$\mathbf{e}(t-\mathbf{d}) = [\mathbf{H}(q)\mathbf{G}(q) - \mathbf{D}(q)]\mathbf{s}(t) + \mathbf{v}(t)$$

and $\mathbf{s}(t)$ and $\mathbf{v}(t)$ are independent, there holds

$$\mathrm{E}\{\|\mathbf{e}(t)\|^2\} = \|(\mathbf{H}\mathbf{G}-\mathbf{D})\Sigma_s^{1/2}\|_2^2 + \mathrm{E}\{\|\mathbf{v}(t)\|^2\}.$$

An MMSE criterion leads to minimization of $\|(\mathbf{H}\mathbf{G}-\mathbf{D})\Sigma_s^{1/2}\|_2^2$. In light of the argument in the previous subsection, $\|\mathbf{G}\Sigma_s^{1/2}\|_2^2$ needs to be minimized as well in order to maximize the SNR, leading to the performance index J_G in (7.41) with $\rho > 0$ a trade-off parameter.

Thus far, it is unclear how J_G can be minimized by synthesizing a causal and stable precoder $\mathbf{G}(z)$. Denote

$$\mathbf{P}(z) = \begin{bmatrix} (\mathbf{H}(z)\mathbf{G}(z) - \mathbf{D})(z)\Sigma_s^{1/2} \\ \rho\mathbf{G}(z)\Sigma_s^{1/2} \end{bmatrix} \implies J_G = \|\mathbf{P}\|_2^2.$$

Moreover, $\mathbf{P}(z) = \mathscr{F}_\ell[\mathbf{T}(z), \mathbf{G}(z)]$ by taking

$$\mathbf{T}(z) = \left[\begin{array}{c|c} \mathbf{T}_{11}(z) & \mathbf{T}_{12}(z) \\ \hline \mathbf{T}_{21}(z) & \mathbf{T}_{22}(z) \end{array}\right] = \left[\begin{array}{c|c} -\mathbf{D}(z)\Sigma_s^{1/2} & \mathbf{H}(z) \\ \mathbf{0} & \rho I \\ \hline \Sigma_s^{1/2} & \mathbf{0} \end{array}\right] \tag{7.42}$$

Hence, the design algorithm in Chap. 6 can be employed. To proceed, a realization is assumed for $\mathbf{T}(z)$ as follows:

$$\mathbf{T}(z) = \left[\begin{array}{c|c} D_1 & D_2 \\ \mathbf{0} & \rho I \\ \hline \Sigma_s^{1/2} & \mathbf{0} \end{array}\right] + \left[\begin{array}{c} C \\ \mathbf{0} \\ \hline \mathbf{0} \end{array}\right] (zI - A)^{-1} \left[\begin{array}{c|c} B_1 & B_2 \end{array}\right]. \tag{7.43}$$

The next result provides the optimal solution to minimization of J_G.

Theorem 7.6. *Consider* $\mathbf{T}(z)$ *in (7.43) of which* A *is a stability matrix and* Σ_s *is nonsingular. Let* $A_R = A - B_2 R_\rho^{-1} D_2^* C$, $R_\rho = \rho^2 I + D_2^* D_2$, $X \geq \mathbf{0}$ *be the stabilizing solution to*

$$X = A_R^* X (I + B_2 R_\rho^{-1} B_2 X)^{-1} + C^*(I - D_2 R_\rho^{-1} D_2^*) C. \tag{7.44}$$

Then the optimal precoder $\mathbf{G}(z)$ *that minimizes* J_G *is given by*

$$\mathbf{G}(z) = \left[\begin{array}{c|c} A + B_2 F & (B_2 F_0 + B_1)\Sigma_s^{-1/2} \\ \hline F & F_0 \Sigma_s^{-1/2} \end{array}\right] \tag{7.45}$$

where $\begin{bmatrix} F & F_0 \end{bmatrix} = -(R_\rho + B_2^* X B_2)^{-1} \left(B_2^* X \begin{bmatrix} A & B_1 \end{bmatrix} + D_2^* \begin{bmatrix} C & D_1 \end{bmatrix}\right)$.

Proof. The hypotheses imply that (A, B_2) is stabilizable and

$$\text{rank}\left\{\begin{bmatrix} A - e^{j\omega} I & B_2 \\ C & D_2 \\ \mathbf{0} & \rho I \end{bmatrix}\right\} = \text{full}$$

for all real ω. Hence, there exists a stabilizing solution $X \geq \mathbf{0}$ to the ARE in (7.44), and the full information feedback gains F and F_0 can be readily obtained. For output estimation, stability of A implies detectability of (C_2, A) even though $C_2 = \mathbf{0}$, and

$$\text{rank}\left\{\begin{bmatrix} A - e^{j\omega} I & B_1 \\ \mathbf{0} & \Sigma_s^{1/2} \end{bmatrix}\right\} = \text{full}$$

holds as well for all real ω. However, the corresponding filtering ARE has $Y = \mathbf{0}$ as the stabilizing solution. It follows that

$$L = -B_1 \Sigma_s^{-1/2}, \quad L_0 = -F_0 \Sigma_s^{-1/2}$$

are the output estimation gains. The expression of the optimal controller in (7.45) follows from Sect. 6.1. □

The performance index J_G of the non-ZF precoder also depends on $\{d_i\}$. Recall that $\mathbf{D}(z) = \text{diag}(z^{-d_1}, z^{-d_2}, \cdots, z^{-d_M})$. Hence, there is again a trade-off between the performance index J_G and the complexity of $\mathbf{G}(z)$.

7.3 Decision Feedback Receivers

Recall the results in the previous section. In the case of ZF precoders, the received signal has the form of $\hat{\mathbf{s}}(t) = \mathbf{s}(t) + \mathbf{v}(t)$ where $\mathbf{v}(t)$ is the additive white Gauss noise. If the noise covariance is given by $\Sigma_v = \sigma_v^2 I$, then each symbol in the symbol vector $\mathbf{s}(t)$ can be detected individually. In fact, optimal symbol detection amounts to quantization for each entry of $\hat{\mathbf{s}}(t)$. See Problem 7.18 in Exercises. However, if $\Sigma_v \neq \sigma_v^2 I$ and symbol vector has large dimension, then optimal detection becomes not only difficult, but also too complex to implement in practice. This section will present a Kalman filtering approach to vector symbol detection using the notion of decision feedback. A more complex situation is when the additive Gauss noise $\mathbf{v}(t)$ is not even temporal white. This is the case for ZF channel equalizers and also for the non-ZF transceivers in the previous subsection. Clearly, the past information can be utilized to improve symbol detection. An interesting method, termed as decision feedback equalization, will be presented in this section to show that significant improvement can be achieved using decision feedback equalization over linear MMSE receivers.

7.3.1 Vector Symbol Detection

Consider the received vector $\hat{\mathbf{s}} = \mathbf{s} + \mathbf{v}$ with dimension M. Assume that $\mathbf{v}$ is Gauss distributed with mean zero and covariance Σ_v, and each entry of $\mathbf{s}$ is from a finite alphabet table. While optimal detection amounts to quantization in the case of $\Sigma_v = \sigma_v^2 I$, it becomes much more complex in the case of $\Sigma_v \neq \sigma_v^2 I$ when M is large. This subsection develops a Kalman filtering method for vector symbol detection augmented by decision feedback.

Consider the following fictitious state-space process:

$$\mathbf{s}(m+1) = \mathbf{s}(m) = \mathbf{s}, \qquad y(m) = \check{\mathbf{s}}(m) = \mathbf{e}_m \mathbf{s} \tag{7.46}$$

for $0 < m < M$ where $\mathbf{e}_k$ is a row vector consisting of zeros except the kth element equal to one and $\check{\mathbf{s}}(m)$ is the detected symbol. The simple idea of decision feedback assumes that the detected symbol $\check{s}(m) = s_m := \mathbf{e}_m \mathbf{s}$ and is fed back in the form of output measurement $y(m) = \check{s}(m)$. Since Σ_v is a full matrix, detected symbol will be beneficial to estimation of the state vector, i.e., the symbol vector $\mathbf{s}$, thereby improving detection of the subsequent symbol. This idea indeed works through successive improvement for estimation of $\mathbf{s}$ based on feedback of each detected symbol. Let $\hat{\Sigma}_0 = \Sigma_v$ and $\hat{\mathbf{s}}_{1|0} = \hat{\mathbf{s}}$ which are prior data. Set $y(1) = \check{s}(1) = \text{Quan}[\mathbf{e}_1 \hat{\mathbf{s}}_{1|0}]$. An application of the state estimator in Theorem 5.6 leads to the spatial Kalman filter:

$$\begin{aligned} \hat{\mathbf{s}}_{m+1|m} &= \hat{\mathbf{s}}_{m|m-1} - K_m[y(m) - \mathbf{e}_m \hat{\mathbf{s}}_{m|m-1}], \\ K_m &= -\hat{\Sigma}_m \mathbf{e}_m^* (\mathbf{e}_m \hat{\Sigma}_{m-1} \mathbf{e}_m^*)^{-1}, \\ \hat{\Sigma}_{m+1} &= (I + K_m \mathbf{e}_m)\hat{\Sigma}_m, \quad y(m+1) = \text{Quan}[\mathbf{e}_{m+1} \hat{\mathbf{s}}_{m+1|m}], \end{aligned} \tag{7.47}$$

for $m = 1, 2, \cdots, M-1$. The above skips the calculation of $\hat{s}_{m|m}$ due to

$$\begin{aligned}\hat{s}_{m|m} &= \mathbf{e}_{m+1}\hat{\mathbf{s}}_{m|m-1} - \mathbf{e}_{m+1}K_m\left(y(m) - \mathbf{e}_m\hat{\mathbf{s}}_{m|m-1}\right)\\ &= \mathbf{e}_{m+1}\left[\hat{\mathbf{s}}_{m|m-1} - K_m\left(y(m) - \mathbf{e}_m\hat{\mathbf{s}}_{m|m-1}\right)\right]\\ &= \mathbf{e}_{m+1}\hat{\mathbf{s}}_{m+1|m}.\end{aligned}$$

The estimator in (7.47) is obtained by noting that $\mathbf{x}(t) \to \mathbf{s}$ is the state to estimate, and $\mathbf{y}(t) \to y(m) = \check{s}_m$ is the observation. Thus, with time index t replaced by spatial index m, Theorem 5.6 can be readily applied using

$$\begin{aligned}A_t &\to A_m = I, \quad B_t \to B_m = \mathbf{0},\\ C_t &\to C_{2m} = \mathbf{e}_m, \quad D_t \to 0,\\ L_t &\to L_m = K_m, \quad R_t \to R_m = 0.\end{aligned}$$

The spatial Kalman filter in (7.47) will be termed as sequential interference cancellation (SIC), or Kalman SIC in short, due to its ability to mitigate the noise effect in vector symbol detection. Its weakness lies in the possible error propagation if a detection error occurs. A moment of reflection prompts the idea of ordering by detecting the symbol with the least error probability first. Two different orderings will be considered. The first is based on SNR. Denote $\kappa(\cdot)$ as the new ordering. Then

$$\kappa(1) = \arg\min\left\{\mathbf{e}_i\Sigma_v\mathbf{e}_i^* : \ 1 \le i \le M\right\}.$$

In addition, the spatial Kalman SIC in (7.47) will be replaced by

$$\begin{aligned}\hat{\mathbf{s}}_{m+1|m} &= \hat{\mathbf{s}}_{m|m-1} - K_m[\check{s}_{\kappa(m)} - \mathbf{e}_{\kappa(m)}\hat{\mathbf{s}}_{m|m-1}],\\ K_m &= -\hat{\Sigma}_m\mathbf{e}_{\kappa(m)}^*\left(\mathbf{e}_{\kappa(m)}\hat{\Sigma}_m\mathbf{e}_{\kappa(m)}^*\right)^{-1},\\ \hat{\Sigma}_{m+1} &= (I + K_m\mathbf{e}_{\kappa(m)})\hat{\Sigma}_m,\\ \kappa(m+1) &= \arg\min\left\{\mathbf{e}_j\hat{\Sigma}_{m+1}\mathbf{e}_j^* : \ j \notin \{\kappa(i)\}_{i=1}^m\right\},\\ \check{s}_{\kappa(m+1)} &= \text{Quan}[\mathbf{e}_{\kappa(m+1)}\hat{\mathbf{s}}_{m+1|m}],\end{aligned} \tag{7.48}$$

initialized by $\hat{\Sigma}_0 = \Sigma_v$, $\hat{\mathbf{s}}_{1|0} = \hat{\mathbf{s}}$, and $\check{s}(1) = \text{Quan}[\mathbf{e}_{\kappa(1)}\hat{\mathbf{s}}_{1|0}]$.

The second ordering is log-likelihood ratio (LLR). Denote $\mathcal{T}_w = \{w_i\}_{i=1}^f$ as the set of finite alphabetical words. Then $s_m \in \mathcal{T}_w$ for each m, and $\hat{s}_m = s_m + v_m$ for some Gauss random variable v_m with mean zero and variance $\mathbf{e}_m\Sigma_v\mathbf{e}_m^*$. At each m, the MAP decision for $s(i)$ given $\hat{s}_{i|m}$ has the expression

$$\check{s}_i = \arg\max_{w_k} \Pr\{s_i = w_k|\hat{s}_{i|m}\}$$

where $\Pr\{\cdot\}$ stands for probability. Denote

$$\gamma_{i|m}(w_k) = \ln\left(\frac{\Pr\{s_i = \check{s}(i)|\hat{s}_{i|m}\}}{\Pr\{s_i = w_k|\hat{s}_{i|m}\}}\right)$$

as the pairwise LLR. It is left as an exercise (Problem 7.19) to show that the probability of symbol error after knowing $\hat{s}_{i|d}$ is given by

$$\Pr\{\check{s}_i \neq s_i\} = 1 - \frac{1}{\sum_{k=1}^{f} e^{-\gamma_{i|m}(w_k)}}. \tag{7.49}$$

The above holds even if the noise is not Gauss distributed. The LLR-based ordering chooses to first detect the symbol that minimizes $\sum_{k=1}^{f} e^{-\gamma_{i|m}(w_k)}$ over $\{w_k\}_{k=1}^{f}$ at each iteration of the spatial Kalman filter. This choice minimizes the symbol error probability and thus the probability of error propagation. In the case of Gauss noise as assumed in this subsection, there holds

$$\gamma_{i|m}(w_k) = \frac{|\hat{s}_{i|m} - w_k| - |\hat{s}_{i|m} - \check{s}_i|}{\mathbf{e}_m \hat{\Sigma}_m \mathbf{e}_m^*}.$$

The spatial Kalman filter remains the same except the index at the mth step in (7.48) that is replaced by

$$\kappa(m+1) = \arg\min_j \left\{\Pr\{\check{s}_j \neq s_j | \hat{s}_{j|m}\},\ j \notin \{\kappa(i)\}_{i=1}^{m}\right\}.$$

Example 7.7. This example considers the output estimate of the LE receiver from Sect. 2.3 as the symbol vector. For simplicity, BPSK is assumed, and thus, the each symbol takes value of $\pm\sigma_s$ with σ_s^2 the symbol power. The estimate of the LE receiver at time t is

$$\hat{\mathbf{s}}(t-d) = \mathbf{s}(t-d) + \hat{\mathbf{v}}(t)$$

with $d = 1$. Note that $\hat{\mathbf{v}}(t)$ represents the estimation error, different from the noise process at the input of the LE receiver. Although the Gauss assumption does not hold, the LE receiver implies approximate Gauss distribution of the estimation error, provided that the channel order is high or M and P are large, in light of Example 2.12 in Chap. 2. This example employs the MIMO channel with $M = P = 8$ (input/output) and $L = 1$ (channel length) to demonstrate the results for both the LE receiver from Sect. 7.2.3 and for the SIC-based vector symbol detection. Whereas such M, P, and L are not high enough, the simulation results do help to illustrate the improvement due to Kalman SIC. A total of 30 channel models are randomly generated in which each coefficient is Rayleigh distributed with the total variance normalized to one. The BER performance are then averaged over these 30 channel models. Combining the LE with the Kalman SIC in this subsection yields the BER curves shown in Fig. 7.17 below.

The SNR is defined as $20\log_{10}(\sigma_s^2/\sigma_v^2)$ with $\sigma_s^2 I$ as the covariance of the symbol vector and $\sigma_v^2 I$ as the covariance of the noise at the input of the receiver (different from $\hat{\mathbf{v}}$ that is the error noise at the output of the receiver). The improvement due to ordering is also obvious.

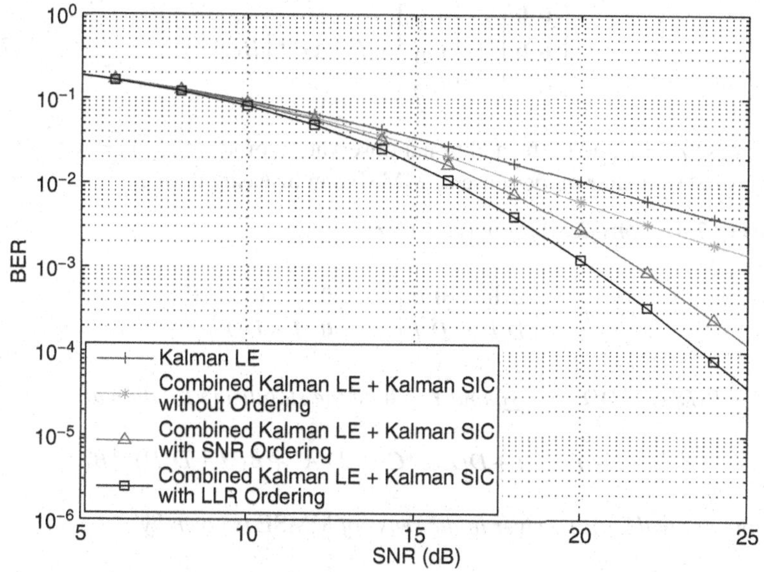

Fig. 7.17 BER performance of the combined Kalman LE and Kalman SIC

Fig. 7.18 Block diagram of the DFE receiver

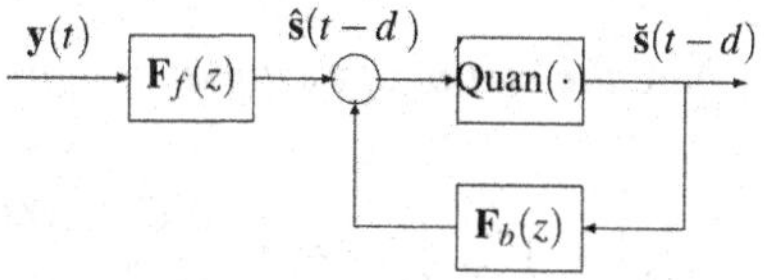

7.3.2 *Decision Feedback Equalization*

The previous subsection shows the BER improvement when the detected symbol is fed back to aid the detection of the subsequent symbol. This improvement is achieved by taking advantage of the correlated nature of the noise vector. Now consider the LE receiver studied in the previous section in which the noise is colored temporally. Hence, there is an incentive to feedback the detected symbol vector in the previous time instant to help estimation of the current symbol vector. This is indeed the case. To be specific, the LE receiver having the structure of the output estimator is shown in Fig. 7.16. When the detected symbol is fed back, the input to the estimator includes $\check{\mathbf{s}}(t-d-1)$, in addition to the received signal $\mathbf{y}(t)$. Figure 7.18 shows the structure of the time-invariant decision feedback equalizer (DFE).

The process model for output estimation in Fig. 7.16 is now changed to

$$\mathbf{T}(z) = \begin{bmatrix} z^{-d}I & \mathbf{0} \\ \hline z^{-(d+1)}I & \mathbf{0} \\ \mathbf{H}(z) & I \end{bmatrix} = \begin{bmatrix} D_1 \\ \hline D_2 \end{bmatrix} + \begin{bmatrix} C_1 \\ \hline C_2 \end{bmatrix} (zI - A)^{-1} B$$

$$= \left[\begin{array}{c} \mathbf{0}\ \mathbf{0} \\ \hline \mathbf{0}\ \mathbf{0} \\ D\ I \end{array}\right] + \left[\begin{array}{c} C_1 \\ \hline C_{21} \\ C_{22} \end{array}\right] (zI - A)^{-1} \left[B_1\ \mathbf{0} \right] \tag{7.50}$$

for the case $d \geq 1$ due to the additional observation of $\check{\mathbf{s}}(t-d-1) = \mathbf{s}(t-d-1)$. The optimal estimator in the sense of MMSE can be easily derived.

Theorem 7.8. *Consider* $\mathbf{T}(z)$ *in (7.50). Let*

$$R = \begin{bmatrix} \mathbf{0}\ \mathbf{0} \\ D\ I \end{bmatrix} \begin{bmatrix} \mathbf{0}\ \mathbf{0} \\ D\ I \end{bmatrix}^* = \begin{bmatrix} \mathbf{0} & \mathbf{0} \\ \mathbf{0} & (I + DD^*) \end{bmatrix},$$

$\tilde{A} = A - B_1 D^*(I + DD^*)^{-1} C_{22}$, *and* $Y \geq \mathbf{0}$ *be the stabilizing solution to*

$$Y = \tilde{A}[I + YC_{22}^*(I + DD^*)^{-1}C_{22}]^{-1} Y \tilde{A}^* + B_1(I + D^*D)^{-1}B_1^*.$$

Then the optimal DFE *receiver in the sense of* MMSE *is given by*

$$\mathbf{F}(z) = \left[\mathbf{F}_b(z)\ \mathbf{F}_f(z) \right] = \left[\begin{array}{c|c} A + L_1C_{21} + L_2C_{22} & L_1\ L_2 \\ \hline C_1 + Q_1C_{21} + Q_2C_{22} & Q_1\ Q_2 \end{array}\right]$$

where the stationary estimation gains are given by

$$\begin{bmatrix} L \\ Q \end{bmatrix} = \begin{bmatrix} L_1\ L_2 \\ Q_1\ Q_2 \end{bmatrix} = \begin{bmatrix} AYC_{21}^* & AYC_{22}^* + B_1D^* \\ C_1YC_{21}^* & \mathbf{0} \end{bmatrix} \begin{bmatrix} C_{21}YC_{21}^* & C_{21}YC_{22}^* \\ C_{22}YC_{21}^* & DD^* + C_{22}YC_{22}^* \end{bmatrix}^{-1}.$$

Proof. Since R is singular, pseudoinverse has to be used leading to

$$R^+ = \begin{bmatrix} \mathbf{0} & \mathbf{0} \\ \mathbf{0} & (I + DD^*)^{-1} \end{bmatrix}.$$

By using the stationary version of the output estimation result in Theorem 5.14,

$$\tilde{A} = A - BD_2^*R^+C_2 = A - B_1D^*(I + DD^*)^{-1}C_{22}, \tag{7.51}$$
$$\tilde{B} = B[I - D_2^*R^+D_2] = B_1(I + D^*D)^{-1} \left[I\ -D^* \right],$$
$$\tilde{B}\tilde{B}^* = B_1(I + D^*D)^{-1}B_1^*, \quad C_2^*R^+C_2 = C_{22}^*(I + DD^*)^{-1}C_{22}. \tag{7.52}$$

Hence, the filtering ARE in the theorem can be verified, and expressions of the state and output estimation gains in (7.51) can be easily derived. □

In light of Theorem 7.8, the estimator $\mathbf{F}(z) = \left[\mathbf{F}_b(z)\ \mathbf{F}_f(z) \right]$ has an IIR structure in general. However, this is not the case for DFE, if the channel model is FIR, which holds in the case for wireless channels. Recall the derivation of the channel model in Chap. 1. For the sake of generality, consider the time-varying channel model of length L with M input/P output:

$$\mathbf{H}_t(q) = \sum_{i=0}^{L} H_i(t)q^{-i}. \tag{7.53}$$

The next result is surprising and shows that the optimal DFE receiver in the sense of MMSE has an FIR structure, even if the channel experiences time-selective fading.

Theorem 7.9. *Let the $P \times M$ channel model be the same as in (7.53) and*

$$\mathbf{y}(t) = \sum_{i=0}^{L} H_i(t)\mathbf{s}(t-i) + \mathbf{v}(t)$$

be the received signal where the symbol vector $\mathbf{s}(t)$ and noise vector $\mathbf{v}(t)$ are temporally white and uncorrelated to each other. Consider the DFE *receiver*

$$\hat{\mathbf{s}}(t-d) = \sum_{i=0}^{L_g} F_i(t)\mathbf{y}(t-i) + \sum_{j=1}^{L_b} G_j(t)\check{\mathbf{s}}(t-d-j). \tag{7.54}$$

If $\check{\mathbf{s}}(t-d-i) = \mathbf{s}(t-d-i)$ for all integer $i \geq 1$, then the MSE *of the estimation error $\mathrm{E}\{\|\mathbf{s}(t-d) - \hat{\mathbf{s}}(t-d)\|^2\}$ is minimized by the* FIR DFE *receiver in form of (7.54) among all linear receivers, provided that $L_g \geq d$ and $L_b \geq L$.*

Proof. Assume $L_g = d$, $L_b = L$, and $\check{\mathbf{s}}(t-d-i) = \mathbf{s}(t-d-i)$ for $i \geq 1$. Then the estimate $\hat{\mathbf{s}}(t-d)$ in (7.54) can be written as

$$\begin{aligned}
\mathbf{r}(t) = \hat{\mathbf{s}}(t-d) &= \sum_{i=0}^{d} F_i(t)\mathbf{y}(t-i) + \sum_{j=1}^{L} G_j(t)\mathbf{s}(t-d-j) \\
&= \sum_{i=0}^{d} F_i(t)\left(\sum_{k=0}^{L} H_k(t-i)\mathbf{s}(t-i-k) + \mathbf{v}(t-i)\right) + \sum_{j=d+1}^{L+d} G_{j-d}(t)\mathbf{s}(t-j) \\
&= \sum_{\ell=0}^{L+d} \Gamma_\ell(t)\mathbf{s}(t-\ell) + \sum_{j=d+1}^{L+d} G_{j-d}(t)\mathbf{s}(t-j) + \sum_{i=0}^{d} F_i(t)\mathbf{v}(t-i) \\
&= \sum_{\ell=0}^{d} \Gamma_\ell(t)\mathbf{s}(t-\ell) + \sum_{j=d+1}^{L+d} [\Gamma_j(t) + G_{j-d}(t)]\mathbf{s}(t-j) + \sum_{i=0}^{d} F_i(t)\mathbf{v}(t-i),
\end{aligned}$$

where $\Gamma_\ell(t) = \sum_{i=0}^{\ell} F_i(t)H_{\ell-i}(t-i)$ by defining $F_i(t) = \mathbf{0}\ \forall\, i > d$. Taking

$$G_j(t) = -\Gamma_{j+d}(t) = -\sum_{i=0}^{d} F_i(t)H_{j+d-i}(t-i) \tag{7.55}$$

for all $j \in \{i\}_{i=1}^{L}$ concludes that

$$\mathbf{r}(t) = \hat{\mathbf{s}}(t-d) = \sum_{\ell=0}^{d} \Gamma_\ell(t)\mathbf{s}(t-\ell) + \sum_{i=0}^{d} F_i(t)\mathbf{v}(t-i).$$

Let $\mathbf{e}_s(t-d) = \mathbf{s}(t-d) - \hat{\mathbf{s}}(t-d)$ be the error signal. The above implies

$$\mathrm{E}\{\mathbf{e}(t-d)\mathbf{s}(t-d-i)^*\} = \mathbf{0} \ \ \forall\, i \geq 1 \tag{7.56}$$

by $\mathrm{E}\{\mathbf{s}(t)\mathbf{s}(\tau)^*\} = \mathbf{0}$ whenever $t \neq \tau$ and $\mathrm{E}\{\mathbf{s}(t)\mathbf{v}(\tau)^*\} = \mathbf{0}$ for all (t, τ). If $\{F_i(t)\}_{i=0}^{d}$ are chosen such that

$$\mathrm{E}\{\mathbf{e}(t-d)\mathbf{y}(t-i)^*\} = \mathbf{0} \ \ \forall\, i \geq 0, \tag{7.57}$$

then the output estimate $\hat{\mathbf{s}}(t-d)$ satisfies the orthogonality condition required for the MMSE estimate. See Problem 5.6 in Chap. 5. Hence, the estimate (7.54) is indeed optimal provided that (7.57) has a unique solution $\{F_i(t)\}_{i=0}^{d}$, since then $\{Q_j(t)\}_{j=1}^{L}$ can be obtained according to (7.55). For this purpose, denote $\Omega = \begin{bmatrix} \mathbf{0} \cdots \mathbf{0} \ I \end{bmatrix}$ and

$$\underline{\mathbf{s}}_d(t) = \begin{bmatrix} \mathbf{s}(t) \\ \vdots \\ \mathbf{s}(t-d) \end{bmatrix}, \quad \underline{\mathbf{v}}_d(t) = \begin{bmatrix} \mathbf{v}(t) \\ \vdots \\ \mathbf{v}(t-d) \end{bmatrix}. \tag{7.58}$$

Thus, $\mathbf{s}(t-d) = \Omega \underline{\mathbf{s}}_d(t)$. In addition, denote $\underline{F}(t) = \begin{bmatrix} F_0(t) \cdots F_d(t) \end{bmatrix}$ and

$$\underline{H}_d(t) = \begin{bmatrix} H_0(t) & H_1(t) & \cdots & H_d(t) \\ \mathbf{0} & H_0(t-1) & \cdots & H_{d-1}(t-1) \\ \vdots & \ddots & \ddots & \vdots \\ \mathbf{0} & \cdots & \mathbf{0} & H_0(t-d) \end{bmatrix}$$

by the convention $H_\ell(k) = 0$ if $\ell > L$. Then under the relation in (7.55) and $\mathbf{e}_s(t-d) = \mathbf{s}(t-d) - \hat{\mathbf{s}}(t-d)$, there holds

$$\begin{aligned} \mathbf{e}_s(t-d) &= \Omega \underline{\mathbf{s}}_d(t) - \underline{F}(t)\left[\underline{H}_d(t)\underline{\mathbf{s}}_d(t) + \underline{\mathbf{v}}_d(t)\right] \\ &= \left[\Omega - \underline{F}(t)\underline{H}_d(t)\right]\underline{\mathbf{s}}_d(t) - \underline{F}(t)\underline{\mathbf{v}}_d(t). \end{aligned} \tag{7.59}$$

In light of the assumption for $\mathbf{s}(t)$ and $\mathbf{v}(t)$,

$$\mathrm{E}\{\mathbf{s}(k)\mathbf{s}(i)^*\} = \Sigma_s \delta(k-i), \quad \mathrm{E}\{\mathbf{v}(k)\mathbf{v}(i)^*\} = \Sigma_v \delta(k-i)$$

where $\delta(\cdot)$ is the Kronecker delta function. Recall the Kronecker product symbol $\otimes$. It is now claimed that (7.57) holds, if and only if

$$\underline{F}(t) = \Omega\,(I_d \otimes \Sigma_s)\,\underline{H}_d(t)^*\,[I_d \otimes \Sigma_v + \underline{H}_d(t)\,(I_d \otimes \Sigma_s)\,\underline{H}_d(t)^*]^{-1}\,. \tag{7.60}$$

Indeed, denoting $\underline{\mathbf{y}}_{\ell+1}(t) = \left[\mathbf{y}(t)^* \cdots \mathbf{y}(t-\ell-1)^*\right]^*$ with $\ell > d$ yields

$$\underline{\mathbf{y}}_{\ell+1}(t) = \begin{bmatrix} \underline{H}_d(t) \\ \mathbf{0} \end{bmatrix} \underline{\mathbf{s}}_d(t) + \begin{bmatrix} \underline{\mathbf{v}}_d(t) \\ \underline{\mathbf{v}}_{\ell-d}(t-d-1) \end{bmatrix} + \underline{H}_R(t)\underline{\mathbf{s}}_{L+\ell-d}(t-d-1)$$

for some $\underline{H}_R(t)$ dependent on the channel coefficients where $\underline{\mathbf{s}}_{L+\ell-d}(t-d-1)$ and $\underline{\mathbf{v}}_{\ell-d}(t-d-1)$ are defined along the lines of (7.58). Hence, the orthogonality condition (7.57) holds, if and only if

$$\mathrm{E}\{\mathbf{e}(t-d)\underline{\mathbf{y}}_{\ell+1}(t)^*\} = \mathbf{0}$$

for each integer $\ell > d$. The above is equivalent to

$$[\Omega - \underline{F}(t)\underline{H}_d(t)]\,(I_d \otimes \Sigma_s)\left[\underline{H}_d(t)^*\ \mathbf{0}\right] = \underline{F}(t)\left[I_d \otimes \Sigma_v\ \mathbf{0}\right]$$

by substituting the expression of $\underline{\mathbf{y}}_{\ell+1}(t)$ and that of $\mathbf{e}_s(t-d)$ in (7.59), and by noting that $\mathrm{E}\{\underline{\mathbf{s}}_d(t)\underline{\mathbf{s}}_d(t)^*\} = I_d \otimes \Sigma_s$, $\mathrm{E}\{\underline{\mathbf{v}}_d(t)\underline{\mathbf{v}}_d(t)^*\} = I_d \otimes \Sigma_v$, and mutual orthogonality of $\underline{\mathbf{s}}_d(t)$, $\underline{\mathbf{v}}_d(t)$, $\underline{\mathbf{s}}_{L+\ell-d}(t-d-1)$, and $\underline{\mathbf{v}}_{\ell-d}(t-d-1)$. The above has a unique solution identical to the one in (7.60). Since $\ell > d$ is an arbitrary integer, (7.57) holds that concludes the proof. □

In the case of the time-invariant channel model, the feedforward filter (FFF) and feedback filter (FBF) are time invariant as well and are given by $\{F_i\}$ and $\{G_i\}$, respectively. The proof of Theorem 7.9 shows that the roles of FFF and FBF are different. The former is designed to satisfy the orthogonality condition (7.57) and the latter to (7.56). Together, they achieve the same MMSE performance as the Kalman DFE in Theorem 7.8, provided that $L_g \geq d$ and $L_b \geq L$. In fact, $L_g = d$ and $L_b = L$ can be chosen which are of the minimum order. Moreover, the Kalman DFE has an FIR structure, if the channel model has an FIR structure.

Example 7.10. Consider a 2-input/2-output ($M = 2$, $P = 2$) system over the normalized Rayleigh fading channels with $L = 2$. Similar to the previous example, BPSK symbols are again assumed and 30 channel models are randomly generated. Figure 7.19 shows the BER performance of the Kalman DFE in Theorem 7.8 and the finite length DFE (FLDFE) in Theorem 7.9. The results are averaged over 30 randomly generated channels. The lower four curves illustrate the BER performance of the Kalman DFE with respect to the detection delay d. The performance improvement is clearly seen as the detection delay increases. It is emphasized that there is only one parameter, the detection delay d, can be adjusted. Since the detection delay is directly related to the dimension of the state vector, better performance (larger delay) is achieved at the expense of higher complexity.

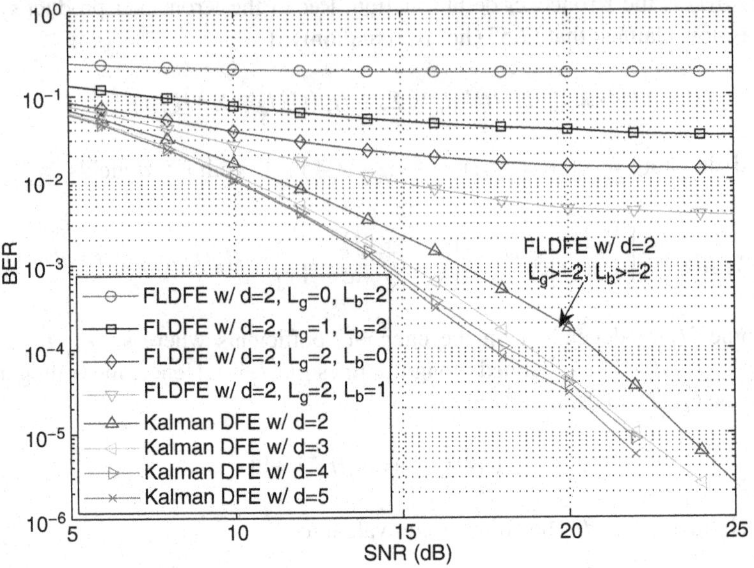

Fig. 7.19 BER performance of the combined Kalman receiver

Fig. 7.20 Combined Kalman DFE and SIC receiver

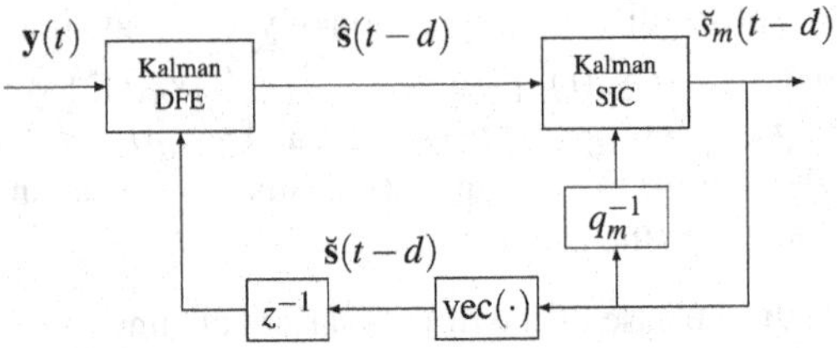

For comparison, Fig. 7.19 also shows the BER performance of the FLDFE. It is seen that the FLDFE never outperforms the Kalman DFE, and its performance improves with the increase of the FFF and the FBF lengths. When $L_g \geq d$ and $L_b \geq L$, the FLDFE has the identical performance as the Kalman DFE, which confirms Theorem 7.9. It is also observed that reducing the FFF length causes much more performance loss than reducing the FBF length. In fact, when the FBF length reduces to zero, a DFE becomes an LE that is still able to suppress the ISI to some extent.

While the FLDFE in (7.54) is attractive, the Kalman DFE in Theorem 7.8 has some advantages. It can be easily generalized to time-varying channels using Theorem 5.14. Moreover, both DRE and ARE associated with Kalman DFE are easy to compute. Finally, Kalman DFE provides the error covariance matrix for $\hat{\mathbf{s}}(t-d)$ that can be utilized in its quantization by using the SIC method in the previous subsection. See the schematic block diagram in Fig. 7.20 in which q_m^{-1} denotes unit spatial delay.

For simplicity, consider time-invariant channel model

$$\mathbf{H}(z) = H_0 + H_1 z^{-1} + \cdots + H_L z^{-L}.$$

A block state-space realization (A, B_1, C, D) is given by

$$A = \begin{bmatrix} \mathbf{0} & \cdots & \cdots & \mathbf{0} \\ I_M & \ddots & & \vdots \\ & \ddots & \ddots & \vdots \\ * & & I_M & \mathbf{0} \end{bmatrix}, \quad B_1 = \begin{bmatrix} I_M \\ \mathbf{0} \\ \vdots \\ \mathbf{0} \end{bmatrix},$$

$$C = \begin{bmatrix} H_1 & \cdots & H_L & \mathbf{0} & \cdots & \mathbf{0} \end{bmatrix}, \quad D = H_0.$$

The associated state vector can be easily identified as

$$\mathbf{x}(t) = \begin{bmatrix} \mathbf{s}(t-1)^* & \mathbf{s}(t-2)^* & \cdots & \mathbf{s}(t-n)^* \end{bmatrix}^*$$

with $n > L$ and $n > d > 0$ giving rise to the state-space equation

$$\mathbf{x}(t+1) = A\mathbf{x}(t) + \begin{bmatrix} B_1 & \mathbf{0} \end{bmatrix} \begin{bmatrix} \mathbf{s}(t) \\ \mathbf{v}(t) \end{bmatrix}, \quad \mathbf{y}(t) = C\mathbf{x}(t) + \begin{bmatrix} D & I \end{bmatrix} \begin{bmatrix} \mathbf{s}(t) \\ \mathbf{v}(t) \end{bmatrix}.$$

The above agrees with $\mathbf{T}(z)$ in (7.50) by taking C_1 and C_{21} such that

$$C_1\mathbf{x}(t) = \mathbf{s}(t-d), \quad C_{21}\mathbf{x}(t) = \mathbf{s}(t-d-1),$$

and by taking $C_{22} = C$. The following result shows more than just the performance improvement when the Kalman DFE and SIC are combined.

Theorem 7.11. *Let the* FFF $\mathbf{F}_f(z)$ *and* FBF $\mathbf{F}_b(z)$ *be specified in* Theorem *7.8 with* $Y \geq \mathbf{0}$ *be the stabilizing solution to the corresponding filtering* ARE. *The linear* MMSE *estimator for* $\hat{\mathbf{s}}(t-d)$ *is given by*

$$\begin{bmatrix} \hat{\mathbf{x}}(t+1) \\ \hat{\mathbf{s}}(t-d) \end{bmatrix} = \begin{bmatrix} A + LC_2 \\ C_1 + QC_2 \end{bmatrix} \hat{\mathbf{x}}(t) + \begin{bmatrix} L_1 \\ Q_1 \end{bmatrix} \check{\mathbf{s}}(t-d-1) + \begin{bmatrix} L_2 \\ Q_2 \end{bmatrix} \mathbf{y}(t), \tag{7.61}$$

where $L = \begin{bmatrix} L_1 & L_2 \end{bmatrix}$, $Q = \begin{bmatrix} Q_1 & Q_2 \end{bmatrix}$, *and* $C_2 = \begin{bmatrix} C_{21}^* & C^* \end{bmatrix}^*$. *Assume that* $\check{s}_m(t-d) =$ Quan$[\hat{s}_m(t-d)] = s_m(t-d)$ *is correct for each m where the subscript m denotes the mth entry of the corresponding vector. Then* $\hat{s}_m(t-d)$ *can be obtained from the spatial* Kalman *filter:*

$$\begin{aligned}
\hat{\mathbf{b}}_{m+1|m} &= \hat{\mathbf{b}}_{m|m-1} - L_m[\breve{b}_m - \mathbf{e}_m\hat{\mathbf{b}}_{m|m-1}],\\
L_m &= -\hat{\Sigma}_m\mathbf{e}_m^*(\mathbf{e}_m\hat{\Sigma}_m\mathbf{e}_m^*)^{-1},\\
\hat{\Sigma}_{m+1} &= (I + L_m\mathbf{e}_m)\hat{\Sigma}_m, \quad \breve{b}_{m+1} = \text{Quan}[\mathbf{e}_{m+1}\hat{\mathbf{b}}_{m+1|m}],
\end{aligned} \tag{7.62}$$

for $m = 1, 2, \cdots, M-1$, *initialized by* $\hat{\Sigma}_0 = C_1YC_1^*$ *and* $\hat{\mathbf{b}}_{1|0} = \hat{\mathbf{s}}(t-d) = C_1\hat{\mathbf{x}}(t)$. *Moreover, the combined* Kalman *receiver in* Fig. *7.20 based on* Kalman DFE *in (7.61) and the above recursive* SIC *is jointly optimal in the sense of achieving* MMSE.

Proof. Because the M symbols of $\hat{\mathbf{s}}_{t-d}$ are detected and fed back sequentially, a fictitious upsampling with ratio of M can be assumed. For each integer m such that $1 \le m < M$, the state vector remains the same, but observations are replaced by $s_m(t-d)$, the mth element of $\mathbf{s}_{t-d}$, in light of the assumption of the correct detection of the previous symbols. Since $\hat{\mathbf{x}}(t) = \mathbf{x}(t) + \hat{\mathbf{v}}_x(t)$ for some error vector $\hat{\mathbf{v}}_x(t)$ with covariance Y being the stabilizing solution to the corresponding ARE in Theorem 7.8, an equivalent MMSE estimation problem arises for the following state-space model:

$$\hat{\mathbf{x}}(t|m+1) = \hat{\mathbf{x}}(t|m), \quad y(t|m) = \mathbf{e}_mC_1\hat{\mathbf{x}}(t|m) - \mathbf{e}_mC_1\hat{\mathbf{v}}_x(t) \tag{7.63}$$

for $m = 1, 2, \cdots, M-1$ that is initialized by a priori estimate $\hat{\mathbf{x}}(t|1) = \hat{\mathbf{x}}(t)$. Similar to (7.47), the above leads to the Kalman SIC:

$$\begin{aligned}
\widehat{\mathbf{x}}_{m+1|m}(t) &= [I + K_m(t)\mathbf{e}_mC_1]\widehat{\mathbf{x}}_{m|m-1}(t) - K_m(t)\breve{b}_m,\\
K_m(t) &= -Y_{m-1}(\mathbf{e}_mC_1)^*[(\mathbf{e}_mC_1)Y_{m-1}(\mathbf{e}_mC_1)^*]^{-1},\\
Y_{m+1} &= [I + K_m(t)\mathbf{e}_mC_1]Y_m,\\
\breve{b}_{m+1} &= \text{Quan}[\mathbf{e}_{m+1}C_1\widehat{\mathbf{x}}_{m+1|m}(t)]
\end{aligned} \tag{7.64}$$

for $m = 1, 2, \cdots, M-1$, initialized by $\widehat{\mathbf{x}}_{1|0} = \hat{\mathbf{x}}(t)$, $\breve{b}_1 = \text{Quan}[\mathbf{e}_1C_1\widehat{\mathbf{x}}_{1|0}]$, and $Y_0 = Y$. Multiplying C_1 from left to the first three equations of (7.64) and multiplying C_1^* from right to the third yield the spatial Kalman filter in (7.62) by taking $L_m = C_1K_m(t)$ and $\hat{\Sigma}_i = C_1Y_iC_1^*$ for $i = m$ and $i = m-1$. Joint optimality follows from the fact that quantized symbols are fed back to the Kalman DFE as input in vector form, and thus, the Kalman DFE has no access of $\breve{s}_m(t-d)$ until $m = M$ which preserves the optimality of the Kalman DFE as well. □

The combined Kalman DFE and SIC will be referred to as combined Kalman receiver. Recall that the SIC part can be improved, if SNR or LLR orderings are employed. The use of ordering will alter the Kalman SIC in Fig. 7.20, but its joint optimality remains for the combined Kalman receiver with ordering. The next example illustrates the performance improvement when the combined Kalman receiver is employed.

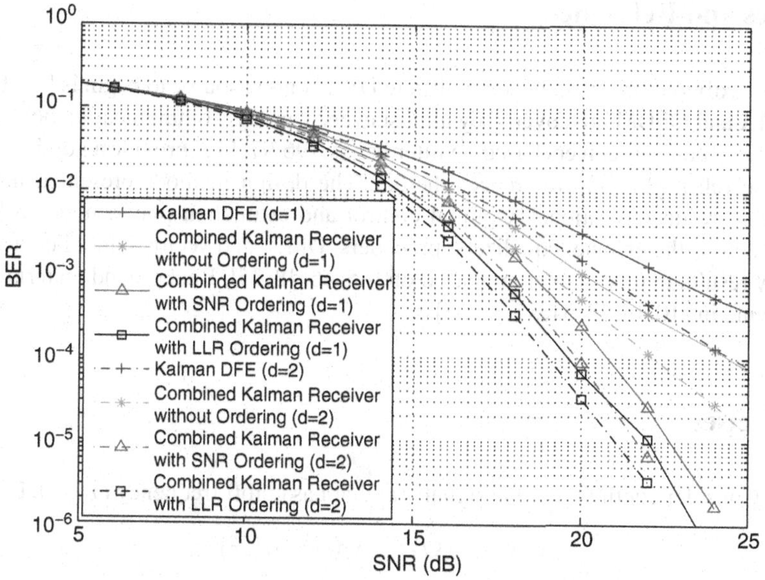

Fig. 7.21 BER performance of the combined Kalman receiver

Example 7.12. This example considers time-varying 8×8 channels with

$$\mathbf{H}_t(q) = H_0(t) + H_1(t)q^{-1}.$$

Thus, the channel length $L = 1$. Each of the channel coefficient, denoted by $h_i(t)$, is assumed to be Rayleigh random variable, generated via

$$h_i(t+1) = \alpha_i h_i(t) + \beta_i \eta_i(t)$$

where $\eta(t)$ has mean 0 and variance 1, $\alpha_i = 0.9999$, and $\alpha_i^2 + \beta_i^2 = 1$. This is the result of the so called WSSUS, and autoregressive model in time evolution, observed from large number of experiments. It is assumed that the full channel information is known at the receiver, in which case the Kalman DFE is a time-varying output estimator (Problem 7.21 in Exercises). Figure 7.21 shows the BER performance averaged again over 30 randomly generated channels.

For any given detection delay, it is observed that the Kalman SIC improves the performance of the Kalman DFE. Detection ordering further improves the performance of the Kalman SIC. It is also observed that the LLR-based detection ordering outperforms the SNR-based detection ordering consistently. As expected, increasing the detection delay d improves the performance of each receiver at the cost of complexity. Our numerical results show that the performance gain due to detection ordering is influenced more by the channel length L than by the detection delay d, but this gain becomes smaller as the channel length L becomes larger.

Notes and References

Many textbooks are available on multirate DSP. A good source to begin is [47, 112]. CDMA and OFDM are covered by almost every wireless communication book [98, 107, 111]. For transceivers based on filter banks, [96, 97, 119] provide a good sample of research papers and references therein. The design methods presented in this chapter are unconventional. Optimal control and estimation play a major role in synthesis of the channel equalizers, precoders, DFEs, and vector symbol detectors. Many results in this chapter are based on [45, 46, 48, 75]. Background material can be found in [10, 14, 15, 31, 37, 77, 93].

Exercises

7.1. Consider continuous-time signal $s(t)$ that has cutoff frequency $\omega_c > 0$. Let

$$x_1(k) = s(kT_1), \quad x_2(t) = s(kT_2)$$

be two sampled version of $s(t)$ where $k = 0, \pm 1, \cdots$. Suppose that $T_1 = LT_2$ with L integer and sampling frequency

$$\omega_{s1} = 2\pi/T_1 > 2\omega_c, \quad \omega_{s2} = L\omega_{s1}.$$

Show that if an ideal interpolator is employed with $\{x_1(k)\}$ as input, then the output is identical to $\{x_2(k)\}$. Also, test this fact in Matlab by using $s(t)$ as a combination of sinusoidal signals.

7.2. Prove the equality in (7.1). (*Hint:* Note first the following representation for the periodic Kroneker delta function:

$$\delta_p(k) = \frac{1}{M}\sum_{i=0}^{M-1} W_M^{ik} = \begin{cases} 1, & k = 0, \pm M, \cdots, \\ 0, & \text{elsewhere.} \end{cases}$$

Note next that the frequency response of $y_D(k)$ is given by

$$Y_D(e^{j\omega}) = \sum_{k=-\infty}^{\infty} \delta_p(k)x(k)e^{-jk\omega/M}$$

from which the equality in (7.1) can be derived.)

7.3. (Resampling) Fig. 7.22 in next page shows the fractional sampling rate converter termed resampling in which L and M are coprime. Show that the low-pass filter in ideal resampler has passband $[-\frac{L\pi}{M}, \frac{L\pi}{M}]$ in the case $M > L$. The Matlab command "resample" can be used to implement the resampler.

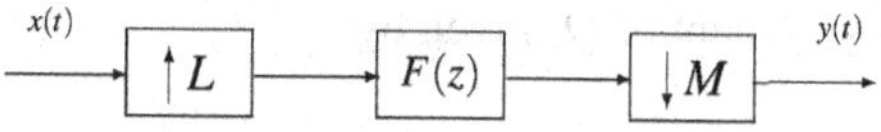

Fig. 7.22 Resampling with expander, low-pass filter, and decimator

Fig. 7.23 Two-channel QMF bank

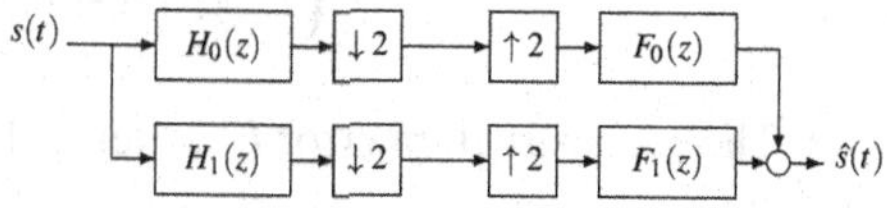

7.4. For the case $M = 2$, the QMF bank is shown in Fig. 7.23.

(i) Assume $H_0(z)$ is a low-pass filter. Show that

$$H_1(z) = H_0(-z), \quad F_1(z) = -2H_0(-z)$$

are both high-pass filters, $F_0(z) = 2H_1(z)$ is low-pass filter, and

$$\hat{S}(z) = [H_0(z)H_0(z) - H_0(-z)H_0(-z)]\, S(z).$$

The above reduces the design of four filters into that of $H_0(z)$ only.

(ii) Let $H_0(z)$ be FIR having the form

$$H_0(z) = \sum_{i=0}^{n} h_0(i)z^{-i}, \quad h_0(i) = h_0(n-i),$$

where $i = 0, 1, \ldots, \frac{n+1}{2}$ with n odd. Show that the DTFT of $\{h_0(i)\}$ has a linear phase and

$$\left[H_0(e^{j\omega})\right]^2 = |H_0(e^{j\omega})|^2 e^{-jn\omega} \quad \Longrightarrow$$

$$\hat{S}(e^{j\omega}) = e^{-jn\omega}\left[|H_0(e^{j\omega})|^2 + |H_0(e^{j(\omega+\pi)})|^2\right] S(e^{j\omega}).$$

(iii) Show that $|H_0(e^{j\omega})| = 2|\underline{h}_0^* C(\omega)|$ with $m = \frac{n+1}{2}$ and

$$\underline{h}_0 = \begin{bmatrix} h_0(0) \\ h_0(1) \\ \vdots \\ h_0(m-1) \end{bmatrix}, \quad C(\omega) = \begin{bmatrix} \cos(m-1/2)\omega \\ \cos(m-3/2)\omega \\ \vdots \\ \cos(1-1/2)\omega \end{bmatrix},$$

and thus, the stopband filtering error is given by

$$E_s := \frac{1}{\pi}\int_{\omega_s}^{\pi} |H_0(e^{j\omega})|^2 \, d\omega = \underline{h}_0^* Q_s \underline{h}_0$$

where the (i,j)th element of Q_s given by

$$Q_s(i,k) = \begin{cases} -\frac{1}{2\pi}\left(\frac{\sin(k-i)\omega_s}{k-i} + \frac{\sin(2m-i-k+1)\omega_s}{2m-i-k+1}\right), & i \neq k, \\ \frac{1}{2\pi}\left(\pi - \omega_s - \frac{\sin 2(m-i+0.5)\omega_s}{2(m-i+0.5)}\right), & i = k. \end{cases}$$

(iv) The reconstruction error $E_{\rm r}$ is defined by

$$E_{\rm r} = \frac{1}{\pi}\int_0^{\pi}\left(\left[|H_0(e^{j\omega})|^2 + |H_0(e^{j(\omega+\pi)})|^2\right] - 1\right)^2 d\omega.$$

Show that $E_{\rm r} = (1-2\underline{h}^*\underline{h})^2 + 8\sum_{i=1}^{m-1}\left(\underline{h}^* S^{2i}\underline{h}\right)^2$ where

$$S = \begin{bmatrix} 0 & 1 & 0 & \cdots & 0 \\ 0 & 0 & 1 & \cdots & 0 \\ \vdots & \vdots & \ddots & \ddots & \vdots \\ \vdots & \vdots & \cdots & 0 & 1 \\ 0 & \cdots & \cdots & \cdots & 0 \end{bmatrix}, \quad \underline{h} = \begin{bmatrix} h_0(0) \\ h_0(1) \\ \vdots \\ h_0(2m-2) \\ h_0(2m-1) \end{bmatrix}.$$

(*Hint:* See Ref. [43].)

7.5. For the two-channel QMF bank in the previous problem, consider the total error $E = (1-\alpha)E_{\rm r} + \alpha E_s$ where $0 < \alpha < 1$ is a design parameter.

(i) Show that

$$E = (1-\alpha)\left[(1-4\underline{h}_0^*\underline{h}_0)^2 + 8\sum_{i=1}^{m-1}\left(\underline{h}_0^* M^* S^{2i} M\underline{h}_0\right)^2\right] + \alpha\underline{h}_0^* Q_s\underline{h}_0$$

where $\underline{h} = M\underline{h}_0$, $M = \begin{bmatrix} I_m & \tilde{I}_m \end{bmatrix}^*$, and

$$\tilde{I}_m = \begin{bmatrix} 0 & \cdots & 0 & 1 \\ \vdots & \cdots & 1 & 0 \\ \vdots & \cdots & \cdots & \vdots \\ 1 & 0 & \cdots & 0 \end{bmatrix}.$$

(ii) Define the design index $J(\underline{h}_0) = E - (1-\alpha)$. Show that

$$J(\underline{h}_0) = \underline{h}_0^* P[\underline{h}_0,\underline{h}_0]\underline{h}_0 + \underline{h}_0^* Q\underline{h}_0, \quad P[\underline{h}_0,\underline{h}_0] = \sum_{k=0}^{m-1} P_k\underline{h}_0\underline{h}_0^* P_k$$

where $Q = Q_+ + Q_-$ with $Q_+ = \alpha Q_s$, $Q_- = -8(1-\alpha)I$, and

$$P_0 = 4\sqrt{1-\alpha}I, \quad P_k = \sqrt{2(1-\alpha)}M^*\left(S^{2k} + (S^*)^{2k}\right)M.$$

(iii) Use Matlab to design $H_0(z)$ that minimizes $E = (1-\alpha)E_r + \alpha E_s$ where $0 < \alpha < 1$ is a design parameter with $\omega_s = 0.55\pi$ and $n = 55$. An iterative algorithm can be adopted. Denote $x = \underline{h}_0$. The algorithm searches for the optimal x as follows:

- Step 1: Choose initial guess $x = x_0 \neq 0$ such that $J_0 = \tilde{J}(x_0) < 0$.
- Step 2: For $i = 0, 1, 2, \ldots$, do the following:
  - Step 2a: Find $x = x_{i+1}$ such that it minimizes the auxiliary index function

$$\tilde{J}_i[x_i](x) = x^* \left(P[x_i, x_i] + \frac{1}{2}Q_+ \right) x + x_i^* Q_- x + \frac{1}{2} x_i^* Q_+ x_i.$$

The optimal solution is given by

$$x = x_{i+1} = -\left(2P[x_i, x_i] + Q_+\right)^{-1} Q_- x_i. \tag{7.65}$$

  - Step 2b: If $\|x_{i+1} - x_i\| \leq \varepsilon$, with ε a prespecified error tolerance, stop; otherwise, set $i := i+1$ and repeat Step 2a.

(*Hint:* See Ref. [43].)

7.6. Two block diagrams are equivalent, if the same input produces the same output. Show the following:

(i) The M-fold decimator followed by a filter $F(z)$ is equivalent to the filter $F(z^M)$ followed by the M-fold decimator.
(ii) The L-fold expander followed by a filter $F(z^L)$ is equivalent to the filter of $F(z)$ followed by the L-fold decimator.

7.7. Prove the relation in (7.11).

7.8. Suppose that the continuous-time channel is flat with impulse response $a_m\delta(t-\tau_m)$ where $\tau_m = (\kappa_m + \varepsilon_m)T_c$ with $\kappa_m \geq 0$ an integer and $0 \leq \varepsilon_m < 1$. For simplicity, assume that $p(t) = [\mathbf{1}(t) - \mathbf{1}(t-T_c)]$ is the chip wave function with $\mathbf{1}(t)$, the unit step function. Show that the impulse response of the discretized channel is given by

$$h_{m,k} = \begin{cases} a_m(1-\varepsilon_m)T_c, & k = \kappa_m, \\ a_m\varepsilon_m T_c, & k = \kappa_m + 1. \end{cases}$$

(*Hint:* Using the matched filter in Chap. 1, the impulse response of the discretized channel can be shown to be

$$h_{m,k} = a_m \int p(\tau - \tau_m)\bar{p}(\tau - k)\, \mathrm{d}\tau$$

from which the expression of $h_{m,k}$ can be derived.)

7.9. For $\underline{\mathbf{G}}_m(z)$, the mth column of $\underline{\mathbf{G}}(z)$ in (7.15), show that

$$\underline{\mathbf{G}}_m(z) = \underline{\mathbf{G}_\mathbf{m}}(z)\underline{c}_m = \underline{\mathbf{C}}_m(z)\underline{\mathbf{G}}_m^{(1)}(z)$$

where $\underline{\mathbf{G}}_m^{(1)}(z)$ is the first column of $\underline{\mathbf{G}_\mathbf{m}}(z)$ and

$$\underline{\mathbf{C}}_m(z) = \begin{bmatrix} c_m(0) & z^{-1}c_m(P-1) & \cdots & z^{-1}c_m(1) \\ c_m(1) & c_m(0) & \ddots & \vdots \\ \vdots & \ddots & \ddots & z^{-1}c_m(P-1) \\ c_m(P-1) & \cdots & c_m(1) & c_m(0) \end{bmatrix}.$$

7.10. Prove the relation in (7.16) where $\{H_k\}_{k=0}^{N-1}$ are the N-point DFT of $\{h_i\}_{i=0}^{n-1}$.

7.11. (i) Let $\{v_i\}_{i=0}^{N-1}$ be i.i.d. Gauss random variables of mean zero and variance σ^2. Show that its normalized DFT, denoted by $\{\widehat{v}_k\}_{i=0}^{N-1}$, is also i.i.d. Gauss with the same mean and variance.

(ii) Repeat (i) for the case when v_i is complex and has i.i.d. real and imaginary parts with mean zero and variance $\sigma^2/2$.

(iii) Consider detection of $\widehat{b}$ based on noisy observation:

$$\widehat{y} = \widehat{H}\widehat{b} + \widehat{v}.$$

Assume that $\widehat{H} \neq 0$ is known, $\widehat{v}$ is the same as in (ii), and both real and imaginary parts of $\widehat{b}$ are binary. Show that the above is the same as

$$\begin{bmatrix} \widehat{y}_r \\ \widehat{y}_i \end{bmatrix} = \begin{bmatrix} H_r & -H_i \\ H_i & H_r \end{bmatrix} \begin{bmatrix} \widehat{b}_r \\ \widehat{b}_i \end{bmatrix}$$

where $\widehat{y} = \widehat{y}_r + j\widehat{y}_i$ and $\widehat{b} = \widehat{b}_r + j\widehat{b}_i$.

7.12. Consider ZF equalization in Sect. 7.2.1 in which $\mathbf{v}(k)$ has mean zero but covariance $\Sigma_v \neq I$. Find the optimal ZF equalizer that minimizes

$$J_F = \mathrm{Tr}\left\{\frac{1}{2\pi}\int_{-\pi}^{\pi} \mathbf{F}(e^{j\omega})\Sigma_v\mathbf{F}(e^{j\omega})^*\, d\omega\right\}.$$

7.13. Consider $\mathbf{H}(z) = D + C(zI - A)^{-1}B$ with size $P \times M$, $P > M$, and A, a stability matrix. Assume that $\mathrm{rank}\{D\} = M$ and $D^+D = I$ for some left inverse D^+. Show that $\mathbf{H}(z)$ is strictly minimum phase, if and only if $(D_\perp^* C, A_0)$ is detectable, where $A_0 = A - BD^+C$ and $D_\perp$ is the same as the one in (7.20).

7.14. Consider $\mathbf{H}(z) = D + C(zI - A)^{-1}B$ with size $P \times M$ and A, a stability matrix. Assume that (A, B, C, D) is strictly minimum phase, except that $D \neq \mathbf{0}$ is not full rank, i.e., $\mathbf{H}(z)$ has zeros at $z = \infty$.

(i) Let $P < M$, and $X \geq \mathbf{0}$ be the stabilizing solution to

$$X = A^*XA + C^*C - (A^*XB + C^*D)(D^*D + B^*XB)^+(B^*XA + D^*C).$$

The stabilizing solution $X \geq \mathbf{0}$ can be obtained as the limit of the iteration:

$$\begin{aligned} F_i &= -(D^*D + B^*XB)^+(B^*XA + D^*C), \\ X_{i+1} &= (A + BF_i)^*X_i(A + BF_i) + (C + DF_i)^*X_i(C + DF_i), \end{aligned}$$

initialized by $X_0 \geq \mathbf{0}$ satisfying $X_0 = A^*X_0A + C^*C$. Let

$$F = -(D^*D + B^*XB)^+(B^*XA + D^*C),$$

and Ω of size $P \times M$ satisfy $\Omega^*\Omega = (D^*D + B^*XB)$. Show that

$$\mathbf{H}_{\rm i}(z) = \left[\begin{array}{c|c} A + BF & B \\ \hline C + DF & D \end{array}\right] \Omega^+, \quad \mathbf{H}_{\rm o}(z) = \Omega \left[\begin{array}{c|c} A & B \\ \hline -F & I \end{array}\right]$$

satisfy inner-outer factorization relation $\mathbf{H}(z) = \mathbf{H}_{\rm i}(z)\mathbf{H}_{\rm o}(z)$ with $\mathbf{H}_{\rm i}(z)$ square inner, consisting of only z^{-d_i} or zeros, and $\mathbf{H}_{\rm o}(z)$ of size $P \times M$ has full rank P at $z = \infty$. Note that Ω has rank M.

(ii) Let $P > M$, and $Y \geq \mathbf{0}$ be the stabilizing solution to

$$Y = AYA^* + BB^* - (AYC^* + BD^*)(DD^* + CYC^*)^+(CYA^* + DB^*).$$

The stabilizing solution $Y \geq \mathbf{0}$ can be obtained as the limit of the iteration:

$$\begin{aligned} L_i &= -(AY_iC^* + BD^*)(DD^* + CY_iC^*)^+, \\ Y_{i+1} &= (A + L_iC)Y_i(A + LC_i)^* + (B + L_iD)Y_i(B + L_iD)^*, \end{aligned}$$

initialized by $Y_0 \geq \mathbf{0}$ satisfying $Y_0 = AYA^* + BB^*$. Let

$$L = -(AYC^* + BD^*)(DD^* + CYC^*)^+,$$

and Ω of size $P \times M$ satisfy $\Omega\Omega^* = (DD^* + CYC^*)$. Show that

$$\mathbf{H}_{\rm i}(z) = \Omega^+ \left[\begin{array}{c|c} A + LC & B + LD \\ \hline C & D \end{array}\right], \quad \mathbf{H}_{\rm o}(z) = \left[\begin{array}{c|c} A & -L \\ \hline C & I \end{array}\right] \Omega$$

satisfy inner-outer factorization relation $\mathbf{H}(z) = \mathbf{H}_o(z)\mathbf{H}_o(z)$ with $\mathbf{H}_i(z)$ square inner satisfying $\det[\mathbf{H}_i(z)] = z^{-(d_1+\cdots+d_M)}$, and $\mathbf{H}_o(z)$ of size $P \times M$ has full rank M at $z = \infty$. Note that $\mathbf{H}_i(z)$ can be a permutation of $\text{diag}(z^{-d_1}, \cdots, z^{-d_M})$.

7.15. Prove Lemma 7.3.

7.16. Prove Theorem 7.3.

7.17. Prove Theorem 7.5 by using the stationary version of Theorem 5.14.

7.18. Suppose that $\hat{s} = s + v$ where s is from a finite alphabet table $\mathscr{T}$ and v is Gauss distributed with mean zero and variance σ_v^2. Show that the optimal detection is given by

$$\breve{s} = \text{Quan}(\hat{s}) := \arg\min_{s \in \mathscr{T}} |\hat{s} - s|.$$

7.19. (i) Prove (7.49) in the general case when the noise is not Gauss distributed.
(ii) For BPSK modulation, $\mathscr{T} = \{\pm\sigma_s\}$ with σ_s^2 the symbol power. Show that (7.49) can be simplified to

$$\Pr\{\breve{s}_i \neq s_i\} = \frac{1}{1 + e^{|\gamma_i|}}, \quad \gamma_i = \frac{4\sigma_s \text{Re}\{\hat{s}_{i|m}\}}{\mathbf{e}_m \hat{\Sigma}_m \mathbf{e}_m^*}.$$

(*Hint*: The proof can be found in Ref. [67].)

7.20. Consider the received signal given by

$$\hat{\mathbf{s}} = H\mathbf{s} + \mathbf{v},$$

where the channel matrix H has the full column rank, and $\mathbf{v}$ is gauss distributed with mean zero and covariance $\sigma_v^2 I$. Show that the SIC in (7.47) can be generalized to the following:

$$\begin{aligned}
\hat{\mathbf{s}}_{m+1|m} &= \hat{\mathbf{s}}_{m|m-1} - K_m[y(m) - \mathbf{e}_m \hat{\mathbf{s}}_{m|m-1}], \\
K_m &= -\hat{\Sigma}_m \mathbf{e}_m^* (\mathbf{e}_m \hat{\Sigma}_{m-1} \mathbf{e}_m^*)^{-1}, \\
\hat{\Sigma}_{m+1} &= (I + K_m \mathbf{e}_m)\hat{\Sigma}_m, \quad y(m+1) = \text{Quan}[\mathbf{e}_{m+1}\hat{\mathbf{s}}_{m+1|m}],
\end{aligned}$$

for $m = 1, 2, \cdots, M-1$, initialized with $\hat{\mathbf{s}}_{1|0} = (H^*H)^{-1}H^*\hat{\mathbf{s}}$ and $\hat{\Sigma}_0 = \sigma_v^2 (H^*H)^{-1}$. (*Hint:* Use the transform $\widehat{\mathbf{s}} = (H^*H)^{-1}H^*\hat{\mathbf{s}}$ to obtain the new model $\widehat{\mathbf{s}} = \mathbf{s} + \widehat{\mathbf{v}}$ that is the same as in Sect. 7.3.1, and then apply Theorem 5.6 to the fictitious state-space model $\mathbf{x}(m+1) = \mathbf{x}(m) = \mathbf{s}$ with $y(m) = \breve{s}_m = s_m$ as the observation. Note that $H^+ = (H^*H)^{-1}H^*$ has the smallest Frobenius norm among all left inverses of H, and thus the smallest covariance for the new noise vector $\widehat{\mathbf{v}}$.)

7.21. Suppose that the channel model is time varying given by

$$\mathbf{H}_t(q) = \sum_{\ell=0}^{L} H_i(t)q^{-i}.$$

Derive the optimal Kalman DFE by generalizing the result in Theorem 7.8.

Chapter 8
System Identification

Mathematical models are essential to modern science and engineering, and have been very successful in advancing the technology that has had profound impact to our society. A serious question to be addressed in this chapter is how to obtain the model for a given physical process. There are two basic approaches. The first one is based on principles in physics or other sciences. The inverted pendulum in Chap. 1 provides an example to this approach. Its advantage lies in its capability to model nonlinear systems and preservation of the physical parameters. However, this approach can be costly and time consuming. The second approach is based on input and output data to extrapolate the underlying system model. This approach treats the system as a black box and is only concerned with its input–output behaviors. While experiments need to be carried out and restrictions on input signals may apply, the second approach overcomes the weakness of the first approach.

This chapter examines the input/output approach to modeling of the physical system which is commonly referred to as system identification. In this approach, the mathematical model is first parameterized and then estimated based on input/output experimental data. Autoregressive moving average (ARMA) models are often used in feedback control systems due to their ability to capture the system behavior with lower order and fewer parameters than the MA models or transversal filters. On the other hand, wireless channels are more suitable to be described by MA models due to the quick die out of the CIR. Many identification algorithms exist, and most of them uses squared error as the identification criterion. The squared error includes energy or mean power of the model matching error that results in least squares (LS), or total LS (TLS), or MMSE algorithms. These algorithms will be presented and analyzed in two different sections. For ease of the presentation, only real matrices and variable are considered, but the results are readily extendable to the case of complex valued signals and systems. A very important result in estimation is the well-known Cramér–Rao lower bound (CRLB). See Sect. 6 in Appendix B for details.

G. Gu, *Discrete-Time Linear Systems: Theory and Design with Applications*,
DOI 10.1007/978-1-4614-2281-5_8,

8.1 Least-Squares-Based Identification

Consider an m-input/p-output system with plant model

$$\mathbf{H}(z) = \mathbf{M}(z)^{-1}\mathbf{N}(z) = \left(I - \sum_{k=1}^{n_\mu} M_k z^{-k}\right)^{-1} \left(\sum_{k=1}^{n_\nu} N_k z^{-k}\right).$$

The parameter matrices are those of $\{M_k\}_{k=1}^{n_\mu}$ with dimension $p \times p$, and of $\{N_k\}_{k=1}^{n_\nu}$ with dimension $p \times m$. Due to the existence of measurement error, the input and output are related through the following difference equation:

$$\mathbf{y}(t) = \sum_{k=1}^{n_\mu} M_k \mathbf{y}(t-k) + \sum_{k=1}^{n_\nu} N_k \mathbf{u}(t-k) + \mathbf{v}(t), \tag{8.1}$$

where $\mathbf{v}(t)$ is a WSS process with mean zero and covariance $\sigma^2 I$. Define

$$\Theta = \begin{bmatrix} M_1 & \cdots & M_{n_\mu} & N_1 & \cdots & N_{n_\nu} \end{bmatrix}$$

as the true parameter matrix of dimension $p \times (n_\mu + n_\nu)m$, and

$$\underline{\phi}(t) = \begin{bmatrix} \mathbf{y}(t-1)' & \cdots & \mathbf{y}(t-n_\mu)' & \mathbf{u}(t-1)' & \cdots & \mathbf{u}(t-n_\nu)' \end{bmatrix}'$$

as the regressor vector of dimension $(n_\mu + n_\nu)m$. There holds

$$\mathbf{y}(t) = \Theta\underline{\phi}(t) + \mathbf{v}(t). \tag{8.2}$$

The linearity is owing to the linearity of the system. Although the above signal model is derived from the input/output model (8.1), this section assumes temporarily that $\underline{\phi}(t)$ is noise-free. This assumption holds for the FIR model, including wireless channels. The dependence of the signal model (8.2) on input/output model (8.1) will be revisited in the next section.

8.1.1 LS and RLS Algorithms

The LS algorithm is perhaps the most widely adopted in the practice of system identification. It has an interpretation of maximum likelihood estimate (MLE), if the observation noise is Gauss distributed. Consider Gauss-distributed $\mathbf{v}(t)$ that is temporally white with mean zero and covariance $\Sigma_v = \sigma^2 I$ that is known. The MLE is equivalent to minimizing the squared error index

$$J_\Theta(t_0, t_f) = \frac{1}{2\sigma^2} \sum_{t=t_0}^{t_f} [\mathbf{y}(t) - \Theta\underline{\phi}(t)]'[\mathbf{y}(t) - \Theta\underline{\phi}(t)]$$

by noting that MLE maximizes the PDF given by

$$f_V(\mathbf{y};\Theta) = \frac{1}{\sqrt{(2\pi\sigma^2)^{(t_f - t_0 + 1)p}}} \exp\left\{-J_\Theta(t_0, t_f)\right\},$$

and by noting the independence of $\{\mathbf{v}(t)\}$. For convenience, denote

$$\begin{aligned} Y_{0,f} &= \left[\mathbf{y}(t_0)\ \mathbf{y}(t_0+1)\ \cdots\ \mathbf{y}(t_f)\right], \\ \Phi_{0,f} &= \left[\underline{\phi}(t_0)\ \underline{\phi}(t_0+1)\ \cdots\ \underline{\phi}(t_f)\right]. \end{aligned} \tag{8.3}$$

Then the squared error index can be rewritten as

$$J_\Theta(t_0, t_f) = \frac{1}{2\sigma^2} \mathrm{Tr}\left\{\left(Y_{0,f} - \Theta\Phi_{0,f}\right)\left(Y_{0,f} - \Theta\Phi_{0,f}\right)'\right\}. \tag{8.4}$$

The LS solution $\Theta = \Theta_{\mathrm{LS}}$ minimizes $J_\Theta(t_0, t_f)$ and is the MLE.

Recall the definition in (B.50). It is left as an exercise to show (Problem 8.1)

$$\begin{aligned} &\frac{\partial \mathrm{Tr}\{AXB\}}{\partial X} = A'B', \quad \frac{\partial \mathrm{Tr}\{AX'B\}}{\partial X} = BA, \\ &\frac{\partial \mathrm{Tr}\{AXBX'\}}{\partial X} = A'XB' + AXB. \end{aligned}$$

See Sect. B.5 in Appendix B. The next result provides the MLE in the general case.

Theorem 8.1. *Consider the signal model in (8.2) where* $\mathbf{v}(t)$ *is* Gauss *distributed with mean zero and covariance* $\Sigma_v = \mathrm{diag}(\sigma_1^2, \cdots, \sigma_p^2)$*. Then the* MLE *estimate based on* $\{(\mathbf{y}(t), \underline{\phi}(t)\}_{t=t_0}^{t_f}$ *is the* LS *solution and given by*

$$\Theta_{\mathrm{LS}} = Y_{0,f}\Phi_{0,f}'(\Phi_{0,f}\Phi_{0,f}')^{-1}, \tag{8.5}$$

provided that $(\Phi_{0,f}\Phi_{0,f}')$ *is invertible. Moreover, let* $\underline{\theta}_i$ *be the ith row of* Θ*. Then* $\sigma_i^{-2}\Phi_{0,f}\Phi_{0,f}'$ *is the* FIM *associated with estimation of* $\underline{\theta}_i'$ *for* $1 \le i \le p$*, and thus,* $\sigma_i^2(\Phi_{0,f}\Phi_{0,f}')^{-1}$ *is the corresponding* CRLB.

Proof. Suppose that Σ_v is known. Then the Gauss assumption and (8.3) imply that the MLE minimizes

$$\begin{aligned} J_\Theta &= \frac{1}{2}\mathrm{Tr}\left\{\Sigma_v^{-1}\left(Y_{0,f} - \Theta\Phi_{0,f}\right)\left(Y_{0,f} - \Theta\Phi_{0,f}\right)'\right\} \\ &= \frac{1}{2}\mathrm{Tr}\left\{\Sigma_v^{-1}\left(Y_{0,f}Y_{0,f}' - \Theta\Phi_{0,f}Y_{0,f}' - Y_{0,f}\Phi_{0,f}'\Theta' + \Theta\Phi_{0,f}\Phi_{0,f}'\Theta'\right)\right\}. \end{aligned}$$

Direct calculation shows

$$\frac{\partial J_\Theta}{\partial \Theta} = \Sigma_v^{-1}\left(\Theta \Phi_{0,f}\Phi_{0,f}' - Y_{0,f}\Phi_{0,f}'\right).$$

Setting the above to zero yields the MLE in (8.5). Since the MLE is independent of Σ_v, $\Theta_{\rm LS}$ is indeed the MLE. With partition row-wise,

$$Y_{0,f} = \begin{bmatrix} \underline{y}_1(t_0,t_f) \\ \vdots \\ \underline{y}_p(t_0,t_f) \end{bmatrix}, \quad \Theta = \begin{bmatrix} \underline{\theta}_1 \\ \vdots \\ \underline{\theta}_p \end{bmatrix},$$

and $\underline{\varepsilon}_i = \underline{y}_i(t_0,t_f) - \underline{\theta}_i \Phi_{0,f}$ for $1 \le i \le p$. There holds

$$\begin{aligned} \ln f_V(\mathbf{y};\Theta) &= -J_\Theta(t_0,t_f) = -\frac{1}{2}\sum_{i=1}^{p} \sigma_i^{-2}\underline{\varepsilon}_i\underline{\varepsilon}_i' \\ &= -\frac{1}{2}\sum_{i=1}^{p} \sigma_i^{-2}\left(\underline{y}_i(t_0,t_f) - \underline{\theta}_i\Phi_{0,f}\right)\left(\underline{y}_i(t_0,t_f) - \underline{\theta}_i\Phi_{0,f}\right)' \end{aligned}$$

by $\Sigma_v = \mathrm{diag}(\sigma_1^2, \cdots, \sigma_p^2)$. It can be easily verified that

$$\frac{\partial \ln f_V(\mathbf{y};\Theta)}{\partial \underline{\theta}_i'} = -\sigma_i^{-2}\Phi_{0,f}\left(\underline{y}_i(t_0,t_f) - \underline{\theta}_i\Phi_{0,f}\right)' = -\sigma_i^{-2}\Phi_{0,f}\underline{\varepsilon}_i'.$$

By recognizing $\mathrm{E}\{(\underline{y}_i(t_0,t_f) - \underline{\theta}_i\Phi_{0,f})'(\underline{y}_i(t_0,t_f) - \underline{\theta}_i\Phi_{0,f})\} = \sigma_i^2 I$, the above yields the FIM for $\underline{\theta}_i'$:

$$\mathrm{FIM}(\underline{\theta}_i') = \mathrm{E}\left\{\frac{\partial \ln f_V(\mathbf{y};\Theta)}{\partial \underline{\theta}_i'}\frac{\partial \ln f_V(\mathbf{y};\Theta)}{\partial \underline{\theta}_i}\right\} = \sigma_i^{-2}\Phi_{0,f}\Phi_{0,f}'.$$

The corresponding CRLB is thus $\sigma_i^2(\Phi_{0,f}\Phi_{0,f}')^{-1}$ that concludes the proof. □

It needs to keep in mind that the LS solution may not have the MLE interpretation, if $\underline{\phi}(t)$ involves random noises. Consider the case when the covariance of $\mathbf{v}(t)$ is $\Sigma_v(t)$, and thus, $\{\mathbf{v}(t)\}$ is not a WSS process. Suppose that $\Sigma_v(t) = \sigma_t^2 I$ that is known for all t. Then J_Θ is replaced by

$$J_\Theta = \frac{1}{2}\sum_{t=t_0}^{t_f} \mathrm{Tr}\left\{\sigma_t^{-2}[\mathbf{y}(t) - \Theta\underline{\phi}(t)][\mathbf{y}(t) - \Theta\underline{\phi}(t)]'\right\}. \tag{8.6}$$

The MLE minimizes the above J_Θ and satisfies

$$\sum_{t=t_0}^{t_f} \sigma_t^{-2}\left[\Theta_{\rm LS}\underline{\phi}(t)\underline{\phi}(t)' - \mathbf{y}(t)\underline{\phi}(t)'\right] = \mathbf{0}.$$

Hence, the MLE is the weighted LS with weighting $\{\sigma_t^{-2}\}$. Without loss of generality, assume that $t_0 = 0$. For $k > 0$, denote

$$P_k = \left[\sum_{t=0}^{k-1} \sigma_t^{-2}\underline{\phi}(t)\underline{\phi}(t)'\right]^{-1}, \quad Q_k = \sum_{t=0}^{k-1} \sigma_t^{-2}\mathbf{y}(t)\underline{\phi}(t)'. \tag{8.7}$$

By the proof of Theorem 8.1, $\widehat{\Theta}_{t_f} = \Theta_{\text{LS}} = Q_{t_f+1}P_{t_f+1}$ is the MLE.

The simplicity form of the LS solution allows its recursive computation with low complexity. To be specific, denote $\widehat{\Theta}_k$ as the LS solution based on the measurement data over the time horizon $[0,\ k)$ for some $k > 0$. Suppose that new input and output measurements are obtained at k. There hold

$$\widehat{\Theta}_k = Q_k P_k, \quad \widehat{\Theta}_{k+1} = Q_{k+1}P_{k+1}.$$

The recursive LS (RLS) algorithm is aimed at computing $\widehat{\Theta}_{k+1}$ based on $\widehat{\Theta}_k$ and the updated regressor $\underline{\phi}(k)$ without explicitly computing $Q_{k+1}P_{k+1}$. In this regard, RLS is similar to Kalman filtering in Theorem 5.6. The key is computation of P_{k+1} based on P_k and $\underline{\phi}(k)$.

First, it is noted that the covariance type matrix P_{k+1} can be written as

$$\begin{aligned} P_{k+1} &= [P_k^{-1} + \sigma_k^{-2}\underline{\phi}(k)\underline{\phi}(k)]^{-1} \\ &= P_k - P_k\underline{\phi}(k)[\sigma_k^2 + \underline{\phi}(k)'P_k\underline{\phi}(k)]^{-1}\underline{\phi}(k)'P_k \end{aligned}$$

by the matrix inversion formula in Appendix A. See also Problem 8.4 in Exercises. The above can be rewritten as

$$P_{k+1} = P_k - P_k\underline{\phi}(k)\mathbf{g}_k, \quad \mathbf{g}_k = [\sigma_k^2 + \underline{\phi}(k)'P_k\underline{\phi}(k)]^{-1}\underline{\phi}(k)'P_k. \tag{8.8}$$

The derivation next shows the relation between $\underline{\phi}(k)'P_{k+1}$ and $\mathbf{g}_k$:

$$\begin{aligned} \underline{\phi}(k)'P_{k+1} &= \underline{\phi}(k)'P_k - \underline{\phi}(k)'P_k\underline{\phi}(k)[\sigma_k^2 + \underline{\phi}(k)'P_k\underline{\phi}(k)]^{-1}\underline{\phi}(k)'P_k \\ &= \{I - \underline{\phi}(k)'P_k\underline{\phi}(k)[\sigma_k^2 + \underline{\phi}(k)'P_k\underline{\phi}(k)]^{-1}\}\underline{\phi}(k)'P_k \\ &= \sigma_k^2[\sigma_k^2 + \underline{\phi}(k)'P_k\underline{\phi}(k)]^{-1}\underline{\phi}(k)'P_k = \sigma_k^2\mathbf{g}_k. \end{aligned}$$

It follows from $Q_{k+1} = Q_k + \sigma_k^{-2}\mathbf{y}(k)\underline{\phi}(k)'$ that

$$\begin{aligned} \widehat{\Theta}_{k+1} &= Q_{k+1}\left[P_k - P_k\underline{\phi}(k)\mathbf{g}_k\right] \\ &= \widehat{\Theta}_k - \widehat{\Theta}_k\underline{\phi}(k)\mathbf{g}_k + \sigma_k^{-2}\mathbf{y}(k)\underline{\phi}(k)'P_{k+1} \\ &= \widehat{\Theta}_k - \widehat{\Theta}_k\underline{\phi}(k)\mathbf{g}_k + \mathbf{y}(k)\mathbf{g}_k = \widehat{\Theta}_k + [\mathbf{y}(k) - \widehat{\mathbf{y}}(k)]\mathbf{g}_k, \end{aligned}$$

where $\widehat{\mathbf{y}}(k) = \widehat{\Theta}_k\underline{\phi}(k)$ can be regarded as the predicted output. The above and (8.8) form the RLS algorithm. If no knowledge is available at $k = 0$, then $\widehat{\Theta}_0 = \mathbf{0}$ and $P_0 = \rho^2 I$ with large ρ can be employed which admit a similar interpretation to that in Kalman filter.

An important problem in parameter estimation is the convergence of the estimate. Consider the case when $\sigma_t^2 = \sigma^2 > 0$ is a constant. Then the signal is called persistent exciting (PE), if $P_t \to \mathbf{0}$ as $t \to \infty$. Accordingly, the PE condition implies $\widehat{\Theta}_t \to \Theta$ asymptotically based on the facts that $\mathbf{g}_t \to \mathbf{0}$ as $t \to \infty$ and that LS algorithm yields the MLE. As a result $\widehat{\Theta}_t$ stops updating eventually. In fact, the LS estimate $\widehat{\Theta}_t$ may stop updating very quickly when the signal is rich in its information content.

While the convergence is welcomed, it does not suit to estimation of the time-varying parameter matrix. Basically, the RLS algorithm fails to track the underlying parameter matrix, when the PE condition holds. One may periodically reset P_t to prevent it from being zero. However, this method is not suggested due to the loss of past information. A more sophisticated method considers the following performance index:

$$J_\Theta(t_f) = \frac{1}{2}\|\mathbf{y}(t_f) - \Theta\underline{\phi}(t_f)\|^2 + \gamma_{t_f} J_\Theta(t_f - 1), \tag{8.9}$$

where $\gamma_t \in (0,\ 1)$ is referred to as the forgetting factor at time t. In the case when $\gamma_t = \gamma$ is a constant and $0 < \gamma < 1$,

$$J_\Theta(t_f) = \frac{1}{2}\sum_{t=0}^{t_f} \gamma^{t_f - t}\|\mathbf{y}(t_f) - \Theta\underline{\phi}(t_f)\|^2.$$

Thus, the terms in the distant past decay exponentially with respect to the time duration. The resultant minimizer is very similar to the RLS algorithm derived earlier with an appropriate modification.

To derive the RLS with the forgetting factor, it is noted that

$$J_\Theta(t-1) = C_t + \Theta P_t^{-1}\Theta' - \Theta Q_t' - Q_t\Theta'$$

for some time-dependent constant matrices P_t, Q_t, and C_t. Consequently,

$$J_\Theta(t) = \frac{\gamma_t}{2}\mathrm{Tr}\left\{C_t + \Theta P_t^{-1}\Theta' - \Theta Q_t' - Q_t\Theta'\right\} + \frac{1}{2}\|\mathbf{y}(t_f) - \Theta\underline{\phi}(t_f)\|^2.$$

Its partial derivative can be easily computed and is given by

$$\left.\frac{\partial J_\Theta(t)}{\partial\Theta}\right|_{\Theta = \widehat{\Theta}_{t+1}} = \widehat{\Theta}_{t+1}[\gamma_t P_t^{-1} + \underline{\phi}(t)\underline{\phi}(t)'] - [\gamma_t Q_t + \mathbf{y}(t)\underline{\phi}(t)'].$$

Setting the above to zero yields $\widehat{\Theta}_{t+1} = \tilde{Q}_{t+1}\tilde{P}_{t+1}$ with

$$\tilde{P}_{t+1} = \gamma_t P_t + \underline{\phi}(t)\underline{\phi}(t)', \quad \tilde{Q}_{t+1} = \gamma_t Q_t + \mathbf{y}(t)\underline{\phi}(t)'.$$

It is interesting to observe that

$$\tilde{P}_{t+1} = \gamma_t P_{t+1}, \quad \tilde{Q}_{t+1} = \gamma_t Q_{t+1}$$

where P_t and Q_t are defined in (8.7) by taking $\gamma_t = \sigma_t^2$. It is left as an exercise (Problem 8.6) to show that the RLS with forgetting factor is given by

$$\widehat{\Theta}_{t+1} = \Theta_t + [\mathbf{y}(t) - \widehat{\mathbf{y}}(t)]\,\mathbf{g}_t, \; \widehat{\mathbf{y}}(t) = \widehat{\Theta}_t \underline{\phi}(t), \tag{8.10}$$

$$P_{t+1} = \gamma_t^{-1}\left[P_t - P_t \underline{\phi}(t)\mathbf{g}_t\right], \; \mathbf{g}_t = [\gamma_t + \underline{\phi}(t)' P_t \underline{\phi}(t)]^{-1} \underline{\phi}(t)' P_t. \tag{8.11}$$

Two examples are used to illustrate the LS algorithm next.

Example 8.2. Let $\{H_i\}_{i=0}^2$ be the CIR with one input and two outputs. Thus, each H_i has dimension 2×1, specified by

$$H_0 = \begin{bmatrix} 0.6360 \\ 0.0636 \end{bmatrix}, \; H_1 = \begin{bmatrix} -0.3552 \\ 0.2439 \end{bmatrix}, \; H_2 = \begin{bmatrix} 0.5149 \\ -0.3737 \end{bmatrix}. \tag{8.12}$$

Hence, $\|\mathbf{H}\|_2 = 1$. The training sequence of binary symbols $\pm P_b$ is transmitted with length ranging from 10 to 20 bits. The received signals are corrupted by i.i.d. Gauss noises with variance σ^2. Since $\|\mathbf{H}\|_2 = 1$, the SNR is the same as the ratio of P_b to σ^2. Numerical simulations are carried out for the cases when SNR $= 10$ dB and when SNR $=20$ dB. A total of 500 ensemble runs are used to evaluate the channel estimation performance. Let $\{\widehat{H}_i^{(k)}\}$ be the estimated CIR at the kth ensemble run. The RMSE is computed according to

$$\text{RMSE} = \sqrt{\frac{1}{T}\sum_{k=1}^{T} \text{Tr}\left\{\sum_{i=0}^{2}\left(H_k - \widehat{H}_i^{(k)}\right)\left(H_k - \widehat{H}_i^{(k)}\right)'\right\}}.$$

The results are plotted in the following figure.

The upper two curves correspond to the case of SNR $= 10$ dB. The dashed line marked with diamond shows the RMSE, while the solid line marked with circle is the corresponding CRLB curve, defined by $\sqrt{\text{Tr}\{\text{FIM}^{-1}\}}$ that is the lower bound for RMSE. These two curves are close to each other. In fact, the two curves overlap for large T (Problem 8.5 in Exercises). The lower two curves in Fig. 8.1 correspond to the case of SNR $= 20$ dB. The dashed line marked with square shows the RMSE, while the solid line marked with $*$ shows the corresponding CRLB.

The simulation result validates the MLE nature of the LS solution in the case of FIR models that hold for the wireless channels (Problem 8.8 in Exercises). However, if the temporary assumption on noise-free $\{\underline{\phi}(t)\}$ is removed that is the case for IIR models (refer to (8.1) in which $M_k \neq \mathbf{0}$ for at least one k), the MLE interpretation of the LS solution will be lost. Specifically, the physical systems in feedback control are generically described by IIR models. Figure 8.2 illustrates the plant model with input/output signals together with observation noises.

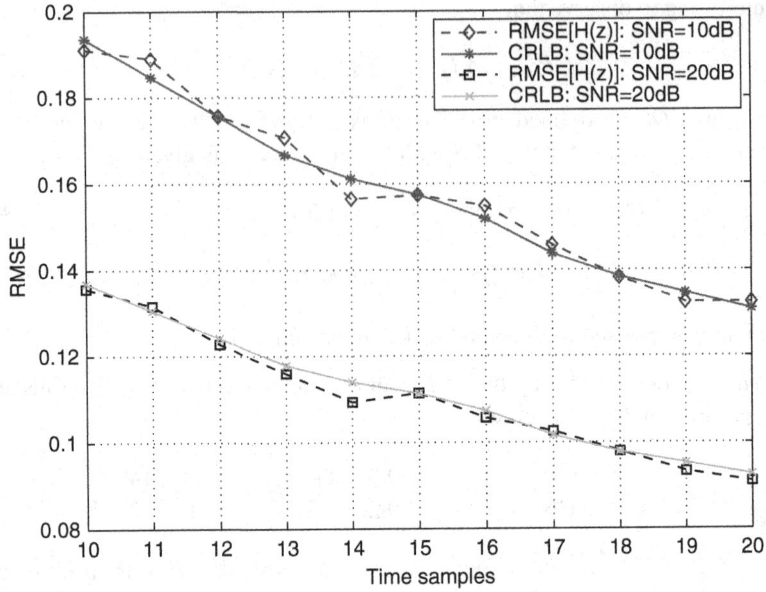

Fig. 8.1 RMSE plots for LS-based channel estimation

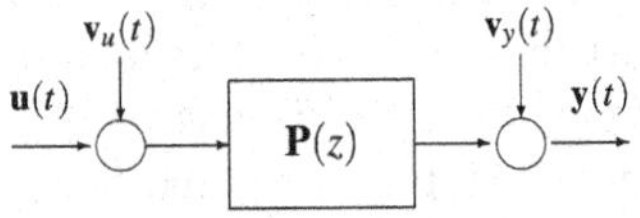

Fig. 8.2 Plant model with noisy measurement data

It is clear that both observations $\mathbf{u}(t)$ and $\mathbf{y}(t)$ are no longer the actual input and output of the plant. In fact, the signal model in (8.1) is replaced by

$$\mathbf{y}(t) = \mathbf{v}_y(t) + \sum_{k=1}^{n_\mu} M_k \mathbf{y}(t-k) + \sum_{k=1}^{n_\nu} N_k[\mathbf{u}(t-k) + \mathbf{v}_u(t-k)]. \tag{8.13}$$

Hence, when the LS algorithm is applied to estimate the system parameters, the estimation performance is different from the case for FIR models. The next example illustrates the identification results.

Example 8.3. Consider identification of a SISO plant represented by its transfer function

$$P(z) = \frac{0.3624z + 0.1812}{z^2 - 1.5z + 0.7}.$$

The input $u(t)$ is chosen as white Gauss with variance 1 that results in output variance about 4 dB. Observation noises are added to input, or output, or both in the same way as shown in Fig. 8.2. The corrupting noises are white and Gauss distributed with variance 0.1, or -20 dB. The corresponding RMSE curves are

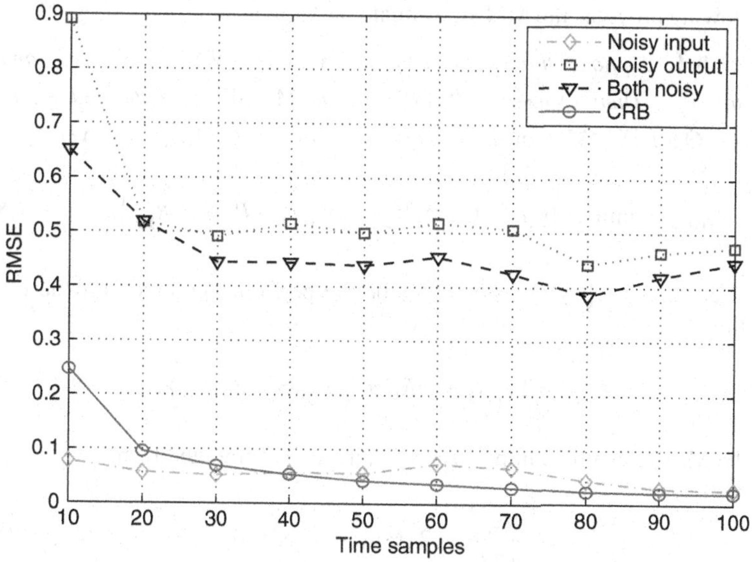

Fig. 8.3 RMSE curves for LS-based IIR model estimation

plotted in Fig. 8.3 with a larger observation interval than the previous example, in order to illustrate the trend of the identification error. It is clearly seen that the RMSE curves do not monotonically decline anymore. The simulation results indicate that the LS solution is biased for identification of IIR models in the presence of output observation noises. In fact, the corruption noise at the output impacts more negatively than at the input in terms of the estimation performance.

It needs to be pointed out that the CRLB curve used in the figure is the same as $\sigma_v\sqrt{\mathrm{Tr}\{(\Phi_{0,f}\Phi_{0,f}')^{-1}\}}$. This expression is actually not the true CRLB anymore due to the observation noise involved in $\Phi_{0,f}$. The derivation of the CRLB for the case of noisy $\Phi_{0,f}$ will be investigated in a later subsection.

8.1.2 MMSE Algorithm

For the signal model $\mathbf{y}(t) = \Theta\underline{\phi}(t) + \mathbf{v}(t)$ studied in the previous section, $\{\underline{\phi}(t)\}$ is likely to involve observation noises, in which case the LS solution is not the MLE, and the MLE solution is difficult to compute in general. An alternative to MLE is the MMSE estimate that minimizes $\mathrm{E}\{\|\mathbf{y}(t) - \Theta\underline{\phi}(t)\|^2\}$. Assume that both $\underline{\phi}(t)$ and $\mathbf{v}(t)$ are WSS processes, and denote

$$R_{\mathbf{y}} = \mathrm{E}\{\mathbf{y}(t)\mathbf{y}(t)'\}, \quad R_\phi = \mathrm{E}\{\underline{\phi}(t)\underline{\phi}(t)'\}, \quad R_{\mathbf{y},\phi} = \mathrm{E}\{\mathbf{y}(t)\underline{\phi}(t)'\}.$$

The following provides the MMSE estimate.

Theorem 8.4. *Suppose that both $\underline{\phi}(t)$ and $\mathbf{v}(t)$ are WSS processes, and R_ϕ is nonsingular. Then $\Theta_{\text{MMSE}} = R_{\mathbf{y}\phi}R_\phi^{-1}$ is the MMSE estimate and minimizes $\mathrm{E}\{\|\mathbf{y}(t) - \Theta\underline{\phi}(t)\|^2\}$. Moreover, the MSE associated with Θ_{MMSE} is given by*

$$\varepsilon_{\text{MMSE}} = \min_\Theta \mathrm{E}\{\|\mathbf{y}(t) - \Theta\underline{\phi}(t)\|^2\} = \mathrm{Tr}\left\{R_\mathbf{y} - R_{\mathbf{y}\phi}R_\phi^{-1}R_{\mathbf{y}\phi}'\right\}. \tag{8.14}$$

Proof. Let $\varepsilon_{\text{MSE}} = \mathrm{E}\{\|\mathbf{y}(t) - \Theta\underline{\phi}(t)\|^2\}$ be the performance index for the MMSE estimation. Then

$$\varepsilon_{\text{MSE}} = \mathrm{Tr}\left\{R_\mathbf{y} + \Theta R_\phi\Theta' - R_{\mathbf{y}\phi}\Theta' - \Theta R_{\mathbf{y}\phi}'\right\}.$$

Since the MMSE estimate minimizes ε_{MSE}, it can be computed from

$$\frac{\partial\varepsilon_{\text{MSE}}}{\partial\Theta} = 2\left(\Theta R_\phi - R_{\mathbf{y}\phi}\right) = \mathbf{0} \tag{8.15}$$

that shows $\Theta_{\text{MMSE}} = R_{\mathbf{y}\phi}R_\phi^{-1}$. Substituting $\Theta = \Theta_{\text{MMSE}}$ into the expression of ε_{MSE} yields (8.14). □

The autocorrelation matrices R_ϕ and $R_{\mathbf{y}\phi}$ are often unavailable in practice. Estimates based on N samples of measurements over $[0,\ t_f]$ with $t_f = N-1$ can be used:

$$R_\phi \approx \frac{1}{N}\sum_{t=0}^{N-1}\underline{\phi}(t)\underline{\phi}(t)' = \frac{P_N^{-1}}{N}, \quad R_{\mathbf{y}\phi} \approx \frac{1}{N}\sum_{t=0}^{N-1}\mathbf{y}(t)\underline{\phi}(t)' = \frac{Q_N}{N} \tag{8.16}$$

where P_N and Q_N are the same as defined in (8.7). If

$$R_\phi = \lim_{N\to\infty}\frac{1}{N}\sum_{t=0}^{N-1}\underline{\phi}(t)\underline{\phi}(t)', \quad R_{\mathbf{y}\phi} = \lim_{N\to\infty}\frac{1}{N}\sum_{t=0}^{N-1}\mathbf{y}(t)\underline{\phi}(t)',$$

then $\{\underline{\phi}(t)\}$ and $\{\mathbf{y}(t)\}$ are called ergodic processes. The PE condition is clearly necessary for R_ϕ to be nonsingular. Under both WSS and ergodic assumptions, the RLS solution approaches the MMSE estimate, i.e.,

$$\widehat{\Theta}_N = P_N Q_N = \left(\frac{1}{N}\sum_{t=0}^{N-1}\underline{\phi}(t)\underline{\phi}(t)'\right)^{-1}\left(\frac{1}{N}\sum_{t=0}^{N-1}\mathbf{y}(t)\underline{\phi}(t)'\right) \to \Theta_{\text{MMSE}}$$

as $N \to \infty$. It follows that (cf. Problem 8.10 in Exercises)

$$\widehat{\Theta}_N \to \Theta + \lim_{N\to\infty}\mathrm{E}\left\{V_{0,f}\Phi_{0,f}'\right\}\left(\mathrm{E}\{\Phi_{0,f}\Phi_{0,f}'\}\right)^{-1} \tag{8.17}$$

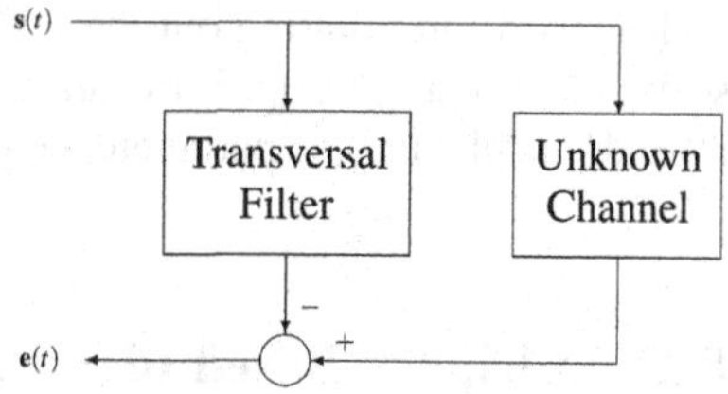

Fig. 8.4 Schematic illustration for channel estimation

where $N = t_f - t_0 + 1$. Hence, if $\mathrm{E}\{V_{0,f}\Phi_{0,f}'\} \to \mathbf{0}$ as $N \to \infty$, then the LS algorithm is asymptotically unbiased. Otherwise, the LS solution is biased in which case large number of samples does not help to eliminate the bias in the LS solution.

The next example illustrates the use of MMSE estimation.

Example 8.5. In wireless communications, channel information is essential for reliable data detection. While pilot tones such as training sequence can be used to estimate the CIR, it is desirable to estimate the CIR based on statistical information of the data sequence. Consider MIMO channel estimation in Fig. 8.4 with $\mathrm{E}\{\mathbf{s}(t)\mathbf{s}(t-k)'\} = R_{\mathbf{s}}(k)$ assumed to be known for each integer k. The received signal at the output of the channel is given by

$$\mathbf{y}(t) = \sum_{k=1}^{L} H_i \mathbf{s}(t-i) + \mathbf{v}(t) = \Theta\underline{\phi}(t) + \mathbf{v}(t)$$

for some white Gauss noise $\mathbf{v}(t)$ where

$$\Theta = \begin{bmatrix} H_1 & \cdots & H_L \end{bmatrix}, \quad \phi(t) = \begin{bmatrix} \mathbf{s}(t-1) \\ \vdots \\ \mathbf{s}(t-L) \end{bmatrix}.$$

It follows from $\mathrm{E}\{\mathbf{s}(t)\mathbf{s}(t-k)'\} = R_{\mathbf{s}}(k)$ that

$$R_\phi = \underline{R}_{\mathbf{s}} = \mathrm{E}\{\underline{\phi}(t)\underline{\phi}(t)'\} = [\, R_{\mathbf{s}}(k-i) \,]_{i,k=1,1}^{L,L}$$

is a block Toeplitz matrix known at the receiver. However, $R_{\mathbf{y}\phi}$ has to be estimated using

$$R_{\mathbf{y}\phi} \approx \frac{1}{N} \sum_{t=0}^{N-1} \mathbf{y}(t)\underline{\phi}(t)'$$

for some integer $N >> 1$. In this case, the MMSE estimate for CIR can be obtained from $\Theta_{\mathrm{MMSE}} = R_{\mathbf{y}\phi} R_\phi^{-1}$. If $\{R_{\mathbf{s}}(k)\}_{k=0}^{L-1}$ are not available at the receiver, they need to be estimated as well. A commonly seen method employs the past detected symbols, assumed to be correct, which yields effective way to estimate R_ϕ.

It is worth to pointing out the difference between the MMSE estimation in this section and that in Chap. 5. Recall that in Chap. 5, the state vector under estimation is random, whereas the parameters under estimation in this section are deterministic.

8.2 Subspace-Based Identification

For the signal model $\mathbf{y}(t) = \Theta\phi(t) + \mathbf{v}(t)$ studied in the previous section, the MLE interpretation for the LS algorithm does not hold in general, if the noise sequence $\{\mathbf{v}(t)\}$ is not Gauss distributed. Although the LS solution can still be used to extrapolate the system model, the estimate is not unbiased anymore for IIR models, because $\phi(t)$ involves $\{\mathbf{y}(k)\}$ for $k < t$ (Problem 8.10 in Exercises). It turns out that it is the TLS algorithm that yields the unbiased estimate asymptotically which will be used for system identification in this section.

8.2.1 TLS Estimation Algorithm

The TLS algorithm arises from the class of error-in-variable (EIV) models. Let Θ be the parameter matrix of interest satisfying $\Theta A_0 = B_0$ where A and B are wide matrices. The precise values of A_0 and B_0 are not available. Instead, only their measurements, denoted by A and B, respectively, are available given by the EIV model

$$\begin{bmatrix} A \\ B \end{bmatrix} = \begin{bmatrix} A_0 \\ B_0 \end{bmatrix} + M\Delta. \tag{8.18}$$

To be specific, the dimensions of A, B, and M are $L \times N$, $p \times N$, and $p \times \ell$ with $N > (L+p)$. Thus, Δ has dimension $\ell \times N$. The elements of Δ are assumed to be independent and identically distributed (i.i.d.) random variables with mean zero and variance σ^2. The goal is to estimate Θ of dimension $p \times L$ based on A, B, and M.

The previous section studied the estimation problem for the case of $A = A_0$. The only measurement error comes from B. Recall the deterministic assumption for $\Phi_{0,f}$ by taking $A = A_0 = \Phi_{0,f}$ and $B = Y_{0,f} \neq B_0$. The LS algorithm finds $\widehat{B}$ closest to B such that

$$\operatorname{rank}\left\{\begin{bmatrix} A \\ \widehat{B} \end{bmatrix}\right\} = \operatorname{rank}\{A\},$$

and then solve for $\widehat{\Theta}$ from $\Theta A = \widehat{B}$. It is noted that $B = \widehat{B} + \widehat{B}_\perp$ of which each row of $\widehat{B}$ lies in the row space of A, and $\widehat{B}_\perp A' = \mathbf{0}$. That is, the row spaces of $\widehat{B}$ and $\widehat{B}_\perp$ are orthogonal to each other. Consequently, $\Theta AA' = BA' = \widehat{B}A'$ yielding the LS solution $\Theta_{\rm LS} = BA'(AA')^{-1}$. If the measurement errors are Gauss distributed, then the LS algorithm yields the MLE estimate.

When $A \neq A_0$ in addition to $B \neq B_0$, both $\widehat{A}$ and $\widehat{B}$ closest to A and B, respectively, are searched for such that

$$\text{rank}\left\{\begin{bmatrix}\widehat{A}\\ \widehat{B}\end{bmatrix}\right\} = \text{rank}\{\widehat{A}\}$$

prior to solving for Θ from $\Theta\widehat{A} = \widehat{B}$ that is the essence of the TLS. Let $M = I$ to begin with. The TLS algorithm is aimed at minimizing

$$J_{A,B} := \left\|\begin{bmatrix}A\\ B\end{bmatrix} - \begin{bmatrix}\widehat{A}\\ \widehat{B}\end{bmatrix}\right\|_{\mathscr{F}} \quad \text{subject to} \;\; \text{rank}\left\{\begin{bmatrix}\widehat{A}\\ \widehat{B}\end{bmatrix}\right\} = L \tag{8.19}$$

with $\|X\|_{\mathscr{F}} = \sqrt{\text{Tr}\{X'X\}}$ the Frobenius norm. A similar problem is encountered in Hankel-norm approximation. See (4.58) in Sect. 4.3 and the discussion therein. Hence, SVD can be used to compute such a pair of $\widehat{A}$ and $\widehat{B}$. As a result, there exists a unique solution pair (X_1, X_2) with X_2 of dimension $p \times p$ to $X_1\widehat{A} = X_2\widehat{B}$. If X_2 is nonsingular, then $\widehat{\Theta} = X_2^{-1}X_1$ is the TLS solution.

A formal procedure for TLS is stated next. Define

$$W := \begin{bmatrix}A\\ B\end{bmatrix}\begin{bmatrix}A' & B'\end{bmatrix}, \tag{8.20}$$

and let $W = G'\Lambda G$ be the eigenvalue decomposition with

$$\Lambda = \text{diag}(\lambda_1, \lambda_2, \cdots, \lambda_{L+p})$$

arranged in descending order. Partition the eigenvector matrix G and eigenvalue matrix Λ according to

$$G = \begin{bmatrix}G_{11} & G_{12}\\ G_{21} & G_{22}\end{bmatrix}, \quad \Lambda = \begin{bmatrix}\Lambda_1 & \mathbf{0}\\ \mathbf{0} & \Lambda_2\end{bmatrix}, \tag{8.21}$$

where G_{11} and Λ_1 have the same dimension of $L \times L$. Then

$$\Theta_{\text{TLS}} = G_{21}G_{11}^{-1} = -(G_{22}')^{-1}G_{12}' \tag{8.22}$$

is the TLS estimate, provided that G_{22} is nonsingular. It can be shown that $\det(G_{22}) \neq 0$ has probability 1, but the proof is skipped because of the involvement of more specialized mathematical background. The next result shows the MLE property when the elements of Δ are Gauss distributed.

Theorem 8.6. *Suppose that $M = I$ and elements of Δ are normal* i.i.d. *with mean zero and variance σ^2. Let W be defined in (8.20), $W = G'\Lambda G$ be the eigenvalue decomposition with diagonal elements of Λ arranged in descending order. Partition*

G and Λ as in (8.21) where G_{11} and Λ_1 are of dimension $L \times L$. Then G_{11} and G_{22} are nonsingular w.p. 1 (*with probability* 1), *and*

$$\widehat{\Theta} = G_{21}G_{11}^{-1} = -(G_{22}')^{-1}G_{12}', \quad \widehat{\sigma}^2 = \frac{\mathrm{Tr}\{\Lambda_2\}}{(L+p)N} \tag{8.23}$$

are MLEs *for Θ and σ^2, respectively.*

Proof. The PDF for the measurement data A and B is given by

$$f_\Delta(\{\delta_{ij}\}) = \frac{1}{\left(\sqrt{2\pi\widehat{\sigma}^2}\right)^{(L+p)N}} \exp\left\{-\frac{1}{2\widehat{\sigma}^2}\left\|\begin{bmatrix} A \\ B \end{bmatrix} - \begin{bmatrix} \widehat{A} \\ \widehat{B} \end{bmatrix}\right\|_{\mathscr{F}}^2\right\}.$$

The MLE searches $\widehat{A}, \widehat{B}$ and $\widehat{\sigma}^2$ which maximize $f_\Delta(\{\delta_{ij}\})$. Since $J_{A,B}$ defined in (8.19) is independent of σ^2, its minimum is $\mathrm{Tr}\{\Lambda_2\}$ by taking $\widehat{\Theta} = G_{21}G_{11}^{-1}$. See Problem 8.13 in Exercises. Hence,

$$\max f_\Delta(\{\delta_{ij}\}) = \max_{\widehat{\sigma}^2} \frac{1}{\left(\sqrt{2\pi\widehat{\sigma}^2}\right)^{(L+p)N}} \exp\left\{-\frac{1}{2\widehat{\sigma}^2}\mathrm{Tr}\{\Lambda_2\}\right\}.$$

Taking derivative with respect to $\widehat{\sigma}$ and setting it to zero yield

$$\frac{(L+p)N}{\widehat{\sigma}} - \frac{\mathrm{Tr}\{\Lambda_2\}}{\widehat{\sigma}^3} = 0.$$

Hence, $\widehat{\sigma}^2$ in (8.23) is the MLE for σ^2 that concludes the proof. □

Theorem 8.6 shows the MLE of the TLS, but whether or not it is unbiased, estimate remains unknown. The following shows that the TLS solution is asymptotically unbiased.

Theorem 8.7. *Suppose that the same hypotheses of* Theorem *8.6 hold, and assume that*

$$\text{(i)}\ \Pi_0 := \lim_{N\to\infty} \frac{A_0 A_0'}{N} > \mathbf{0}, \quad \text{(ii)}\ \lim_{N\to\infty} \frac{\Delta\begin{bmatrix} A_0' & B_0' \end{bmatrix}}{N} = \mathbf{0}. \tag{8.24}$$

Then the TLS *estimate $\Theta_{\mathrm{TLS}} \to \Theta$ as $N \to \infty$. In addition, there holds*

$$\lim_{N\to\infty} \frac{W}{N} = \sigma^2 I + \begin{bmatrix} I \\ \Theta \end{bmatrix} \Pi_0 \begin{bmatrix} I & \Theta' \end{bmatrix}. \tag{8.25}$$

Proof. By the EIV model (8.18), $M = I$, W in (8.20), and $\Theta A_0 = B_0$,

$$\frac{W}{N} = \frac{1}{N}\begin{bmatrix} A_0 \\ B_0 \end{bmatrix}\begin{bmatrix} A_0' & B_0' \end{bmatrix} + \frac{\Delta\Delta'}{N} + \frac{\Delta}{N}\begin{bmatrix} A_0' & B_0' \end{bmatrix} + \frac{1}{N}\begin{bmatrix} A_0 & B_0 \end{bmatrix}\Delta'.$$

Taking limit $N \to \infty$ and using the two assumptions in (8.24) arrive at

$$\lim_{N\to\infty} \frac{W}{N} = \lim_{N\to\infty} \frac{1}{N} \begin{bmatrix} A_0 \\ B_0 \end{bmatrix} \begin{bmatrix} A_0' & B_0' \end{bmatrix} + \frac{\Delta\Delta'}{N}$$
$$= \sigma^2 I + \begin{bmatrix} I \\ \Theta \end{bmatrix} \Pi_0 \begin{bmatrix} I & \Theta' \end{bmatrix}$$

by the assumption on the i.i.d. of elements of Δ, which verifies (8.25). Hence, $\lambda_{L+i} \to \sigma^2$ as $N \to \infty$ for $1 \le i \le p$. Denote $\mathscr{R}(\cdot)$ for the range space. Then

$$\mathscr{R}\left(\begin{bmatrix} G_{11} \\ G_{12} \end{bmatrix}\right) \longrightarrow \mathscr{R}\left(\begin{bmatrix} I \\ \Theta \end{bmatrix}\right)$$

as $N \to \infty$. The asymptotic convergence of $\widehat{\Theta}_{\text{TLS}}$ to Θ thus follows. □

Two comments are made. The first regards the assumption in (8.24): Statement (ii) is in fact implied by (i). But because the proof is more involved, it is skipped. The second is the convergence of the MLE for the noise variance σ^2. The proof of Theorem 8.7 indicates that

$$\lim_{N\to\infty} \frac{(L+p)\widehat{\sigma}^2}{p} = \lim_{N\to\infty} \frac{\text{Tr}\{\Lambda_2\}}{pN} = \frac{\text{Tr}\{\sigma^2 I_p\}}{p} = \sigma^2 \tag{8.26}$$

where $\widehat{\sigma}^2$ is the MLE for σ^2 in Theorem 8.6. Therefore, the MLE for the σ^2 is not an asymptotically unbiased estimate. The regularity condition breaks down for estimation of σ^2.

Theorems 8.6 and 8.7 address the estimation problem for the EIV model in the case of $M = I$. If $M \neq I$ is a full rank and possibly wide matrix, it can be converted to the estimation problem of $M = I$.

Corollary 8.1. *Under the same conditions and hypotheses of* Theorem 8.6 *except that $M(\neq I)$ has the full row rank, the expressions of* MLEs *in (8.23) hold, provided that the eigenvalue decomposition of W is replaced by that of $W_0 = \Sigma_0^{-1/2} W \Sigma_0^{-1/2}$ where $MM' = \Sigma_0$. In addition, the* MLE $\widehat{\Theta}$ *is an asymptotically unbiased estimate for Θ, and $\frac{(L+p)\widehat{\sigma}^2}{p}$ is an asymptotically unbiased estimate for σ^2.*

Proof. It is important to note that $\sigma^2 \Sigma_0 = \sigma^2 MM'$ can be regarded as the common covariance for each column of ΔM, and thus,

$$\Sigma_0^{-1/2} \begin{bmatrix} A \\ B \end{bmatrix} = \Sigma_0^{-1/2} \begin{bmatrix} A_0 \\ B_0 \end{bmatrix} + U\Delta \tag{8.27}$$

with $\Sigma_0^{1/2}$ symmetric and $U = \Sigma_0^{-1/2} M$ satisfying $UU' = I$. Theorem 8.6 can now be applied, leading to

$$\left(\lambda_i I - \Sigma_0^{-1/2} W \Sigma_0^{-1/2}\right) \mathbf{v}_i = 0 \tag{8.28}$$

for $i = 1, 2, \cdots, L+p$. Hence, by setting

$$G = \begin{bmatrix} \mathbf{v}_1 & \cdots & \mathbf{v}_{L+p} \end{bmatrix}, \quad \Lambda = \text{diag}(\lambda_1, \cdots, \lambda_{L+p}),$$

the proof of the corollary can be concluded. □

It is noted that the eigenvalue/eigenvector equation in (8.28) is the same as the following generalized eigenvalue/eigenvector equation:

$$(\lambda_i \Sigma_0 - W)\,\mathbf{g}_i = \mathbf{0} \tag{8.29}$$

by taking $\mathbf{g}_i = \Sigma_0^{-1/2} \mathbf{v}_i$ for $i = 1, 2, \cdots, L+p$. The above is more convenient to compute than (8.28), and avoids the potential numerical problem associated with the inversion of $\Sigma_0^{1/2}$.

In the case when $M (\neq I)$ is a wide and full rank matrix, the MLE for Θ corresponds to the generalized TLS solution. It requires to compute the p smallest (generalized) eigenvalues and their respective eigenvectors in (8.29) in order to obtain the MLE. Let $\text{diag}(X)$ be a diagonal matrix using diagonal elements of X. It is interesting to observe that the p eigenvectors associated with the p smallest (generalized) eigenvalues in (8.29) solve the following minimization problem (Problem 8.12 in Exercises):

$$\min_H \left\{ \text{Tr}\{H'WH\} : \ \text{diag}\left(H'\Sigma_0 H\right) = I_p \right\}. \tag{8.30}$$

Indeed, by denoting $\mathbf{h}_i$ as the ith column of H, and by setting the cost index

$$J = \sum_{i=1}^{p} \left[\mathbf{h}_i' W \mathbf{h}_i + \gamma_i (1 - \mathbf{h}_i' \Sigma_0 \mathbf{h}_i) \right] \tag{8.31}$$

with $\{\gamma_i\}_{i=1}^p$ Lagrange multipliers, the constrained minimization in (8.30) is equivalent to the unconstrained minimization of J in (8.31). Carrying out computation of the necessary condition leads to $(\gamma_i \Sigma_0 - W)\mathbf{h}_i = 0$ that has the same form as (8.29). Hence, the optimality is achieved by taking $\gamma_i = \lambda_{L+i}$ and $\mathbf{h}_i = \mathbf{g}_{L+i}$ for $1 \le i \le p$ that are the p smallest (generalized) eigenvalues and their respective eigenvectors in (8.29). In the next two subsections, the results of the TLS solution will be applied to channel estimation in wireless communications, and also to system identification in feedback control systems.

8.2.2 Subspace Method

In wireless communications, the CIR has finite duration. Let $\{H_i\}$ be the CIR of a MIMO channel with M input and P output. The received signal is mathematically described by

$$\mathbf{y}(t) = \sum_{i=0}^{L} H_i \mathbf{s}(t-i) + \mathbf{v}(t), \tag{8.32}$$

where $\{\mathbf{s}(k)\}$ is the sequence of the transmitted symbols and $\{\mathbf{v}(k)\}$ is the sequence of i.i.d. with normal distribution. Denote

$$\underline{\mathbf{y}}(t) = \begin{bmatrix} \mathbf{y}(t) \\ \mathbf{y}(t-1) \\ \vdots \\ \mathbf{y}(t-q) \end{bmatrix}, \quad \underline{\mathbf{v}}(t) = \begin{bmatrix} \mathbf{v}(t) \\ \mathbf{v}(t-1) \\ \vdots \\ \mathbf{v}(t-q) \end{bmatrix}, \quad \underline{\mathbf{s}}(t) = \begin{bmatrix} \mathbf{s}(t) \\ \mathbf{s}(t-1) \\ \vdots \\ \mathbf{s}(t-L_q) \end{bmatrix}$$

with $L_N = L + q$. Let $T_{\mathscr{H}}$ be a block Toeplitz matrix defined by

$$T_{\mathscr{H}} = \begin{bmatrix} H_0 & \cdots & H_L & \mathbf{0} & \cdots & \mathbf{0} \\ \mathbf{0} & \ddots & & \ddots & \ddots & \vdots \\ \vdots & \ddots & \ddots & & \ddots & \mathbf{0} \\ \mathbf{0} & \cdots & \mathbf{0} & H_0 & \cdots & H_L \end{bmatrix} \tag{8.33}$$

that has dimension $(q+1)P \times (q+L+1)M$. There holds

$$\underline{\mathbf{y}}(t) = T_{\mathscr{H}} \underline{\mathbf{s}}(t) + \underline{\mathbf{v}}(t). \tag{8.34}$$

Training signals are often employed to estimate the CIR. However, the use of training signals consume precious channel bandwidth. There is thus a strong incentive to estimate the channel blindly, given the statistics of the symbol sequence $\{\mathbf{s}(t)\}$. A common assumption is that $\mathbf{s}(t)$ is a WSS process and has mean zero and covariance Σ_s that is known at the receiver site. This subsection considers the subspace method for blind channel estimation.

Suppose that $P > M$. Then $T_{\mathscr{H}}$ is a strictly tall matrix, if $(q+1)(P-M) > LM$. Since L and M are fixed for each MIMO channel, the block Toeplitz matrix $T_{\mathscr{H}}$ can be made strictly tall by taking large q.

Lemma 8.1. *Let $T_{\mathscr{H}}$ of dimension $(q+1)P \times (q+L+1)M$ be the block Toeplitz matrix defined in (8.33) with $P > M$ and $(q+1)(P-M) > LM$. Suppose that both H_0 and H_L have the full column rank. Then $T_{\mathscr{H}}$ has the full column rank, if*

$$\text{rank}\{\ \mathbf{H}(z)\ \} = M \ \forall\, z \in \mathbb{C} \tag{8.35}$$

where $\mathbf{H}(z) = H_0 + H_1 z^{-1} + \cdots + H_L z^{-L}$ is the channel transfer matrix.

Proof. The contrapositive argument will be used for the proof. Suppose that $T_{\mathscr{H}}$ has the full column rank, but the rank condition (8.35) fails. Since $\mathbf{H}(z)$ losses its rank for some $z = z_0$, there exists $\mathbf{s} \neq \mathbf{0}$ such that $\mathbf{H}(z_0)\mathbf{s} = 0$. The hypothesis that both H_0 and H_L have the full column rank implies that $z_0 \neq 0$ and $z_0 \neq \infty$. As a result,

$$\sum_{i=0}^{L} H_i z_0^{-(i+k)} \mathbf{s} = \mathbf{0}$$

for each positive integer k. By taking $\mathbf{s}(t-k) = z_0^{-k}\mathbf{s}$ for each element of $\underline{\mathbf{s}}(t)$ in (8.34) yields $T_{\mathscr{H}}\underline{\mathbf{s}}(t) = \mathbf{0}$, contradicting to the full column rank assumption for $T_{\mathscr{H}}$ at the beginning. The proof is now complete. $\square$

A common assumption on the measurement noise is that it is not only temporally but also spatially white. Hence, $\mathrm{E}\{\mathbf{v}(t)\mathbf{v}(t)'\} = \sigma_v^2 I$. Under the condition that $\{\mathbf{s}(t)\}$ and $\{\mathbf{v}(t)\}$ are independent random processes,

$$\Sigma_y = \mathrm{E}\left\{\mathbf{y}(t)\mathbf{y}(t)'\right\} = T_{\mathscr{H}}\Sigma_s T_{\mathscr{H}}' + \sigma^2 I.$$

Both $\{\mathbf{s}(t)\}$ and $\{\mathbf{v}(t)\}$ are assumed to be not only WSS, but also ergodic. Consequently, there holds

$$\hat{\Sigma}_y = \frac{1}{N}\sum_{t=0}^{N-1} \mathbf{y}(t)\mathbf{y}(t)' \longrightarrow \Sigma_y = T_{\mathscr{H}}\Sigma_s T_{\mathscr{H}}' + \sigma^2 I, \tag{8.36}$$

as $N \to \infty$. Applying eigenvalue decomposition to Σ_y yields

$$\Sigma_y = \begin{bmatrix} G_{11} & G_{12} \\ G_{21} & G_{22} \end{bmatrix} \begin{bmatrix} \Lambda_1 & \mathbf{0} \\ \mathbf{0} & \sigma_v^2 I_\nu \end{bmatrix} \begin{bmatrix} G_{11}' & G_{21}' \\ G_{12}' & G_{22}' \end{bmatrix}$$

where the partitions are compatible and $\nu = (q+1)(P-M) - LM > 0$. Recall $\mathscr{R}\{\cdot\}$ for the range space and $\mathscr{N}\{\cdot\}$ for the null space. There hold

$$\mathscr{R}\{T_{\mathscr{H}}\} = \mathscr{R}\left\{\begin{bmatrix} G_{11} \\ G_{21} \end{bmatrix}\right\}, \quad \mathscr{N}\{T_{\mathscr{H}}'\} = \mathscr{R}\left\{\begin{bmatrix} G_{12} \\ G_{22} \end{bmatrix}\right\}. \tag{8.37}$$

The former is termed signal subspace and the latter termed noise subspace. The orthogonality of the two subspaces leads to the subspace method for channel estimation. Specifically, $T_{\mathscr{H}} = T_{\mathscr{H}}(\{H_i\})$, and there holds

$$\begin{bmatrix} G_{12}' & G_{22}' \end{bmatrix} T_{\mathscr{H}}(\{H_i\}) = \mathbf{0}. \tag{8.38}$$

However, the precise Σ_y is not available due to finitely many samples of the received signal and the existence of the measurement error. Hence, the relation in (8.38) does not hold, if the eigen-matrix G is computed based on the estimated Σ_y. Nevertheless,

(8.38) suggests an effective way for channel estimation by searching for $\{\widehat{H}_i\}$ to minimize

$$J_{\mathscr{F}} = \left\| \left[G_{12}' \; G_{22}' \right] T_{\mathscr{H}}(\{\widehat{H}_i\}) \right\|_{\mathscr{F}} \quad \text{subject to} \quad \sum_{i=0}^{L} \widehat{H}_i' \widehat{H}_i = I. \tag{8.39}$$

The normalization constraint is necessary to prevent the CIR estimates $\{\widehat{H}_i\}$ from being the meaningless zero. Although other norms and normalization constraints can be adopted, the constrained minimization of $J_{\mathscr{F}}$ in (8.39) leads to a simpler solution for the CIR estimates $\{\widehat{H}_i\}$.

Specifically, let $C = \left[G_{12}' \; G_{22}' \right]$ that has dimension $v \times (q+1)P$. It can be partitioned into $(q+1)$ blocks of the same size as follows:

$$\left[G_{12}' \; G_{22}' \right] = \left[C_0 \; C_1 \; \cdots \; C_q \right].$$

Thus, each C_i has dimension $v \times P$. Let $F = [G_{12}' \; G_{22}']T_{\mathscr{H}}(\{\widehat{H}_i\})$. Recall that $T_{\mathscr{H}}(\{\widehat{H}_i\})$ has dimension $(q+1)P \times (q+L+1)M$. There exists a partition

$$\left[G_{12}' \; G_{22}' \right] T_{\mathscr{H}}(\{\widehat{H}_i\}) = \left[F_0 \; F_1 \; \cdots \; F_{q+L+1} \right] \tag{8.40}$$

with $\{F_i\}$ of the same dimension of $v \times M$. Denote Θ as the parameter matrix of dimension $(L+1)P \times M$ with H_i being the $(i+1)$th block, $\underline{F}$ as a $(q+L+1)v \times M$ matrix with F_i as the $(i+1)$th block, and $T_{\mathscr{C}}$ as a block Toeplitz matrix consisting of $\{C_i\}$ as shown next:

$$\widehat{\Theta} = \begin{bmatrix} \widehat{H}_0 \\ \widehat{H}_1 \\ \vdots \\ \widehat{H}_L \end{bmatrix}, \quad \underline{F} = \begin{bmatrix} F_0 \\ F_1 \\ \vdots \\ F_{N+L+1} \end{bmatrix}, \quad T_{\mathscr{C}} = \begin{bmatrix} C_0 & \mathbf{0} & \cdots & \mathbf{0} \\ \vdots & \ddots & \ddots & \vdots \\ \vdots & & \ddots & \mathbf{0} \\ \vdots & & & C_0 \\ C_N & & & \vdots \\ \mathbf{0} & \ddots & & \vdots \\ \vdots & \ddots & \ddots & \vdots \\ \mathbf{0} & \cdots & \mathbf{0} & C_N \end{bmatrix},$$

assuming that $q \geq L$. It can be verified that $\underline{F} = T_{\mathscr{C}}\widehat{\Theta}$ and

$$J_{\mathscr{F}} = \left\| \left[G_{12}' \; G_{22}' \right] T_{\mathscr{H}}\left(\{\widehat{H}_i\} \right) \right\|_{\mathscr{F}} = \sum_{i=0}^{q+L+1} \operatorname{Tr}\{F_i' F_i\} = \underline{F}'\underline{F}.$$

Therefore, the constrained minimization of $J_{\mathscr{F}}$ is the same as minimization of

$$J_{\mathscr{F}} = \operatorname{Tr}\left\{ \widehat{\Theta}' T_{\mathscr{C}}' T_{\mathscr{C}} \widehat{\Theta} \right\} \quad \text{subject to} \quad \widehat{\Theta}'\widehat{\Theta} = I. \tag{8.41}$$

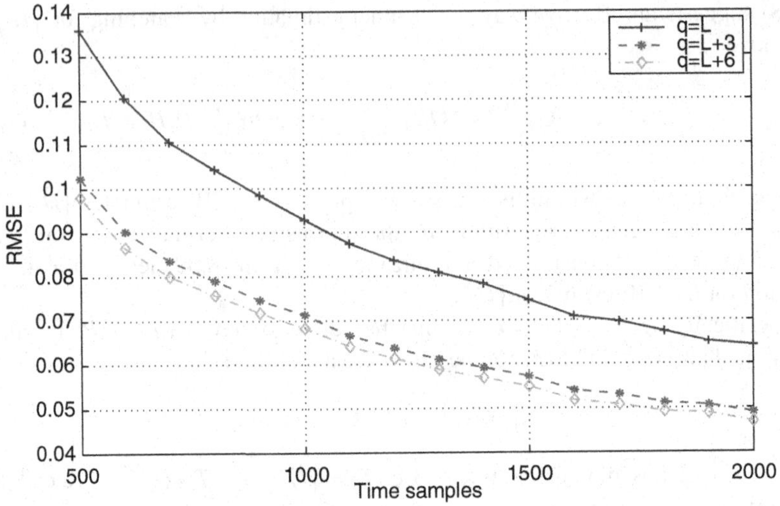

Fig. 8.5 RMSE for blind channel estimation

It is important to observe that $T_{\mathscr{C}}'T_{\mathscr{C}}$ is a block Toeplitz matrix. The minimizer consists of the M right singular vectors $\{\mathbf{v}_i\}$ corresponding to the M smallest nonzero singular values $\{\sigma_i\}$ that can be obtained via SVD of $T_{\mathscr{C}} = USV'$ and $V = \begin{bmatrix} \mathbf{v}_1 & \mathbf{v}_2 & \cdots & \mathbf{v}_{(L+1)M} \end{bmatrix}$. Since the true second-order statistics Σ_y is not available, its sampled version $\hat{\Sigma}_y$ defined in (8.36) has to be used. As a result, $\widehat{\Theta}_{\text{opt}} \neq \Theta$ in general with $\widehat{\Theta}_{\text{opt}}$ the solution to the constrained minimization in (8.41). The performance of the subspace algorithm depends on SNR and on the estimation of the noise subspace.

Example 8.8. Consider the same CIR in Example 8.2 with $P = 2$, $M = 1$, and $L = 2$. The transmitted signal consists of binary symbols that is white. The SNR, defined as the ratio of the signal power to the noise power, is taken to be 0 dB. The received signal is measured at 2000 times samples and used to compute the sampled second-order statistics Σ_y. The constrained minimization of (8.39) is employed to compute the estimated CIR. Let $\{\widehat{H}_i^{(k)}\}$ be the estimated CIR for the kth emsemble run. The RMSE is computed according to

$$\text{RMSE} = \sqrt{\frac{1}{T}\sum_{k=1}^{T} \text{Tr}\left\{\sum_{i=0}^{L}\left(H_i - \widehat{H}_i^{(k)}\right)'\left(H_i - \widehat{H}_i^{(k)}\right)\right\}}$$

with a total of $T = 2500$ ensemble runs. The simulation results are plotted in the Fig. 8.5 that shows the improvement when N increases. However the improvement due to large N diminishes as SNR increases that is why a small SNR is used in this example.

The subspace algorithm for blind channel estimation is closely related to the TLS algorithm studied in the previous subsection. Both compute the sampled second-order statistics, and both use eigenvalue decomposition. The difference lies in the Toeplitz structure of the second-order statistics for blind channel estimation that prevents the subspace algorithm from being MLE. Nonetheless, the following result is true.

Theorem 8.9. *Let $\widehat{\Theta}_{\rm opt}$ be the channel estimate for blind channel estimation based on the subspace algorithm. If the input signal is both temporally with nonsingular Σ_s, then the estimate $\widehat{\Theta}_{\rm opt}$ converges to the true Θ as the number of time samples approaches infinity* w.p.1.

Proof. The assumption on the input signal implies that (8.36) holds. In addition, the convergence has probability 1 which is the same as that for the TLS algorithm. Hence, the strong convergence holds. □

Thus far, the optimality of the subspace method is not addressed. The main hurdle lies in the structure of the signal model that is not in the same form as the EIV model. It will be shown in the next subsection that the subspace is asymptotically optimal in the sense of MLE.

8.2.3 Graph Space Method

For identification of the plant model in feedback control, the subspace method can also be used to estimate the plant parameters. Let $\mathbf{P}(z)$ be the transfer matrix with m input/p output. It is assumed that $\mathbf{P}(z) = \mathbf{B}(z)\mathbf{A}(z)^{-1}$ with

$$\mathbf{A}(z) = I + \sum_{k=1}^{L} A_k z^{-k}, \quad \mathbf{B}(z) = \sum_{k=0}^{L} B_k z^{-k}. \tag{8.42}$$

Even though the physical system is strictly causal, $B_0 \neq \mathbf{0}$ is assumed which can help to reduce the modeling error. It is further assumed that

$$\text{rank}\left\{\begin{bmatrix}\mathbf{A}(z)\\ \mathbf{B}(z)\end{bmatrix}\right\} = m \quad \forall\, z \in \mathbb{C}.$$

The above ensures that $\{\mathbf{A}(z), \mathbf{B}(z)\}$ are right coprime. Because physical systems in practice are more complex than the linear finite dimensional models, this assumption holds for most real systems.

Let $\{\mathbf{u}(t)\}$, and $\{\mathbf{y}(t)\}$ be the input and output of the system, respectively. The graph space associated with the plant model $\mathbf{P}(z)$ is defined as

$$\underline{\mathbf{G}}_P := \left\{\mathbf{z}(t) = \begin{bmatrix}\mathbf{u}(t)\\ \mathbf{y}(t)\end{bmatrix} \middle| : \exists\, \mathbf{w}(t) : \ \mathbf{z}(t) = \begin{bmatrix}\mathbf{A}(q)\\ \mathbf{B}(q)\end{bmatrix}\mathbf{w}(t)\right\}. \tag{8.43}$$

The unknown signal $\{\mathbf{w}(t)\}$ will be referred to auxiliary input. For this reason, an FIR transfer matrix $\mathbf{G}(z)$ can be defined via

$$\mathbf{G}(z) = \sum_{k=0}^{L} G_k z^{-k}, \quad G_k = \begin{bmatrix} A_k \\ B_k \end{bmatrix}, \tag{8.44}$$

where $A_0 = I_m$ is taken. By taking $\mathbf{z}(t)$ as the observation and $\mathbf{w}(t)$ as the unknown input, system identification for the plant model $\mathbf{P}(z)$ is converted to parameter estimation for $\{G_k\}$. As a result, the subspace method from the previous subsection can be employed to estimate the system parameters. To emphasize the graph space of the system and to distinguish it from blind channel estimation, the subspace method used for control system identification is termed as the graph space method.

Denote $\Theta = \begin{bmatrix} G_0' \; G_1' \; \cdots \; G_L' \end{bmatrix}'$ as the parameter matrix of the system. The constraint $\Theta'\Theta = I$ from the subspace method is replace by the first square block of Θ being $A_0 = I$. However, this constraint does not change the estimation algorithm. Let $\mathbf{v}(t)$ be the noise vector comprising both measurement errors at the plant input and output. There holds

$$\mathbf{z}(t) = \sum_{k=0}^{L} G_k \mathbf{w}(t-k) + \mathbf{v}(t) \tag{8.45}$$

that is almost identical to (8.32). Define the block Toeplitz matrix $T_{\mathscr{G}}$ of dimension $(q+1)(p+m) \times (L+q+1)m$ as

$$T_{\mathscr{G}} = \begin{bmatrix} G_0 & \cdots & G_L & \mathbf{0} & \cdots & \mathbf{0} \\ \mathbf{0} & \ddots & & \ddots & \ddots & \vdots \\ \vdots & \ddots & \ddots & & \ddots & \mathbf{0} \\ \mathbf{0} & \cdots & \mathbf{0} & G_0 & \cdots & G_L \end{bmatrix} \tag{8.46}$$

that is identical to (8.33), except that H_i is replaced by G_i for $0 \le i \le L$. Similarly, denote

$$\underline{\mathbf{z}}(t) = \begin{bmatrix} \mathbf{z}(t) \\ \mathbf{z}(t-1) \\ \vdots \\ \mathbf{z}(t-q) \end{bmatrix}, \quad \underline{\mathbf{v}}(t) = \begin{bmatrix} \mathbf{v}(t) \\ \mathbf{v}(t-1) \\ \vdots \\ \mathbf{v}(t-q) \end{bmatrix}, \quad \underline{\mathbf{w}}(t) = \begin{bmatrix} \mathbf{w}(t) \\ \mathbf{w}(t-1) \\ \vdots \\ \mathbf{w}(t-L_q) \end{bmatrix}$$

with $L_q = L + q$. It follows that at each time sample t,

$$\underline{\mathbf{z}}(t) = T_{\mathscr{G}} \underline{\mathbf{w}}(t) + \underline{\mathbf{v}}(t). \tag{8.47}$$

The observation noise vector $\mathbf{v}(t)$ is assumed to be both spatially and temporally white with noise variance σ^2. The sampled second-order statistics can be computed via

$$\hat{\Sigma}_z = \frac{1}{N}\sum_{t=0}^{N-1} \mathbf{z}(t)\mathbf{z}(t)' \longrightarrow \Sigma_z = T_{\mathscr{G}}\Sigma_w T_{\mathscr{G}}' + \sigma^2 I$$

as $N \to \infty$. The above convergence has probability 1, and is similar to that for blind channel estimation.

For the graph space method, the persistent excitation (PE) for the input signal $\{\mathbf{u}(t)\}$ needs to be assumed which ensures strictly positivity of Σ_w. Hence, under the PE condition and $\mu = (q+1)p - Lm > 0$, the covariance matrix Σ_z has precisely μ zero eigenvalues. Let $\{\mathbf{x}_i\}_{i=1}^{\mu}$ be the corresponding eigenvectors that span the noise subspace of the sampled second-order statistics. Denote

$$X = \begin{bmatrix} \mathbf{x}_1 \ \mathbf{x}_2 \ \cdots \ \mathbf{x}_\mu \end{bmatrix}.$$

The graph space method is aimed at searching for $\{\widehat{G}_i\}$ to minimize

$$J_{\mathscr{F}} = \left\| X' T_{\mathscr{G}}\left(\{\widehat{G}_i\}\right)\right\|_{\mathscr{F}}^2 \quad \text{subject to} \quad \begin{bmatrix} I_m \ \mathbf{0} \end{bmatrix} \widehat{G}_0 = I_m. \tag{8.48}$$

The matrices X and $F = X'T_{\mathscr{G}}(\{\widehat{G}_i\})$ have dimensions of $(q+1)(p+m) \times \mu$ and $\mu \times (L+q+1)m$, respectively. Partition these two matrices in accordance with

$$X' = \begin{bmatrix} X_0 \ X_1 \ \cdots \ X_q \end{bmatrix},$$
$$F = \begin{bmatrix} F_0 \ F_1 \ \cdots \ F_{L+q+1} \end{bmatrix},$$

of which each X_i has the dimension $\mu \times (p+m)$ and each F_i has the dimension $\mu \times m$. There holds $F = T_{\mathscr{X}}\widehat{\Theta}$ where

$$\widehat{\Theta} = \begin{bmatrix} \widehat{G}_0 \\ \widehat{G}_1 \\ \vdots \\ \widehat{G}_L \end{bmatrix}, \quad F = \begin{bmatrix} F_0 \\ F_1 \\ \vdots \\ F_{q+L+1} \end{bmatrix}, \quad T_{\mathscr{X}} = \begin{bmatrix} X_0 & \mathbf{0} & \cdots & \mathbf{0} \\ \vdots & \ddots & \ddots & \vdots \\ \vdots & & \ddots & \mathbf{0} \\ \vdots & & & X_0 \\ X_q & & & \vdots \\ \mathbf{0} & \ddots & & \vdots \\ \vdots & \ddots & \ddots & \vdots \\ \mathbf{0} & \cdots & \mathbf{0} & X_q \end{bmatrix},$$

assuming that $q \geq L$. As a result,

$$J_{\mathscr{F}} = \left\| X' T_{\mathscr{G}}\left(\{\widehat{G}_i\}\right) \right\|_{\mathscr{F}}^2 = \sum_{i=0}^{q+L+1} \operatorname{Tr}\{F_i' F_i\} = F'F.$$

Therefore, the constrained minimization of $J_{\mathscr{F}}$ in (8.48) is the same as minimization of

$$J_{\mathscr{F}} = \operatorname{Tr}\left\{\widehat{\Theta}' T_{\mathscr{X}}' T_{\mathscr{X}} \widehat{\Theta}\right\} \text{ subject to } \begin{bmatrix} I_m & \mathbf{0} \end{bmatrix} \widehat{\Theta} = I_m. \tag{8.49}$$

The minimizer is given by $\widehat{\Theta}_{\text{opt}} = \begin{bmatrix} \mathbf{v}_{Lm+1} & \mathbf{v}_{Lm+2} & \cdots & \mathbf{v}_{(L+1)m} \end{bmatrix} \Omega^{-1}$, consisting of the m right singular vectors $\{\mathbf{v}_i\}_{i=Lm+1}^{(L+1)m}$ corresponding to the m smallest nonzero singular values $\{\sigma_i\}_{i=Lm+1}^{(L+1)m}$. The normalization matrix Ω is used to satisfy the constraint $\begin{bmatrix} I_m & \mathbf{0} \end{bmatrix} \widehat{\Theta} = I_m$. Similar to the subspace algorithm, the right singular vectors can be computed via SVD.

Theorem 8.10. *Consider square transfer matrix* $\mathbf{P}(z) = \mathbf{B}(z)\mathbf{A}(z)^{-1}$ *with* $\{\mathbf{A}(z), \mathbf{B}(z)\}$ *specified in (8.42). Suppose that* $\mathbf{G}(z)$ *defined in (8.44) is right coprime, the input* $\{\mathbf{u}(t)\}$ *is* PE, *and the noise vectors* $\{\mathbf{v}(t)\}$ *in (8.45) are both spatially and temporally white and Gauss with variance* σ^2. *If the auxiliary input* $\{\mathbf{w}(t)\}$ *is also spatially and temporally white with covariance identity, then the graph space algorithm in this subsection is asymptotically optimal in the sense of* MLE, *and the estimate* $\widehat{\Theta}_{\text{opt}}$ *converges to the true system parameter matrix* Θ *with probability* 1.

Proof. By the hypothesis, $p = m$, although the result is true for $p \neq m$. The right coprime assumption implies the existence of

$$\mathbf{G}_\ell(z) = \sum_{i=0}^{L} \begin{bmatrix} A_{\ell_k} & B_{\ell_k} \end{bmatrix} z^{-k}$$

such that $\mathbf{G}_\ell(z) J \mathbf{G}(z) \equiv \mathbf{0}$ where

$$J = \begin{bmatrix} \mathbf{0} & I_p \\ -I_m & \mathbf{0} \end{bmatrix}, \quad A_{\ell_0} = I.$$

That is, $\mathbf{A}_\ell(z)\mathbf{B}(z) = \mathbf{B}_\ell(z)\mathbf{A}(z)$ with

$$\mathbf{A}_\ell(z) = I + \sum_{i=1}^{L} A_{\ell_k} z^{-k}, \quad \mathbf{B}_\ell(z) = \sum_{i=0}^{L} B_{\ell_k} z^{-k},$$

and thus, $\{\mathbf{A}_\ell(z), \mathbf{B}_\ell(z)\}$ are left coprime and $\mathbf{P}(z) = \mathbf{A}_\ell(z)^{-1}\mathbf{B}_\ell(z)$. Since $A_{\ell_0} = A_0$ by $p = m$, the left coprime factors $\{\mathbf{A}_\ell(z), \mathbf{B}_\ell(z)\}$ are uniquely determined by right coprime factors $\{\mathbf{A}(z), \mathbf{B}(z)\}$ up to a unitary matrix, and vice versa. As a result, identification of the right coprime factors is the same as that of the left coprime factors. Consider the case $q = L$. Denote

$$\mathbf{z}_J(t) = J\mathbf{z}(t) = \begin{bmatrix} \mathbf{y}(t) \\ -\mathbf{u}(t) \end{bmatrix}.$$

In the noise-free case, there holds the relation

$$\begin{bmatrix} I \ \Theta_\ell \end{bmatrix} \underline{\mathbf{z}}_J(t) = \mathbf{0}, \quad \Theta_\ell = \begin{bmatrix} B_{\ell_0} \ A_{\ell_1} \ B_{\ell_1} \ \cdots \ A_{\ell_L} \ B_{\ell_L} \end{bmatrix}$$

where $\underline{\mathbf{z}}_J(t)$ is the blocked column vector of $\mathbf{z}_J(t)$ with size $(L+1)(p+m)$. Clearly, $\underline{\mathbf{z}}_J(t)$ is permutation of $\underline{\mathbf{z}}(t)$. In the noisy case, an EIV model is resulted in but the elements of the noise matrix are not i.i.d. anymore.

Without loss of generality, the measurements $\underline{\mathbf{z}}_J(t)\}$ at times samples $[t_0,\ t_f]$ are assumed. Hence, the corresponding EIV model is given by

$$\begin{bmatrix} \underline{\mathbf{z}}_J(t_0) \cdots \underline{\mathbf{z}}_J(t_f) \end{bmatrix} = \begin{bmatrix} \underline{\mathbf{z}}_J^{(0)}(t_0) \cdots \underline{\mathbf{z}}_J^{(0)}(t_f) \end{bmatrix} + \begin{bmatrix} \underline{\mathbf{v}}_J(t_0) \cdots \underline{\mathbf{v}}_J(t_f) \end{bmatrix}$$

with $\underline{\mathbf{z}}_J^{(0)}(t)$ the noise-free blocked graph signal at time t. Indeed, the elements of the noise matrix are not i.i.d., because $\underline{\mathbf{v}}_J(t)$ is a blocked column vector of $\mathbf{v}_J(t) = J\mathbf{v}(t)$ consisting of $\{\mathbf{v}(t-i)\}_{i=0}^N$. It follows that the TLS solution is not an MLE for Θ_ℓ. On the other hand, let

$$\varepsilon_N^2 = \frac{1}{N} \sum_{t=t_0}^{t_f} \left\| \underline{\mathbf{z}}_J(t) - \underline{\mathbf{z}}_J^{(0)}(t) \right\|^2, \quad N = t_f - t_0 + 1.$$

Since $\mathbf{v}_J(t) = \underline{\mathbf{z}}_J(t) - \underline{\mathbf{z}}_J^{(0)}(t)$, it can be verified that

$$\varepsilon_N^2 = \frac{q}{N} \sum_{t=t_0}^{t_f - q} \|\mathbf{v}_J(t)\|^2 + \frac{1}{N} \sum_{i=1}^{q-1} \left[(q-i)\|\mathbf{v}_J(t_0 - i)\|^2 + i\|\mathbf{v}_J(t_f - i + 1)\|^2 \right].$$

Recall $q = L$ that is fixed and finite. The second summation on the right-hand side of the above equation approaches zero as $N \to \infty$. Hence, the TLS solution minimizes ε_T^2 asymptotically in the same spirit of MLE. In addition, the white assumption on $\mathbf{w}(t)$ leads to

$$\frac{1}{N} \sum_{t=t_0}^{t_f} \underline{\mathbf{z}}(t)\underline{\mathbf{z}}(t)' \longrightarrow Q_J T_{\mathscr{G}} T_{\mathscr{G}}' Q_J' + \sigma^2 I$$

for some permutation matrix Q_J dependent on J. The right-hand side is deterministic. Consequently, the TLS solution for $\widehat{\Theta}_\ell$, and thus the graph space estimate $\widehat{\Theta}_{\rm opt}$, are indeed the asymptotic MLE. The convergence with probability 1 follows from the PE condition and identity covariance of $\{\mathbf{w}(t)\}$. The proof for the case of $q > L$ can be covered by adding zero blocks to Θ_ℓ. The proof is now complete. □

It is known that the error covariance for MLE approaches the CRLB asymptotically under certain regularity condition. To compute the CRLB, it is necessary to obtain first the corresponding FIM. Denote $f_V(\{\mathbf{z}(t)\})$ as the joint PDF. By the Gauss assumption and the signal model in (8.45),

$$\ln f_V(\{\mathbf{z}(t)\}) \sim J = -\frac{1}{2\sigma^2}\sum_{t=t_0}^{t_f} \mathrm{Tr}\left\{\left[\mathbf{z}(t) - \tilde{\Theta}\mathbf{w}(t)\right]\left[\mathbf{z}(t) - \tilde{\Theta}\mathbf{w}(t)\right]'\right\}, \tag{8.50}$$

where $\tilde{\Theta} = \begin{bmatrix} G_0 \; G_1 \; \cdots \; G_L \end{bmatrix}$. Denote

$$\vartheta = \begin{bmatrix} \mathrm{vec}(G_0) \\ \vdots \\ \mathrm{vec}(G_L) \end{bmatrix}, \quad \underline{\mathbf{z}}_N = \begin{bmatrix} \mathbf{z}(t_f) \\ \vdots \\ \mathbf{z}(t_0) \end{bmatrix},$$

$$\underline{\mathbf{w}}_{T+L} = \begin{bmatrix} \mathbf{w}(t_f) \; \vdots \\ \mathbf{w}(t_0 - L) \end{bmatrix}, \quad \underline{\mathbf{v}}_N = \begin{bmatrix} \mathbf{v}(t_f) \\ \vdots \\ \mathbf{v}(t_0) \end{bmatrix}.$$

Direct calculation yields

$$\begin{aligned} \frac{\partial J}{\partial \vartheta} &= \frac{1}{\sigma^2}\begin{bmatrix} \mathbf{w}_1(t_f) I_{p+m} & \cdots & \mathbf{w}_1(t_0) I_{p+m} \\ \vdots & \cdots & \vdots \\ \mathbf{w}_{(L+1)m}(t_f) I_{p+m} & \cdots & \mathbf{w}_{(L+1)m}(t_0) I_{p+m} \end{bmatrix}\begin{bmatrix} \mathbf{v}(t_f) \\ \vdots \\ \mathbf{v}(t_0) \end{bmatrix} \\ &=: \frac{1}{\sigma^2} M_{\mathbf{w}} \underline{\mathbf{v}}_N \end{aligned} \tag{8.51}$$

where $\mathbf{w}_k(t)$ is the kth component of $\mathbf{w}(t)$. A caution needs to be taken to that the first m rows of ϑ are the same as I_m which are known. Let $\tilde{\vartheta}$ be obtained from ϑ by deleting the m^2 known elements. Then

$$\frac{\partial J(\tilde{\Theta})}{\partial \tilde{\vartheta}} = \frac{1}{\sigma^2}\tilde{M}_{\mathbf{w}}\underline{\mathbf{v}}_N, \quad \tilde{M}_{\mathbf{w}} = \begin{bmatrix} Z & 0 \\ 0 & I_{mpL} \end{bmatrix} M_{\mathbf{w}}, \tag{8.52}$$

and $Z = I_m \otimes \begin{bmatrix} \mathbf{0} \; I_p \end{bmatrix}$.

It is important to notice that the auxiliary input $\{\mathbf{w}(t)\}$ is also unknown. Its impact to the CRLB needs to be taken into account. Let $T_{\mathscr{G}}(\Theta)$ be the same as in (8.46) but with blocking size $q = N$. There holds

$$\underline{\mathbf{z}}_N = T_{\mathscr{G}}(\Theta)\underline{\mathbf{w}}_{N+L} + \underline{\mathbf{v}}_N.$$

Then the likelihood function in (8.50) can be written alternatively as

$$J(\Theta, \mathbf{w}) = -\frac{1}{2\sigma^2}\left[\underline{\mathbf{z}}_N - T_{\mathscr{G}}(\Theta)\underline{\mathbf{w}}_{N+L}\right]'\left[\underline{\mathbf{z}}_N - T_{\mathscr{G}}(\Theta)\underline{\mathbf{w}}_{N+L}\right]. \quad (8.53)$$

Thus, the partial derivative of $J(\Theta, \mathbf{w})$ with respect to $\underline{\mathbf{w}}_{N+L}$ is given as

$$\frac{\partial J(\Theta, \mathbf{w})}{\partial \underline{\mathbf{w}}_{N+L}} = \frac{1}{\sigma^2}T_{\mathscr{G}}(\Theta)'\left[\underline{\mathbf{z}}_N - T_{\mathscr{G}}(\Theta)\underline{\mathbf{w}}_{N+L}\right] = \frac{1}{\sigma^2}T_{\mathscr{G}}(\Theta)'\underline{\mathbf{v}}_N. \quad (8.54)$$

To compute the FIM, the following matrices

$$\begin{aligned}
\text{FIM}(\Theta) = \text{E}\left\{\frac{\partial J(\Theta,\mathbf{w})}{\partial \tilde{\vartheta}}\frac{\partial J(\Theta,\mathbf{w})}{\partial \tilde{\vartheta}'}\right\} &= \frac{1}{\sigma^2}\tilde{M}_{\mathbf{w}}\tilde{M}_{\mathbf{w}}', \\
\text{E}\left\{\frac{\partial J(\Theta,\mathbf{w})}{\partial \tilde{\vartheta}}\frac{\partial J(\Theta,\mathbf{w})}{\partial \mathbf{w}'}\right\} &= \frac{1}{\sigma^2}\tilde{M}_{\mathbf{w}}T_{\mathscr{G}}(\Theta), \\
\text{FIM}(\mathbf{w}) = \text{E}\left\{\frac{\partial J(\Theta,\mathbf{w})}{\partial \mathbf{w}}\frac{\partial J(\Theta,\mathbf{w})}{\partial \mathbf{w}'}\right\} &= \frac{1}{\sigma^2}T_{\mathscr{G}}(\Theta)'T_{\mathscr{G}}(\Theta)
\end{aligned}$$

are useful. The CRLB for estimation of Θ can be obtained according to

$$\text{CRB}(\Theta) = \sigma^2\left(\text{FIM}(\Theta) - \tilde{M}_{\mathbf{w}}T_{\mathscr{G}}(\Theta)\left[T_{\mathscr{G}}(\Theta)'T_{\mathscr{G}}(\Theta)\right]^{-1}T_{\mathscr{G}}(\Theta)'\tilde{M}_{\mathbf{w}}'\right)^{-1}. \quad (8.55)$$

The above CRLB can be difficult to compute, if N, the number of time samples, is large, in light of the fact that the blocked Toeplitz matrix $T_{\mathscr{G}}(\Theta)$ has size $(N+1)(p+m) \times (N+L+1)m$. It is left as an exercise to derive an efficient algorithm for computing the CRLB (Problem 8.16).

Example 8.11. Consider the SISO plant model given by

$$P(z) = \frac{1.4496z^{-1} + 0.7248z^{-2}}{1 - 1.5z^{-1} + 0.7z^{-2}}.$$

This plant model has the same poles and zeros as the one in Example 8.3. The difference lies in the gain factor of 4. A total of $N = 3{,}000$ input and output measurements are generated by taking the auxiliary input $\{\mathbf{w}(t)\}$ as white Gauss of variance one. The resulting $\{u(t)\}$ and $\{y(t)\}$ are WSS and admit variance of 5.672 and 4.264 dB, respectively. The measurement error $\{v(t)\}$ is also taken as white Gauss with variance one, implying that the SNR for the input and output signals is 5.672 and 4.264 dB, respectively. Using the graph space method, the estimation errors are plotted against the CRLB similar to that in Example 8.8. A total of 2500 ensemble runs are used to compute the RMSE value of the estimation error. The RMSE is seen to converge to the CRLB, albeit slowly. In fact, a larger number of time samples are used in order to see such a convergence.

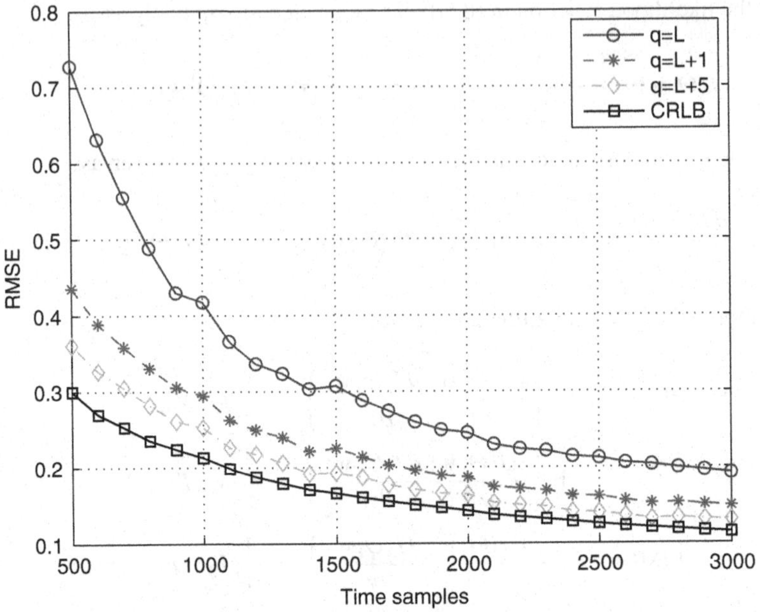

Fig. 8.6 RMSE curves for identification using the graph space method

The simulation results show that as q increases, the RMSE value decreases. However, the largest drop of the RMSE value occurs at $q = L+1$. As N increases beyond $L+1$, the decrease of the RMSE values slows down dramatically that is why the RMSE values are shown for only the cases of $q = L, L+1, L+5$. It needs to be pointed out that the white assumption for auxiliary input $\{\mathbf{w}(t)\}$ is important in order to achieve asymptotic MLE. Figure 8.6 shows the case when the input signal $\{u(t)\}$ is white with variance one, and both $\{y(t)\}$ and $\{w(t)\}$ become colored signals with variances 15.78 and 9.28, respectively. Under the same statistical noise $\{v(t)\}$, the total SNR is greater than the previous case. However, the simulation results in Fig. 8.7 shows that the RMSE values resulted from the graph space method do not converge to the CRLB. In fact, the larger the q, the worse the RMSE performance. The simulation results in this example indicates the importance of the auxiliary input being white, in order to obtain asymptotic MLE, which is consistent with the theoretical result in Theorem 8.10. Since white auxiliary input is not possible to generate prior to system identification, it is suggested to apply the LS algorithm first for system identification. After a reasonably good plant model is identified, white $\{\mathbf{w}(t)\}$ can be generated, and the plant input $\mathbf{u}(t) = \widehat{\mathbf{A}}(q)\mathbf{w}(t)$ can be obtained using the estimate model, which can then be applied as the exogenous input. Once the output measurements are available, the graph space algorithm can be applied to obtain more accurate identification results.

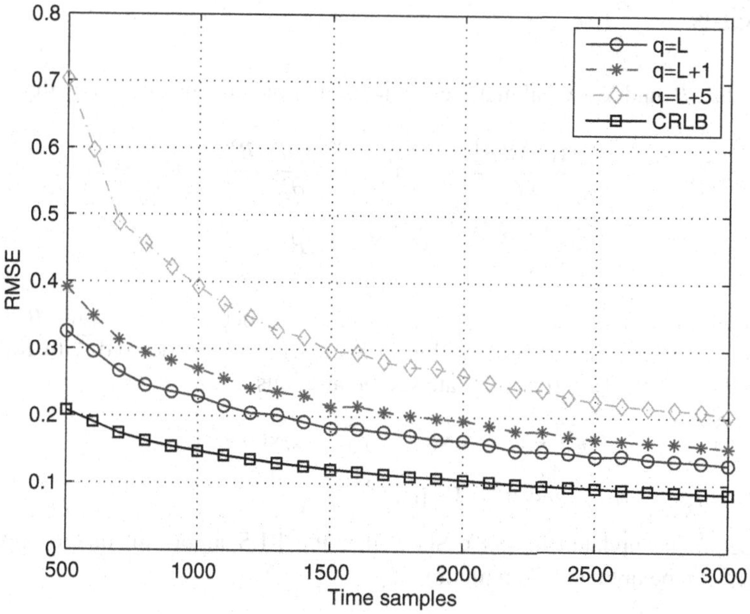

Fig. 8.7 RMSE curves for (graph space) identification with white input

Let σ_u^2 and σ_y^2 be the noise variances of the input and output, respectively. This subsection assumes $\sigma_u = \sigma_y = \sigma$ thus far. In the case when $\sigma_u \neq \sigma_y$,

$$\mathbf{z}_r(t) = \begin{bmatrix} r\mathbf{u}(t) \\ \mathbf{y}(t) \end{bmatrix}, \quad r = \frac{\sigma_y}{\sigma_u},$$

can be employed to replace $\mathbf{z}(t)$ in (8.45). The noise vectors associated with $\{\mathbf{z}_r(t)\}$ admit covariance $\sigma_y^2 I$, and thus, the graph space method can be applied to $\{\mathbf{z}_r(t)\}$ which estimates $r\mathbf{A}(z)$ and $\mathbf{B}(z)$. A more serious issue is how to estimate the variances σ_u^2 and σ_y^2. Methods have been developed in the research literature, and are not pursued in this book.

Notes and References

LS algorithm is presented in almost every textbook on system identification. See for instance [78, 102]. The RLS algorithm can not only be found in system identification books but also in adaptive control books [7, 38]. The TLS algorithm has a shorter history. A good source is [35, 113]. It is basically the same as the bias elimination LS [49, 103, 104]. Blind channel estimation based on subspace in [87] is connected to TLS. See also [51, 110] for blind channel estimation. The graph space method in this chapter is generalized from the subspace method in [87].

Exercises

8.1. Let A, B, and X be real matrices with compatible dimensions. Show that

$$\frac{\partial \operatorname{Tr}\{AXB\}}{\partial X} = A'B', \quad \frac{\partial \operatorname{Tr}\{AX'B\}}{\partial X} = BA,$$
$$\frac{\partial \operatorname{Tr}\{AXBX'\}}{\partial X} = A'XB' + AXB.$$

8.2. Suppose that $p = m = 1$, and thus, (8.2) is reduced to $y(t) = \underline{\phi}(t)'\theta$ where $\Theta = \theta'$. Show that the RLS algorithm in Sect. 8.1.1 can be derived with Kalman filter. (*Hint:* Use $\mathbf{x}(t) = \theta$ as the state vector, and thus,

$$\mathbf{x}(t+1) = A_t\mathbf{x}(t), \quad y(t) = \mathbf{c}_t\mathbf{x}(t) + v(t)$$

with $A_t = I$, $\mathbf{c}_t = \underline{\phi}(t)'$, and $\sigma_t^2 = \mathrm{E}\{|v(t)|^2\}$.)

8.3. Let P_t be updated in (8.8). Show that the RLS algorithm in the previous problem can be obtained from minimizing

$$J_\theta = \left(\widehat{\theta}_{t+1} - \widehat{\theta}_t\right)' P_t^{-1}\left(\widehat{\theta}_{t+1} - \widehat{\theta}_t\right) + \sigma_t^{-2}\left[y(t) - \underline{\phi}(t)'\widehat{\theta}_{t+1}\right]^2.$$

8.4. Show that $(D - CA^{-1}B)^{-1} = D^{-1} + D^{-1}C(A - BD^{-1}C)^{-1}BD^{-1}$. (*Hint:* Recall the inverse of the square transfer matrix:

$$[D + C(zI - A)^{-1}B]^{-1} = D^{-1} - D^{-1}C(zI - A + BD^{-1}C)^{-1}BD^{-1}$$

and then set $z = 0$.)

8.5. For the LS solution $\Theta_{\rm LS} = Y_{0,f}\Phi_{0,f}'(\Phi_{0,f}\Phi_{0,f}')^{-1}$ in Theorem 8.1, denote $\underline{\theta}_i$ and $\widehat{\underline{\theta}}_i$ as the ith row of Θ and $\Theta_{\rm LS}$, respectively. Assume that $\Phi_{0,f}$ is noise-free. Show that

$$\text{(i) } \mathrm{E}\left\{\widehat{\underline{\theta}}_i\right\} = \underline{\theta}_i, \quad \text{(ii) } \mathrm{E}\left\{\left(\widehat{\underline{\theta}}_i - \underline{\theta}_i\right)'\left(\underline{\theta}_i - \widehat{\underline{\theta}}_i\right)\right\} = \sigma_i^2(\Phi_{0,f}\Phi_{0,f}')^{-1}.$$

8.6. Prove the RLS algorithm with the forgetting factor in (8.10) and (8.11).

8.7. Program the RLS algorithm, and the RLS algorithm with forgetting factor. Use the following transfer function as a testing example:

$$P(z) = \frac{1.4496z^{-1} + b_t z^{-2}}{1 - 1.5z^{-1} + a_t z^{-2}}$$

where both a_t and b_t lie in the interval of $[0.6,\ 0.8]$ and change slowly. The forgetting factor can be generated via $\gamma_t = \gamma_0\gamma_{t-1} + (1 - \gamma_0)$ with $\gamma_0 \in [0.95,\ 0.99]$.

8.8. Consider the input/output measurement model in Fig. 8.2. Show that the LS solution to system identification is MLE, if the plant model has an FIR structure and input is noise-free. What happens when the plant input is not noise-free?

8.9. Consider the input/output measurement model in Fig. 8.2. (i) If the plant input is noise-free, show that the LS solution is asymptotically unbiased, and (ii) if the plant input involves noise, show that the corresponding RMSE depends on the system parameters.

8.10. For the signal model in (8.2) for $t \in [t_0,\ t_f]$, arising from the system input/output description in (8.1) in which the observation noises corrupt both input and output signals, show that

1. The matrix $\Phi_{0,f}$ in $Y_{0,f} = \Theta\Phi_{0,f} + V_{0,f}$ involves observation noises if $M_k \neq \mathbf{0}$ for at least one $k > 0$.
2. The LS solution can be written as

$$\Theta_{\rm LS} = \Theta + V_{0,f}\Phi_{0,f}'(\Phi_{0,f}\Phi_{0,f}')^{-1}.$$

3. Show that $\Theta_{\rm LS}$ is a biased estimate of Θ, if $M_k \neq \mathbf{0}$ for at least one $k > 0$.

(*Hint:* $\mathrm{E}\{V_{0,f}\Phi_{0,f}'(\Phi_{0,f}\Phi_{0,f}')^{-1}\} \neq \mathbf{0}$, because the noise components of $\{\mathbf{v}(t)\}$ corrupted to $\{\mathbf{y}(t)\}$ cannot be removed from $\Phi_{0,f}$, if $M_k \neq \mathbf{0}$ for at least one $k > 0$.)

8.11. Consider partition of the eigenvector matrix G and eigenvalue matrix Λ in (8.21).

(i) Show that

$$G_{12}'G_{11} + G_{22}'G_{21} = \mathbf{0}.$$

(ii) Show that G_{11} is nonsingular, if and only if G_{22} is nonsingular.
(iii) Show that $\Theta_{\rm TLS}$ in (8.22) is indeed the TLS solution.

8.12. Show that the optimal solution to (8.30) is the generalized TLS solution.

8.13. Consider minimization of $J_{A,B}$ in (8.19). Show that $\min J_{A,B} = \mathrm{Tr}\{\Lambda_2\}$ where Λ_2 of dimension $p \times p$ is defined in (8.21).

8.14. For the case $p = 1$, the bias-eliminating LS estimate is given by

$$\widehat{\Theta} = Y_{0,f}\Phi_{0,f}'(\Phi_{0,f}\Phi_{0,f}' - \widehat{\lambda}_{\min}^2 I)^{-1}$$

where $\widehat{\lambda}_{\min}$ is the minimum eigenvalue of

$$\hat{\Sigma}_z = \frac{1}{N}\sum_{t=0}^{N-1}\mathbf{z}(t)\mathbf{z}(t)'$$

with $\mathbf{z}(t)$ the same as in (8.43). Show that the bias-eliminating LS estimate is the same as the TLS solution.

8.15. Show that the TLS solution to $Y_{0,f} \approx \Theta\Phi_{0,f}$ minimizes

$$J_{\text{TLS}}(T) := \text{Tr}\left\{ \left(Y_{0,f} - \Theta\Phi_{0,f}\right)' \left(I + \Theta\Theta'\right)^{-1} \left(Y_{0,f} - \Theta\Phi_{0,f}\right) \right\}.$$

8.16. The CRLB in (8.55) is difficult to compute, if the time horizon $[t_0,\ t_f]$ is large. This exercise provides a guideline on an efficient algorithm for computing the CRLB in (8.55) in the case of large time horizon:

1. Show that $T_{\mathscr{G}}(\Theta)'T_{\mathscr{G}}(\Theta) = \tilde{T}_{\mathscr{G}}(\Theta)'\tilde{T}_{\mathscr{G}}(\Theta) - T_G'T_G$ where

$$\tilde{T}_{\mathscr{G}}(\Theta) = \begin{bmatrix} G_L & & \\ \vdots & \ddots & \\ \vdots & & G_L \\ G_0 & & \vdots \\ & \ddots & \vdots \\ & & G_0 \end{bmatrix},$$

$$T_G = \begin{bmatrix} G_L & & & & & \\ \vdots & \ddots & & & & \\ G_1 & \cdots & G_L & & & \\ & & & G_0 & \cdots & G_{L-1} \\ & & & & \ddots & \vdots \\ & & & & & \tilde{g}_0 \end{bmatrix}$$

   are of dimension $(t_f - t_0 + 2L + 1)(p+m) \times (t_f - t_0 + L + 1)m$ and $2(p+m)L \times (t_f - t_0 + L + 1)m$, respectively.
2. Denote $\Psi = \tilde{T}_{\mathscr{G}}(\Theta)'\tilde{T}_{\mathscr{G}}(\Theta)$. Show that

$$\begin{aligned}\left[T_{\mathscr{G}}(\Theta)'T_{\mathscr{G}}(\Theta)\right]^{-1} &= (\Psi - T_G'T_G)^{-1} \\ &= \Psi^{-1} + \Psi^{-1}T_G'(I_{2(p+m)L} - T_G\Psi^{-1}T_G')^{-1}T_G\Psi^{-1}.\end{aligned}$$

3. Let $\Psi^{-1} = \tilde{T}_{-1}'\tilde{T}_{-1}$ be Cholesky factorization such that $\tilde{T}_{-1}$ is block lower triangular with block size $m \times m$. Denote $\Omega = T_G\tilde{T}_{-1}'$ and $\Gamma = \tilde{M}_w T_{\mathscr{G}}(\Theta)\tilde{T}_{-1}'$. Show that

$$\text{CRB}(\Theta) = \sigma^2 \left(\text{FIM}(\Theta) - \Gamma\Gamma' - \Gamma\Omega'(I - \Omega\Omega')^{-1}\Omega\Gamma'\right)^{-1}.$$

   The efficient computation of $\text{CRB}(\Theta)$ is hinged to the Cholesky factorization of $\Psi^{-1} = \tilde{T}_{-1}'\tilde{T}_{-1}$ that will be worked out in the next problem.

8.17. (i) For $\{\mathbf{A}(z), \mathbf{B}(z)\}$ in (8.42) which are right coprime, show that the existence of the spectral factorization

$$\mathbf{A}(z)^{\sim}\mathbf{A}(z) + \mathbf{B}(z)^{\sim}\mathbf{B}(z) = \mathbf{C}(z)^{\sim}\mathbf{C}(z)$$

where $\mathbf{C}(z)$ is the right spectral factor with size $m \times m$ and given by

$$\mathbf{C}(z) = \sum_{k=0}^{L} C_k z^{-k}.$$

(ii) Show that the Toeplitz matrix $\Psi = T_{\mathscr{G}}(\Theta)' T_{\mathscr{G}}(\Theta)$ in the previous problem corresponds to spectral factorization of (Sect. C.3 in Appendix C)

$$\left[z^{-L}\mathbf{G}\left(z^{-1}\right)\right]^{\sim}\left[z^{-L}\mathbf{G}\left(z^{-1}\right)\right] = \left[z^{-L}\mathbf{C}\left(z^{-1}\right)\right]^{\sim}\left[z^{-L}\mathbf{C}\left(z^{-1}\right)\right].$$

(iii) Show that $\tilde{T}_{-1}$ in the previous problem is lower block Toeplitz, and consists of the impulse response of $[z^{-L}\mathbf{C}(z^{-1})]^{-1}$.

Appendix A
Linear Algebra

The purpose of this appendix is to provide a quick review on the related materials in linear algebra used in the text. The exposition will be concise and sketchy. Readers are encouraged to read the more complete mathematical texts in the subject area.

A.1 Vectors and Matrices

A vector of dimension n is denoted by boldfaced letter, say, $\mathbf{x}$, which has n elements in $\mathbb{F}$. The field $\mathbb{F}$ is either $\mathbb{R}$, the collection of all real numbers, or $\mathbb{C}$, the collection of all complex numbers. A vector space, also called linear space, over $\mathbb{F}$ is a nonempty set $\mathscr{V}$ for which addition and scalar multiplications (these are familiar element-wise operations) are closed:

$$\mathbf{x}, \mathbf{y} \in \mathscr{V} \implies \mathbf{x} + \mathbf{y} \in \mathscr{V}, \quad \alpha \mathbf{x} \in \mathscr{V},$$

where α is a scalar in $\mathbb{F}$. The "length" of the vector $\mathbf{x}$ is measured by the Euclidean norm: $\|\mathbf{x}\| = \sqrt{\mathbf{x}^*\mathbf{x}}$. Its location in the vector space $\mathscr{V}$ is uniquely determined by its n components. For convenience, $\mathscr{V}$ is denoted by $\mathbb{F}^n$.

Two vectors, $\mathbf{x}$ and $\mathbf{y}$, are linearly independent, if $\alpha\mathbf{x} + \beta\mathbf{y} \neq \mathbf{0}$ whenever $\alpha\beta \neq 0$. The m vectors $\{\mathbf{x}_k\}_{k=1}^m$ with $m \leq n$ and n the dimension of $\mathbf{x}_k$ are linearly independent, if no nontrivial linear combination of $\mathbf{x}_k$'s is a zero vector. A set of n linearly independent vectors $\{\mathbf{x}_k\}_{k=1}^n$ forms a basis for all $n \times 1$ vectors, in which each is a unique linear combination of $\{\mathbf{x}_k\}_{k=1}^n$.

A matrix A of size $n \times m$ over $\mathbb{F}$ can be viewed as a linear map from $\mathbb{F}^m$ to $\mathbb{F}^n$. That is, $\mathbf{y} = A\mathbf{x} \in \mathbb{F}^n$ for $\mathbf{x} \in \mathbb{F}^m$, and the map: $\mathbf{x} \mapsto \mathbf{y}$ is linear. It is noted that $\mathbf{y} = A\mathbf{x}$ is a set of n linear equations:

$$\begin{bmatrix} y_1 \\ \vdots \\ y_n \end{bmatrix} = \begin{bmatrix} a_{1,1} & \cdots & a_{1,m} \\ \vdots & \ddots & \vdots \\ a_{n,1} & \cdots & a_{n,m} \end{bmatrix} \begin{bmatrix} x_1 \\ \vdots \\ x_m \end{bmatrix}. \tag{A.1}$$

G. Gu, *Discrete-Time Linear Systems: Theory and Design with Applications*,
DOI 10.1007/978-1-4614-2281-5, © Springer Science+Business Media, LLC 2012

A natural question is: Given $\mathbf{y} \in \mathbb{F}^n$, does there exist a (unique) solution to (A.1)? The following notions are introduced.

Definition A.1. For $A \in \mathbb{F}^{n\times m}$, let $\mathbf{a}_k$ be its kth column. Its range space, denoted by $\mathscr{R}(A)$, and the null space, denoted by $\mathscr{N}(A)$, are defined as

$$\mathscr{R}(A) = \left\{ \mathbf{y} \in \mathbb{F}^n : \mathbf{y} = \sum_{k=1}^{m} \alpha_k \mathbf{a}_k \right\}, \quad \mathscr{N}(A) = \{ \mathbf{x} \in \mathbb{F}^m : A\mathbf{x} = \mathbf{0} \},$$

respectively, where $\{\alpha_k\}$ range over $\mathbb{F}$.

By definition, $\mathscr{R}(A)$ is a vector space spanned by the linearly independent column vectors of A as the basis, which is a closed subspace of $\mathbb{F}^n$. On the other hand $\mathscr{N}(A)$ is a vector space composed of all vectors $\mathbf{x} \in \mathbb{F}^m$ such that $A\mathbf{x} = \mathbf{0}$, which is a closed subspace of $\mathbb{F}^m$ too (Problem A.1 in Exercises). The next result answers the question regarding the solution to $\mathbf{y} = A\mathbf{x}$, or (A.1). Its proof is left as an exercise (Problem A.4).

Lemma A.1. *Given $\mathbf{y} \in \mathbb{F}^n$, the linear equation $\mathbf{y} = A\mathbf{x}$, or (A.1), admits a solution $\mathbf{x} \in \mathbb{F}^m$, if and only if $\mathbf{y} \in \mathscr{R}(A)$. In addition, the solution is unique, if and only if $\mathscr{N}(A) = \emptyset$.*

The collection of all matrices of the same dimensions is also a vector space. One may associate A with vector vec(A), which stacks the column vectors of A sequentially into a single column. Different from vectors, multiplication of two different matrices is possible with $C = AB$ well defined, if the number of columns of A is the same as the number of rows of B. In general, $AB \neq BA$, even if both multiplications make sense. The *Kronecker product* of the two matrices $A \in \mathbb{F}^{n\times m}$ and B of any other dimension is defined as

$$A \otimes B = \begin{bmatrix} a_{1,1}B & \cdots & a_{1,m}B \\ \vdots & \ddots & \vdots \\ a_{n,1}B & \cdots & a_{n,m}B \end{bmatrix}.$$

If A and B are square of sizes n and m, respectively, then the *Kronecker sum* of A and B is defined as

$$A \oplus B := (A \otimes I_m) + (I_n \otimes B) \in \mathbb{F}^{nm\times nm}.$$

For square matrices, A^k for positive integer k defines the power of matrices with the convention $A^0 = I$. If $A^k = \mathbf{0}$ for some integer $k > 0$, then A is called *nilpotent*. The trace of A, denoted by Tr$\{A\}$, is defined as the sum of the diagonal elements of A. If AB and BA are both square, then

$$\text{Tr}\{AB\} = \text{Tr}\{BA\}. \tag{A.2}$$

In addition, $\det(A)$, the determinant of A, can be defined, and A^{-1} exists, if and only if $\det(A) \neq 0$, in which case A is said to be nonsingular. If A and B are both square of the same dimensions, there holds:

$$\det(AB) = \det(A)\det(B) = \det(BA). \tag{A.3}$$

A matrix norm, termed *Frobenius norm* can be defined by

$$\|A\|_F := \|\text{vec}(A)\| = \sqrt{\text{Tr}\{A^*A\}} = \sqrt{\text{Tr}\{AA^*\}} \tag{A.4}$$

which measures the "size" of A.

Let $\text{rank}\{A\}$ be the maximum number of linearly independent column vectors of A, or the maximum number of linearly independent row vectors of A. The next result is concerned with the rank of the product of two matrices.

Lemma A.2 (Sylvester inequality). *Let $A \in \mathbb{F}^{n\times m}$ and $B \in \mathbb{F}^{m\times \ell}$. Then*

$$\text{rank}\{A\} + \text{rank}\{B\} - m \leq \text{rank}\{AB\} \leq \min\{\text{rank}[A],\ \text{rank}[B]\}.$$

For a given subspace $\mathscr{S}$, the dimension of $\mathscr{S}$ is denoted by $\dim\{\mathscr{S}\}$, which is defined as the maximum number of linearly independent vectors in $\mathscr{S}$. Lemma A.1 indicates that $\mathscr{N}(A) = \emptyset$ implies that $\text{rank}\{A\} = m$, and thus, $A\mathbf{x} = \mathbf{0}$ admits no solution $\mathbf{x} \neq \mathbf{0}$. In fact, there holds

$$\dim\{\mathscr{R}(A)\} = \text{rank}\{A\}, \quad \dim\{\mathscr{N}(A)\} = m - \text{rank}\{A\}. \tag{A.5}$$

If $\text{rank}\{A\} = r$, then the matrix A has r linearly independent column vectors $\{\mathbf{a}_{k_i}\}_{i=1}^r$, where $1 \leq k_i \leq r$. If $\dim\{\mathscr{N}(A)\} = \rho$, then there exits a maximum number of ρ linearly independent vectors $\{\mathbf{x}_k\}_{k=1}^{\rho} \in \mathscr{N}(A)$ and $\rho = m - \text{rank}\{A\}$. The following result is true.

Lemma A.3. *Let $A \in \mathbb{F}^{n\times m}$. Suppose that $\text{rank}\{A\} = r \leq \min\{n,m\}$. Then $A = TF$ where T has size $n \times r$, F has size $r \times m$, and there hold*

$$\mathscr{R}(A) = \mathscr{R}(T), \mathscr{N}(A) = \mathscr{N}(F),$$
$$\mathscr{R}(A^*) = \mathscr{R}(F^*), \mathscr{N}(A^*) = \mathscr{N}(T^*).$$

Proof. Since A has rank r, there exist r column vectors of A, denoted by $\{\mathbf{a}_{k_i}\}_{i=1}^r$, which are linearly independent. It follows that

$$A = \begin{bmatrix} \mathbf{a}_{k_1} & \cdots & \mathbf{a}_{k_r} \end{bmatrix} \begin{bmatrix} \alpha_{1,1} & \cdots & \alpha_{1,m} \\ \vdots & \ddots & \vdots \\ \alpha_{r,1} & \cdots & \alpha_{r,m} \end{bmatrix} = TF,$$

where T is a "tall" matrix with size $n \times r$ and F a "fat" matrix with size $r \times m$. Hence, $\mathscr{R}(A) = \mathscr{R}(T)$. Because T has full column rank, $A\mathbf{x} = \mathbf{0}$, if and only if $F\mathbf{x} = \mathbf{0}$. Thus, $\mathscr{N}(A) = \mathscr{N}(F)$. By noting that $A^* = F^*T^*$, $\mathscr{R}(A^*) = \mathscr{R}(F^*)$ and $\mathscr{N}(A^*) = \mathscr{N}(T^*)$ follow accordingly. □

Let $\mathscr{S}_1$ and $\mathscr{S}_2$ be two closed subspaces of the same vector space $\mathscr{V}$, which do not have common elements, except $\mathbf{0}$. The direct sum, denoted by $\oplus$, is defined by:

$$\mathscr{S}_1 \oplus \mathscr{S}_2 := \{\mathbf{x} = \mathbf{x}_1 + \mathbf{x}_2 : \mathbf{x}_1 \in \mathscr{S}_1, \mathbf{x}_2 \in \mathscr{S}_2\}. \tag{A.6}$$

Theorem A.1. *For $A \in \mathbb{F}^{n\times m}$, there hold*

$$\mathscr{R}(A) \oplus \mathscr{N}(A^*) = \mathbb{F}^n, \quad \mathscr{R}(A^*) \oplus \mathscr{N}(A) = \mathbb{F}^m.$$

Proof. By Lemma A.3, $A = TF$ with T and F full column and row ranks, respectively, and $\mathscr{R}(A) = \mathscr{R}(T)$. Hence, there exists $T_\perp$ such that $\begin{bmatrix} T & T_\perp \end{bmatrix}$ is a square nonsingular matrix with size $n \times n$ and $T^*T_\perp = \mathbf{0}$. It follows that $\mathscr{R}(T_\perp) = \mathscr{N}(T^*) = \mathscr{N}(A^*)$, and thus, $\mathscr{R}(T) \oplus \mathscr{R}(T_\perp) = \mathscr{R}(A) \oplus \mathscr{N}(A^*) = \mathbb{F}^n$, in light of the fact that columns of T and $T_\perp$ are all linearly independent, and form the basis of $\mathbb{F}^n$. Applying the same procedure to A^* shows that $\mathscr{R}(A^*) \oplus \mathscr{N}(A) = \mathbb{F}^n$ is true as well. □

A.2 Projections and Generalized Inverses

Let $\mathscr{S}$ be a closed subspace of $\mathbb{F}^n$. That is, $\mathscr{S}$ is itself a vector space over the filed $\mathbb{F}$. Let $\dim(\mathscr{S}) = r < n$. Then it has r linearly independent vectors $\{\mathbf{s}_i\}_{i=1}^r$. Any element of $\mathscr{S}$ is a linear combination of $\mathbf{s}_i$. Hence, $\{\mathbf{s}_i\}_{i=1}^r$ form a basis for $\mathscr{S}$. It is a fact that the basis can be chosen such that

$$\mathbf{s}_i^*\mathbf{s}_k = \delta(k-i), \quad \|\mathbf{s}_k\| = 1.$$

(Problem A.14 in Exercises). Such a basis is called orthonormal basis of $\mathscr{S}$.

Since $r < n$, there exists an orthogonal complement of the subspace $\mathscr{S}$ defined by

$$\mathscr{S}_\perp := \{\mathbf{x} \in \mathbb{F}^n : \mathbf{x}^*\mathbf{s}_i = 0 \text{ for } 1 \le i \le r\}.$$

It follows that $\mathbb{F} = \mathscr{S} \oplus \mathscr{S}_\perp$ and $\dim(\mathscr{S}_\perp) = n - r$. As such, there exist linearly independent vectors $\{\mathbf{s}_{r+k}\}_{k=1}^{n-r}$ which form a basis of $\mathscr{S}_\perp$ and which can be chosen as an orthonormal basis of $\mathscr{S}_\perp$. Define

$$S = \begin{bmatrix} \mathbf{s}_1 & \cdots & \mathbf{s}_r \end{bmatrix}, \quad S_\perp = \begin{bmatrix} \mathbf{s}_{r+1} & \cdots & \mathbf{s}_n \end{bmatrix}.$$

Both S and $S_\perp$ are *orthogonal* matrices satisfying

$$\begin{bmatrix} S & S_\perp \end{bmatrix} \begin{bmatrix} S^* \\ S_\perp^* \end{bmatrix} = \begin{bmatrix} S^* \\ S_\perp^* \end{bmatrix} \begin{bmatrix} S & S_\perp \end{bmatrix} = I_n. \tag{A.7}$$

Therefore, $S_a = \begin{bmatrix} S \ S_\perp \end{bmatrix}$ is a unitary matrix, and $S_a^{-1} = S_a^*$.

A matrix $P \in \mathbb{F}^{n\times n}$ is a projection, if $P^2 = P$. Let $r < n$ be the rank of P. Then $P = TF$ with $T \in \mathbb{F}^{n\times r}$ and $F \in \mathbb{F}^{r\times m}$ full rank matrices by Lemma A.3. Hence,

$$P = P^2 \iff TFTF = TF \implies FT = I_r. \tag{A.8}$$

That is, F is a *left inverse* of T and T is a *right inverse* of F. Moreover, there exists $T_\perp$ such that $FT_\perp = \mathbf{0}$ and $\begin{bmatrix} T \ T_\perp \end{bmatrix}$ is both square and nonsingular. As a result, there exists $F_\perp$ with an appropriate dimension such that

$$\begin{bmatrix} F \\ F_\perp \end{bmatrix} \begin{bmatrix} T \ T_\perp \end{bmatrix} = I \iff TF + T_\perp F_\perp = I. \tag{A.9}$$

Thus, $F_\perp$ is a left inverse of $T_\perp$. It follows that

$$\mathscr{R}(T) \oplus \mathscr{N}(F) = \mathscr{R}(P) \oplus \mathscr{N}(P) = \mathbb{F}^n.$$

Geometrically, P projects $\mathbf{x} \in \mathbb{F}^n$ into the subspace $\mathscr{R}(P)$ along the subspace $\mathscr{N}(P)$: Any $\mathbf{x} \in \mathbb{F}^n$ can be decomposed into $\mathbf{x} = \mathbf{x}_1 + \mathbf{x}_2$ with $\mathbf{x}_1 \in \mathscr{R}(P)$ and $\mathbf{x}_2 \in \mathscr{N}(P) = \mathscr{N}(F)$, and

$$P\mathbf{x} = \mathbf{x}_1 \in \mathscr{R}(P), \quad P\mathbf{x}_2 = \mathbf{0}.$$

In other words, the four tuples $(\mathbf{0}, \mathbf{x}_1, \mathbf{x}_2, \mathbf{x})$ form a parallelogram. If, in addition, $\mathscr{R}(P)$ and $\mathscr{N}(P)$ are orthogonal and complement to each other, then P is an orthogonal projection. Thus, the four tuples $(\mathbf{0}, \mathbf{x}_1, \mathbf{x}_2, \mathbf{x})$ form a rectangle. Note that $P = TF$ satisfying (A.8) implies that

$$Q = I - P = T_\perp F_\perp$$

is also a projection. On the other hand, for S and $S_\perp$ satisfying (A.7), both $P_{\mathscr{S}} = SS^*$ and $P_{\mathscr{S}_\perp} = S_\perp S_\perp^*$ are orthogonal projections, and $P_{\mathscr{S}} + P_{\mathscr{S}_\perp} = I_n$.

Let $A \in \mathbb{F}^{n\times m}$ and $B \in \mathbb{F}^{n\times r}$ be given, and $X \in \mathbb{F}^{m\times r}$ be unknown. The more general linear equation $AX = B$ admits a solution X, if and only if

$$\mathscr{R}(B) \subseteq \mathscr{R}(A) \iff \mathscr{N}(A^*) \subseteq \mathscr{N}(B^*),$$

in light of Lemma A.1 and Theorem A.1. Two scenarios arise: The solution X does not exits, or there are more than one solution X. In the former, it is natural to seek X such that $\|AX - B\|_F$ is minimized, and in the latter, there is an interest in searching for X which has the minimum Frobenius norm among all possible solutions.

Lemma A.4. *Let $A \in \mathbb{F}^{n\times m}$ and $B \in \mathbb{F}^{n\times r}$.* (i) *Suppose that $\mathscr{R}(B)$ is not a subset of $\mathscr{R}(A)$ and* rank$\{A\} = m \le n$. *Then $X = X_m := (A^*A)^{-1}A^*B$ minimizes $\|AX - B\|_F$ over all $X \in \mathbb{F}^{m\times r}$.* (ii) *Suppose that $\mathscr{R}(B) \subseteq \mathscr{R}(A)$, $\mathscr{N}(A) \ne \emptyset$, and* rank$\{A\} = n \le m$. *Then $X = X_n := A^*(AA^*)^{-1}B$ has the minimum Frobenius norm over all solutions to $AX = B$.*

Proof. For (i), let S be an orthogonal matrix of size $n \times m$, whose columns form an orthonormal basis of $\mathscr{R}(A)$. Then

$$A = SF, \quad B = SB_1 + S_\perp B_2$$

for some nonsingular matrix F of size $m \times m$, where $S_\perp$ satisfies (A.7). By the fact that unitary matrices do not change the Frobenius norm, there holds

$$\begin{aligned}\|AX - B\|_F &= \left\| \begin{bmatrix} S^* \\ S_\perp^* \end{bmatrix} (SFX - SB_1 - S_\perp B_2) \right\|_F \\ &= \sqrt{\|FX - B_1\|_F^2 + \|B_2\|^2} \geq \|B_2\|_F.\end{aligned}$$

On the other hand, there holds

$$\begin{aligned}\|AX_m - B\|_F &= \left\| \left[I - A(A^*A)^{-1}A^*\right] B \right\|_F \\ &= \|(I_n - SS^*)B\|_F = \|S_\perp S_\perp^* B\|_F = \|B_2\|_F.\end{aligned}$$

That is, $X = X_m = A^*(AA^*)^{-1}B$ minimizes the error $\|AX - B\|_F$. For (ii), it is noted that $AX_n = AA^*(AA^*)^{-1}B = B$. So, X_n is a solution to $AX = B$. Because $\mathscr{N}(A) \neq \mathbf{0}$, there exists an orthogonal matrix U whose columns form an orthonormal basis of $\mathscr{N}(A)$. Moreover, all solutions to $AX = B$ is parameterized by $X = X_n + UR$ with R an arbitrary matrix of the appropriate dimensions. Now direct calculation shows that (by noting that $AU = \mathbf{0}$)

$$\begin{aligned}\|X\|_F^2 &= \|A^*(AA^*)^{-1}B + UR\|_F^2 = \operatorname{Tr}\left\{B^*(AA^*)^{-1}B + R^*R\right\} \\ &\geq \operatorname{Tr}\left\{B^*(AA^*)^{-1}B\right\} = \|X_n\|_F^2.\end{aligned}$$

Hence, X_n is the minimum norm solution. The lemma is true. □

In (i) of Lemma A.4, A admits a left inverse $A^+ = (A^*A)^{-1}A^*$, and all left inverses of A are parameterized by

$$A^\dagger = (A^*A)^{-1}A^* + \Theta S_\perp^* = A^+ + \Theta S_\perp^*, \tag{A.10}$$

where $\Theta \in \mathbb{F}^{m\times(n-m)}$ is arbitrary and $\mathscr{R}(S_\perp)$ is the orthogonal complement of $\mathscr{R}(S) = \mathscr{R}(A)$. It is noted that $P = AA^\dagger$ is a projection and $P = AA^+$ is an orthogonal projection. Hence, orthogonal projection matrices are always hermitian, but projection matrices are nonhermitian in general. The proof of Lemma A.4 also shows that

$$\|A^\dagger\|_F = \sqrt{\|A^+\|_F^2 + \left\|\Theta S_\perp^*\right\|_F^2} \geq \|A^+\|_F.$$

In (ii) of Lemma A.4, A admits a right inverse $A^+ = A^*(AA^*)^{-1}$, and all right inverses of A are parameterized by

$$A^\dagger = A^*(AA^*)^{-1} + U^*R = A^+ + U^*R, \tag{A.11}$$

where $R \in \mathbb{F}^{(n-m)\times m}$ is arbitrary and U is an orthogonal matrix whose columns form an orthonormal basis of $\mathscr{N}(A)$. Again, $P = A^\dagger A$ is a projection which is nonhermitian in general, and $P = A^+A$ is an orthogonal projection which is always a hermitian matrix. Similarly,

$$\|A^\dagger\|_F = \sqrt{\|A^+\|_F^2 + \|UR\|_F^2} \geq \|A^+\|_F.$$

More generally for any matrix A of arbitrary rank, generalized inverses or *pseudoinverses* of A are denoted by $A^\dagger$ and defined as satisfying

$$\text{(a) } AA^\dagger A = A, \quad \text{(b) } A^\dagger AA^\dagger = A^\dagger. \tag{A.12}$$

Clearly, $A^\dagger$ is not unique. It is left as an exercise to derive a general expression of $A^\dagger$ (Problem A.16). The one that minimizes $\|A^\dagger\|_F$ among all possible $A^\dagger$ is called Moore–Penrose pseudoinverse and is given by

$$A^+ = F^*(FF^*)^{-1}(T^*T)^{-1}T^*,$$

where $A = TF$ is the canonical factorization as in Lemma A.3. Again, both $AA^\dagger$ and $A^\dagger A$ are projections, and AA^+ and A^+A are orthogonal projections. Often A^+ is computed via the SVD to be discussed in the next section.

A.3 Decompositions of Matrices

A.3.1 Eigenvalues and Eigenvectors

For $A \in \mathbb{F}^{n\times n}$, its eigenvalues are the n roots of the characteristic polynomial

$$\mathbf{a}(\lambda) = \det(\lambda I_n - A) = \lambda^n + a_1\lambda^{n-1} + \cdots + a_n. \tag{A.13}$$

The set of n eigenvalues, denoted by $\{\lambda_k\}_{k=1}^n$, is called *spectrum* of A, and

$$\rho(A) = \max_{1\leq k\leq n} |\lambda_k|$$

is the *spectral radius* of A. Let λ be an eigenvalue of A. Then there exist a row vector $\mathbf{q}$ and a column vector $\mathbf{v}$ such that

$$\mathbf{q}A = \lambda\mathbf{q}, \quad A\mathbf{v} = \lambda\mathbf{v},$$

and $\mathbf{q}$ and $\mathbf{v}$ are left and right eigenvectors of A, respectively. Moreover, A admits the decomposition termed as *Jordan canonical form* as follows. However, its proof is nontrivial and skipped in this text.

Theorem A.2. *For $A \in \mathbb{F}^{n\times n}$, there exists a nonsingular matrix T such that $A = TJT^{-1}$ and $J = \mathrm{diag}(J_1,\ldots,J_\ell)$ where,*

$$J_i = \begin{bmatrix} \lambda_{k_i} & 1 & & \\ & \ddots & \ddots & \\ & & \ddots & 1 \\ & & & \lambda_{k_i} \end{bmatrix}, \quad 1 \le k_i \le n.$$

The Jordan canonical form in Theorem A.2 does not rule out the possibility that $\lambda_{k_i} \ne \lambda_{k_j}$ for $i \ne j$. On the other hand, if $\{\lambda_{k_i}\}_{i=1}^{\ell}$ are all distinct, then A is called *cyclic*, and J_i the Jordan block. If in addition $\{\lambda_k\}_{k=1}^{n}$ are all distinct, then each Jordan block has size one, and A is *diagonalizable* by a nonsingular matrix T with $J = \Lambda = \mathrm{diag}(\lambda_1,\ldots,\lambda_n)$.

The following Cayley–Hamilton theorem is useful in linear system theory.

Theorem A.3. *Let $A \in \mathbb{F}^{n\times n}$ and $\det(\lambda I_n - A)$ be as in (A.13). Then*

$$A^n + a_1 A^{n-1} + \cdots + a_n I_n = \mathbf{0}.$$

The proof is left as an exercise (Problem A.17).

A.3.2 QR Factorizations

Let $\mathbf{v} \in \mathbb{R}^n$ be nonzero. A square matrix

$$P = I_n - 2\mathbf{v}\mathbf{v}^T/\|\mathbf{v}\|^2 \tag{A.14}$$

is both symmetric and unitary, i.e., $P = P^T$ and $P^2 = I_n$. Such a matrix is called *Householder reflection*, or *Householder transformation* and $\mathbf{v}$ the *Householder vector*. Let $\mathbf{x} \in \mathbb{R}^n$, and denote $\mathbf{e}_1$ as the vector with the first entry 1 and the rest zeros. If $P\mathbf{x} \in \mathscr{R}(\mathbf{e}_1)$, then $\mathbf{v} \in \mathscr{R}([\,\mathbf{x}\ \ \mathbf{e}_1\,])$, and

$$P\mathbf{x} = \left(I_n - 2\frac{\mathbf{v}\mathbf{v}^T}{\|\mathbf{v}\|^2}\right)\mathbf{x} = \mp\|\mathbf{x}\|\mathbf{e}_1. \tag{A.15}$$

The Householder reflection sets the elements of $P\mathbf{x}$ to zeros except the first one. It sets $\mathbf{v} = \mathbf{x} \pm \|\mathbf{x}\|\mathbf{e}_1$ with $\pm$ sign so chosen that $\|\mathbf{v}\|$ has a larger value to avoid the possible rounding-off error. Furthermore, for $A \in \mathbb{F}^{n\times m}$,

$$PA = \left(I_n - 2\frac{\mathbf{v}\mathbf{v}^T}{\|\mathbf{v}\|^2}\right)A = A + \mathbf{v}\mathbf{q}^T, \tag{A.16}$$

where $\mathbf{q} = \alpha A^T \mathbf{v}$ with $\alpha = -2/\|\mathbf{v}\|^2$. That is, a Householder update of a matrix involves a matrix–vector multiplication followed by an outer product update, which avoids the multiplication of P and A directly. For this reason, Householder transformation is widely used in triangularization of matrices such as QR factorizations.

Lemma A.5. *Let $A \in \mathbb{R}^{n\times m}$. Then there exists QR factorization $A = QR$ where Q is an orthogonal matrix and R is an upper triangular matrix.*

Proof. Applying (A.15) with the first column of A as $\mathbf{x}$ eliminates the elements of the first column of PA, except the first one. Record such a Householder transformation as H_1. Now applying (A.15) with the last $(n-1)$ elements of the second column of H_1A as $\mathbf{x}$ eliminates the second column of H_2H_1A except the first two elements. By induction,

$$H_m \cdots H_1 A = HA = \begin{bmatrix} R \\ \mathbf{0} \end{bmatrix}, \quad H_k = \begin{bmatrix} I_{k-1} & \mathbf{0} \\ \mathbf{0} & P_k \end{bmatrix} \implies Q = H^* \begin{bmatrix} I \\ \mathbf{0} \end{bmatrix},$$

where P_k is a Householder matrix. Thus, R is upper triangular. Since product of unitary matrices is unitary, $Q^TQ = I$ or Q is an orthogonal matrix. □

In light of (A.16), multiplication of a Householder matrix with another matrix can be easily implemented. The use of the unitary matrices improves the numerical property of QR factorizations. Moreover, R can be made full row rank by taking Q comprised of the minimum number of columns. It is noted that the QR factorization also holds for complex matrices.

A.3.3 Schur Decompositions

For the Jordan canonical form as in Theorem A.2, partition

$$T = \begin{bmatrix} T_{k_1} & T_{k_2} & \cdots & T_{k_\ell} \end{bmatrix}$$

compatibly with that of J. The decomposition $A = TJT^{-1}$ gives the rise of $AT = TJ$, implying that

$$AT_{k_i} = T_{k_i}J_i, \quad 1 \le i \le \ell. \tag{A.17}$$

It follows that for any vector $\mathbf{v} \in \mathscr{R}(T_{k_i})$, $A\mathbf{v} \in \mathscr{R}(T_{k_i})$. In other words, $\mathscr{R}(T_{k_i})$ is A invariant for each i. Moreover, the direct sum of any pair of $\{\mathscr{R}(T_{k_i}\}_{i=1}^{\ell}$ is also A invariant. It is left as an exercise to estimate the number of distinct A-invariant subspaces (Problem A.20).

Suppose that J_1 has size $m > 1$. Then

$$A\mathbf{v}_1 = \lambda_1\mathbf{v}_1, \quad A\mathbf{v}_k = \lambda_1\mathbf{v}_k + \mathbf{v}_{k-1}, \quad 1 < k \le m.$$

Hence, both $\mathscr{R}(\mathbf{v}_1)$ and $\mathscr{R}\left(\begin{bmatrix}\mathbf{v}_1 & \cdots & \mathbf{v}_m\end{bmatrix}\right)$ are invariant subspaces of A. The same holds true for other J_i's. Let $\mathscr{S} \subset \mathbb{F}^n$ be an A-invariant subspace with $\dim\{\mathscr{S}\} > 0$. Then there is at least one $\mathbf{v} \in \mathscr{S}$ such that $A\mathbf{v} = \lambda\mathbf{v}$. If all eigenvalues of A constrained into $\mathscr{S}$ are strictly inside the unit circle, then $\mathscr{S}$ is a stable A-invariant subspace. The following Schur decomposition is useful.

Theorem A.4. *For $A \in \mathbb{C}^{n\times n}$, there exists a unitary matrix Q such that $A = QR_AQ^*$, where R_A is an upper triangular matrix with eigenvalues of A on the diagonal.*

Proof. By Theorem A.2, $AT = TJ$ with J in the Jordan canonical form, and T nonsingular. Applying QR factorization to T yields

$$T = QR \implies AQR = QRJ \iff Q^*AQ = RJR^{-1} = R_A.$$

Since both R and J are upper triangular matrices, R_A is also an upper triangular matrix (Problem A.19 in Exercises). The fact that Q is orthogonal and square implies that Q is unitary. Thus, A and R_A have the same eigenvalues which are on the diagonal of R_A and can be arranged in any order. □

Schur decomposition can be computed directly without going through the Jordan form. In fact, due to the use of unitary matrices, the Schur decomposition is numerically reliable and efficient. It is a preferred method in computing eigenvalues as well as invariant subspaces.

In general, eigenvalues are complex, even if A is a real matrix. However, if A is real and has a real eigenvalue λ, then its corresponding eigenvector can be chosen as real too. Moreover, if A is real, then real Schur form exists in which Q and R_A are both real, but R_A is now block upper triangular to account for the complex eigenvalues. For the case when all eigenvalues of A are real, $\lambda_k(A)$ denotes the kth largest eigenvalue of A, and the real Schur form can place eigenvalues of A in the diagonal of R_A arranged in descending order.

A.3.4 Singular Value Decompositions

Singular value decomposition (SVD) is widely used in matrix analysis and linear algebra. Roughly speaking, singular values measure the "size," and the corresponding singular vectors the "directions" for the underlying matrix.

Theorem A.5. *Let $A \in \mathbb{F}^{n\times m}$. There exists* SVD*: $A = USV^*$ where $U \in \mathbb{F}^{n\times n}$ and $V \in \mathbb{F}^{m\times m}$ are unitary matrices and $S \in \mathbb{F}^{n\times m}$ contains nonzero elements $\{\sigma_k\}_{k=1}^{\rho}$ on the diagonal in descending order where $\rho = \min\{n, m\}$.*

Proof. Applying Schur decomposition to AA^* and A^*A gives

$$AA^* = Q_nD_n^2Q_n^*, \quad A^*A = Q_mD_m^2Q_m^*,$$

where $Q_n \in \mathbb{F}^{n\times n}$ and $Q_m \in \mathbb{F}^{m\times m}$ are unitary matrices, and D_n and D_m are upper triangular matrices. By the hermitian of AA^* and A^*A, and the fact that nonzero eigenvalues of AA^* and A^*A are identical, D_n^2 and D_m^2 are in fact diagonal and positive real of the form

$$D_k^2 = \text{diag}\left(d_1^2, d_2^2, \ldots, d_k^2\right), \quad k = n, m.$$

If $n \geq m$, then there exists an orthogonal matrix $Q \in \mathbb{F}^{n\times m}$ such that

$$AQ_m = QD_m = \begin{bmatrix} Q & Q_\perp \end{bmatrix} \begin{bmatrix} D_m \\ \mathbf{0} \end{bmatrix} = Q_a D,$$

by $Q_m^* A^* A Q_m = D_m^2$ and thus $\mathcal{N}(AQ_m) = \mathcal{N}(D_m)$. Hence, the SVD holds with $U = Q_a$, $S = D$, and $V = Q_m$. If $n \leq m$, a similar procedure leads to

$$Q_n^* A = D_n Q^* = \begin{bmatrix} D_n & \mathbf{0} \end{bmatrix} \begin{bmatrix} Q & Q_\perp \end{bmatrix}^* = DQ_a^*$$

with Q orthogonal. Hence, the SVD holds with $U = Q_n$, $V = Q_a$, and $S = D$. The nonzero diagonal elements of S are $\{|d_k|\}_{k=1}^{\rho} = \{\sigma_k\}_{k=1}^{\rho}$, which are called singular values of A and which can be arranged in descending order. □

Denote $\mathbf{u}_k$ and $\mathbf{v}_i$ as the kth and ith column of U and V, respectively. Then the SVD of A implies that

$$A\mathbf{v}_k = \sigma_k \mathbf{u}_k, \quad A^*\mathbf{u}_k = \sigma_k \mathbf{v}_k \tag{A.18}$$

for $1 \leq k \leq r = \text{rank}\{A\}$. For this reason, $\mathbf{u}_k$ and $\mathbf{v}_k$ are called kth left and right singular vectors of A. The relations in (A.18) also show that

$$A^*A\mathbf{v}_k = \sigma_k^2 \mathbf{v}_k, \quad AA^*\mathbf{u}_k = \sigma_k^2 \mathbf{u}_k.$$

It follows that $\sigma_k = \sqrt{\lambda_k(AA^*)} = \sqrt{\lambda_k(A^*A)}$ for $1 \leq k \leq \min\{n, m\}$.

The next result is concerned with the *induced norm* for matrices, and shows that the induced norm is the maximum achievable "gain."

Corollary A.1. *Let $\overline{\sigma}(A) = \sigma_1$ be the maximum singular value of A. Then*

$$\|A\| := \sup_{\mathbf{x}\neq \mathbf{0}} \frac{\|A\mathbf{x}\|}{\|\mathbf{x}\|} = \sup_{\|\mathbf{x}\|=1} \|A\mathbf{x}\| = \overline{\sigma}(A).$$

Proof. By SVD, $A = USV^*$ with V a unitary matrix. Hence, any nonzero vector is a linear combination of $\mathbf{v}_k$, the kth column of V. That is,

$$\mathbf{x} = \sum_{k=1}^{m} \beta_k \mathbf{v}_k = V\mathbf{b}, \quad \mathbf{b} = \begin{bmatrix} \beta_1 & \cdots & \beta_m \end{bmatrix}^T,$$

where $A \in \mathbb{F}^{n\times m}$. It follows that $\|\mathbf{x}\| = \|\mathbf{b}\|$. Let $\rho = \min\{n,m\}$. Then

$$A\mathbf{x} = AV\mathbf{b} = US\mathbf{b} = \sum_{k=1}^{n} \beta_k\sigma_k\mathbf{u}_k,$$

with $\mathbf{u}_k$ the kth column of U. The fact that U is unitary implies that

$$\|A\mathbf{x}\|^2 = \sum_{k=1}^{\rho} \sigma_k^2\beta_k^2 \leq \sigma_1^2\|\mathbf{b}\|^2 = \sigma_1^2\|\mathbf{x}\|^2.$$

Thus, $\overline{\sigma}(A) = \sigma_1$ is the upper bound for the induced norm $\|A\|$. Choosing $\mathbf{x} = \mathbf{v}_1$ shows that $\|\mathbf{x}\| = 1$ and $\|A\mathbf{x}\| = \overline{\sigma}(A)$. □

A.4 Hermitian Matrices

In light of the Schur decomposition $A = QR_AQ^*$, there hold

$$AA^* = QR_AR_A^*Q^*, \quad A^*A = QR_A^*R_AQ^*.$$

Hence, if a square matrix A of size n is *normal*, i.e., $AA^* = A^*A$, then

$$R_A^*R_A = R_AR_A^* = \text{diag}\left(\sigma_1^2,\ldots,\sigma_n^2\right).$$

It follows that for a normal matrix A, it is necessary that R_A be diagonal. That is, normal matrices are diagonalizable with unitary matrices.

Hermitian matrices are clearly normal matrices. If A is hermitian, then

$$A = QR_AQ^* = A^* = QR_A^*Q^*,$$

leading to the conclusion that R_A is a real diagonal matrix. Consequently, eigenvalues of a hermitian matrix are all real. If in addition all eigenvalues of a hermitian matrix are positive or strictly positive, then A is called positive semidefinite or positive definite, denoted by $A \geq \mathbf{0}$ and $A > 0$, respectively.

If $A = A^* \geq 0$, then there exists a "square root" $R = A^{1/2}$ such that $A = R^2$. One may take the square root of A as $R = US^{1/2}U^*$ with $A = USU^*$ as the SVD of A. Suppose that $A > B \geq 0$. Then

$$\mathbf{0} < I - A^{-1/2}BA^{-1/2} = I - A^{-1/2}BA^{-1}A^{1/2}.$$

Hence, $\rho\left(BA^{-1}\right) < 1$, and all eigenvalues of BA^{-1} are strictly positive real.

Two results will be presented for positive semidefinite matrices with one on Cholesky factorizations, and the other on simultaneous diagonalization of two positive semidefinite matrices.

Proposition A.1. *Let $\Theta_1 = \Theta_1^* \geq 0$ and $\Theta_3 = \Theta_3^* \geq 0$, which may not have the same dimension. If*

$$\Theta = \begin{bmatrix} \Theta_1 & \Theta_2^* \\ \Theta_2 & \Theta_3 \end{bmatrix} \geq 0,$$

then $\nabla_1 = \Theta_1 - \Theta_2^* \Theta_3^+ \Theta_2 \geq 0$, $\nabla_3 = \Theta_3 - \Theta_2 \Theta_1^+ \Theta_2^* \geq 0$, and

$$\text{(i) } \mathscr{R}(\Theta_2) \subseteq \mathscr{R}(\Theta_3), \quad \text{(ii) } \mathscr{R}(\Theta_2^*) \subseteq \mathscr{R}(\Theta_1). \tag{A.19}$$

There hold factorizations of the form

$$\Theta = \begin{bmatrix} I & \Theta_2^* \Theta_3^+ \\ 0 & I \end{bmatrix} \begin{bmatrix} \nabla_1 & 0 \\ 0 & \Theta_3 \end{bmatrix} \begin{bmatrix} I & 0 \\ \Theta_3^+ \Theta_2 & I \end{bmatrix} \tag{A.20}$$

$$= \begin{bmatrix} I & 0 \\ \Theta_2 \Theta_1^+ & I \end{bmatrix} \begin{bmatrix} \Theta_1 & 0 \\ 0 & \nabla_3 \end{bmatrix} \begin{bmatrix} I & \Theta_1^+ \Theta_2^* \\ 0 & I \end{bmatrix}. \tag{A.21}$$

Proof. As (ii) of (A.19) and (A.21) are dual to (i) of (A.19) and (A.20), respectively, only (i) of (A.19) and (A.20) will be proven. Because $\Theta \geq 0$, choose a nonzero vector $\mathbf{v}$ and set $\mathbf{w} = -\Theta_3^+ \Theta_2 \mathbf{v}$. Then

$$\begin{bmatrix} \mathbf{v}^* & \mathbf{w}^* \end{bmatrix} \begin{bmatrix} \Theta_1 & \Theta_2^* \\ \Theta_2 & \Theta_3 \end{bmatrix} \begin{bmatrix} \mathbf{v} \\ \mathbf{w} \end{bmatrix} = \mathbf{v}^* \Theta_1 \mathbf{v} + \mathbf{w}^* \Theta_3 \mathbf{w} + 2\mathbf{v}^* \Theta_2^* \mathbf{w}$$

$$= \mathbf{v}^* (\Theta_1 - \Theta_2^* \Theta_3^+ \Theta_2) \mathbf{v} = \mathbf{v}^* \nabla_1 \mathbf{v}.$$

Since $\Theta \geq 0$, $\nabla_1 \geq 0$. If $\mathbf{w} = -\Theta_1^+ \Theta_2^* \mathbf{v}$ is chosen, then similar calculations show that $\nabla_3 \geq 0$. Now if $\mathscr{R}(\Theta_2) \subseteq \mathscr{R}(\Theta_3)$ is false, then there exists a nonzero vector $\mathbf{v}$ such that $\mathbf{v}^* \Theta_2 \neq 0$ and $\mathbf{v}^* \Theta_3 = 0$. Hence, $\mathbf{v}^* \nabla_3 \mathbf{v} = -\mathbf{v}^* \Theta_2 \Theta_1^+ \Theta_2^* \mathbf{v} < 0$, contradicting to $\nabla_3 \geq 0$. As a result, $\mathscr{R}(\Theta_2) \subseteq \mathscr{R}(\Theta_3)$, and $\Theta_2 = \Theta_3 \Gamma$ for some matrix Γ of appropriate size. It follows that

$$\Theta_3 \Theta_3^+ \Theta_2 = \Theta_3 \Theta_3^+ \Theta_3 \Gamma = \Theta_3 \Gamma = \Theta_2.$$

The above implies the decomposition in (A.20). The proof is complete. □

The factorizations in (A.20) or (A.21) have the form $\Theta = LDL^*$ or $\Theta = L^*DL$ with L lower block triangular and $D \geq 0$ block diagonal. Hence, it is the Cholesky factorization in block form. Note that $D = I$ can be taken in Cholesky factorizations.

Theorem A.6. *Let P and Q be two positive semidefinite matrices. Then there exists a nonsingular matrix T such that*

$$TPT^* = \begin{bmatrix} \Sigma_1 & & & \\ & \Sigma_2 & & \\ & & \mathbf{0} & \\ & & & \mathbf{0} \end{bmatrix}, \quad (T^*)^{-1} Q T^{-1} = \begin{bmatrix} \Sigma_1 & & & \\ & \mathbf{0} & & \\ & & \Sigma_3 & \\ & & & \mathbf{0} \end{bmatrix},$$

where Σ_1, Σ_2, and Σ_3 are diagonal and positive definite.

Proof. Since $P \geq 0$, there exists a transformation T_1 such that

$$T_1 P T_1^* = \begin{bmatrix} I & \mathbf{0} \\ \mathbf{0} & \mathbf{0} \end{bmatrix}.$$

Compute $(T_1^*)^{-1} Q T_1^{-1}$, and partition compatibly as

$$(T_1^*)^{-1} Q T_1^{-1} = \begin{bmatrix} Q_1 & Q_2 \\ Q_2^* & Q_3 \end{bmatrix}.$$

By the SVD, there exists a unitary matrix U_1 such that

$$U_1 Q_1 U_1^* = \begin{bmatrix} \Sigma_1^2 & \mathbf{0} \\ \mathbf{0} & \mathbf{0} \end{bmatrix}, \quad \Sigma_1 > 0.$$

Setting $T_2^{-1} = \operatorname{diag}(U_1^*, I)$ implies that

$$(T_2^*)^{-1} (T_1^*)^{-1} Q T_1^{-1} T_2^{-1} = \begin{bmatrix} \Sigma_1^2 & \mathbf{0} & \hat{Q}_{2,1} \\ \mathbf{0} & \mathbf{0} & \hat{Q}_{2,2} \\ \hat{Q}_{2,1}^* & \hat{Q}_{2,2}^* & Q_3 \end{bmatrix}.$$

Since $Q \geq \mathbf{0}$, $\hat{Q}_{2,2} = \mathbf{0}$. Let

$$(T_3^*)^{-1} = \begin{bmatrix} I & \mathbf{0} & \mathbf{0} \\ \mathbf{0} & I & \mathbf{0} \\ -\hat{Q}_{2,1}^* \Sigma_1^{-2} & \mathbf{0} & I \end{bmatrix}.$$

Direct computation yields

$$T_3 T_2 T_1 P T_1^* T_2^* T_3^* = \begin{bmatrix} \Sigma_1^2 & \mathbf{0} & \mathbf{0} \\ \mathbf{0} & I & \mathbf{0} \\ \mathbf{0} & \mathbf{0} & \mathbf{0} \end{bmatrix},$$

$$(T_3^*)^{-1} (T_2^*)^{-1} (T_1^*)^{-1} Q T_1^{-1} T_2^{-1} T_3^{-1} = \begin{bmatrix} \Sigma_1^2 & \mathbf{0} & \mathbf{0} \\ \mathbf{0} & \mathbf{0} & \mathbf{0} \\ \mathbf{0} & \mathbf{0} & Q_3 - \hat{Q}_{2,1}^* \Sigma_1^{-2} \hat{Q}_{2,1} \end{bmatrix}.$$

Finally, a unitary matrix U_2 can be obtained such that

$$U_2 \left(Q_3 - \hat{Q}_{2,1}^* \Sigma_1^{-2} \hat{Q}_{2,1} \right) U_2^* = \begin{bmatrix} \Sigma_3 & \mathbf{0} \\ \mathbf{0} & \mathbf{0} \end{bmatrix}, \quad \Sigma_3 > \mathbf{0}.$$

Setting $T = T_4 T_3 T_2 T_1$ yields the desired expressions for TPT^* as well as for $(T^*)^{-1} Q T^{-1}$ with $\Sigma_2 = I$. □

The proof of Theorem A.6 shows that if Q is nonsingular, then P and Q^{-1} are simultaneously diagonalizable in the sense that there exists a nonsingular matrix M such that M^*PM and $M^*Q^{-1}M$ are both diagonal. In fact the positivity is not required. See Problem A.28 in Exercises.

A.5 Algebraic Riccati Equations

In Chap. 5, the necessary and sufficient conditions to the ARE (5.96) are obtained for the existence of the stabilizing solution. This section will present a numerical algorithm to obtain the unique stabilizing solution when the existence conditions hold. For simplicity, it is assumed that $\det(A) \neq 0$. The case of $\det(A) = 0$ will be commented at the end of the section.

Let $R = R^* > \mathbf{0}$, $G = BR^{-1}B^*$, and $Q = Q^* \geq \mathbf{0}$. Denote

$$S = \begin{bmatrix} A + G(A^*)^{-1}Q & -G(A^*)^{-1} \\ -(A^*)^{-1}Q & (A^*)^{-1} \end{bmatrix}, \quad J = \begin{bmatrix} 0 & -I_n \\ I_n & 0 \end{bmatrix}. \tag{A.22}$$

Then X is a solution to the ARE (5.96) such that $\det(I_n + XG) \neq 0$, if and only if (refer to Problem A.29 in Exercises)

$$\begin{bmatrix} -X & I_n \end{bmatrix} S \begin{bmatrix} I_n \\ X \end{bmatrix} = \mathbf{0}. \tag{A.23}$$

It can be verified that $J^{-1}S^*J = S^{-1}$. Such matrices are called *simplectic*. Thus, $\lambda = re^{j\theta}$ is an eigenvalue of S implies that $\bar{\lambda}^{-1} = r^{-1}e^{j\theta}$ is also an eigenvalue of S. As a result, the $2n$ eigenvalue of S form a mirror image pattern about the unit circle. Moreover, the following result is true.

Lemma A.6. *Let $G = BR^{-1}B^*$, and $Q = C^*C$. If (A,B) is stabilizable and (C,A) has no unobservable modes on the unit circle, i.e., (5.100) holds, then the simplectic matrix S in (A.23) has no eigenvalues on the unit circle.*

Proof. By the contradiction argument, assume that S has an eigenvalue $e^{j\theta}$. Then there exist vectors $\mathbf{q}$ and $\mathbf{v}$ of dimension n such that

$$S \begin{bmatrix} \mathbf{q} \\ \mathbf{v} \end{bmatrix} = e^{j\theta} \begin{bmatrix} \mathbf{q} \\ \mathbf{v} \end{bmatrix} \neq \begin{bmatrix} \mathbf{0} \\ \mathbf{0} \end{bmatrix}.$$

The above is equivalent to

$$[A + G(A^*)^{-1}Q]\mathbf{q} - G(A^*)^{-1}\mathbf{v} = e^{j\theta}\mathbf{q}, \tag{A.24}$$

$$-(A^*)^{-1}Q\mathbf{q} + (A^*)^{-1}\mathbf{v} = e^{j\theta}\mathbf{v}. \tag{A.25}$$

Multiplying (A.25) by G from left and adding it to (A.24) give

$$A\mathbf{q} - e^{j\theta}G\mathbf{v} = e^{j\theta}\mathbf{q}, \tag{A.26}$$

$$-Q\mathbf{q} + \mathbf{v} = e^{j\theta}A^*\mathbf{v}. \tag{A.27}$$

Premultiplying (A.26) by $e^{-j\theta}\mathbf{v}^*$ and (A.27) by $\mathbf{q}^*$ yield

$$e^{-j\theta}\mathbf{v}^*A\mathbf{q} = \mathbf{v}^*G\mathbf{v} + \mathbf{v}^*\mathbf{q},$$

$$-\mathbf{q}^*Q\mathbf{q} + \mathbf{q}^*\mathbf{v} = e^{j\theta}\mathbf{q}^*A^*\mathbf{v}.$$

It follows that $\mathbf{q}^*Q\mathbf{q} + \mathbf{v}^*G\mathbf{v} = 0$, and thus,

$$\mathbf{v}^*B = \mathbf{0} \implies \mathbf{v}^*G = \mathbf{0}, \quad C\mathbf{q} = \mathbf{0} \implies Q\mathbf{q} = \mathbf{0}.$$

Substituting the above into (A.26) and (A.27) leads to

$$A\mathbf{q} = e^{j\theta}\mathbf{q}, \quad A^*\mathbf{v} = e^{-j\theta}\mathbf{v}.$$

Since $\mathbf{q}$ and $\mathbf{v}$ are not zero simultaneously, $e^{j\theta}$ is either an unobservable mode of (C,A), contradicting to (5.100), or uncontrollable mode of (A,B), contradicting to the stabilizability of (A,B). Hence, the lemma is true. □

If S has no eigenvalues on the unit circle, then its $2n$ eigenvalues are split into evenly inside and outside the unit circle, in light of the property of the simplectic matrix. Let T be the $2n \times n$ matrix, whose columns consist of the basis vectors of the stable eigensubspace of S, i.e., eigensubspace of S corresponding to the eigenvalues of S inside the unit circle. Then

$$ST = T\Lambda_-, \quad T = \begin{bmatrix} T_1 \\ T_2 \end{bmatrix}, \tag{A.28}$$

where T_1, T_2, and Λ_- are all $n \times n$ matrices, and eigenvalues of Λ_- coincide with the stable eigenvalues of S.

Lemma A.7. *Let S be the simplectic matrix in (A.23). Under the same hypotheses as in Lemma A.6, the stable eigenspace represented by T as in (A.28) satisfies $T_1^*T_2 \geq 0$ and* $\det(T_1) \neq 0$.

Proof. Define $\{U_k\} = \{T\Lambda_-^k\}$ where T is defined as in (A.28). Then

$$U_0 = T, \quad U_{k+1} = (T\Lambda_-)\Lambda_-^k = ST\Lambda_-^k = SU_k.$$

With T_1 and T_2 as in (A.28),

$$T_1^*T_2 = U_0^*VU_0, \quad V = \begin{bmatrix} \mathbf{0} & I \\ \mathbf{0} & \mathbf{0} \end{bmatrix}.$$

Hence, recursive computations and $U_{k+1} = SU_k$ yield

$$Y_k = U_0^* V U_0 - U_k^* V U_k = \sum_{i=0}^{k-1} \left(U_i^* V U_i - U_{i+1}^* V U_{i+1}\right) = \sum_{i=0}^{k-1} U_i^* (V - S^* V S) U_i.$$

Because $Q \geq \mathbf{0}$ and $G \geq \mathbf{0}$,

$$V - S^* V S = \begin{bmatrix} I & -Q \\ 0 & I \end{bmatrix} \begin{bmatrix} Q & 0 \\ 0 & A^{-1}G(A^*)^{-1} \end{bmatrix} \begin{bmatrix} I & 0 \\ -Q & I \end{bmatrix} \geq \mathbf{0}$$

from which $Y_k \geq \mathbf{0}$ follows. Stability of Λ_- implies that $U_k = T\Lambda_-^k \to \mathbf{0}$ as $k \to \infty$, and thus, $U_0' V U_0 = Y_\infty = T_1^* T_2 \geq 0$. Now write (A.28) as

$$\begin{bmatrix} A + G(A^*)^{-1}Q & -G(A^*)^{-1} \\ -(A^*)^{-1}Q & (A^*)^{-1} \end{bmatrix} \begin{bmatrix} T_1 \\ T_2 \end{bmatrix} = \begin{bmatrix} T_1 \\ T_2 \end{bmatrix} \Lambda_-,$$

which is in turn the same as

$$AT_1 + G\left[(A^*)^{-1}QT_1 - (A^*)^{-1}T_2\right] = T_1\Lambda_-, \tag{A.29}$$

$$-(A^*)^{-1}QT_1 + (A^*)^{-1}T_2 = T_2\Lambda_-. \tag{A.30}$$

The expression of (A.30) implies that (A.29) is the same as

$$AT_1 - GT_2\Lambda_- = T_1\Lambda_-. \tag{A.31}$$

Using contradiction argument, assume that T_1 is singular. Let $\mathbf{v} \in \mathscr{N}(T_1)$. Premultiplying (A.31) by $\mathbf{v}^*\Lambda_-^* T_2^*$ and postmultiplying by $\mathbf{v}$ give

$$-\mathbf{v}^*\Lambda_-^* T_2^* G T_2 \Lambda_- \mathbf{v} = \mathbf{v}^*\Lambda_-^* T_2^* T_1 \Lambda_- \mathbf{v},$$

where $T_1\mathbf{v} = \mathbf{0}$ is used. Since $\Lambda_-^* T_2^* G T_2 \Lambda_- \geq 0$, and $T_2^* T_1 \geq 0$,

$$GT_2\Lambda_-\mathbf{v} = 0 \iff B^* T_2 \Lambda_- \mathbf{v} = 0, \tag{A.32}$$

by $G = BR^{-1}B^*$. Postmultiplying (A.31) by $\mathbf{v}$ concludes that $T_1\Lambda_-\mathbf{v} = 0$, by $GT_2\Lambda_-\mathbf{v} = 0$. That is, $\mathbf{v} \in \mathscr{N}(T_1)$ implies that $\Lambda_-\mathbf{v} \in \mathscr{N}(T_1)$, and thus, $\mathscr{N}(T_1)$ is an Λ_--invariant subspace. As a result, there exists an eigenvector $\mathbf{0} \neq \mathbf{v} \in \mathscr{N}(T_1)$ such that $\Lambda_-\mathbf{v} = \lambda\mathbf{v}$ and $|\lambda| < 1$ due to stability of Λ_-. Rewrite (A.30) into $-QT_1 + T_2 = A^* T_2 \Lambda_-$. Then multiplication by $\mathbf{v}$ from right yields

$$T_2\mathbf{v} = \lambda A^* T_2 \mathbf{v} \iff A^* T_2 \mathbf{v} = \lambda^{-1} T_2 \mathbf{v}.$$

In connection with (A.32), $\mathbf{0} = B^* T_2 \Lambda_- \mathbf{v} = \lambda B^* T_2 \mathbf{v}$. By the PBH test with $\mathbf{q} = T_2 \mathbf{w}$ and the fact that $|\lambda^{-1}| > 1$, (A,B) is not stabilizable, contradicting to the hypothesis of the lemma. Hence, $\det(T_1) \neq 0$ is true. □

With the established results, it is ready to show that $X = T_2 T_1^{-1}$ is the stabilizing solution to the ARE in (5.96) which is the ARE

$$X = A^*(I + XG)^{-1} XA + Q \tag{A.33}$$

under the notations in (A.23). Indeed, under the condition that S has no eigenvalues on the unit circle and that (A,B) is stabilizable, T_1 is nonsingular and $T_1^* T_2 \geq \mathbf{0}$, which implies that

$$X = T_2 T_1^{-1} = T_1 (T_1^* T_1)^{-1} (T_1^* T_2)(T_1^* T_1)^{-1} T_1^* \geq 0.$$

By multiplying, T_1^{-1} from right on both sides of (A.28) obtains

$$S \begin{bmatrix} T_1 \\ T_2 \end{bmatrix} = \begin{bmatrix} T_1 \\ T_2 \end{bmatrix} \Lambda_- \iff S \begin{bmatrix} I \\ X \end{bmatrix} = \begin{bmatrix} I \\ X \end{bmatrix} A_c, \tag{A.34}$$

where $A_c = T_1 \Lambda_- T_1^{-1}$ is a stability matrix. If premultiplying the equation on the right of (A.34) by $\begin{bmatrix} I & 0 \end{bmatrix}$ from left, then

$$\begin{aligned} A_c &= A + G(A^*)^{-1} Q - G(A^*)^{-1} X = A + G(A^*)^{-1}(Q - X) \\ &= A - G(I + XG)^{-1} XA = A + BF, \end{aligned}$$

where $(Q - X) = -A^*(I + XG)^{-1} XA$ from the ARE (A.33) is used and:

$$\begin{aligned} F &= -R^{-1} B (I + XBR^{-1} B)^{-1} XA = -R^{-1} (I + B^* XBR^{-1})^{-1} BXA \\ &= -(R + B^* XB)^{-1} BXA \end{aligned}$$

is the state-feedback gain. Stability of A_c implies that F is a stabilizing feedback gain, and thus, $X = T_2 T_1^{-1}$ is the stabilizing solution to the ARE (5.96), i.e., (A.33).

In summary, the stabilizing solution to the ARE (A.33) can be obtain by computing the stable eigensubspace of S as in (A.28) under the condition that (A,B) is stabilizable and (C,A) has no unobservable modes on the unit circle. Clearly, the Schur decomposition is well suited for such a task, which is numerically reliable and efficient. The stabilizing solution $X = T_2 T_1^{-1}$ can then be obtained. If A is singular, then the generalized eigenvalue/eigenvector problem (which is also referred to as matrix pencil)

$$\left(\lambda \begin{bmatrix} I & G \\ 0 & A^* \end{bmatrix} - \begin{bmatrix} A & 0 \\ -Q & I \end{bmatrix} \right) \mathbf{x} = 0 \tag{A.35}$$

needs be solved for each stable eigenvalue λ and its corresponding eigenvector $\mathbf{x}$. If λ has multiplicity μ, then (A.35) needs be added by

$$\left(\lambda\begin{bmatrix} I & G \\ 0 & A^* \end{bmatrix} - \begin{bmatrix} A & 0 \\ -Q & I \end{bmatrix}\right)\mathbf{x}_i + \begin{bmatrix} I & G \\ 0 & A^* \end{bmatrix}\mathbf{x}_{i-1} = 0,$$

for $i = 2, \ldots, \mu$ with $\mathbf{x} = \mathbf{x}_1$. Such generalized eigenvectors are called *generalized principle vectors.*

Exercises

A.1. A subset $\mathscr{S} \subset \mathbb{F}^n$ is said to be a closed subspace, if $\mathscr{S}$ is itself a vector space. Show that both range space and null space associated with a given matrix are closed subspaces.

A.2. Show that the Euclidean norm $\|\cdot\|$ satisfies (i) $\|\mathbf{x}\| \geq 0$, (ii) $\|\mathbf{x}\| = 0$, if and only if $\mathbf{x} = \mathbf{0}$, (iii) $\|\alpha\mathbf{x}\| = |\alpha|\|\mathbf{x}\|$, and (iv) $\|\mathbf{x}+\mathbf{y}\| \leq \|\mathbf{x}\| + \|\mathbf{y}\|$. *Note:* Any map: $\mathbf{x} \mapsto \mathbb{R}_+$ satisfies the properties (i)–(iv) defines a norm, which can be different from the Euclidean norm.

A.3. For $\mathbf{x} \in \mathbb{F}^n$ and $p \geq 1$, define

$$\|\mathbf{x}\|_p = \left(\sum_{k=1}^{n} |x_k|^p\right)^{1/p}, \quad \|\mathbf{x}\|_\infty = \sup_{k\geq 1} |x_k|.$$

Show that $\|\cdot\|_p$ defines the norm, i.e., $\|\cdot\|_p$ satisfies properties (i)–(iv) as in Problem A.2. Draw the unit circle in the case $p = 1, 2, \infty$ for $\mathbb{R}^2$.

A.4. Prove Lemma A.1.

A.5. (i) Show that

$$\text{rank}\{AB\} = \begin{cases} \text{rank}\{A\}, & \text{if } \det(BB') \neq 0 \\ \text{rank}\{B\}, & \text{if } \det(A'A) \neq 0 \end{cases}$$

(ii) Show that $\text{rank}\{ARB\} = \text{rank}\{R\}$ if $\det(A'A) \neq 0$ and $\det(BB') \neq 0$.

A.6. Prove Lemma A.2.

A.7. (i) Prove (A.2). (ii) Let $A \in \mathbb{F}^{n\times m}$ and $B \in \mathbb{F}^{m\times n}$ and show that $A(I_m + BA)^{-1} = (I_n + AB)^{-1}A$.

A.8. Show that (A.3) is true for the case when both A and B are square of the same dimension. (*Hint:* If E be an elementary matrix which performs some row operations to A, then $\det(EA) = \det(E)\det(A)$.)

A.9. (Laplace expansion) Let $A = \{a_{i,k}\}_{i,k=1,1}^{n,n}$. Show that

$$\det(A) = \sum_{k=1}^{n} a_{i,k}\gamma_{i,k}, \quad \gamma_{i,k} = (-1)^{i+k}\det(M_{i,k}),$$

where $M_{i,k}$ is the $(n-1)\times(n-1)$ matrix obtained by deleting the ith row and kth column of A and $1 \le i \le n$. (*Note:* $\gamma_{i,k}$ is called cofactor corresponding to $a_{i,k}$, and $\det(M_{i,k})$ is called the (i,k)th minor of the matrix; if $i = k$, $\det(M_{i,k})$ is called leading or principle minor.)

A.10. (i) If A is nonsingular, show that

$$\det\left(\begin{bmatrix} A & B \\ C & D \end{bmatrix}\right) = \det(A)\det(D - CA^{-1}B).$$

(ii) If D is nonsingular, show that

$$\det\left(\begin{bmatrix} A & B \\ C & D \end{bmatrix}\right) = \det(D)\det(A - BD^{-1}C).$$

(iii) Show that $\det(I + AB) = \det(I + BA)$, if AB and BA are both square.

A.11. Suppose that $A = A^* > \mathbf{0}$. Show that

$$\mathbf{x}^*A\mathbf{x} - \mathbf{x}^*\mathbf{b} - \mathbf{b}^*\mathbf{x} + c \ge c - \mathbf{b}^*A^{-1}\mathbf{b}.$$

A.12. (i) Suppose that A^{-1} exists. Show that

$$\begin{bmatrix} A & B \\ C & D \end{bmatrix}^{-1} = \begin{bmatrix} A^{-1} + A^{-1}B\Delta^{-1}CA^{-1} & -A^{-1}B\Delta^{-1} \\ -\Delta^{-1}CA^{-1} & \Delta^{-1} \end{bmatrix},$$

where $\Delta = D - CA^{-1}B$ is known as the Schur complement of A.

(ii) Suppose that D^{-1} exists. Show that

$$\begin{bmatrix} A & B \\ C & D \end{bmatrix}^{-1} = \begin{bmatrix} \nabla^{-1} & -\nabla^{-1}BD^{-1} \\ -D^{-1}C\nabla^{-1} & D^{-1} + D^{-1}C\nabla^{-1}BD^{-1} \end{bmatrix},$$

where $\nabla = A - BD^{-1}C$ is known as the Schur complement of D.

(iii) If both A^{-1} and D^{-1} exist, prove the matrix inversion formula:

$$(A - BD^{-1}C)^{-1} = A^{-1} + A^{-1}B(D - CA^{-1}B)^{-1}CA^{-1}.$$

A.13. Let $A \in \mathbb{C}^{n\times r}$, $B \in \mathbb{C}^{m\times k}$, and $X \in \mathbb{C}^{r\times m}$. Show that

$$\text{vec}(AXB) = (B^T \otimes A)\,\text{vec}(X),$$

and thus, $\text{vec}(AX + XB) = (B^T \oplus A)\text{vec}(X)$.

A.14. (Gram–Schmidt orthogonalization) Given m linearly independent vectors $\{\mathbf{x}_k\}_{k=1}^m$, develop a procedure to obtain an orthonormal basis $\{\mathbf{v}_k\}_{k=1}^m$ which span the same range space of $\{\mathbf{x}_k\}_{k=1}^m$. (*Hint*: One may proceed as follows. Step 1: Choose $\mathbf{v}_1 = \mathbf{x}_1/\|\mathbf{x}_1\|$. Step 2: Set

$$\mathbf{v}_2 = \frac{\mathbf{x}_2 - \alpha\mathbf{v}_1}{\|\mathbf{x}_2 - \alpha\mathbf{v}_1\|}, \qquad \alpha = \mathbf{v}_1^*\mathbf{x}_2.$$

Step k ($2 < k \le m$): For $1 \le i < k$, compute $\alpha_{i,k} = \mathbf{v}_i^*\mathbf{x}_k$. Set

$$\mathbf{v}_k = \left(\mathbf{x}_k - \sum_{i=1}^{k-1}\alpha_{i,k}\mathbf{v}_i\right) \Bigg/ \left\|\mathbf{x}_k - \sum_{i=1}^{k-1}\alpha_{i,k}\mathbf{v}_i\right\|.$$

Show that such a set of $\{\mathbf{v}_k\}_{k=1}^m$ is indeed an orthonormal basis.)

A.15. Let S be an orthogonal matrix, whose columns form an orthonormal basis for the projection P. Let $S_\perp$ be the complement of S, i.e., satisfying (A.7). Show that $P = S\left(S^* + \Theta S_\perp^*\right)$ for some Θ, and

$$PP^* = S(I + \Theta\Theta^*)S^*, \quad (I-P)^*(I-P) = S_\perp(I + \Theta^*\Theta)S_\perp^*.$$

Show also that $Q = I - P$ is a projection, and $\|Q\|_F = \|P\|_F$.

A.16. Let $A \in \mathbb{C}^{n\times m}$ have rank r. Show that $A = SRV^*$ where R of size $r \times r$ is nonsingular, and both S and V are orthogonal matrices. Show also that all pseudo-inverses of A, satisfying (A.12), are parameterized by

$$A^\dagger = (V + V_\perp\Theta_v)R^{-1}\left(S^* + \Theta_s S_\perp^*\right),$$

and the Moore-Penrose pseudoinverse are given by $A^+ = VR^{-1}S^*$ where Θ_v and Θ_s are arbitrary, and

$$\begin{bmatrix} S^* \\ S_\perp^* \end{bmatrix}\begin{bmatrix} S & S_\perp \end{bmatrix} = I_n, \quad \begin{bmatrix} V^* \\ V_\perp^* \end{bmatrix}\begin{bmatrix} V & V_\perp \end{bmatrix} = I_m.$$

A.17. Prove Theorem A.3. (*Hint:* Use (A.13) and

$$(\lambda I_n - A)^{-1} = \frac{1}{\det(\lambda I_n - A)}\left(R_1\lambda^{n-1} + R_2\lambda^{n-2} + \cdots + R_n\right)$$

to show that $\det(\lambda I_n - A)I_n = \left(R_1\lambda^{n-1} + R_2\lambda^{n-2} + \cdots + R_n\right)(\lambda I_n - A)$. By matching the coefficient matrices of λ^k on both sides, show that

$$\begin{aligned}
a_1 &= -\mathrm{Tr}\{A\}, & R_1 &= I_n, \\
a_2 &= -\frac{1}{2}\mathrm{Tr}\{R_2A\}, & R_2 &= R_1A + a_1I_n, \\
&\ \vdots & &\ \vdots \\
a_{n-1} &= -\frac{1}{n-1}\mathrm{Tr}\{R_{n-1}A\}, & R_n &= R_{n-1}A + a_{n-1}I_n, \\
a_n &= -\frac{1}{n}\mathrm{Tr}\{R_nA\}, & \mathbf{0} &= R_nA + a_nI.
\end{aligned}$$

Finally, show that the recursive formulas on the right-hand side lead to

$$\mathbf{0} = R_n A + a_n I_n = A^n + a_1 A^{n-1} + \cdots + a_n I_n,$$

which concludes the proof for the Cayley–Hamilton theorem.)

A.18. Let P be a Householder reflection matrix in (A.14). Verify (A.15) and (A.16).

A.19. For square matrices, show that (i) the product of upper triangular matrices is upper triangular, (ii) the inverse of a triangular matrix is upper triangular, and (iii) the equality in (A.17) can be converted into $AQ_i = Q_i R_i$, where Q_i is an orthogonal, and R_i is a triangular matrix.

A.20. Consider Theorem A.3. (i) Show that there are at least 2^ℓ distinct A-invariant subspaces. (ii) Show that if κ of $\{J_i\}_{i=1}^{\ell}$ has sizes greater than 1, then there are at more $(2^{\ell+\kappa})$ distinct A-invariant subspaces. (*Note*: $\mathbb{F}^n$ and $\mathbf{0}$ are also A invariant).

A.21. Suppose that $A, B \in \mathbb{C}^{n\times n}$ and A has n distinct eigenvalues. Show that if A and B are commute, i.e., $AB = BA$, then A and B are simultaneously diagonalizable, i.e., there exists a nonsingular matrix T such that both $\Lambda_A = T^{-1}AT$ and $\Lambda_B = T^{-1}BT$ are diagonal.

A.22. Suppose that $A, B \in \mathbb{C}^{n\times n}$ are diagonalizable. Show that A and B are simultaneously diagonalizable, if and only if A and B are commute. (*Note:* When A and B are not diagonalizable, eigenvalue decomposition can be replaced by Schur decomposition as follows: There exists a unitary matrix Q such that $T_A = Q^*AQ$ and $T_B = Q^*BQ$ are both upper triangular, if and only if A and B commute. See p. 81 in [50].)

A.23. Let $A \in \mathbb{C}^{n\times n}$ be a hermitian matrix. Then all eigenvalues of A are real. Denote n_+, n_-, and n_0 as the number of strictly positive, strictly negative, and zero eigenvalues of A, respectively. Then (n_+, n_-, n_0) is called the inertia of A. (i) Show that S^*AS does not change the inertia of A whenever S is nonsingular. (ii) Assume that A is nonsingular and $B \in \mathbb{C}^{n\times n}$ is also a hermitian. Show that A and B are simultaneously diagonalizable, if and only if $A^{-1}B$ is diagonalizable. (*Note:* Hermitian matrices A and B are simultaneously diagonalizable, if there exists a nonsingular matrix S such that both S^*AS and S^*BS are diagonal.)

A.24. Given $A \in \mathbb{C}^{n\times n}$, use the Schur decomposition to show that for every $\varepsilon > 0$, there exists a diagonalizable matrix B such that $\|A - B\|_2 \le \varepsilon$. This shows that the set of diagonalizable matrices is dense in $\mathbb{C}^{n\times n}$ and that the Jordan canonical form is not a continuous matrix decomposition.

A.25. Let A and $A+E$ be in $\mathbb{R}^{n\times m}$ with $m \ge n$. Show that

$$\text{(i)} \quad |\sigma_k(A+E) - \sigma_k(A)| \le \sigma_1(E) = \|E\|_2,$$

$$\text{(ii)} \quad \sum_{k=1}^{n} [\sigma_k(A+E) - \sigma_k(A)]^2 \le \|E\|_F^2.$$

Let $\mathbf{a}_k$ be the kth column of A and $A_r = \begin{bmatrix} \mathbf{a}_1 & \cdots & \mathbf{a}_r \end{bmatrix}$. Show that

$$\sigma_1(A_{r+1}) \geq \sigma_1(A_r) \geq \sigma_2(A_{r+1}) \geq \cdots \geq \sigma_r(A_{r+1}) \geq \sigma_r(A_r) \geq \sigma_{r+1}(A_{r+1}),$$

where $1 < r < n$.

A.26. (Condition number) Let $A \in \mathbb{R}^{n\times n}$ be nonsingular. Consider

$$(A+\varepsilon E)\mathbf{x}(\varepsilon) = \mathbf{y} + \varepsilon \mathbf{v}, \quad \mathbf{x} = \mathbf{x}(0).$$

(i) Show that $\mathbf{x}(\varepsilon)$ is differentiable at $\varepsilon = 0$ and

$$\dot{\mathbf{x}}(0) = A^{-1}(\mathbf{v} - E\mathbf{x}).$$

(ii) Let $\mathbf{x}(\varepsilon) = \mathbf{x} + \varepsilon\dot{\mathbf{x}}(0) + \mathscr{O}(\varepsilon^2)$. Show that

$$\frac{\|\mathbf{x}(\varepsilon)-\mathbf{x}\|}{\|\mathbf{x}\|} \leq \varepsilon \left\|A^{-1}\right\| \left(\frac{\|\mathbf{v}\|}{\|\mathbf{x}\|} + \|E\|\right) + \mathscr{O}(\varepsilon^2).$$

(iii) Define the condition number $\kappa(A) = \|A\| \left\|A^{-1}\right\|$. Show that

$$\kappa(A) = \|A\| \left\|A^{-1}\right\| = \frac{\sigma_{\max}(A)}{\sigma_{\min}(A)}.$$

(iv) Denote $\rho_A = \varepsilon\|E\|/\|A\|$ and $\rho_{\mathbf{y}} = \varepsilon\|\mathbf{v}\|/\|\mathbf{y}\|$. Show that

$$\frac{\|\mathbf{x}(\varepsilon)-\mathbf{x}\|}{\|\mathbf{x}\|} \leq \kappa(A)(\rho_A + \rho_{\mathbf{y}}) + \mathscr{O}(\varepsilon^2).$$

A.27. Suppose that Θ is a hermitian matrix. Partition Θ as

$$\Theta = \begin{bmatrix} \Theta_1 & \Theta_2^* \\ \Theta_2 & \Theta_3 \end{bmatrix}, \quad \Theta_1 \in \mathbb{F}^{n_1\times n_1}, \quad \Theta_3 \in \mathbb{F}^{n_3\times n_3}.$$

(i) If $\Theta_1 > \mathbf{0}$, and $\Theta_3 < \mathbf{0}$, show that Θ has n_1 strictly positive and n_3 strictly negative eigenvalues. (ii) If $\Theta_1 \geq \mathbf{0}$ and $\Theta_3 \leq \mathbf{0}$ with one of them singular, show that the numbers of positive, and negative eigenvalues of Θ cannot be concluded based on Θ_1 and Θ_3 alone.

A.28. Let A and B be hermitian matrices of the same size. Suppose that $P = \alpha A + \beta B$ is positive definite for some real α and β. Show that A and B are simultaneously diagonalizable. (*Hint:* Hermitian matrices A and B are simultaneously diagonalizable, if and only if A and P or B and P are simultaneously diagonalizable.)

A.29. Assume that $\det(A) \neq 0$. Denote

$$S = \begin{bmatrix} A + G(A^*)^{-1}Q & -G(A^*)^{-1} \\ -(A^*)^{-1}Q & (A^*)^{-1} \end{bmatrix}, \quad J = \begin{bmatrix} 0 & -I_n \\ I_n & 0 \end{bmatrix},$$

where $G = BR^{-1}B^*$, $Q = C^*C$, and $R > 0$. (i) Show that $J^{-1}S^*J = S^{-1}$. Such a matrix is called simplectic. (ii) Show that X is a solution to the ARE (5.96) such that $(I_n + XG)^{-1}$ exists, if and only if X satisfies

$$\begin{bmatrix} -X & I_n \end{bmatrix} S \begin{bmatrix} I_n \\ X \end{bmatrix} = \mathbf{0}.$$

Appendix B
Random Variables and Processes

This appendix gives an overview for some of the materials from probability theory and theory of random processes which are used in this text. The focus is on real random variables. This appendix serves to refresh the reader's memory, or to fill in a few gaps in the reader's knowledge, but, by no means, replaces a formal course on probability theory and random processes. Readers are advised to consult more extensive textbooks on this subject.

B.1 Probability

An experiment may result in several possible *outcomes*. The collection of all possible outcomes from an experiment forms a *sample space* Ω. An *event* A is a subset of Ω. If $A = \Omega$, then it is the *sure event*. If A is empty, denoted by $\emptyset$, then it is the *impossible event*. Let $\mathscr{F}$ be the collection of all events. It is assumed that $\mathscr{F}$ is a σ-filed, meaning that it contains $\emptyset$ and is closed under complements and countable unions. A probability measure $P(\cdot)$ is a mapping from $\mathscr{F}$ into $\mathbb{R}$ satisfying the following three axioms:

$$\begin{array}{ll} \text{(i)} \quad P(A) \geq 0, & \text{(ii)} \; P(\Omega) = 1, \\ \text{(iii)} \; P(A_i \cup A_k) = P(A_i) + P(A_k) \; \text{ if } A_i \cap A_k = \emptyset. \end{array} \tag{B.1}$$

Furthermore, it is required in the infinity countable case,

$$P\left(\bigcup_{k=1}^{\infty} A_k\right) = \sum_{k=1}^{\infty} P(A_k) \tag{B.2}$$

provided that the events are pairwise disjoint. The triplet $(\Omega, \mathscr{F}, P)$ is called a probability space.

G. Gu, *Discrete-Time Linear Systems: Theory and Design with Applications*,
DOI 10.1007/978-1-4614-2281-5,

The joint probability of two events A and B is $P(A \cap B)$. By the elementary set operations and the axioms of probability, there holds

$$P(A \cup B) = P(A) + P(B) - P(A \cap B) \leq P(A) + P(B). \tag{B.3}$$

More generally, there holds union bound

$$P\left(\bigcup_{k=1}^{\infty} A_k\right) \leq \sum_{k=1}^{\infty} P(A_k). \tag{B.4}$$

The conditional probability of A given the occurrence of B is defined as

$$P(A|B) = \frac{P(A \cap B)}{P(B)}, \quad \text{if } P(B) \neq 0.$$

It follows that the joint probability of two events A and B is given by

$$P(A \cap B) = P(A|B)P(B). \tag{B.5}$$

In fact, the above holds even if $P(B) = 0$. Since $P(A \cap B) = P(B \cap A) = P(B|A)P(A)$,

$$P(A|B) = \frac{P(B|A)P(A)}{P(B)} \tag{B.6}$$

which is the well-known Bayes rule. By repeated applications of (B.5),

$$P(A_1 \cap A_2 \cdots \cap A_n) = P(A_1|A_2 \cdots A_n) \cdots P(A_{n-1}|A_n)P(A_n)$$

can be obtained which is the chain's rule for probabilities.

The events $\{A_i\}_{i=1}^m$ are said to form a partition of B, if

$$B = \cup_{i=1}^m A_i \cap B, \quad A_i \cap A_j = \emptyset \ \forall \, i \neq j.$$

In this case, the *law of total probability* states that

$$P(B) = \sum_{i=1}^{m} P(B \cap A_i) = \sum_{i=1}^{m} P(B|A_i)P(A_i). \tag{B.7}$$

This law enables the computation of the probability of an event by analyzing disjoint constituent events and allows Bayes rule to be written as

$$P(A_i|B) = \frac{P(B|A_i)P(A_i)}{\sum_{j=1}^{m} P(B|A_j)P(A_j)}. \tag{B.8}$$

Two events A and B are *independent*, if and only if

$$P(A \cap B) = P(A)P(B) \tag{B.9}$$

implying that $P(A|B) = P(A)$ by (B.6), i.e., the conditioning upon B does not alter the probability of the occurrence of A, if A and B are independent. The multiple events $\{A_i\}_{i=1}^m$ are *jointly independent*, if for every choice of subsets of these events $\{A_{k_i}\}_{i=1}^{\mu}$,

$$P\left(A_{k_1} \cap A_{k_2} \cdots \cap A_{k_\mu}\right) = P\left(A_{k_1}\right) P\left(A_{k_2}\right) P\left(A_{k_\mu}\right). \tag{B.10}$$

Joint independence implies pairwise independence of the underlying events, but the converse is not true in general.

B.2 Random Variables and Vectors

For a discrete r.v. (random variable), its sample space Ω is discrete in the sense that there exists an invertible map between Ω and a discrete subset of $\mathbb{R}$. As such, it can be represented by X with values $x_i \in \mathbb{R}$ and specified by the *probability mass function* (PMF):

$$P[X = x_i] := P[\omega \in \Omega : X(\omega) = x_i], \quad i = 1, 2, \ldots. \tag{B.11}$$

Clearly, $P[X = x_i] \geq 0$, and $\sum_i P[X = x_i] = 1$. It can also be described by the *cumulative distribution function* (CDF):

$$P_X(x) = P[X \leq x] = \sum_{i: x_i \leq x} P[X = x_i] = \sum_i P[X = x_i] \mathbf{1}(x - x_i), \tag{B.12}$$

where $\mathbf{1}(x)$ is the unit step function. Thus, $P_X(-\infty) = 0$, and $P_X(\infty) = 1$.

For the continuous r.v., its CDF is the same as $P_X(x) = P[X \leq x]$ but is continuous and nondecreasing in x. If $P_X(x)$ is differentiable at all x, then

$$p_X(x) = \frac{\mathrm{d}P_X(x)}{\mathrm{d}x} \geq 0 \tag{B.13}$$

is the *PDF* and satisfies

$$P[a < X \leq b] = \int_a^b p_X(x)\, \mathrm{d}x = P_X(b) - P_X(a) \tag{B.14}$$

for $b > a$. In particular, $p_X(x)\mathrm{d}x$ is the probability for $x \in (x,\ x + \mathrm{d}x)$.

Example B.1. Consider the toss of a coin. One side of the coin is mapped to 0, and the other side to 1. The sample space is taken as

$$\Omega = \{A_i\}_{i=0}^3 = \{(0,0), (0,1), (1,0), (1,1)\}.$$

Under the equal probable assumption, $P(A_i) = 0.25$. Let X be a discrete r.v. with $x_i = i$ for $i = 0,1,2,3$, the elements of the corresponding sample space, and $P[X = x_i] = 0.25$. Then this experiment of tossing a coin can be represented by X. Its CDF is given by

$$P_X(x) = 0.25\left[\mathbf{1}(x) + \mathbf{1}(x-1) + \mathbf{1}(x-2) + \mathbf{1}(x-3)\right].$$

Next, consider the continuous r.v. with Gaussian distribution or the PDF

$$p_X(x) = \frac{1}{\sqrt{2\pi}\sigma_X}\exp\left\{-\frac{(x-m_X)^2}{2\sigma_X^2}\right\}.$$

The PDF $p_X(x)$ has the shape of a bell, and is symmetric with respect to $x = m_X$ which is the peak of $p_X(x)$. If $m_X = 0$, and $\sigma_X^2 = 1$, then

$$P_X(x) = \int_{-\infty}^{x} p_X(x)\,\mathrm{d}x = 1 - \int_{x}^{\infty} \frac{1}{\sqrt{2\pi}} \mathrm{e}^{-y^2/2}\,\mathrm{d}y = 1 - Q(x), \tag{B.15}$$

where $Q(\cdot)$ is the Q function or the tail integration.

It is possible for outcomes to be a collection of several r.v.'s. In this case, X denotes random vector of dimension $n > 1$ with components $\{X_i\}_{i=1}^{n}$ random variables. Its joint CDF is defined by

$$P_X(\mathbf{x}) = P[X \le \mathbf{x}] = P[X_1 \le x_1, \ldots, X_n \le x_n]. \tag{B.16}$$

Clearly, the joint CDF is nonnegative, nondecreasing in each of its n arguments, and bounded above by 1. The joint PDF is given by

$$p_X(\mathbf{x}) = p_{X_1,\ldots,X_n}(x_1,\ldots,x_n) = \frac{\partial^n P_{X_1,\ldots,X_n}(x_1,\ldots,x_n)}{\partial x_1 \cdots \partial x_n}, \tag{B.17}$$

provided that the partial derivatives exist.

The marginal distributions can be introduced. In the case of $n = 2$, the marginal PDF and CDF for X_1 are given, respectively, by

$$p_{X_1}(x_1) = \int_{-\infty}^{\infty} p_{X_1,X_2}(x_1,x_2)\,\mathrm{d}x_2, \tag{B.18}$$

$$P_{X_1}(x_1) = P_{X_1,X_2}(x_1,\infty) = P[X \le x_1, X_2 < \infty]. \tag{B.19}$$

In addition, the conditional CDF for X_1, given $X_2 \le x_2$, is

$$P_{X_1|X_2 \le x_2}(x_1) = \frac{P[X_1 \le x_1, X_2 \le x_2]}{P[X_2 \le x_2]} = \frac{P_{X_1,X_2}(x_1,x_2)}{P_{X_2}(x_2)}. \tag{B.20}$$

The conditional CDF $P_{X_1|X_2 \le x_2}(x_1)$ is itself CDF, satisfying

$$P_{X_1|X_2 \le x_2}(-\infty) = 0, \quad P_{X_1|X_2 \le x_2}(\infty) = 1,$$

and the nondecreasing property in x_1. Taking derivative with respect to x_1 in (B.20) yields the conditional PDF

$$p_{X_1|X_2 \le x_2}(x_1) = \frac{\partial P_{X_1|X_2 \le x_2}(x_1)}{\partial x_1}. \tag{B.21}$$

On the other hand, the conditional PDF for X_1, given $X_2 = x_2$, is defined by

$$p_{X_1|X_2}(x_1|x_2) := p_{X_1|X_2 = x_2}(x_1) = \frac{p_{X_1,X_2}(x_1,x_2)}{p_{X_2}(x_2)}, \quad \text{if } p_{X_2}(x_2) \ne 0. \tag{B.22}$$

Note that $p_{X_1|X_2}(x_1|x_2)$ is a PDF of X_1 with $X_2 = x_2$ fixed. The above marginal and conditional distributions can be extended easily to random vectors of dimension $n > 2$. The details are skipped.

The n random variables X_i are independent, if and only if

$$P_{X_1,\dots,X_n}(x_1,\dots,x_n) = P_{X_1}(x_1)\cdots P_{X_n}(x_n).$$

The above is equivalent to

$$p_{X_1,\dots,X_n}(x_1,\dots,x_n) = p_{X_1}(x_1)\cdots p_{X_n}(x_n).$$

The independence of n random variables implies that any pair or subset of these are independent, but the converse is not true in general.

Often the random vector of interest is a transformation of another. Its general form is $Y = f(X)$ with X dimension of n and Y dimension of m. Given the distribution of X, how to find the distribution of Y is the question to be answered next. A general procedure is computing the CDF of Y via

$$P_Y(y) = P[Y \le \mathbf{y}] = P[X : f(\mathbf{x}) \le \mathbf{y}]. \tag{B.23}$$

Example B.2. Let X_1 and X_2 be two identical and independent Gaussian r.v.'s of zero mean and variance σ^2. Define transformation

$$Y = f(X) = \sqrt{X_1^2 + X_2^2}.$$

By $P_Y(y) = P[Y \le y] = P\left[\mathbf{x} : \sqrt{x_1^2 + x_2^2} \le y\right]$, the CDF for Y is

$$P_Y(y) = \frac{1}{2\pi\sigma^2} \iint \exp\left\{-\frac{x_1^2 + x_2^2}{2\sigma^2}\right\} \mathrm{d}x_1 \mathrm{d}x_2$$

over $\sqrt{x_1^2+x_2^2} \leq y$. The polar transform $x_1 = r\cos(\theta)$ and $x_2 = r\sin(\theta)$ gives

$$P_Y(y) = \frac{1}{2\pi\sigma^2}\int_0^{2\pi}\int_0^{y} r\mathrm{e}^{-\frac{r^2}{2\sigma^2}}\,\mathrm{d}r\mathrm{d}\theta = 1-\mathrm{e}^{-\frac{y^2}{2\sigma^2}}.$$

It follows that the PDF for Y is $p_Y(y) = 0$ for $y < 0$ and

$$p_Y(y) = \frac{\mathrm{d}F_Y(y)}{\mathrm{d}y} = \frac{y}{\sigma^2}\mathrm{e}^{-\frac{y^2}{2\sigma^2}}, \quad \text{for } y \geq 0,$$

which is termed Rayleigh distribution. For the case when X_1 or X_2 have nonzero mean, Y is Rician distributed (Problem B.4 in Exercises).

B.3 Expectations

For a given r.v. X, its expectation is defined as

$$\mathrm{E}\{X\} := \begin{cases} \int_{-\infty}^{\infty} x p_X(x)\,\mathrm{d}x, & \text{if } X \text{ is a continuous r.v.,} \\ \sum_i x_i P[X = x_i], & \text{if } X \text{ is a discrete r.v..} \end{cases} \tag{B.24}$$

Expected values are simply probabilistic averages of random variables in an experiment, which are also termed mean values. For transformed r.v.'s, their expectations are defined by

$$\mathrm{E}\{f(X)\} := \begin{cases} \int_{-\infty}^{\infty} f(X) p_X(x)\,\mathrm{d}x, & \text{for continuous r.v.'s,} \\ \sum_i f(x_i) P[X = x_i], & \text{for discrete r.v.'s.} \end{cases} \tag{B.25}$$

B.3.1 Moments and Central Moments

The expected value in (B.24) is also called the first moment of r.v. X and denoted by $\overline{X}$. The nth moment for positive integer n is defined by

$$\overline{X^n} := \mathrm{E}\{f(X)\} = \mathrm{E}\{X^n\}.$$

The second moment $\overline{X^2}$ is commonly referred to as *mean-squared value* of X.

Often r.v.'s may not have zero means. Hence, it makes sense to define central moments for the case $n > 1$:

$$\overline{|X-\overline{X}|^n} = \mathrm{E}\{|X-\overline{X}|^n\}. \tag{B.26}$$

The second central moment reduces to the familiar *variance*

$$\mathrm{var}\{X\} = \mathrm{E}\left\{|X - \overline{X}|^2\right\} = \overline{X^2} - \overline{X}^2. \tag{B.27}$$

The operation of expectation is linear in the sense that

$$\mathrm{E}\{X_1 + X_2 + \cdots + X_n\} = \mathrm{E}\{X_1\} + \mathrm{E}\{X_2\} + \cdots + \mathrm{E}\{X_n\}. \tag{B.28}$$

However, var$\{\cdot\}$ is not linear in general. But if X_i's are independent, then there holds

$$\mathrm{var}\{X_1 + \cdots + X_n\} = \mathrm{var}\{X_1\} + \cdots + \mathrm{var}\{X_n\}. \tag{B.29}$$

Let X be a continuous random vector and $Y = f(X)$. Then the expected value of Y is given by (the discrete counterpart is left as an exercise)

$$\mathrm{E}\{Y\} = \int_{-\infty}^{\infty} \cdots \int_{-\infty}^{\infty} f(\mathbf{x}) p_X(\mathbf{x})\, d\mathbf{x}. \tag{B.30}$$

The conditional expectation can be defined similarly:

$$\mathrm{E}\{f(X)|Y = \mathbf{y}\} = \int_{-\infty}^{\infty} \cdots \int_{-\infty}^{\infty} f(\mathbf{x}) p_{X|Y}(\mathbf{x}|\mathbf{y})\, d\mathbf{x}. \tag{B.31}$$

The quantity $\mathrm{E}\{X|Y = y\}$ can be randomized to $\mathrm{E}\{X|Y\}$ by taking value $\mathrm{E}\{X|Y(\omega) = y\}$ as the experimental outcome ω leads to $Y(\omega) = y$. Hence, $\mathrm{E}\{X|Y\}$ is a function of r.v. Y, and there holds

$$\mathrm{E}\{\mathrm{E}[X|Y]\} = \mathrm{E}\{X\}.$$

If X and Y are conditionally independent for a given Z, then

$$\mathrm{E}\{XY|Z\} = \mathrm{E}\{X|Z\}\mathrm{E}\{Y|Z\}.$$

B.3.2 Correlation and Covariance

If two r.v.'s X and Y have zero mean and are not independent, then

$$\mathrm{cor}\{X,Y\} = \mathrm{E}\{XY\} = \int_{-\infty}^{\infty}\int_{-\infty}^{\infty} xy p_{XY}(x,y)\, dxdy \tag{B.32}$$

defines the *correlation* between the two. If $\mathrm{cor}\{X,Y\} = \mathrm{E}\{X\}\mathrm{E}\{Y\}$, then X and Y are termed *uncorrelated*, and if $\mathrm{cor}\{X,Y\} = 0$, then they are termed *orthogonal*. A more accurate quantity is the *covariance* between X and Y:

$$\mathrm{cov}\{X,Y\} = \mathrm{E}\{(X - m_X)(Y - m_Y)\} = \mathrm{cor}\{X,Y\} - m_X m_Y, \tag{B.33}$$

where $m_X = \mathrm{E}\{X\}$ and $m_Y = \mathrm{E}\{Y\}$ may not be zero. Clearly, $\mathrm{cov}\{X,Y\} = 0$ implies that X and Y are *uncorrelated*. The degree of the correlation between X and Y is measured by *correlation coefficient*

$$\rho_{XY} = \frac{\mathrm{cov}\{X,Y\}}{\sqrt{\mathrm{cov}\{X\}\mathrm{cov}\{Y\}}} = \frac{\mathrm{E}\{(X-m_X)(Y-m_Y)\}}{\sqrt{\mathrm{E}\{(X-m_X)^2\}\mathrm{E}\{(Y-m_Y)^2\}}}. \tag{B.34}$$

It can be easily shown that $0 \leq |\rho_{XY}| \leq 1$. Independent r.v.'s are uncorrelated, but the converse does not hold in general except Gaussian r.v.'s for which independence is equivalent to being uncorrelated.

Example B.3. The random vector X with n tuple is Gaussian, if the joint PDF is given by

$$p_X(\mathbf{x}) = \frac{1}{\sqrt{(2\pi)^n \det(\Sigma)}} \exp\left\{ -\frac{(\mathbf{x}-m_X)^* \Sigma^{-1} (\mathbf{x}-m_X)}{2} \right\}, \tag{B.35}$$

where m_X is the mean and Σ is the covariance matrix. The (i,k)th element of Σ represents the correlation between X_i and X_k, and has the expression

$$\Sigma_{i,k} = \mathrm{E}\{(X_i - m_{X_i})(Y_k - m_{Y_k})\}.$$

If $\Sigma_{i,k} = 0$ for all $i \neq k$, then $\Sigma_{k,k} = \mathrm{diag}(\sigma_{X_1}^2, \ldots, \sigma_{X_n}^2)$, and thus,

$$p_X(\mathbf{x}) = \prod_{k=1}^{n} \frac{1}{\sqrt{2\pi\sigma_{X_k}^2}} \exp\left\{ -\frac{(x_k - m_{X_k})^2}{2\sigma_{X_k}^2} \right\} p_{X_1}(x_1) \cdots p_{X_n}(x_n).$$

That is, for Gaussian r.v.'s, being uncorrelated is equivalent to independence. It is left as an exercise (Problem B.8) to show that linear transform of Gaussian random vectors and conditional (Gaussian) r.v.'s conditioned on Gaussian r.v.'s are also Gaussian-distributed.

Let $Y = X_1 X_2 \cdots X_n$. Then its expectation is

$$\mathrm{E}\{Z\} = \mathrm{E}\{X_1 X_2 \cdots X_n\} = \int_{-\infty}^{\infty} \cdots \int_{-\infty}^{\infty} x_1 x_2 \cdots x_n p_X(\mathbf{x}) \mathrm{d}\mathbf{x}.$$

If X_i's are independent, then $\mathrm{E}\{Z\} = \mathrm{E}\{X_1\} \cdots \mathrm{E}\{X_n\}$.

B.3.3 Characteristic Functions

For r.v. X, its characteristic function is defined by:

$$\Phi_X(\omega) = \mathrm{E}\left\{ \mathrm{e}^{j\omega X} \right\} = \int_{\infty}^{\infty} \mathrm{e}^{j\omega x} p_X(x) \,\mathrm{d}x. \tag{B.36}$$

So, $\Phi_X(\omega)$ is the (conjugate of the) Fourier transform of the PDF. For n r.v.'s $X_1,\ldots,X_n$, the joint characteristic function is

$$\Phi_X(\omega_1,\ldots,\omega_n) = \mathrm{E}\left\{\exp\left[j\sum_{i=1}^{n}\omega_i X_i\right]\right\}.$$

It follows that $\Phi(0) = 1$ and $|\Phi_X(\omega_1,\ldots,\omega_n)| \leq 1$. Moreover, the PDF can be recovered via inverse Fourier transform of the characteristic function.

In regard to moments, it is noted that

$$\frac{\mathrm{d}\Phi_X(\omega)}{\mathrm{d}\omega} = \int_{\infty}^{\infty} jx\mathrm{e}^{j\omega x} p_X(x)\,\mathrm{d}x.$$

Multiplying $-j$ to both sides yields

$$\overline{X} = -j\frac{\mathrm{d}\Phi_X(\omega)}{\mathrm{d}\omega}\bigg|_{\omega=0} = \int_{\infty}^{\infty} x\mathrm{e}^{j\omega x} p_X(x)\,\mathrm{d}x. \tag{B.37}$$

By an induction argument, the nth moment of X is obtained as

$$\overline{X^n} = (-j)^n\frac{\mathrm{d}^n\Phi_X(\omega)}{\mathrm{d}\omega^n}\bigg|_{\omega=0} = \int_{\infty}^{\infty} x^n\mathrm{e}^{j\omega x} p_X(x)\,\mathrm{d}x. \tag{B.38}$$

If X and Y are jointly distributed, then $\Phi_X(\omega) = \Phi_{X,Y}(\omega,0)$. They are independent, if and only if

$$\Phi_{X,Y}(\omega_1,\omega_2) = \Phi_X(\omega_1)\Phi_Y(\omega_2).$$

Let $\{X_i\}$ be a set of n independent r.v.'s and $Z = X_1 + X_2 + \cdots + X_n$. Then

$$\Phi_Z(\omega) = \Phi_{X_1}(\omega)\Phi_{X_2}(\omega)\cdots\Phi_{X_n}(\omega). \tag{B.39}$$

It is left as an exercise to show that the PDF of Z is convolution of $\{p_{X_i}(x_i)\}$'s.

B.4 Sequences of Random Variables

Let $\{X_k\}$ be a sequence of random vectors. If $X_n(\omega) \to X(\omega)$ for all $\omega \in \Omega$, then X_k is said to be convergent to X everywhere as $n \to \infty$. Convergence everywhere is usually too restrictive. The commonly used notions are:

(C1) X_k converges to X almost surely or with probability 1, if $X_n(\omega) \to X(\omega)$ for almost all ω (i.e., for all $\omega \in A \subset \Omega$ such that $P(A) = 1$).
(C2) X_k converges to X in mean square, if $\mathrm{E}\{\|X_k - X\|^2\} \to 0$.
(C3) X_k converges to X in probability, if $P[\|X_k - X\| > \varepsilon] \to 0\ \forall \varepsilon > 0$.

Both (C1) and (C2) imply (C3). As a result, (C1) and (C2) are called the *strong law of convergence*, while (C3) is called the *weak law of convergence*. Conversely, (C3) implies that a subsequence of $\{X_k\}$ satisfies (C1). If, in addition, $\|X_k\| < \alpha$ for some $\alpha > 0$ and all $n \geq n_0$ for some $n_0 > 0$, and almost all ω, then (C2) holds.

Example B.4. Let $\{X_i\}_{i=1}^{\infty}$ be independent and identically distributed (i.i.d.) r.v.'s with PDF $f_X(x)$, mean m_X, and variance σ_X^2. Then

$$S_n = \frac{1}{n}\sum_{i=1}^{n} X_i, \quad n = 1, 2, \ldots, \tag{B.40}$$

form a sequence of r.v.'s although S_k are now strongly correlated. By the Chebyshev inequality (Problem B.11 in Exercises),

$$P[|S_n - m_X| \geq \varepsilon] \leq \frac{\sigma_X^2}{n\varepsilon^2} \to 0 \quad \text{as } n \to \infty,$$

for all $\varepsilon > 0$. Hence, S_n converges to the constant m_X in probability. It is also noted that the assumption on X_i gives

$$\mathrm{E}\{S_n\} = \frac{1}{n}\mathrm{E}\{X_i\} = m_X, \quad \mathrm{var}\{S_n\} = \frac{1}{n}\mathrm{var}\{X_i\} = \frac{\sigma_X^2}{n} \to 0$$

as $n \to \infty$. Thus, S_n converges to m_X in mean square as well. It can be shown that S_n also converges to m_X with probability 1.

The next result is the well-known central limit theorem.

Theorem B.5. *Suppose that $\{X_i\}_{i=1}^{\infty}$ are i.i.d. r.v.'s with mean m_X and variance σ_X^2, and higher-order moments are all finite. Then the random variable*

$$Z_n = \sum_{i=1}^{n} \frac{X_i - m_X}{\sqrt{n}\sigma_X} \tag{B.41}$$

converges in distribution to the Gaussian r.v. having zero mean and unit variance, i.e., $F_{Z_n}(z) \to [1 - Q(z)]$ as $n \to \infty$ with $Q(x)$ as in (B.15).

The proof is left as an exercise (Problem B.12). Let

$$Y_i = \frac{X_i - m_X}{\sigma_X}, \quad i = 1, 2, \ldots. \tag{B.42}$$

Then $\{Y_i\}$ are i.i.d. with zero mean and unit variance, which imply that

$$Z_n = \frac{Y_1 + Y_2 + \cdots + Y_n}{\sqrt{n}}.$$

Thus, Z_n does not degenerate into a constant, as in Example B.4. In fact, Z_n has a unit variance for each n. Because Z_n converges to the standard Gaussian r.v., it is approximately Gaussian distributed for large n.

B.5 Random Processes

Discrete-time random processes are basically sequences of r.v.'s or random vectors of the same dimension, indexed by discretized time. One has a mapping from $\omega \in \Omega$ to a sequence of values $X_\omega(t)$ where time index $t = 0, \pm 1, \ldots$. Normally, the notation $\{X(t)\}$ denotes the process in general. As such $\{\mathbf{x}(t)\}$ is a sample or realization of the random process $\{X(t)\}$. However, $\{\mathbf{x}(t)\}$ is also used to denote the random process in the main body of the text in order to avoid the cumbersome notations.

Clearly, $X(t)$ is a random vector at each time instant t. Assume that any k random vectors $X(t_1), X(t_2), \ldots, X(t_k)$ at k distinct time indexes are jointly distributed. Then the set of all joint PDFs (at all possible k distinct time indexes) $p_{X(t_1)X(t_2)\cdots X(t_k)}(\mathbf{x}(t_1), \mathbf{x}(t_2), \ldots, \mathbf{x}(t_k))$ defines the kth order densities of $\{X(t)\}$. The set of densities of all orders serves to define the probability structure of the random process. The conditional PDFs can be obtained in the usual way.

Markov processes are examples of random processes. Roughly speaking, a process is called Markov, if given that the present is known, the past has no influence on the future. That is, if $t_1 > t_2 > \cdots > t_k$, then

$$p_{X(t_1)|X(t_2)\cdots X(t_k)}[X(t_1)|X(t_2)\cdots X(t_k)] = p_{X(t_1)|X(t_2)}[X(t_1)|X(t_2)].$$

For a given random process $\{X(t)\}$, its expectation and autocovariance are defined respectively by

$$\overline{X}(t) = \mathrm{E}\{X(t)\}, \quad R_X(t;k) := \mathrm{E}\left\{\left[X(t) - \overline{X}(t)\right]\left[X(t-k) - \overline{X}(t-k)\right]\right\}, \tag{B.43}$$

which are functions of time t in general. If $\{X(t)\}$ admit constant mean $\overline{X}$ and covariance $R_X(k)$, then $\{X(t)\}$ is called *WSS*. If the kth densities satisfy

$$p_{X(t_1),\ldots,X(t_k)}(\mathbf{x}(t_1), \ldots, \mathbf{x}(t_k)) = p_{X(t_1+n),\ldots,X(t_k+n)}(\mathbf{x}(t_1+n), \ldots, \mathbf{x}(t_k+n))$$

for any integers k and n, then $\{X(t)\}$ is called *strict-sense stationary*.

For WSS processes, the PSDs are defined as the DTFT of the ACS:

$$\Psi_X(\omega) = \sum_{k=-\infty}^{\infty} R_X(k) \mathrm{e}^{-j\omega k}. \tag{B.44}$$

If $X(t)$ has zero mean for all t, then the power of the random process is

$$P_X = \mathrm{E}\{X(t)X^*(t)\} = R_X(0) = \frac{1}{2\pi}\int_0^{2\pi} \Psi_X(\omega)\, \mathrm{d}\omega \tag{B.45}$$

by the inverse DTFT. Hence ,$\Psi_X(\omega)$ is indeed the power density function of the random process distributed over frequency.

A WSS process $\{X(t)\}$ is called white process, if its mean is zero and its autocovariance is $R_X(k) = R_X\delta(k)$ with $R_X \geq \mathbf{0}$ and $\delta(\cdot)$ the Kronecker delta function. In this case, its PSD is R_X for all frequency. The notion of white processes can be generalized to nonstationary processes. A random process $\{X(t)\}$ is called white process, if

$$\overline{X}(t) = \mathrm{E}\{X(t)\} = \mathbf{0}, \quad R_X(t;k) := \mathrm{E}\{X(t)X^*(t-k)\} = R_X(t)\delta(k). \tag{B.46}$$

So far, random processes are characterized by their *ensemble averages*. That is, a random process is treated as collection of random vectors at various time instants. A natural question is whether the ensemble averages are equivalent to time averages for WSS processes. Unfortunately, not all stationary processes possess such equivalence properties, but those that do so are called *ergodic processes*. If $\{X(t)\}$ is ergodic, then there holds

$$\mathrm{E}\{f[X(t)]\} = \lim_{T\to\infty} \frac{1}{2T+1} \sum_{t=-T}^{T} f[\mathbf{x}(t)]. \tag{B.47}$$

In the case, WSS processes, one may take $f[X(t)] = X(t)$ and

$$f[X(t)] = \left[X(t) - \overline{X}\right]\left[X(t) - \overline{X}\right]^*.$$

If $\{X(t)\}$ is Gaussian process with $R_X(k)$ as covariance, then

$$\sum_{k=-\infty}^{\infty} \|R_X(k)\| < \infty$$

is sufficient for ergodicity.

Suppose that a LTI system represented by its impulse response $\{H(t)\}$ is excited by a WSS process $\{X(t)\}$. Then

$$\mathbf{y}(t) = H(t) \star \mathbf{x}(t) = \sum_{k=-\infty}^{\infty} H(k)\mathbf{x}(t-k). \tag{B.48}$$

Thus, the output is also a random process with mean

$$\overline{Y} = \mathrm{E}\{\mathbf{y}(t)\} = \left[\sum_{k=-\infty}^{\infty} H(k)\right]\overline{X} = \mathbf{H}(1)\overline{X},$$

where $\mathbf{H}(z)$ is the $\mathscr{Z}$ transform of $\{H(t)\}$ and $\overline{X}$ is the mean of $\{X(t)\}$. It is left as an exercise to show that $\{Y(t)\}$ is also a WSS process (Problem B.13) with PSD given by

$$\Psi_Y(\omega) = H\left(e^{j\omega}\right)\Psi_X(\omega)\left[H\left(e^{j\omega}\right)\right]^*. \tag{B.49}$$

B.6 Cramér–Rao Lower Bound

This last section presents a well-known result in estimation, namely, the CRLB. The CRLB characterizes the minimum error covariance for unbiased estimates. Before proceeding, partial derivative with respect to matrix variables is defined first. Let X be a real matrix of dimension $n \times m$, and $g(X)$ be a scalar function of X. The partial derivative of $g(X)$ with respect to X is defined as

$$\frac{\partial g}{\partial X} = g_X := \left[\frac{\partial g}{\partial x_{ij}}\right]_{i,j=1,1}^{n,m} \tag{B.50}$$

that is a matrix of dimension $n \times m$ where x_{ij} is the (i,j)th entry of X. If

$$F(X) = \begin{bmatrix} a(X) & b(X) \\ c(X) & d(X) \end{bmatrix}$$

with $a(\cdot)$, $b(\cdot)$, $c(\cdot)$, $d(\cdot)$ all scalar functions of X, then

$$F_X = \frac{\partial F}{\partial X} = \begin{bmatrix} a_X & b_X \\ c_X & d_X \end{bmatrix}.$$

Hence, if $F(X)$ is a column vector of dimension α with each element a function of X, and X is a row vector of dimension β, then, F_X is a matrix function of X with dimension $\alpha \times \beta$.

Generically, the unknown parameters of the given system are deterministic. However, the measurements obtained from experiments are random due to observation errors and unpredictable disturbances. Let $\mathbf{y}$ be noisy observations. Then $\mathbf{y}$ is a random vector. Denote the corresponding PDF by $f(\mathbf{y};\theta)$ with θ the parameter vector under estimation. As a result, the estimate $\widehat{\theta}$ is also random. The estimate $\widehat{\theta}$ is said to be *unbiased*, if $\mathrm{E}\left\{\widehat{\theta}\right\} = \theta$ that is the true parameter vector.

Unbiased estimates are clearly more preferred. Let the error covariance associated with the unbiased estimate $\widehat{\theta}$ be defined by

$$C_e := \mathrm{E}\left\{\left(\widehat{\theta}-\theta\right)\left(\widehat{\theta}-\theta\right)'\right\} \tag{B.51}$$

where $'$ denotes transpose. The CRLB claims that

$$C_e \geq \text{FIM}(\theta)^{-1}, \quad \text{FIM}(\theta) := \text{E}\left\{\frac{\partial \ln f(\mathbf{y};\theta)}{\partial \theta}\frac{\partial \ln f(\mathbf{y};\theta)}{\partial \theta'}\right\}. \tag{B.52}$$

The matrix $\text{FIM}(\theta)$ is called *Fisher information matrix* (FIM).

To prove the CRLB in (B.52), it is first noted that if $\mathbf{y}$ has dimension N with y_i being the ith entry of $\mathbf{y}$, then

$$\int_{-\infty}^{\infty}\cdots\int_{-\infty}^{\infty} f(\mathbf{y};\theta)\, \mathrm{d}y_1\cdots \mathrm{d}y_N = 1$$

in light of the well-known property of PDF. Denote $\int\cdots\int$ by $\int$ and $\mathrm{d}y_1\cdots\mathrm{d}y_N$ by $\mathrm{d}\mathbf{y}$. Taking partial derivative of the above equality with respect to θ yields

$$\int_{-\infty}^{\infty}\frac{\partial f(\mathbf{y};\theta)}{\partial \theta}\,\mathrm{d}\mathbf{y} = \int_{-\infty}^{\infty} f(\mathbf{y};\theta)\frac{\partial \ln f(\mathbf{y};\theta)}{\partial \theta}\,\mathrm{d}\mathbf{y} = \text{E}\left\{\frac{\partial \ln f(\mathbf{y};\theta)}{\partial \theta}\right\} = \mathbf{0}. \tag{B.53}$$

Next, the hypothesis on unbiased estimate $\widehat{\theta}$ implies that

$$\text{E}\left\{\widehat{\theta}\right\} = \int_{-\infty}^{\infty}\widehat{\theta} f(\mathbf{y};\theta)\,\mathrm{d}\mathbf{y} = \theta.$$

Taking partial derivative of the above equality with respect to θ yields

$$\int_{-\infty}^{\infty}\widehat{\theta}\frac{\partial f(\mathbf{y};\theta)}{\partial \theta'}\,\mathrm{d}\mathbf{y} = \int_{-\infty}^{\infty}\widehat{\theta} f(\mathbf{y};\theta)\frac{\partial \ln f(\mathbf{y};\theta)}{\partial \theta'}\,\mathrm{d}\mathbf{y} = \text{E}\left\{\widehat{\theta}\frac{\partial \ln f(\mathbf{y};\theta)}{\partial \theta'}\right\} = I.$$

It follows from (B.53) and the above equation that

$$\text{E}\left\{\left(\widehat{\theta}-\theta\right)\frac{\partial \ln f(\mathbf{y};\theta)}{\partial \theta'}\right\} = I.$$

Consequently, there holds the following matrix equality:

$$\text{E}\left\{\begin{bmatrix}\left(\widehat{\theta}-\theta\right)\\ \dfrac{\partial \ln f(\mathbf{y};\theta)}{\partial \theta}\end{bmatrix}\begin{bmatrix}\left(\widehat{\theta}-\theta\right)' & \dfrac{\partial \ln f(\mathbf{y};\theta)}{\partial \theta'}\end{bmatrix}\right\} = \begin{bmatrix} C_e & I \\ I & \text{FIM}(\theta)\end{bmatrix} \tag{B.54}$$

that is nonnegative definite from which the CRLB in (B.52) follows.

The simplest case to compute the FIM and the corresponding CRLB is when $\mathbf{y} = A\theta + \mathbf{v}$ where $\mathbf{v}$ is Gauss-distributed. See Problem B.15 in Exercises. The next example illustrates computation of the FIM and CRLB for a more complicated estimation problem.

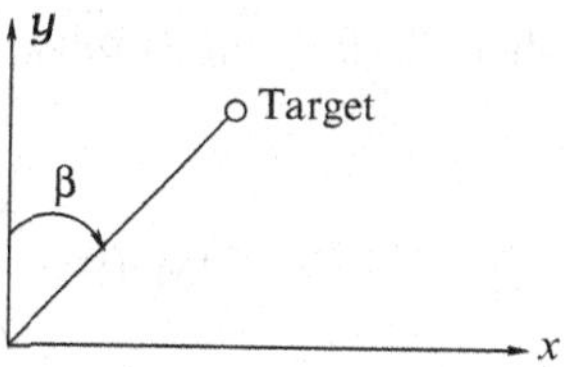

Fig. B.1 Illustration of bearing in the two-dimensional plane

Example B.6. Target motion analysis (TMA) is an important problem in localization and tracking of the moving target where the target moves at the constant speed and in fixed direction. Hence, the motion of the target is uniquely determined by the initial position (x_0, y_0), and velocity (v_x, v_y), assuming two-dimensional motion. It follows that the target coordinate at time t_k is given by $(x_k, y_k) = (x_0 + k\delta_T v_x, y_0 + k\delta_T v_y)$ with δ_T the sampling period. The bearing β is defined as the angle between the positive y-axis and the straight line from the observer at the origin to the target where the angle is positive to the clockwise direction and negative to the counterclockwise direction. See Fig. B.1.

The bearing measurements $\{\hat{\beta}_k\}_{k=0}^{N-1}$ can be obtained with the aid of the antenna array. It is assumed that for $0 \le k < N$,

$$\hat{\beta}_k = \tan^{-1}\left(\frac{x_0 + k\delta_T v_x}{y_0 + k\delta_T v_y}\right) + \eta_k,$$

where $\{\eta_k\}$ are uncorrelated Gaussian noises having mean zero and variance σ_n^2. Bearing-only TMA aims at estimation of the parameter vector $\theta = \begin{bmatrix} x_0 & y_0 & v_x & v_y \end{bmatrix}'$ based on N bearing measurements $\{\hat{\beta}_k\}_{k=0}^{N-1}$. The uncorrelated Gauss assumption on $\{\eta_k\}_{k=0}^{N-1}$ gives the likelihood function

$$\ln f(\beta;\theta) = C_0 - \frac{1}{2\sigma_n^2}\sum_{k=0}^{N-1}\left[\hat{\beta}_k - \tan^{-1}\left(\frac{x_0 + k\delta_T v_x}{y_0 + k\delta_T v_y}\right)\right]^2$$

for some constant C_0. It can be verified by direct calculation that

$$\frac{\partial \ln f(\beta;\theta)}{\partial x_0} = \frac{1}{\sigma_n^2}\sum_{k=0}^{N-1}\frac{\eta_k \sin\beta_k}{R_k},$$

$$\frac{\partial \ln f(\beta;\theta)}{\partial y_0} = -\frac{1}{\sigma_n^2}\sum_{k=0}^{N-1}\frac{\eta_k \cos\beta_k}{R_k},$$

$$\frac{\partial \ln f(\beta;\theta)}{\partial v_x} = \frac{1}{\sigma_n^2}\sum_{k=0}^{N-1}\frac{\eta_k k\delta_T \sin\beta_k}{R_k},$$

$$\frac{\partial \ln f(\beta;\theta)}{\partial v_y} = -\frac{1}{\sigma_n^2}\sum_{k=0}^{N-1}\frac{\eta_k k\delta_T \cos\beta_k}{R_k},$$

where $R_k = \sqrt{(x_0 + k\delta_T v_x)^2 + (y_0 + k\delta_T v_y)^2}$. By the definition in (B.52),

$$\mathrm{FIM}(\theta) = \sum_{k=0}^{N-1} \frac{1}{\sigma_n^2 R_k^2} \begin{bmatrix} \sin\beta_k \\ -\cos\beta_k \\ k\delta_T \sin\beta_k \\ k\delta_T \cos\beta_k \end{bmatrix} \begin{bmatrix} \sin\beta_k \\ -\cos\beta_k \\ k\delta_T \sin\beta_k \\ k\delta_T \cos\beta_k \end{bmatrix}',$$

and thus, $\mathrm{E}\{(\widehat{\theta} - \theta)(\widehat{\theta} - \theta)'\} \geq \mathrm{FIM}(\theta)^{-1}$ can be explicitly calculated.

A natural question regarding the CRLB is whether or not the CRLB is achievable. For the linear signal model with Gauss noise such as in Problem B.15, the answer is affirmative. However, the CRLB is not achievable for most of the other models and estimation algorithms. An exception is the maximum likelihood estimation (MLE). Under certain regularity condition, its associated error covariance converges to the CRLB asymptotically as the number of measurements approaches infinity. This fact indicates that the MLE is asymptotically unbiased as well. While MLE algorithms are important, they are difficult to develop in practice except for some simple signal models. Readers are referred to Chap. 8 for further reading.

Exercises

B.1. Show that the probability measure satisfies

$$P(A) \leq 1, \quad P(\emptyset) = 0, \quad P(A^c) = 1 - P(A), \quad P\left(\bigcup_{i=1}^{\infty} A_i\right) \leq \sum_{i=1}^{\infty} P(A_i)$$

with A^c the compliment of A. Show also that

$$P(A \cup B) = P(A) + P(B) - P(A \cap B).$$

B.2. For a binary symmetric channel with the signal set $\{0, 1\}$,

$$P(B_0|A_0) = P(B_1|A_1) = 0.9, \quad P(B_0|A_1) = P(B_1|A_0) = 0.1,$$

where A_i denotes the event that i is transmitted and B_j the event of j that is received. Suppose that $P(A_0) = P(A_1) = 0.5$. Compute the a posteriori probability $P(A_0|B_0) = P(A_1|B_1)$ and $P(A_1|B_0) = P(A_0|B_1)$.

B.3. Let A and B be two independent and uniformly distributed r.v.'s over $[0,\ 1]$. Consider the transformation

$$X = \sqrt{-2\log(A)}\cos(2\pi B), \quad Y = \sqrt{-2\log(A)}\sin(2\pi B).$$

Show that X and Y are independent, zero mean, and unity variance Gaussian random variables.

B.4. Suppose that X_1 and X_2 in Example B.2 have nonzero mean μ. Let $I_0(\cdot)$ be the modified Bessel function of zeroth order. Show that

$$f_Y(y) = \frac{y}{\sigma^2} I_0\left(\frac{\mu y}{\sigma^2}\right) \exp\left\{-\frac{y^2+\mu^2}{2\sigma^2}\right\}, \quad y \geq 0.$$

B.5. Let X be a discrete random vector and $Y = g(X)$. Find the expression of $\mathrm{E}\{g(X)\}$ which is the counterpart of (B.30).

B.6. Verify (B.28) for the general case and (B.29) for the case when X_i's are independent.

B.7. Let X be a Gaussian random variable with mean zero and variance σ^2. Let $k > 0$ is an integer. Show that

$$\mathrm{E}\left\{X^{2k}\right\} = 1 \times 3 \times 5 \times \cdots \times (2k-1)\sigma^{2k}.$$

B.8. (i) Suppose that

$$\begin{bmatrix} \Psi & \Omega^* \\ \Omega & \Theta \end{bmatrix}$$

is a positive definite matrix with Ψ and Θ positive definite. Show

$$\begin{aligned}\begin{bmatrix} \Psi & \Omega^* \\ \Omega & \Theta \end{bmatrix}^{-1} &= \begin{bmatrix} I & -\Psi^{-1}\Omega^* \\ \mathbf{0} & I \end{bmatrix}\begin{bmatrix} \Psi^{-1} & \mathbf{0} \\ \mathbf{0} & \nabla_\ell^{-1} \end{bmatrix}\begin{bmatrix} I & \mathbf{0} \\ -\Omega\Psi^{-1} & I \end{bmatrix} \\ &= \begin{bmatrix} I & \mathbf{0} \\ -\Theta^{-1}\Omega & I \end{bmatrix}\begin{bmatrix} \nabla_u^{-1} & \mathbf{0} \\ \mathbf{0} & \Theta^{-1} \end{bmatrix}\begin{bmatrix} I & -\Omega^*\Theta^{-1} \\ \mathbf{0} & I \end{bmatrix},\end{aligned}$$

where $\nabla_\ell = \Psi - \Omega^*\Theta^{-1}\Omega$ and $\nabla_u = \Theta - \Omega\Psi^{-1}\Omega^*$.

(ii) Show that if X and Y are jointly Gaussian, the marginal PDF of X or Y is Gaussian distributed, and the conditional PDF of X conditioned on Y is also Gaussian distributed. (*Hint*: Use (i) for the proof.)

(iii) Suppose that X is a Gaussian random vector. Show that $Y = AX + \mathbf{b}$ is also Gaussian random vector where A and $\mathbf{b}$ are constant matrix and vector, respectively.

B.9. Let X be Gaussian r.v. with mean m_X and variance σ_X^2. Show that its characteristic function is

$$\Phi_X(\omega) = \mathrm{e}^{jm_X\omega - \sigma_X^2\omega^2/2}.$$

B.10. Let $\{X_i\}$ be a set of n independent r.v.'s, and $Z = X_1 + X_2 + \cdots + X_n$. Show that

$$p_Z(z) = p_{X_1}(z) \star p_{X_2}(z) \star \cdots \star p_{X_n}(z),$$

where the convolution is defined by

$$p_Z(z) = p_X(z) \star p_Y(z) = \int_{-\infty}^{\infty} p_X(u) p_Y(z-u)\, du.$$

B.11. Let X be a nonnegative r.v., and $a > 0$. (i) Show that

$$P[X \geq a] \leq \frac{\mathrm{E}\{X\}}{a}.$$

The above is called Markov inequality. (ii) Show also that

$$P[|X - m_X| \geq b] \leq \frac{\mathrm{var}\{X\}}{b^2},$$

where $m_X = \mathrm{E}\{X\}$ and $b > 0$. The above is termed Chebyshev inequality.

B.12. Prove Theorem B.5. (*Hint:* Use characteristic function

$$\Phi_{Z_n}(\omega) = \mathrm{E}\left\{e^{j\omega Z_n}\right\} = \prod_{i=1}^{n} \mathrm{E}\left\{e^{jY_i/\sqrt{n}}\right\} = \left[\Phi_Y\left(\omega/\sqrt{n}\right)\right]^n,$$

where Y_i is the same as in (B.42). Next, use Taylor series to show that

$$\Phi_{Z_n}(\omega) = 1 - \frac{\omega^2}{n} + \frac{1}{\sqrt{n^3}} r(n),$$

where $r(n)$ involves third- or higher-order moments of Y_i.)

B.13. Show that the output of an LTI system is WSS process, provided that the input is WSS process, and prove the relation in (B.49).

B.14. Let $\mathrm{FIM}(\theta)$ be defined in (B.52). Show that

$$\mathrm{FIM}(\theta) = -\mathrm{E}\left\{\frac{\partial^2 \ln f(\mathbf{y};\theta)}{\partial\theta\partial\theta'}\right\}.$$

B.15. Consider $\mathbf{y} = A\theta + \mathbf{v}$ where $\mathbf{y}$ is the observation and $\mathbf{v}$ is Gauss-distributed noise with mean zero and covariance Σ_v:

(i) Show that $\mathrm{FIM}(\theta) = A'\Sigma_v^{-1}A$.
(ii) Show that the MLE is given by $\theta_{\mathrm{LS}} = \left(A'\Sigma_v^{-1}A\right)^{-1}\left(A'\Sigma_v^{-1}\mathbf{y}\right)$.
(iii) Show that the CRLB is achievable by proving that:

$$\mathrm{E}\left\{\left(\widehat{\theta}_{\mathrm{LS}} - \theta\right)\left(\widehat{\theta}_{\mathrm{LS}} - \theta\right)'\right\} = \left(A'\Sigma_v^{-1}A\right)^{-1}.$$

Appendix C
Transfer Function Matrices

Multivariable LTI systems can be represented by transfer matrices. The most important class of such matrices are the rational ones which are fractions of polynomial matrices. This appendix covers various forms of polynomial and rational matrices, introduces poles and zeros of the systems, and proves the existence of spectral factorizations for positive para-hermitian matrices. Only the basic elements on transfer matrices are presented which are the integral part of this text.

C.1 Polynomial Matrices

A polynomial matrix of z is represented by

$$\mathbf{P}(z) = P_0 z^{\ell} + P_1 z^{\ell-1} + \cdots + P_{\ell} = \sum_{k=0}^{\ell} P_{\ell-k} z^k, \tag{C.1}$$

where $\{P_k\}$ are matrices of the same dimension. Each element of $\mathbf{P}(z)$ is a polynomial of z with degree no more than ℓ with $(\ell+1)$ referred to as *length* of the polynomial. A square polynomial matrix is *singular*, if $\det[\mathbf{P}(z)] \equiv 0$.

Example C.1. Consider polynomial matrices

$$\mathbf{P}_1(z) = \begin{bmatrix} z^2+3z+2 & z+1 \\ z^2+z-2 & z+1 \end{bmatrix},$$

$$\mathbf{P}_2(z) = \begin{bmatrix} z^2+3z+2 & z+1 \\ z^2+z+2 & z+1 \end{bmatrix}.$$

G. Gu, *Discrete-Time Linear Systems: Theory and Design with Applications*, DOI 10.1007/978-1-4614-2281-5, © Springer Science+Business Media, LLC 2012

The coefficient matrices of $\mathbf{P}_1(z)$ and $\mathbf{P}_2(z)$ can be easily obtained by the following alternate expressions:

$$\mathbf{P}_1(z) = \begin{bmatrix} 1 & 0 \\ 1 & 0 \end{bmatrix} z^2 + \begin{bmatrix} 3 & 1 \\ 1 & 1 \end{bmatrix} z + \begin{bmatrix} 2 & 1 \\ -2 & 1 \end{bmatrix},$$

$$\mathbf{P}_2(z) = \begin{bmatrix} 1 & 0 \\ 1 & 0 \end{bmatrix} z^2 + \begin{bmatrix} 3 & 1 \\ 1 & 1 \end{bmatrix} z + \begin{bmatrix} 2 & 1 \\ 2 & 1 \end{bmatrix}.$$

It can be verified that $\det[\mathbf{P}_1(z)] = 0$ for all z, but $\det[\mathbf{P}_2(z)] = -4(z+1)$. Even though $\det[\mathbf{P}_2(z)] = 0$ at $z = -1$, it is nevertheless nonzero for all other $z \in \mathbb{C}$.

Example C.1 indicates a difference between $\mathbf{P}_1(z)$ and $\mathbf{P}_2(z)$ termed as *normal rank*. For a given polynomial matrix, its normal rank is defined as the rank for almost all z except at some finitely isolated points in $\mathbb{C}$. Thus, $\mathbf{P}_1(z)$ has normal rank 1 while $\mathbf{P}_2(z)$ has normal rank 2.

Suppose that $\mathbf{Q}_1(z)$ and $\mathbf{Q}_2(z)$ are two polynomial matrices with the same number of columns. A polynomial matrix $\mathbf{R}(z)$ is called right divisor of $\mathbf{Q}_1(z)$, if $\mathbf{Q}_1(z) = \tilde{\mathbf{Q}}_1(z)\mathbf{R}(z)$ for some polynomial matrix $\tilde{\mathbf{Q}}_1(z)$. It is called a common right divisor of $\mathbf{Q}_1(z)$ and $\mathbf{Q}_2(z)$, if, in addition, $\mathbf{Q}_2(z) = \tilde{\mathbf{Q}}_2(z)\mathbf{R}(z)$ holds as well for some polynomial matrix $\tilde{\mathbf{Q}}_2(z)$. Moreover, $\mathbf{R}(z)$ is called the greatest common right divisor (GCRD), if for any other right divisor $\tilde{\mathbf{R}}(z)$, $\mathbf{R}(z) = \mathbf{P}(z)\tilde{\mathbf{R}}(z)$ with $\mathbf{P}(z)$ a polynomial matrix.

A convenient way to obtain the GCRD is through some elementary row operations which can be represented by a special class of square matrices, called *unimodular* matrices. For instance, given a polynomial matrix $\mathbf{P}(z)$ of dimension 3×3, multiplication of $\mathbf{P}(z)$ by either of the following matrices

$$U_1 = \begin{bmatrix} 0 & 1 & 0 \\ 0 & 0 & 1 \\ 1 & 0 & 0 \end{bmatrix}, \quad U_2(z) = \begin{bmatrix} 1 & 0 & 0 \\ 0 & 1 & a(z) \\ 0 & 0 & 1 \end{bmatrix}$$

from left carries out an elementary row operation. Specifically, multiplying $\mathbf{P}(z)$ by U_1 from left places the first row of $\mathbf{P}(z)$ to the last and moves the second and third rows up by one. On the other hand, $U_2(z)\mathbf{P}(z)$ adds the product of the third row of $\mathbf{P}(z)$ and $a(z)$ to the second row of $\mathbf{P}(z)$. Elementary row operations also include scaling of some rows by nonzero real or complex numbers that can be represented by a diagonal matrix. It is noted that $\det(U_1) = \det[U_2(z)] = 1$. Any square polynomial matrix with determinant a nonzero constant is called *unimodular* matrix. It is left as an exercise to show that a polynomial matrix is unimodular, if and only if its inverse is a polynomial matrix. Thus, matrices representing elementary row operations are unimodular. A consequence of the elementary row operations is the column hermite form, to be used later for computing the GCRD.

Theorem C.2. *Any $p \times m$ polynomial matrix $\mathbf{P}(z)$ of normal rank r can be reduced by elementary row operations to a quasi-triangular form in which:*

(a) *If $p > r$, then the last $(p-r)$ rows are identically zero.*
(b) *If the diagonal element is unity, then all the elements above it are zero.*
(c) *In column k, $1 \le k \le r$, the diagonal element is monic (i.e., unity leading coefficient) and of higher degree than any element above it.*

Proof. Assume that the first column of $\mathbf{P}(z)$ is not identically zero. Then by exchanging rows, the element of the lowest degree in the first column, denoted by $\tilde{p}_{1,1}(z)$, can be moved to the $(1,1)$ position. It follows that any other element $\tilde{p}_{1,k}(z)$ for $k > 1$ in the first column can be written as

$$\tilde{p}_{1,k}(z) = q_k(z)\tilde{p}_{1,1}(z) + r_k(z)$$

uniquely for some polynomials $q_k(z)$ and $r_k(z)$ with degree of $r_k(z)$ strictly smaller than that of $\tilde{p}_{1,1}(z)$ in light of the Euclidean division algorithm. Applying again elementary row operations removes $q_k(z)\tilde{p}_{1,1}(z)$ for $k > 1$, and thus, none of the elements in the first column has degree higher or equal to that of $\tilde{p}_{1,1}(z)$. Now similar elementary row operations can be applied to the first column for the second time and repeatedly until all the elements, except the first one, in the first column are zero. After then, the same procedure can be used for the second column without affecting the element in the first row that will eliminate the elements below the $(2,2)$ position. Continuing this procedure with the next column and so on leads to the upper triangular hermite form. The statements (a) and (b) thus hold. If (c) is not true for some column, then elementary row operations can be applied to eliminate the higher degree terms above the element on the diagonal via the Euclidean division algorithm. Hence, (c) can be made true as well through elementary row operations. □

Lower triangular hermite form can also be obtained by interexchanging the rows of the upper triangular hermite form which is skipped. The next is an illustrative example for Theorem C.2.

Example C.3. The following shows the steps in reduction of a given $\mathbf{P}(z)$ to the column hermite form:

$$\mathbf{P}(z) = \begin{bmatrix} 1-z & z^2-z+1 \\ -z^2 & z^3-z \\ 1 & -z+1 \end{bmatrix} \rightarrow \begin{bmatrix} 1 & -z+1 \\ -z^2 & z^3-z \\ 1-z & z^2-z+1 \end{bmatrix}$$

$$\rightarrow \begin{bmatrix} 1 & -z+1 \\ 0 & z^2-z \\ 0 & z \end{bmatrix} \rightarrow \begin{bmatrix} 1 & -z+1 \\ 0 & z \\ 0 & z^2-z \end{bmatrix} \rightarrow \begin{bmatrix} 1 & 1 \\ 0 & z \\ 0 & 0 \end{bmatrix}.$$

The corresponding unimodular matrix is

$$\mathbf{U}(z) = \begin{bmatrix} 1 & 1 & 0 \\ 0 & 1 & 0 \\ 0 & 1-z & 1 \end{bmatrix} \begin{bmatrix} 1 & 0 & 0 \\ 0 & 0 & 1 \\ 0 & 1 & 0 \end{bmatrix} \begin{bmatrix} 1 & 0 & 0 \\ z^2 & 1 & 0 \\ z-1 & 0 & 1 \end{bmatrix} \begin{bmatrix} 0 & 0 & 1 \\ 0 & 1 & 0 \\ 1 & 0 & 0 \end{bmatrix}$$

$$= \begin{bmatrix} 1 & 0 & z \\ 1 & 0 & z-1 \\ 1-z & 1 & 2z-1 \end{bmatrix},$$

that represents four elementary row operations on $\mathbf{P}(z)$ and has a determinant 1 for any value of z.

The row hermite form can be obtained similarly by elementary column operations represented by multiplication of some unimodular matrices from right and by simply interchanging the roles of rows and columns in Theorem C.2. Another way to compute the row hermite form is to compute the column hermite form for the transpose of the given polynomial matrix, and then, transpose it back. An exercise is given (Problem C.3) for computing the row hermite form.

For given two polynomial matrices $\mathbf{P}_1(z)$ and $\mathbf{P}_2(z)$ with the same number of columns, a unimodular matrix $\mathbf{U}(z)$ can be obtained to compute the column hermite form as

$$\mathbf{U}(z)\mathbf{P}(z) = \begin{bmatrix} \mathbf{U}_{1,1}(z) & \mathbf{U}_{1,2}(z) \\ \mathbf{U}_{2,1}(z) & \mathbf{U}_{2,2}(z) \end{bmatrix} \begin{bmatrix} \mathbf{P}_1(z) \\ \mathbf{P}_2(z) \end{bmatrix} = \begin{bmatrix} \mathbf{R}(z) \\ \mathbf{0} \end{bmatrix} \tag{C.2}$$

with compatible partition where $\mathbf{R}(z)$ is upper triangular. If $\mathbf{P}(z)$ has full normal column rank, then $\mathbf{R}(z)$ is both square and nonsingular. It is claimed that $\mathbf{R}(z)$ is a GCRD of $\mathbf{P}_1(z)$ and $\mathbf{P}_2(z)$. Indeed, since $\mathbf{U}(z)$ is unimodular,

$$\begin{bmatrix} \mathbf{U}_{1,1}(z) & \mathbf{U}_{1,2}(z) \\ \mathbf{U}_{2,1}(z) & \mathbf{U}_{2,2}(z) \end{bmatrix}^{-1} = \begin{bmatrix} \mathbf{V}_{1,1}(z) & \mathbf{V}_{1,2}(z) \\ \mathbf{V}_{2,1}(z) & \mathbf{V}_{2,2}(z) \end{bmatrix}$$

is a polynomial matrix. It follows that

$$\begin{bmatrix} \mathbf{P}_1(z) \\ \mathbf{P}_2(z) \end{bmatrix} = \begin{bmatrix} \mathbf{V}_{1,1}(z) & \mathbf{V}_{1,2}(z) \\ \mathbf{V}_{2,1}(z) & \mathbf{V}_{2,2}(z) \end{bmatrix} \begin{bmatrix} \mathbf{R}(z) \\ \mathbf{0} \end{bmatrix}.$$

The above implies the relation

$$\mathbf{P}_1(z) = \mathbf{V}_{1,1}(z)\mathbf{R}(z), \quad \mathbf{P}_2(z) = \mathbf{V}_{2,1}(z)\mathbf{R}(z).$$

Hence, $\mathbf{R}(z)$ is a common right divisor of $\mathbf{P}_1(z)$ and $\mathbf{P}_2(z)$. To see that $\mathbf{R}(z)$ is a GCRD, assume that $\tilde{\mathbf{R}}(z)$ is any other common right divisor of $\mathbf{P}_1(z)$ and $\mathbf{P}_2(z)$. Then

$$\mathbf{P}_1(z) = \tilde{\mathbf{P}}_1(z)\tilde{\mathbf{R}}(z), \quad \mathbf{P}_2(z) = \tilde{\mathbf{P}}_2(z)\tilde{\mathbf{R}}(z) \tag{C.3}$$

for some polynomial matrices $\tilde{\mathbf{P}}_1(z)$ and $\tilde{\mathbf{P}}_2(z)$. In light of (C.2), there holds

$$\begin{aligned} \mathbf{R}(z) &= \mathbf{U}_{1,1}(z)\mathbf{P}_1(z) + \mathbf{U}_{1,2}(z)\mathbf{P}_2(z) \\ &= \left[\mathbf{U}_{1,1}(z)\tilde{\mathbf{P}}_1(z) + \mathbf{U}_{1,2}(z)\tilde{\mathbf{P}}_2(z)\right]\tilde{\mathbf{R}}(z), \end{aligned} \tag{C.4}$$

and thus, $\tilde{\mathbf{R}}(z)$ is a right divisor of $\mathbf{R}(z)$, concluding the fact that $\mathbf{R}(z)$ is a GCRD of $\mathbf{P}_1(z)$ and $\mathbf{P}_2(z)$.

It is noted that GCRDs are not unique. In particular, $\mathbf{R}(z)$ does not have to be in column hermite form. Moreover, any two GCRDs $\mathbf{R}_1(z)$ and $\mathbf{R}_2(z)$ are related by

$$\mathbf{R}_1(z) = \mathbf{Q}_2(z)\mathbf{R}_2(z), \quad \mathbf{R}_2(z) = \mathbf{Q}_1(z)\mathbf{R}_1(z),$$

where $\mathbf{Q}_1(z)$ and $\mathbf{Q}_2(z)$ are unimodular.

Polynomial matrices $\mathbf{P}_1(z)$ and $\mathbf{P}_2(z)$ in (C.2) are said to be relatively right prime, or right coprime, if any of their GCRDs is unimodular. If in addition $\mathbf{P}(z)$ in (C.2) has full column rank for all $z \in \mathbb{C}$, then the polynomial matrix $\mathbf{P}(z)$ is called *irreducible*. The next result regards *Bezout identity*.

Lemma C.1. *Polynomial matrices $\mathbf{P}_1(z)$ and $\mathbf{P}_2(z)$ are right coprime, if and only if there exist polynomial matrices $\mathbf{X}_1(z)$ and $\mathbf{X}_2(z)$ such that*

$$\mathbf{X}_1(z)\mathbf{P}_1(z) + \mathbf{X}_2(z)\mathbf{P}_2(z) = I. \tag{C.5}$$

Proof. Given $\mathbf{P}_1(z)$ and $\mathbf{P}_2(z)$, (C.4) holds in light of (C.2). If $\mathbf{P}_1(z)$ and $\mathbf{P}_2(z)$ are right coprime, then $\mathbf{R}(z)$ is unimodular. Multiplying (C.4) by $\mathbf{R}(z)^{-1}$ from left yields

$$\mathbf{R}(z)^{-1}\mathbf{U}_{1,1}(z)\mathbf{P}_1(z) + \mathbf{R}(z)^{-1}\mathbf{U}_{1,2}(z)\mathbf{P}_2(z) = I.$$

Hence, $\mathbf{X}_k(z) = \mathbf{R}(z)^{-1}\mathbf{U}_{1,k}(z)$ is a polynomial matrix for $k = 1,2$ that verifies the Bezout identity (C.5). Conversely, assume that (C.5) is true with $\mathbf{X}_1(z)$ and $\mathbf{X}_2(z)$ polynomial matrices. Let $\tilde{\mathbf{R}}(z)$ be any GCRD of $\mathbf{P}_1(z)$ and $\mathbf{P}_2(z)$, i.e., (C.3) holds for some polynomial matrices $\tilde{\mathbf{P}}_1(z)$ and $\tilde{\mathbf{P}}_2(z)$. Then

$$\begin{aligned} I = \mathbf{X}_1(z)\mathbf{P}_1(z) + \mathbf{X}_2(z)\mathbf{P}_2(z) &= \left[\mathbf{X}_1(z)\tilde{\mathbf{P}}_1(z) + \mathbf{X}_2(z)\tilde{\mathbf{P}}_2(z)\right]\tilde{\mathbf{R}}(z) \\ \Longrightarrow \quad \tilde{\mathbf{R}}(z)^{-1} &= \mathbf{X}_1(z)\tilde{\mathbf{P}}_1(z) + \mathbf{X}_2(z)\tilde{\mathbf{P}}_2(z) \end{aligned}$$

which is a polynomial matrix. Hence, $\tilde{\mathbf{R}}(z)$ is unimodular, implying that $\mathbf{P}_1(z)$ and $\mathbf{P}_2(z)$ are right coprime. □

A square polynomial matrix $\mathbf{R}(z)$ is said to be a common right divisor of polynomial matrices $\{\mathbf{P}_k(z)\}_{k=1}^L$, each having the same number of columns, if $\mathbf{P}_k(z) = \tilde{\mathbf{P}}_k(z)\mathbf{R}(z)$ for $k = 1,2,\ldots,L$ where $\left\{\tilde{\mathbf{P}}_k(z)\right\}_{k=1}^L$ are also polynomial matrices. It is a GCRD of $\{\mathbf{P}_k(z)\}_{k=1}^L$, if for any other right divisor $\tilde{\mathbf{R}}(z)$, $\mathbf{R}(z) = \mathbf{Q}(z)\tilde{\mathbf{R}}(z)$ with $\mathbf{Q}(z)$ a polynomial matrix. The set of polynomial matrices $\{\mathbf{P}_k(z)\}_{k=1}^L$ are said to be right coprime, if any of their GCRDs is unimodular. In particular, a polynomial matrix $\mathbf{P}(z)$ is said to be irreducible, if all its rows are right coprime.

Common left divisor, the greatest common left divisor (GCLD), and left coprime can be defined dually, by interchanging the roles of columns and rows as well as left and right. Problems C.4, C.5, and C.6 are left as exercises.

C.2 Rational Matrices

C.2.1 Matrix Fractional Descriptions

A $p \times m$ rational transfer matrix $\mathbf{T}(z)$ has all its its elements rational functions of z. It can be represented by fractions of two polynomial matrices

$$\mathbf{T}(z) = \mathbf{M}(z)^{-1}\mathbf{N}(z) = \tilde{\mathbf{N}}(z)\tilde{\mathbf{M}}(z)^{-1}. \tag{C.6}$$

Because $\mathbf{T}(z)$ is a matrix, left fraction matrices $\{\mathbf{M}(z),\mathbf{N}(z)\}$ are different from right fraction matrices $\{\tilde{\mathbf{M}}(z),\tilde{\mathbf{N}}(z)\}$ in general. Specifically, $\mathbf{N}(z)$ and $\tilde{\mathbf{N}}(z)$ have the same dimension as $\mathbf{T}(z)$, but $\mathbf{M}(z)$ and $\tilde{\mathbf{M}}(z)$ are square of size $p \times p$ and $m \times m$, respectively. The representation in (C.6) is termed as *matrix fractional descriptions* (MFDs).

Clearly, $\mathbf{M}(z)$ and $\tilde{\mathbf{M}}(z)$ are nonsingular. The left MFD $\{\mathbf{M}(z),\mathbf{N}(z)\}$ are called *irreducible* left MFD, if they are left coprime. Dually, $\{\tilde{\mathbf{M}}(z),\tilde{\mathbf{N}}(z)\}$ are called irreducible right MFD, if the right coprime condition holds. Given a transfer matrix $\mathbf{T}(z)$, how to obtain an irreducible left and right MFDs is partially answered by the following lemma.

Lemma C.2. *Suppose that* $\{\mathbf{M}(z),\mathbf{N}(z)\}$ *is a left* MFD *for* $\mathbf{T}(z)$. *Then there exists a unimodular matrix* $\mathbf{V}(z)$ *such that*

$$\begin{bmatrix}\mathbf{M}(z) & \mathbf{N}(z)\end{bmatrix}\begin{bmatrix}\mathbf{V}_{1,1}(z) & \mathbf{V}_{1,2}(z)\\ \mathbf{V}_{2,1}(z) & \mathbf{V}_{2,2}(z)\end{bmatrix} = \begin{bmatrix}\mathbf{R}(z) & \mathbf{0}\end{bmatrix} \tag{C.7}$$

with compatible partitions where $\mathbf{R}(z)$ *is square. Moreover, the following hold:*

(a) *The polynomial submatrix* $\mathbf{V}_{2,2}(z)$ *is square and nonsingular.*
(b) *The polynomial matrices* $\mathbf{V}_{2,2}(z)$ *and* $\mathbf{V}_{2,1}(z)$ *are right coprime, and*

$$\mathbf{T}(z) = \mathbf{M}(z)^{-1}\mathbf{N}(z) = -\mathbf{V}_{2,1}(z)\mathbf{V}_{2,2}(z)^{-1}. \tag{C.8}$$

(c) *If* $\mathbf{M}(z)$ *and* $\mathbf{N}(z)$ *are left coprime, then*

$$\deg\{\det[\mathbf{M}(z)]\} = \deg\{\det[\mathbf{V}_{2,2}(z)]\}. \tag{C.9}$$

Proof. Applying elementary column operations leads to equality (C.7) that in turn yields $\mathbf{N}(z)\mathbf{V}_{2,2}(z) = -\mathbf{M}(z)\mathbf{V}_{2,1}(z)$ or

$$\mathbf{M}(z)^{-1}\mathbf{N}(z)\mathbf{V}_{2,2}(z) = -\mathbf{V}_{2,1}(z).$$

Thus, for each $z \in \mathbb{C}$, the range space of $\mathbf{V}_{2,2}(z)$ is the same as that of $\mathbf{V}_{2,2}(z)$. Because all columns of $\mathbf{V}(z)$ are linearly independent for all $z \in \mathbb{C}$, (a) holds, and (b) follows. Since $\mathbf{V}(z)$ is unimodular

$$\mathbf{U}(z) = \begin{bmatrix} \mathbf{U}_{1,1}(z) & \mathbf{U}_{1,2}(z) \\ \mathbf{U}_{2,1}(z) & \mathbf{U}_{2,2}(z) \end{bmatrix} = \begin{bmatrix} \mathbf{V}_{1,1}(z) & \mathbf{V}_{1,2}(z) \\ \mathbf{V}_{2,1}(z) & \mathbf{V}_{2,2}(z) \end{bmatrix}^{-1} \tag{C.10}$$

is also a polynomial matrix. By (ii) of Problem A.12 in Appendix A,

$$\mathbf{U}_{1,1}(z) = \left[\mathbf{V}_{1,1}(z) - \mathbf{V}_{1,2}(z)\mathbf{V}_{2,2}(z)^{-1}\mathbf{V}_{2,1}(z)\right]^{-1}.$$

It follows from unimodular of $\mathbf{V}(z)$ and (ii) of Problem A.10 in Appendix A,

$$\begin{aligned} \det[\mathbf{V}(z)] &= \det[\mathbf{V}_{2,2}(z)]\det\left[\mathbf{V}_{1,1}(z) - \mathbf{V}_{1,2}(z)\mathbf{V}_{2,2}(z)^{-1}\mathbf{V}_{2,1}(z)\right] \\ &= \det[\mathbf{V}_{2,2}(z)]/\det[\mathbf{U}_{1,1}(z)] = \text{constant}. \end{aligned}$$

If $\{\mathbf{M}(z), \mathbf{N}(z)\}$ are left coprime, then $\mathbf{R}(z)$ is unimodular. Multiplying (C.7) by $\mathbf{U}(z)$ from right leads to $\mathbf{M}(z) = \mathbf{R}(z)\mathbf{U}_{1,1}(z)$, and thus,

$$\deg\{\det[\mathbf{M}(z)]\} = \deg\{\det[\mathbf{U}_{1,1}(z)]\} = \deg\{\det[\mathbf{V}_{2,2}(z)]\},$$

that concludes (c). □

Lemma C.2 provides a procedure for computing irreducible left and right MFDs for a given transfer matrix $\mathbf{T}(z)$. It begins with any left MFD and elementary column operations embodied in (C.7). Then $\{\mathbf{V}_{2,2}(z), -\mathbf{V}_{1,2}(z)\}$ constitute an irreducible right MFD as shown in (C.8). The computation of $\mathbf{V}(z)^{-1}$ as in (C.10) gives an irreducible left MFD $\{\mathbf{U}_{1,1}(z), \mathbf{U}_{1,2}(z)\}$ by $\mathbf{M}(z) = \mathbf{R}(z)\mathbf{U}_{1,1}(z)$, $\mathbf{N}(z) = \mathbf{R}(z)\mathbf{U}_{1,2}(z)$, and the fact that $\mathbf{R}(z)$ is a GCLD of $\mathbf{M}(z)$ and $\mathbf{N}(z)$. The next result states the generalized Bezout identity.

Theorem C.4. *Let* $\mathbf{M}(z)^{-1}\mathbf{N}(z) = \tilde{\mathbf{N}}(z)\tilde{\mathbf{M}}(z)^{-1}$ *be irreducible left and right* MFDs *of* $p \times m$ *transfer matrix* $\mathbf{T}(z)$. *Then there exist polynomial matrices* $\{\mathbf{X}(z), \mathbf{Y}(z), \tilde{\mathbf{X}}(z), \tilde{\mathbf{Y}}(z)\}$ *such that the generalized Bezout identity*

$$\begin{bmatrix} \mathbf{M}(z) & \mathbf{N}(z) \\ \mathbf{Y}(z) & -\mathbf{X}(z) \end{bmatrix} \begin{bmatrix} \tilde{\mathbf{X}}(z) & \tilde{\mathbf{N}}(z) \\ \tilde{\mathbf{Y}}(z) & -\tilde{\mathbf{M}}(z) \end{bmatrix} = \begin{bmatrix} I_p & \mathbf{0} \\ \mathbf{0} & I_m \end{bmatrix} \tag{C.11}$$

holds in which both of the 2×2 *block matrices on the left are unimodular.*

Proof. The hypothesis on the left and right MFDs implies the existence of polynomial matrices $\{\mathbf{X}(z), \mathbf{Y}(z), \hat{\mathbf{X}}(z), \hat{\mathbf{Y}}(z)\}$ such that

$$\mathbf{M}(z)\tilde{\mathbf{X}}(z) + \mathbf{N}(z)\tilde{\mathbf{Y}}(z) = I_p, \quad \hat{\mathbf{X}}(z)\tilde{\mathbf{M}}(z) + \hat{\mathbf{Y}}(z)\tilde{\mathbf{N}}(z) = I_m,$$

and $\hat{\mathbf{Y}}(z)\tilde{\mathbf{X}}(z) - \hat{\mathbf{X}}(z)\tilde{\mathbf{Y}}(z) = \mathbf{Q}(z)$ is a polynomial matrix, leading to

$$\begin{bmatrix} \mathbf{M}(z) & \mathbf{N}(z) \\ \hat{\mathbf{Y}}(z) & -\hat{\mathbf{X}}(z) \end{bmatrix} \begin{bmatrix} \tilde{\mathbf{X}}(z) & \tilde{\mathbf{N}}(z) \\ \tilde{\mathbf{Y}}(z) & -\tilde{\mathbf{M}}(z) \end{bmatrix} = \begin{bmatrix} I_p & \mathbf{0} \\ \mathbf{Q}(z) & I_m \end{bmatrix}.$$

Multiplying the above equation from left by the inverse of the right-hand side matrix verifies the generalized Bezout identity in (C.11) by taking

$$\mathbf{X}(z) = \hat{\mathbf{X}}(z) + \mathbf{Q}(z)\mathbf{N}(z), \quad \mathbf{Y}(z) = \hat{\mathbf{Y}}(z) - \mathbf{Q}(z)\mathbf{M}(z). \tag{C.12}$$

Because all submatrices in (C.11) are polynomials with the right-hand side identify, both of the 2×2 block matrices are unimodular. □

C.2.2 Row- and Column-Reduced Matrices

A $p \times m$ transfer matrix $\mathbf{T}(z)$ is said to be *proper*, or *strictly proper*, if

$$\lim_{z\to\infty} \mathbf{T}(z) < \infty \quad \text{or} \quad \lim_{z\to\infty} \mathbf{T}(z) = \mathbf{0},$$

respectively. A proper or strictly proper transfer matrix represents a causal or strictly causal multivariable discrete-time system, respectively. Suppose that $\mathbf{T}(z)$ is given by its left MFD, i.e., $\mathbf{T}(z) = \mathbf{M}(z)^{-1}\mathbf{N}(z)$. How can we determine if it is proper or strictly proper? In order to answer this question, a couple of new notions is needed.

The first is the degree of a polynomial row vector, which is defined as the highest degree of all elements in the vector. The second is the row-reduced matrices. Let k_i be the degree of the ith row vector of $\mathbf{M}(z)$. Then the square polynomial matrix $\mathbf{M}(z)$ of size $p \times p$ is said to be row-reduced, if

$$\deg\{\det[\mathbf{M}(z)]\} = \sum_{i=1}^{p} k_i = n. \tag{C.13}$$

Column-reduced form can be defined similarly by replacing rows with columns. However, row-reduced matrices may not be column-reduced and vice-versa.

Let $\mathbf{H}(z) = \text{diag}\left(z^{k_1}, z^{k_2}, \ldots, z^{k_p}\right)$. There exists a matrix M_0 such that

$$\mathbf{M}(z) = \mathbf{H}(z)M_0 + \text{remaining terms of lower degrees.} \tag{C.14}$$

It can be easily shown that $\mathbf{M}(z)$ is row-reduced, if and only if $\det(M_0) \neq 0$. Moreover, $\det[\mathbf{M}(z)] = \det(M_0)z^n +$ terms of lower degrees in z. It needs to be pointed out that any square polynomial matrix can be made row-reduced through elementary row operations.

Example C.5. Consider 2×2 polynomial matrix $\mathbf{M}(z)$ given by

$$\begin{bmatrix} 2 & z^4+3z \\ z^2+1 & z^3 \end{bmatrix} = \begin{bmatrix} z^4 & 0 \\ 0 & z^3 \end{bmatrix}\begin{bmatrix} 0 & 1 \\ 0 & 1 \end{bmatrix} + \begin{bmatrix} 2 & 3z \\ z^2+1 & 0 \end{bmatrix}$$
$$= \begin{bmatrix} 0 & 1 \\ 1 & 0 \end{bmatrix}\begin{bmatrix} z^2 & 0 \\ 0 & z^4 \end{bmatrix} + \begin{bmatrix} 2 & 3z \\ 1 & z^3 \end{bmatrix}.$$

In the first expression, z^4 and z^3 are the highest degree in the first and second row of $\mathbf{M}(z)$, respectively. Since the leading coefficient matrix is singular, $\mathbf{M}(z)$ is not in the row-reduced form. On the other hand, the leading coefficient matrix in the second expression is nonsingular, and z^2 and z^4 are the highest degree in the first and second column of $\mathbf{M}(z)$, respectively. It is thus concluded that $\mathbf{M}(z)$ is in the column-reduced form. A row elementary operation can be applied to obtain

$$\mathbf{U}(z)\mathbf{M}(z) = \begin{bmatrix} -1 & z \\ 0 & 1 \end{bmatrix}\begin{bmatrix} 2 & z^4+3z \\ z^2+1 & z^3 \end{bmatrix}$$
$$= \begin{bmatrix} z^3+z^2-2 & -3z \\ z^2+1 & z^3 \end{bmatrix} = \begin{bmatrix} z^3 & 0 \\ 0 & z^3 \end{bmatrix} + \begin{bmatrix} z^2-2 & -3z \\ z^2+1 & 0 \end{bmatrix}.$$

Hence, the polynomial matrix $\mathbf{U}(z)\mathbf{M}(z)$ is both row- and column-reduced.

Lemma C.3. *If $\mathbf{M}(z)$ is row-reduced, then $\mathbf{T}(z) = \mathbf{M}(z)^{-1}\mathbf{N}(z)$ is proper (strictly proper), if and only if each row of $\mathbf{N}(z)$ has degree less (strictly less) than the degree of the corresponding row of $\mathbf{M}(z)$.*

Proof. Only the case of strictly proper $\mathbf{T}(z)$ will be considered, as the other case is similar and thus omitted. Let $\mathbf{T}_k(z)$ and $\mathbf{N}_k(z)$ be the kth column of $\mathbf{T}(z)$ and $\mathbf{N}(z)$, respectively. Then $\mathbf{M}(z)\mathbf{T}(z) = \mathbf{N}(z)$, and thus,

$$\mathbf{M}(z)\mathbf{T}_k(z) = \mathbf{N}_k(z), \quad k = 1, 2, \ldots, m. \tag{C.15}$$

For the "if" part, the ith element of $\mathbf{T}_k(z)$ has the expression

$$\mathbf{T}_{i,k}(z) = \det\left[\mathbf{M}^{(i,k)}(z)\right] \Big/ \det[\mathbf{M}(z)],$$

where $\mathbf{M}^{(i,k)}(z)$ is the matrix obtained by replacing the ith column of $\mathbf{M}(z)$ with $\mathbf{N}_k(z)$ in light of Cramer's rule (Problem C.7 in Exercises). By (C.14)

$$\mathbf{M}(z) = \mathbf{H}(z)M_0 + \text{remaining terms of lower degrees},$$
$$\mathbf{M}^{(i,k)}(z) = \mathbf{H}(z)M_0^{(i,k)} + \text{remaining terms of lower degrees},$$

where M_0 is nonsingular due to the assumption on row-reduced form for $\mathbf{M}(z)$ and $M_0^{(i,k)}$ is the same as M_0 except ith column which is now zero vector by the assumption that each row of $\mathbf{N}(z)$ has degree strictly less than the degree of the corresponding row of $\mathbf{M}(z)$. It follows that for $k = 1,2,\ldots,m$,

$$\deg\{\det[\mathbf{M}(z)]\} > \deg\left\{\det\left[\mathbf{M}^{(i,k)}(z)\right]\right\},$$

where $i = 1,2,\ldots,p$. Or $\mathbf{T}_{i,k}(z)$ is strictly proper. Conversely, there holds

$$\mathbf{N}_{i,k}(z) = \sum_{\ell=1}^{p} \mathbf{M}_{i,\ell}(z)\mathbf{T}_{\ell,k}(z)$$

by (C.15). Thus, if $\mathbf{T}(z)$ is strictly proper, then each element $\mathbf{T}_{\ell,k}(z)$ is strictly proper, and therefore, the degree of $\mathbf{N}_{i,k}(z)$ is strictly smaller than the highest degree of $\{\mathbf{M}_{i,\ell}(z)\}_{\ell=1}^{p}$ for $k = 1,2,\ldots,m$. It is now concluded that the ith row of $\mathbf{N}(z)$ has degree strictly less than that of the ith row of $\mathbf{M}(z)$. Since $1 \le i \le p$, the "only if" part of the proof is completed. □

C.2.3 Smith–McMillan Form

Given a proper transfer matrix $\mathbf{T}(z) = \mathbf{M}(z)^{-1}\mathbf{N}(z)$ which is in the form of left MFD. Lemma C.2 provides a procedure for computing a unimodular matrix $\mathbf{V}(z)$ to obtain an irreducible left MFD. Moreover, a unimodular matrix $\mathbf{U}(z)$ can be obtained to compute a row-reduced form for irreducible left MFD. In this subsection, Smith–McMillan form will be introduced to give further insight into multivariable systems described by their transfer matrices. More importantly, poles and zeros can be defined for multivariable systems.

Theorem C.6. *Suppose that the $p \times m$ rational transfer matrix $\mathbf{T}(z)$ has normal rank r. Then there exist unimodular matrices $\mathbf{U}(z)$ of size $p \times p$ and $\mathbf{V}(z)$ of size $m \times m$ such that*

$$\mathbf{U}(z)\mathbf{T}(z)\mathbf{V}(z) = \begin{bmatrix} \frac{\beta_1(z)}{\alpha_1(z)} & & & \mathbf{0} \\ & \ddots & & \vdots \\ & & \frac{\beta_r(z)}{\alpha_r(z)} & \vdots \\ \mathbf{0} & \cdots & \cdots & \mathbf{0} \end{bmatrix}, \qquad \text{(C.16)}$$

where $\alpha_{k+1}(z)$ divides $\alpha_k(z)$, $\beta_k(z)$ divides $\beta_{k+1}(z)$ for $1 \le k < r$, and $\{\beta_i(z), \alpha_i(z)\}$ are coprime for $1 \le i \le r$.

Proof. Let $a(z)$ be the monic least common multiple of the denominators of the entries of $\mathbf{T}(z)$. Then $\mathbf{N}(z) = a(z)\mathbf{T}(z)$ is a polynomial matrix. It is claimed that there exist unimodular matrices $\mathbf{U}_1(z)$ of size $p \times p$ and $\mathbf{V}_1(z)$ of size $m \times m$ such that

$$\mathbf{U}_1(z)\mathbf{N}(z)\mathbf{V}_1(z) = \begin{bmatrix} \lambda_1(z) & 0 & \cdots & 0 \\ 0 & & & \\ \vdots & & \mathbf{N}_1(z) & \\ 0 & & & \end{bmatrix},$$

where $\lambda_1(z)$ divides every element of $\mathbf{N}_1(z)$. If not, the division algorithm and row and column interchanges can always bring a lower-degree element to the $(1,1)$ position and then applying elementary row and column operations to zero out all other elements in the first row and first column. Repeating the same procedure will eventually yield $\lambda_1(z)$ in the $(1,1)$ position that divides the rest of the nonzero elements. Now repeat the same steps for $\mathbf{N}_1(z)$ and continuing in this way ultimately leads to the *Smith form*

$$\mathbf{U}(z)\mathbf{N}(z)\mathbf{V}(z) = \begin{bmatrix} \lambda_1(z) & & & \mathbf{0} \\ & \ddots & & \vdots \\ & & \lambda_r(z) & \vdots \\ \mathbf{0} & \cdots & \cdots & \mathbf{0} \end{bmatrix}, \tag{C.17}$$

where $\lambda_{k-1}(z)$ clearly divides $\lambda_k(z)$ for $1 < k \leq r$. The expression in (C.16) then follows from

$$\mathbf{U}(z)\mathbf{T}(z)\mathbf{V}(z) = \mathbf{U}(z)\mathbf{N}(z)\mathbf{V}(z)/a(z)$$

and from the Smith form in (C.17) by setting $\lambda_i(z)/a(z) = \beta_i(z)/\alpha_i(z)$ with $\{\beta_i(z), \alpha_i(z)\}$ coprime for $i = 1, 2, \ldots, r$. The division property for $\{\lambda_k(z)\}$ implies those for $\{\beta_k(z)\}$ and $\{\alpha_k(z)\}$. The proof is now complete. □

In Theorem C.6, the polynomial $\alpha_i(z)$ can be made monic. In addition, $\alpha_1(z) = a(z)$. If not, then $\beta_1(z)/\alpha_1(z) = \lambda_1(z)/a(z)$ implying that $\lambda_1(z)$ and $a(z)$ have at least one common root that is also a common root of $\lambda_k(z)$ and $a(z)$ because $\lambda_1(z)$ divides all other $\lambda_k(z)$ for $1 < k \leq r$. This contradicts the assumption on $a(z)$. Consequently, $\alpha_1(z) = a(z)$ holds.

Zeros and poles of multivariable systems are difficult to describe. As a matter of fact, zeros of the system are dependent on the "direction" of the input or output vectors. However, Theorem C.6 gives a clean view for finite zeros and poles of the multivariable systems without referring to the "direction" of the input/output vectors. To be specific, zeros are roots of $\beta_i(z) = 0$ and poles are roots of $\alpha_i(z) = 0$. Suppose that $r = \min\{p, m\}$. Then z_0 is called a *transmission zero* of $\mathbf{T}(z)$, if $(z - z_0)$ is a divisor of $\beta_i(z)$ for some $i < r$. It is a *block zero* of $\mathbf{T}(z)$, if $(z - z_0)$ is a divisor of $\beta_r(z)$ in light of the fact that $\lambda_k(z)$ divides $\lambda_{k+1}(z)$ for $1 \leq k < r$. The roots of $\alpha_i(z) = 0$ are the poles of $\mathbf{T}(z)$ for $1 \leq i \leq r$. It is thus possible for poles and zeros of $\mathbf{T}(z)$ to have the same value without causing pole/zero cancellations, which is very different from scalar transfer functions. The number of poles of $\mathbf{T}(z)$ is called

McMillan degree that is the sum of the degrees of $\alpha_i(z)$ over index i. For this reason, the diagonal matrix in (C.16) is called Smith–McMillan form.

Example C.7. Consider a 2×2 transfer matrix

$$\mathbf{T}(z) = \frac{1}{z^2-1}\begin{bmatrix} z & z(z+2) \\ -z^2 & z \end{bmatrix}.$$

It can be verified easily that its Smith–McMillan form is given by

$$\begin{bmatrix} 1 & 0 \\ z & 1 \end{bmatrix}\mathbf{T}(z)\begin{bmatrix} 1 & -(z+2) \\ 0 & 1 \end{bmatrix} = \begin{bmatrix} \frac{z}{z^2-1} & 0 \\ 0 & \frac{z(z+1)}{z-1} \end{bmatrix}.$$

Hence, $z = 0$ is a block zero, while $z = -1$ is a transmission zero. It is noted that a multivariable system can have the same pole and zero without causing cancellation. The corresponding system has three poles with one at -1 and two at $+1$. Thus, its McMillan degree is 3.

It needs to be pointed out that Smith–McMillan form does not contain any information about the infinity poles and zeros of the system. As a point of fact, unimodular matrices introduce poles and zeros at infinity, and thus, the Smith–McMillan form is often nonproper, even though the given transfer matrix is proper or strictly proper.

C.3 Spectral Factorizations

Let $\boldsymbol{\Phi}(z)$ be a para-hermitian matrix of size $n \times n$. Then it has the form

$$\boldsymbol{\Phi}(z) = \sum_{k=-\infty}^{\infty} \Phi_k z^{-k}, \quad \Phi_k = \Phi_k^* \ \forall\, k. \tag{C.18}$$

If $\boldsymbol{\Phi}\left(e^{j\omega}\right)$ is continuous and positive definite for all $\omega \in \mathbb{R}$, then $\boldsymbol{\Phi}(z)$ qualifies to be a PSD matrix. This section will show the existence and computation of the spectral factorization

$$\boldsymbol{\Phi}(z) = \mathbf{G}(z)^{\sim}\mathbf{G}(z), \quad \mathbf{G}(z)^{\sim} = \left[\mathbf{G}(\bar{z}^{-1})\right]^*, \tag{C.19}$$

where $\mathbf{G}(z)$ has all its poles and zeros strictly inside the unit circle. The results apply to the case when $\boldsymbol{\Phi}\left(e^{j\omega}\right)$ is positive semidefinite.

Several inequalities are needed in studying the spectral factorization. The first inequality is the arithmetic-geometry mean (AGM) inequality:

$$\frac{1}{n}\sum_{k=1}^{n} x_k \ge \left(\prod_{k=1}^{n} x_k\right)^{1/n}, \quad x_k \ge 0. \tag{C.20}$$

Its proof makes the use of the convexity of $-\log(x)$ over $x > 0$. Recall that a function $f(\cdot)$ is convex over the interval $[a,\ b]$ with $b > a$, if for any $x, y \in [a,\ b]$,

$$f(\alpha x + (1-\alpha)y) \le \alpha f(x) + (1-\alpha) f(y), \quad 0 \le \alpha \le 1.$$

By induction, $f(\cdot)$ being convex implies

$$f\left(\sum_{k=1}^{n} \alpha_k x_k\right) \le \sum_{k=1}^{n} \alpha_k f(x_k), \tag{C.21}$$

where $x_k \in [a,\ b]$, $\alpha_k \ge 0$ and $\sum_k \alpha_k = 1$. Since $f(x) = -\log(x)$ is convex for $x > 0$, the inequality (C.21) leads to the weighted AGM inequality

$$\sum_{k=1}^{n} \alpha_k x_k \ge \prod_{k=1}^{n} x_k^{\alpha_k}.$$

Hence, the AGM inequality (C.20) follows by taking $\alpha_k = 1/n$ for each k.

The second inequality is Minkowski inequality as stated next.

Lemma C.4. *If two square matrices A and B are positive definite with the same size $n \times n$, then*

$$[\det(A+B)]^{1/n} \ge [\det(A)]^{1/n} + [\det(B)]^{1/n}. \tag{C.22}$$

Proof. Let $A = D^* D$. Multiplying both sides of (C.22) by

$$[\det(A)]^{-1/n} = \left(|\det(D)|^{-1/n}\right)^2$$

yields an equivalent inequality:

$$[\det(I+C)]^{1/n} \ge [\det(I)]^{1/n} + [\det(C)]^{1/n} = 1 + [\det(C)]^{1/n},$$

where $C = [D^*]^{-1} B D^{-1}$. Let eigenvalues of C be $\{\lambda_k\}_{k=1}^n$. Then $\lambda_k > 0$ and

$$\prod_{k=1}^{n} (1+\lambda_k)^{1/n} \ge 1 + \left(\prod_{k=1}^{n} \lambda_k\right)^{1/n}. \tag{C.23}$$

Hence, Minkowski inequality is equivalent to (C.23) which can be verified directly by explicit multiplication of both sides and term-by-term comparisons using the AGM inequality. □

By induction, Minkowski inequality can also be extended to

$$\left[\det\left(\sum_{k=1}^{N} A_k\right)\right]^{1/n} \ge \sum_{k=1}^{N} [\det(A_k)]^{1/n} \tag{C.24}$$

with A_k's all positive definite. For spectral factorizations of para-hermitian matrices, inequalities for functions on the unit circle are needed. The next one is the Master inequality.

Lemma C.5. *Let $\Psi(\omega) = \Phi\left(e^{j\omega}\right)$ be hermitian positive definite with size $n \times n$ that is continuous on $[-\pi, \pi]$. Then*

$$\log\left(\det\left[\frac{1}{2\pi}\int_{-\pi}^{\pi}\Psi(\omega)\, d\omega\right]\right) \geq \frac{1}{2\pi}\int_{-\pi}^{\pi}\log\left(\det\left[\Psi(\omega)\right]\right)\, d\omega. \tag{C.25}$$

Proof. Since integration can be replaced by summation, (C.24) implies

$$\left(\det\left[\frac{1}{2\pi}\int_{-\pi}^{\pi}\Psi(\omega)\, d\omega\right]\right)^{1/n} \geq \frac{1}{2\pi}\int_{-\pi}^{\pi}\left(\det\left[\Psi(\omega)\right]\right)^{1/n}\, d\omega.$$

Taking natural logarithm on both sides for the above inequality yields

$$\begin{aligned}\frac{1}{n}\log\left(\det\left[\frac{1}{2\pi}\int_{-\pi}^{\pi}\Psi(\omega)\, d\omega\right]\right) &\geq \log\left[\frac{1}{2\pi}\int_{-\pi}^{\pi}\left(\det\left[\Psi(\omega)\right]\right)^{1/n}\, d\omega\right] \\ &\geq \frac{1}{2\pi}\int_{-\pi}^{\pi}\frac{1}{n}\log\left(\det\left[\Psi(\omega)\right]\right)\, d\omega\end{aligned}$$

in light of the inequality (C.44) in Exercises (Problem C.11) by taking $f(\omega) = \left(\det\left[\Psi(\omega)\right]\right)^{1/n}$. Hence, the Master inequality (C.25) is true. □

The last inequality is Jensen inequality. Denote $\lambda = z^{-1}$. It concerns an analytic function $F(\lambda)$ on the unit disk of the form

$$F(\lambda) = \sum_{k=0}^{\infty} f_k\lambda^k, \quad \sum_{k=0}^{\infty}|f_k|^2 < \infty. \tag{C.26}$$

It follows from the Parserval theorem that $|F(\lambda)|$ is square integrable on the unit circle. For application to spectral factorizations, continuity of $F(\lambda)$ on the unit circle is assumed, even though Jensen inequality holds for any function $F(\lambda)$ in (C.26). As a result, Cauchy integral

$$\frac{1}{j2\pi}\oint_{|\lambda|=r}\lambda^{-1}F(\lambda)\, d\lambda = F(0) \quad \forall\, r \leq 1 \tag{C.27}$$

holds. Let $u(\lambda) = \mathrm{Re}\{F(\lambda)\}$ and $v(\lambda) = \mathrm{Im}\{F(\lambda)\}$. That is, $F(\lambda) = u(\lambda) + jv(\lambda)$. It is left as an exercise (Problem C.12) to show that

$$\frac{1}{2\pi}\int_{-\pi}^{\pi}u\left(re^{j\theta}\right)\, d\theta = u(0), \quad \frac{1}{2\pi}\int_{-\pi}^{\pi}v\left(re^{j\theta}\right)\, d\theta = v(0) \tag{C.28}$$

for all r such that $0 < r \leq 1$. The following is the Jensen inequality.

Lemma C.6. *Suppose that $F(\lambda)$ as in (C.26) is continuous and nonzero on the unit circle. Then*

$$\frac{1}{2\pi}\int_{-\pi}^{\pi}\log\left(\left|F\left(e^{j\omega}\right)\right|\right)\,d\omega \ge \log(|F(0)|). \qquad (C.29)$$

Proof. Let $\{z_k\}_{k=1}^{m}$ be zeros of $F(\lambda)$ on the unit disk, i.e., $|z_k| < 1$. Then

$$F_1(\lambda) = F(\lambda)\prod_{k=1}^{m}\left(\frac{1-\bar{z}_k\lambda}{\lambda - z_k}\right)$$

has no zeros on the unit disk. The first-order function

$$B_k(\lambda) = \frac{1-\bar{z}_k\lambda}{\lambda - z_k}, \quad \left|B_k\left(e^{j\omega}\right)\right| = 1 \;\forall\, \omega \in \mathbb{R},$$

is called Blaschke factor that has a zero at z_k and a pole at $\bar{z}_k^{-1}$. It follows that $\log[F_1(\lambda)]$ is analytic on the unit disk and its real part is

$$u_1(\lambda) = \log(|F_1(\lambda)|), \quad \log\left(\left|F_1\left(e^{j\omega}\right)\right|\right) = \log\left(\left|F\left(e^{j\omega}\right)\right|\right).$$

Applying the first equality in (C.28) to $u_1(\lambda) = \log(|F_1(\lambda)|)$ gives

$$\begin{aligned}\frac{1}{2\pi}\int_{-\pi}^{\pi}\log\left(\left|F\left(e^{j\omega}\right)\right|\right)\,d\omega &= \frac{1}{2\pi}\int_{-\pi}^{\pi}\log\left(\left|F_1\left(e^{j\omega}\right)\right|\right)\,d\omega = \log(|F_1(0)|)\\ &= \log(|F(0)|) + \sum_{k=1}^{m}\log\left(\frac{1}{|z_k|}\right) \ge \log(|F(0)|),\end{aligned}$$

by $|z_k| < 1$ for all k. Hence, Jensen inequality (C.29) is true. □

Although m, the number of zeros for $F(\lambda)$ on the unit disk, is implicitly assumed finite in the proof, Jensen inequality is applicable to the case when m is unbounded. Moreover, the proof of Jensen inequality shows that the equality holds for (C.29), if and only if $F(\lambda)$ has no zeros on the unit disk.

Spectral factorization is rather easy if the given PSD is a scalar function and has finitely many poles/zeros. However, it becomes much harder if neither is true that is the case for the PSD matrix $\Phi(z)$ in (C.18). A feasible approach is through approximation by considering approximant

$$\hat{\Phi}_N\left(e^{j\omega}\right) = \sum_{k=-N}^{N}\left(1-\frac{|k|}{N}\right)\Phi_k e^{-jk\omega} = \frac{1}{2\pi}\int_{-\pi}^{\pi}F_N(\theta)\Phi\left(e^{jk(\omega-\theta)}\right)\,d\theta,$$

where $F_N(\omega)$ is the Nth order Fejér kernel defined in (2.18) in Chap. 2. In light of the properties of Fejér kernel in Lemma 2.1 and Theorem 2.2,

$$\mathbf{0} \le \hat{\Phi}_N(z) \le \Phi(z) \;\text{ and }\; \lim_{N\to\infty}\hat{\Phi}_N(z) = \Phi(z) \;\; \forall\, |z| = 1. \qquad (C.30)$$

Hence, spectral factorization of $\Phi(z)$ can be studied via $\hat{\Phi}_N(z)$. Denote

$$T_m = \begin{bmatrix} \Phi_0 & \Phi_1 & \cdots & \Phi_{m-1} \\ \Phi_{-1} & \ddots & \ddots & \vdots \\ \vdots & \ddots & \ddots & \Phi_1 \\ \Phi_{1-m} & \cdots & \Phi_{-1} & \Phi_0 \end{bmatrix}, \quad Z_m = \begin{bmatrix} I \\ z^{-1}I \\ \vdots \\ z^{-(m-1)}I \end{bmatrix} \frac{1}{\sqrt{m}}. \tag{C.31}$$

Then T_m is Toeplitz, and the approximant $\hat{\Phi}_N(z)$ can be expressed as

$$\hat{\Phi}_N(z) = Z_N^{\sim} T_N Z_N. \tag{C.32}$$

The following describe a procedure for computing the spectral factorization:

- Step 1: Compute Cholesky factorization $T_N = L_N^* L_N$ where

$$L_N = L_N^{(m)} := \begin{bmatrix} L_{0,0}^{(N)} & 0 & \cdots & 0 \\ L_{1,0}^{(N)} & L_{1,1}^{(N)} & \ddots & \vdots \\ \vdots & \ddots & \ddots & 0 \\ L_{m,0}^{(N)} & \cdots & L_{m,m-1}^{(N)} & L_{m,m}^{(N)} \end{bmatrix}, \quad m = N-1, \tag{C.33}$$

  is square and lower triangular with all diagonal entries positive.
- Step 2: Set $G_k^{(N)} = L_{k,0}^{(N)}$ and

$$\hat{\mathbf{G}}_N(z) = \sum_{k=0}^{N-1} G_k^{(N)} z^{-k}. \tag{C.34}$$

The next theorem is the main result of this section and shows that $\hat{\mathbf{G}}_N(z)$ converges to $\mathbf{G}(z)$, the spectral factor of $\Phi(z)$ as $N \to \infty$.

Theorem C.8. *Let $\Phi(z)$ as in (C.18) be a* PSD *matrix that is continuous and strictly positive on the unit circle. Let $\hat{\mathbf{G}}_N(z)$ be constructed with Step 1 and Step 2. Then*

$$\lim_{N\to\infty} \hat{\mathbf{G}}_N(z)^{\sim} \hat{\mathbf{G}}_N(z) = \mathbf{G}(z)^{\sim}\mathbf{G}(z) = \Phi(z)$$

is the spectral factorization. That is, both $\mathbf{G}(z)$ and $\mathbf{G}(z)^{-1}$ are stable.

Proof. It can be shown that the limit of L_N in (C.33) is Toeplitz. Specifically, by noting that the $nN \times nN$ block of T_{N+1} on the lower right corner is the same as T_N and the Cholesky factorization of T_{N+1} can be computed based on T_N, it is seen that

$$L_{k,i}^{(N)} = L_{k+1,i+1}^{(N+1)} \quad \forall\, i \le k.$$

The same argument can be used to conclude that given integer $q \geq 0$,

$$L_{k,i}^{(N)} = L_{k+q,i+q}^{(N+q)} \quad \forall\, i \leq k \tag{C.35}$$

as long as N is sufficiently large. In view of the above relation,

$$\lim_{N\to\infty} L_{k+q,i+q}^{(N+q)} = G_{k+q,i+q} = \lim_{N\to\infty} L_{k,i}^{(N)} = G_{k,i}.$$

Hence, for finite index values, $G_{k,i} = G_{k-i}$ for $k \geq i$, and thus,

$$\lim_{N\to\infty} L_N^{(m)} = \begin{bmatrix} G_0 & 0 & \cdots & 0 \\ G_1 & \ddots & \ddots & \vdots \\ \vdots & \ddots & \ddots & 0 \\ G_m & \cdots & G_1 & G_0 \end{bmatrix} \tag{C.36}$$

is lower triangular Toeplitz for any fixed integer $m \geq 0$. It remains to show that $\mathbf{G}(z)$, the limit of $\hat{\mathbf{G}}_N(z)$, has all its zeros strictly inside the unit circle.

Let $N > m \geq 0$. Partition L_N in a conformable manner:

$$L_N = \begin{bmatrix} L_N^{(m)} & \\ L_{N,a} & L_{N,b} \end{bmatrix}. \tag{C.37}$$

Consider the case $m = 0$. Then $L_N^{(m)} = L_{0,0}^{(N)}$. Let $\mathbf{P}(\omega) = \mathbf{Q}\left(\mathrm{e}^{-j\omega}\right)$ be the boundary value of

$$\mathbf{Q}(\lambda) = I + \sum_{k=1}^{N-1} Q_k \lambda^k. \tag{C.38}$$

It can be verified that (Problem C.13 in Exercises) the integral matrix

$$M(Q) = \frac{1}{2\pi}\int_{-\pi}^{\pi} \mathbf{P}(\omega)^* \Phi\left(\mathrm{e}^{j\omega}\right) \mathbf{P}(\omega)\, \mathrm{d}\omega = \underline{Q}^* T_N \underline{Q}, \tag{C.39}$$

where $\underline{Q}^* = \begin{bmatrix} I & \underline{Q}_1^* \end{bmatrix}$ and $\underline{Q}_1^* = \begin{bmatrix} Q_1^* & \cdots & Q_{N-1}^* \end{bmatrix}$. In light of Cholesky factorization of T_N and the partition in (C.37) with $m = 0$,

$$\begin{aligned} M(Q) &= \left[L_{0,0}^{(N)}\right]^* L_{0,0}^{(N)} + \left(L_{N,a} + L_{N,b}\underline{Q}_1\right)^* \left(L_{N,a} + L_{N,b}\underline{Q}_1\right) \\ &\geq \left[L_{0,0}^{(N)}\right]^* L_{0,0}^{(N)} =: \Gamma_{N,0}. \end{aligned}$$

The lower bound is achieved by taking $\underline{Q}_1 = -L_{N,b}^{-1}L_{N,a}$ because T_N is nonsingular for each $N > 0$ (Problem C.13 in Exercises). Let $\mathbf{P}_{\min}(\omega)$ be the extremizing polynomial matrix. It follows that

$$\left[L_{0,0}^{(N)}\right]^* L_{0,0}^{(N)} = \Gamma_{N,0} = \frac{1}{2\pi}\int_{-\pi}^{\pi} \mathbf{P}_{\min}(\omega)^* \Phi\left(e^{j\omega}\right)\mathbf{P}_{\min}(\omega)\, d\omega. \tag{C.40}$$

Applying the Master inequality and Jensen inequality in succession yields

$$\begin{aligned}
\log[\det(\Gamma_{N,0})] &= \log\left(\det\left[\frac{1}{2\pi}\int_{-\pi}^{\pi} \mathbf{P}_{\min}(\omega)^* \Phi\left(e^{j\omega}\right)\mathbf{P}_{\min}(\omega)\, d\omega\right]\right) \\
&\geq \frac{1}{2\pi}\int_{-\pi}^{\pi} \log\left(\det\left[\mathbf{P}_{\min}(\omega)^* \Phi\left(e^{j\omega}\right)\mathbf{P}_{\min}(\omega)\right]\right) d\omega \\
&\geq \frac{1}{2\pi}\int_{-\pi}^{\pi} \log\left(\det\left[\Phi\left(e^{j\omega}\right)\right]\right) + \log\left(\left|\det\left[\mathbf{P}_{\min}(\omega)\right|^2\right]\right) d\omega \\
&\geq \frac{1}{2\pi}\int_{-\pi}^{\pi} \log\left(\det\left[\Phi\left(e^{j\omega}\right)\right]\right) d\omega
\end{aligned}$$

by $\log\left(\det\left[\mathbf{Q}_{\min}(0)\right]\right) = \log\left(\det[I]\right) = 0$. Taking limit $N \to \infty$, $\Gamma_{N,0} \to G_0^* G_0$, $\hat{\mathbf{G}}_N(z) \to \mathbf{G}(z)$, and hence the important inequality

$$\log\left[\det\left(G_0^* G_0\right)\right] \geq \frac{1}{2\pi}\int_{-\pi}^{\pi} \log\left(\det\left[\Phi\left(e^{j\omega}\right)\right]\right) d\omega \tag{C.41}$$

holds. Moreover, inequalities (C.30) in limit and (C.46) in Exercises imply

$$\log\left(\det[\Phi(z)]\right) \geq \log\left(\left|\det\left[G(z)\right]\right|^2\right) \quad \forall\, |z| = 1. \tag{C.42}$$

By invoking Jensen inequality plus (C.41) gives

$$\begin{aligned}
\frac{1}{2\pi}\int_{-\pi}^{\pi} \log\left(\det\left[\Phi\left(e^{j\omega}\right)\right]\right) d\omega &\geq \frac{1}{2\pi}\int_{-\pi}^{\pi} \log\left(\det\left[\left|\mathbf{G}\left(e^{j\omega}\right)\right|^2\right]\right) d\omega \\
&\geq \log[\det(G_0^* G_0)] \\
&\geq \frac{1}{2\pi}\int_{-\pi}^{\pi} \log\left(\det\left[\Phi\left(e^{j\omega}\right)\right]\right) d\omega.
\end{aligned}$$

Consequently, the above four terms are all equal, and thus,

$$\frac{1}{2\pi}\int_{-\pi}^{\pi} \log\left(\det\left[\left|\mathbf{G}\left(e^{j\omega}\right)\right|\right]\right) d\omega = \log[\det(G_0)].$$

Jensen equality concludes that $\mathbf{G}(z)$ is indeed free of zeros for $|z| \geq 1$. Recall that $z = \lambda^{-1}$. □

The condition on continuity and strict positivity of the PSD matrix $\Phi(z)$ on the unit circle is not necessary that can be weakened to positive semidefinite $\forall |z| = 1$ and to inclusion of the singular component (Dirac delta functions on the unit circle) for $\Phi(z)$, provided that

$$\frac{1}{2\pi}\int_{-\pi}^{\pi} \log\left(\det\left[\Phi\left(e^{j\omega}\right)\right]\right) d\omega > -\infty. \tag{C.43}$$

Interested readers may consult with the research literature to expand further understanding on spectral factorizations. Furthermore, spectral factorization result holds for the case when $\Phi(z)$ has nonfull normal rank by adding $\varepsilon^2 I$ before computing the spectral factorization and then taking limit $\varepsilon \to 0$. Although a stronger assumption is used for the PSD matrix in this section, it is adequate for engineering applications, and more importantly, it prevents the advanced mathematics from overwhelming the engineering significance of the spectral factorization results.

Exercises

C.1. Show that a square polynomial matrix is unimodular, if and only if its inverse is a polynomial matrix.

C.2. Compute column hermite form for

$$\mathbf{P}_1(z) = \begin{bmatrix} z^2 & 0 \\ 0 & z^2 \\ 1 & z+1 \end{bmatrix}, \quad \mathbf{P}_2(z) = \begin{bmatrix} 1 & z & -(z+1) \\ z^2 & z^3+z & -z^3 \\ z-1 & z^2+z & -2z+1 \end{bmatrix}$$

and the unimodular matrices that represent the elementary row operations.

C.3. Compute row hermite form for

$$\mathbf{P}_1(z) = \begin{bmatrix} 1+z & z^2 & 1 \\ z^2+2z & z^3+1 & z \end{bmatrix}, \quad \mathbf{P}_2(z) = \begin{bmatrix} z^2 & -z^3+1 & 1 \\ 1 & z & z+1 \end{bmatrix}$$

and the corresponding unimodular matrix.

C.4. Let $\mathbf{Q}_1(z)$ and $\mathbf{Q}_2(z)$ be polynomial matrices having the same number of rows. Show that $\mathbf{Q}_1(z)$ and $\mathbf{Q}_2(z)$ are left coprime, if and only if there exist polynomial matrices $\mathbf{Y}_1(z)$ and $\mathbf{Y}_2(z)$ such that

$$\mathbf{Q}_1(z)\mathbf{Y}_1(z) + \mathbf{Q}_2(z)\mathbf{Y}_2(z) = I.$$

C.5. Suppose that polynomial matrix $\mathbf{R}(z)$ is a GCLD of polynomial matrices $\{\mathbf{P}_k(z)\}_{k=1}^{L}$, each having the same number of rows. Show that any other common left divisor $\tilde{\mathbf{R}}(z)$ of $\{\mathbf{P}_k(z)\}_{k=1}^{L}$ is related to $\mathbf{R}(z)$ via $\mathbf{R}(z) = \tilde{\mathbf{R}}(z)\mathbf{Q}(z)$ for some polynomial matrix $\mathbf{Q}(z)$. Show that if $\tilde{\mathbf{R}}(z)$ is also a GCLD, then $\mathbf{Q}(z)$ is unimodular.

C.6. Suppose that all columns of $\mathbf{Q}(z)$ are left coprime. Show that row vectors of $\mathbf{Q}(z)$ are all linearly independent for any $z \in \mathbb{C}$. What happens if all columns of $\mathbf{Q}(z)$ are not left coprime?

C.7. (Cramer's rule) Consider linear equation $\mathbf{y} = A\mathbf{x}$ with A square and nonsingular. Show that x_k, the kth element of $\mathbf{x}$, can be obtained as $x_k = \det(A_k)/\det(A)$ for $1 \le k \le n$ where A_k is the same as A except that its kth column is replaced by $\mathbf{y}$.

C.8. (i) Let $\{\mathbf{M}(z), \mathbf{N}(z)\}$ be left MFD for transfer matrix $\mathbf{T}(z)$ of size $p \times m$. Show that there exist unique polynomial matrices $\{\mathbf{Q}(z), \mathbf{R}(z)\}$ such that

$$\mathbf{N}(z) = \mathbf{M}(z)\mathbf{Q}(z) + \mathbf{R}(z),$$

where $\mathbf{M}(z)^{-1}\mathbf{R}(z)$ is strictly proper. (ii) Develop a similar result for right MFD of the transfer matrix $\mathbf{T}(z)$. (*Note:* This is the division theorem for polynomial matrices, extended from the Euclidean division algorithm.)

C.9. Referring to the previous problem, compute $\mathbf{Q}(z)$ and $\mathbf{R}(z)$ with $\{\mathbf{M}(z), \mathbf{N}(z)\}$ given by

$$\mathbf{M}(z) = \begin{bmatrix} z+2 & z^3 \\ z^3 & z^2+1 \end{bmatrix}, \quad \mathbf{N}(z) = \begin{bmatrix} 2 & z^4+3z & z-1 \\ z^2+1 & z^3 & 2z^2 \end{bmatrix}.$$

C.10. Show that $f(x) = -\log(x)$ is convex for $x > 0$.

C.11. Suppose that $f(\theta)$ is continuous and positive on $[-\pi, \pi]$. Show that

$$\log\left[\frac{1}{2\pi}\int_{-\pi}^{\pi} f(\theta)\, d\theta\right] \ge \frac{1}{2\pi}\int_{-\pi}^{\pi} \log[f(\theta)]\, d\theta. \tag{C.44}$$

(*Hint:* Use summation in place of integration and then apply AGM inequality to conclude the proof.)

C.12. Prove (C.27) and mean (C.28).

C.13. Let the PSD matrix $\Phi(z)$ be given as in (C.18). Define

$$\mathbf{H}(z) = \sum_{k=0}^{m-1} H_k z^{-k}$$

as a polynomial matrix with n rows. Show that

$$\mathbf{0} \le \frac{1}{2\pi}\int_{-\pi}^{\pi} \mathbf{H}\left(e^{j\omega}\right)^* \Phi\left(e^{j\omega}\right) \mathbf{H}\left(e^{j\omega}\right)\, d\omega = \underline{H}^* T_m \underline{H},$$

where T_m is the Toeplize matrix as in (C.31) and $\underline{H}^* = \left[H_0^* \cdots H_{m-1}^*\right]$. Hence, T_m is positive definite for any integer $m > 0$, if $\Phi(z)$ is positive definite $\forall |z| = 1$.

C.14. Poisson's kernel is defined by

$$P(\omega) = \sum_{k=-\infty}^{\infty} \rho^{|k|} e^{-jk\omega}, \quad 0 < \rho < 1. \tag{C.45}$$

(i) Show that $P(\omega) \geq 0$ for all $\omega \in \mathbb{R}$ and

$$\frac{1}{2\pi} \int_{-\pi}^{\pi} \frac{1-\rho^2}{1-2\rho\cos(\omega)+\rho^2}\, d\omega = 1.$$

(ii) For $\Phi(z)$ in (C.18), continuous on the unit circle, show that

$$\sum_{k=-\infty}^{\infty} \rho^{|k|} \Phi_k e^{-jk\omega} = \frac{1}{2\pi} \int_{-\pi}^{\pi} \Phi\left(e^{j\theta}\right) \frac{1-\rho^2}{1-2\rho\cos(\omega-\theta)+\rho^2}\, d\theta.$$

C.15. Suppose that both matrices A and B are hermitian nonnegative definite with one of them positive definite. Show that $A - B \geq \mathbf{0}$ and $\det(A) = \det(B)$ imply $A = B$.

C.16. (i) For $A \geq B \geq \mathbf{0}$ with $A > \mathbf{0}$, show that

$$\det(A+B) \geq \det(A) + \det(B). \tag{C.46}$$

(ii) For hermitian matrix C of size $n \times n$ that is nonnegative definite, show that

$$[\det(I+C)]^{1/n} \geq [\det(I)]^{1/n} + [\det(C)]^{1/n} = 1 + [\det(C)]^{1/n}$$

References

1. Ackermann J (1985) Sampled-data control systems: analysis and synthesis, robust system design. Springer, Berlin
2. Adamjan VM, Arov DZ, Krein MG (1978) Infinite block Hankel matrices and related extension problem.' AMS Transl 111:133–156
3. Agarwal RC, Burrus CS (1975) New recursive digital filter structure having very low sensitivity and roundoff noise. IEEE Trans Circuits Syst CAS-22:921–927
4. Anderson BD, Moore JB (1981) Detectability and stabilizability of time-varying discrete-time linear systems. SIAM J Contr Optimiz 19:20–32
5. Anderson, BDO, More JB (1990) Optimal control: linear quadratic methods. Prentice-Hall, Inc., NJ
6. Anderson BDO, Vongpanitlerd S (1973) Network analysis and synthesis, a modern system theory approach. Prentice-Hall, Englewood Cliffs, NJ
7. Åström KJ, Wittermark B (1984) Computer controlled systems: theory and design. Prentice-Hall, Englewood Cliffs, NJ
8. Anderson BD, Moore JB (1979) Optimal filtering. Printice-Hall, Englewood Cliffs, NJ
9. Athans M (1971) The role and use of the stochastic linear-quadratic-gaussian problem in control system design. IEEE Trans Automat Contr AC-16:529–552
10. Ariyavisitakul SL, Winters JH, Lee I (1999) Optimum space-time processors with dispersive interference – unified analysis and required filter span. IEEE Trans Commun 47(7): 1073–1083
11. Başar T, Bernhard P (1991) $\mathscr{H}_\infty$ Optimal control and related minimax design problems: a dynamic game approach, systems and control: foundations and applications. Birkhäuser, Boston, MA
12. Bamieh B, Pearson JB (1992) The $\mathscr{H}_2$ problem for sampled-data systems. Syst Contr Lett 19:1–12
13. Ball JA, Ran ACM (1987) Optimal Hankel norm model reductions and Wiener-Hopf factorization I: the canonical case. SIAM J Contr Optimiz SICON-25:362–382
14. Banaban P, Salz J (1992) Optimum diversity combining and equalization in digital data transmission with applications to cellular mobile radio – part I: theoretical considerations. IEEE Trans Commun TC-40:885–894
15. Banaban P, Salz J (1992) Optimum diversity combining and equalization in digital data transmission with applications to cellular mobile radio – part II: numerical results. IEEE Trans Commun TC-40:895–907
16. Bellman RE, Dreyfus SE (1962) Applied dynamic programming. Princeton University Press, NJ
17. Brizard AJ (2007) Introduction to Lagrange mechnics, Lecture notes. http://academics.smcvt.edu/abrizard/Classical_Mechanics/Notes_070707.pdf

G. Gu, *Discrete-Time Linear Systems: Theory and Design with Applications*,
DOI 10.1007/978-1-4614-2281-5,

18. Brokett RW (1970) Finite dimensional linear systems. Wiley, New York
19. Bullen KE (1971) An introduction to the theory of mechanics, 8th edn. Cambridge University Press, Cambridge, England
20. Chen BM, Saberi A, Sannuti P (1991) Necessary and sufficient conditions for a non-minimum plant to have a recoverable target loop C a stable compensator design for LTR. Automatica 28: 493–507
21. Chen CT (1984) Linear system theory and design. Holt, Rinehart and Windton, New York
22. Chen T, Francis BA (1991) $\mathscr{H}_2$-optimal sampled-data control. IEEE Trans Automat Contr AC-36:387–397
23. Chen T, Francis BA (1995) Optimal sampled-data control systems. Springer, New York
24. Chui CK, Chen G (1997) Discrete $\mathscr{H}_\infty$ optimization. Springer, Berlin
25. Dorato P, Levis AH (1971) Optimal linear regulators: the discrete-time case. IEEE Trans Automat Contr AC-16:613–620
26. Doyle JC, Stein G (1981) Multivariable feedback design: concepts for a classical/modern sunthesis. IEEE Trans Automat Contr AC-26:4–16
27. Doyle JC, Glover K, Khargonekar PP, Francis BA (1989) State-space solutions to standard $\mathscr{H}_2$ and $\mathscr{H}_\infty$ control problems. IEEE Trans Automat Contr AC-34:831–846
28. Enns DF (1984) Model reduction with balanced realizations: an error bound and a frequency weighted generalization. In: Proceedings of the 23rd IEEE Conference on Decision and Control, 127–132
29. Fortmann TE, Hitz KL (1977) An introduction to lienar control systems. Marcel Dekker, New York
30. Feintuch A, Saeks R (1982) System theory: a Hilbert space approach. Academic, New York
31. Foschini GJ, Golden GD, Wolniansky PW Valenzuela RA (1999) Simplified processing for wireless communication at high spectral efficiency. IEEE J Sel Areas Commun 17: 1841–1852
32. Franklin GF, Powell JD, Workman ML (1990) Digital control of dynamic systems. Addison-Wesley, Boston, MA
33. Gantmacher FR (1959) The theory of matrices. Chelsea, New York
34. Gilbert EG (1963) Controllability and observability in multivariable control systems. SIAM J Contr Optimiz SICON-1:128–152
35. Gleser LJ (981) Estimation in a multivariate 'errors in variables' regression model: large sample results. Ann Statist 9:24-44
36. Glover K (1984) All optimal Hankel-norm approximations of linear multivariable systems and their $\mathscr{L}_\infty$-error bounds. Int J Contr IJC-39:1115–1193
37. Golden GD, Foschini GJ, Valenzuela RA, Wolniansky PW (1999) Detection algorithm and initial laboratory results using the V-BLAST space-time communication architecture. Electron Lett EL-35:14–15
38. Goodwin GC, Sin KS (1984)Adaptive filtering, prediction and control. Prentice-Hall, New Jersey
39. Green M, Limebeer D (1985) Linear robust control. Prentice-Hall, Englewood Cliffs, NJ
40. Gu DW, Tsia MV, O'Young SD, Postlethwaite I (1989) State space formulae for discrete-time $\mathscr{H}_\infty$ optimization. Int J Contr IJC-49:1683–1723
41. Gu G (1995) Model reduction with relative/multiplicative error bounds and relations to controller reduction. IEEE Trans Automat Contr AC-40:1478–1485
42. Gu G (2005) All optimal Hankel-norm approximations and their $\mathscr{L}_\infty$ error bounds in discrete-time. Int J Contr IJC-78:408–423
43. Gu G, Huang J (1998) Convergence results on the design of QMF banks. IEEE Trans Signal Process SP-46:758–761
44. Gu G, Cao X-R, Badr H (2006) Generalized LQR control and Kalman filtering with relations to computations of inner-outer and spectral factorizations. IEEE Trans Automat Contr AC-51:595–605
45. Gu G, Badran E (2004) Optimal design for channel equalization via filterbank approach. IEEE Trans Signal Process SP-52:536–545

46. Gu G, Li L (2003) Worst-case design for optimal channel equalization in filterbank transceivers. IEEE Trans Signal Process SP-51:2424–2435
47. Harris FJ (2004) Multirate signal processing for communication systems. Prentice Hall, NJ
48. He J, Gu G, Wu Z (2008) MMSE interference suppression in MIMO frequency selective and time-varying fading channels. IEEE Trans Signal Process SP-56:3638–3651
49. Hong M, Söderström T, Zheng WX (2006) Accuracy analysis of bias-eliminating least squares estimates for errors-in-variables identification. In: Proceedings of the 14th IFAC Symposium on System Identification. Newcastle, Australia, March 29-31, pp 190–195
50. Horn RA, Johnson CR (1999) Matrix analysis. Cambridge University Press, Cambridge, UK
51. Hua Y (1996) Fast maximum likelihood for blind identification of multiple FIR channels. IEEE Trans Signal Process SP-44:661:672
52. Ionescu V, Oara C, Weiss M (1999) Generalized riccati theory and robust control. Wiley, New York
53. Kailath T (1980), Linear systems. Prentice-Hall, Englewood Cliffs, NJ
54. Kailath T, Sayed A, Hassibi B (2000) Linear estimation. Prentice-Hall, New Jersey
55. Kalman RE (1963) Mathematical description of linear systems. SIAM J Contr Optimiz SICON-1:152:192
56. Kalman RE (1982) On the computation of the reachable/observable canonical form. SIAM J Contr Optimiz SICON-20:258–260
57. Kalman RE (1960) Contribution to the theory of optimal control. Bol Soc Matem Mex 5: 102–119
58. Kalman RE (1963) The theory of optimal control and the calculus of variations. In: Bellman RE (ed) Mathematical optimization techniques, Chapter 6. University of california Press, Berkeley, California
59. Kalman RE (1964) When is a linear control system optimal. Trans ASME J Basic Eng 86:1–10
60. Kalman RE, Bucy RS (1961) New results in linear filtering and prediction theory. Trans ASME J Basic Eng 83:95-108
61. Kalman RE, Bertram JE (1960) Control system analsyis and design via the second method of Lyapunov. Trans ASME J. Basic Eng 82:371–392
62. Kalman R, Ho YC, Narendra K (1963) Controllability of linear dynamical systems. In: Contributions to differential equations, vol 1, Interscience, New York
63. Kamen E (2000) Fundamentals of signals and systems, 2nd edn. Prentice-Hall, NJ
64. Keller JP, Anderson BDO (1992) A new approach to the discretization of continuous-time controllers. IEEE Trans Automat Contr AC-37:214–223
65. Khargonekar PP, Sontag E (1982) On the relation between stable matrix fraction factorizations and regulable realizations of linear systems over rings. IEEE Trans Automat Contr AC-37:627–638
66. Khargonekar PP, Sivashankar N (1992) $\mathscr{H}_2$ optimal control for sampled-data systems. Syst Contr Lett 18:627–631
67. Kim SW, Kim K (2006) Log-likelihood-ratio-based detection ordering in V-BLAST. IEEE Trans Commun TC-54:302–307
68. Kucera V (1972) The discrete Riccati equation of optimal control. Kybernetica 8:430–447
69. Kucera V (1979) Discrete linear control: the polynomical equation approach. Wiley, Prague
70. Kung SY (1980) Optimal Hankel-norm model reductions: scalar systems. In: Proceedings of 1980 Joint Automatic Control Conference. San Francisco, CA, Paper FA8.A
71. Kung SY, Lin DW (1981) Optimal Hankel-norm model reductions: multivariable systems. IEEE Trans Automat Contr AC-26:832–852
72. Kung SY, Lin DW (1981) A state-space formulation for optimal Hankel-norm approximations. IEEE Trans Automat Contr AC-26:942–946
73. Kuo BC (1980) Digital control systems. Holt, Rinehart and Windton, New York
74. Lewis FL, Syrmos VL (1995) Optimal control. Wiely, New York
75. Li L, Gu G (2005) Design of optimal zero-forcing precoder for MIMO channels via optimal full information control. IEEE Trans Signal Process SP-53:3238–3246

76. Li Y, Anderson BDO, Ly U-L (1990) Coprime factorization controller reduction with Bezout identity induced frequency weighting. Automatica 26:233–249
77. Li Y(Geoffrey), Ding Z (1995) A simplified approach to optimum diversity combining and equalization in digital data transmission. IEEE Trans Commun 43:47–53
78. Ljung L (1987) System identification: theory for the user, Prentice-Hall, NJ
79. Maciejowski JM (1985) Asymptotic recovery for discrete-time systems. IEEE Trans Automat Contr AC-30:602–605
80. Maciejowski JM (1989) Multivariable feedback design. Addition-Wesley Publishing Company, Inc., England
81. McFarlane DC, Glover K (1990) Robust controller design using normalized coprime factor plant description. Lecture notes in control and information sciences, vol 138. Springer, Berlin
82. Maddock RJ (1982) Poles and zeros in electrical and control engineering. Holt, Rinehart and Windton, New York
83. Middleton RH, GC Goodwin (1986) Improved finite word length characteristics in digital control using delta operator. IEEE Trans Automat Contr AC-31:1015–1021
84. Middleton RH, Goodwin GC (1990) Digital control and estimation: a unified approach. Prentice-Hall, Englewood Cliffs, NJ
85. Mitra SK (2001) Digital signal processing, 3rd edn. McGraw-Hill, New York
86. Moore BC (1981) Principal component analysis in linear systems: controllability, observability, and model reduction. IEEE Trans Automat Contr AC-26:17–32
87. Moulines E, Duhamelm P, Cardoso J-F, Mayrargue S (1995) Subspace methods for the blind identification of multichannel FIR filters. IEEE Trans Signal Process SP-43:516–616
88. CT Mullis, Robers RA (1975) Synthesis of minimum roundoff noise fixed point digital filters. IEEE Trans Circuits Syst CAS-23:551–562
89. Nett CN, Jacobson CA, Balas MJ (1984) A connection between state-space and doubly coprime fractional representations. IEEE Trans Automat Contr AC39:831–832
90. Oppenheim AV, Schafer RW (1989) Discrete-time signal processing. Prentice-Hall, Englewood Cliffs, NJ
91. Pappas T, Laub AJ, Sandell NR Jr (1980) On the numerical solution of the discrete-time algebraic Riccati equation. IEEE Trans Automat Contr AC 25: 631–641
92. Proakis J (2001) Digital communications, 4th edn. McGraw-Hill, New York
93. Roy S (2003) Optimum infinite-length MMSE multi-user decision-feedback space-time processing in broadband cellular radio. Wireless Personal Communications WPC27:1–32
94. Rugh WJ (1996) Linear system theory, 2nd edn. Prentice-Hall, NJ
95. Salgado ME, Middleton RH, Goodwin GC (1987) Connection between continuous and discrete Riccati equations with applications to Kalman filtering. Proc IEE Part D 135:28–34
96. Scaglione A, Giannakis GB, Barbarossa S (1999) Redundant filterbank precoders and equalizers, part I: unification and optimal designs. IEEE Trans Signal Process SP47: 1988–2006
97. Stamoulis A, Giannakis GB, Scaglione A (2001) Block FIR decision-feedback equalizers for filterbank precoded transmissions with blind channel estimation capabilities. IEEE Trans Commun 49(1):69–83
98. Schwartz M (2005) Mobile wireless communications. University Press, Cambridge
99. Shaked U (1985) Explicit solution to the singular discrete-time stationary linear filtering problem. IEEE Trans Automat Contr AC30:34–47
100. Stoica P, Moses R (1997) Introduction to spectral analysis. Prentice Hall, Upper Saddle River, New Jersey
101. Stein EM, Weiss G (1971) Introduction to fourier analysis on euclidean spaces. Princeton University Press, Princeton, NJ
102. Söderström T, Stoica P (1989) System identification. Prentice-Hall International, Hemel Hempstead, UK
103. Söderström T (2007) Accuracy analysis of the Frisch scheme for identifying errors-in-variables systems. IEEE Trans Automat Contr AC52:985–997

104. Söderström T, Soverini U, Mahata K (2002) Perspectives on errors-invariables estimation for dynamic systems. Signal Process 82:1139–1154
105. Stein G, Athans M (1987) The LQG/LTR procedure for multivariable feedback control design. IEEE Trans Automat Contr AC32: 105–114
106. Stevens BL, Lewis FL (1992) Aircraft control and simulation. Wiely, New York
107. G. Stuber (2001) Principles of mobile communication. 2nd edn. Springer, Berlin
108. Synge JL,Griffith BA (1949) Principles of mechanics. McGraw-Hill Book Company, Inc., New York
109. Tadjine M, M'Saad M, Duard L (1994) Discrete-time compensators with loop transfer recovery. IEEE Trans Automat Contr AC 39:1259–1262
110. Tong L, Xu G, Kailath T (1994) Blind identification and equalization based on second-order stistitics: a time domain approach. IEEE Trans Inform Theory IT 40:340–349
111. Tse D, Viswanath P (2005) Fundamentals of wireless communication. Cambridge University Press, Cambridge
112. Vaidyanathan PP (1993) Multirate systems and filter banks. Prentice-Hall, Englewood Cliffs, NJ
113. van Huffel S, Vandewalle J (1991) The total least squares problem: computational aspects and analysis. SIAM Publisher, Philadelphia
114. Vidyasagar M (1984) Graph metric for unstable plants. IEEE Trans Automat Contr AC 29:403–418
115. Vinnicombe G (2001) Uncertainty and feedback – $\mathscr{H}_\infty$ loopshaping and the ν-Gap Metric. Imperial College Press, London, UK
116. Wilson SG (1996) Digital modulation and coding. Prentice-Hall, NJ
117. Wolniansky PW, Foschini GJ, Golden GD, Valenzuela RA (1998) V-BLAST: an architecture for realizing very high data rates over the rich-scattering wireless channel. Proceedings of the 1998 URSI International Symposium on Signals, Systems, and Electronics, ISSSE 98
118. Wonham WM (1985) Linear multiovariable control: a geometric approach. Springer, New York
119. Xia X-G (1997) New precoding for intersymbol interference cancellation using nonmaximally decimated multirate filterbanks with ideal FIR equalizers. IEEE Trans Signal Process SP 45:2431–2441
120. Yamamoto Y (1994) A function space approach to sampled-data control systems and tracking problems. IEEE Trans Automat Contr AC 39:703–713
121. Youla DC (1961) On the factorizations of rational matrices. IRE Trans Inform Theory IT 7:172–189
122. Youla DC, Jabr HA, Bongiorno JJ, Jr (1976) Modern Wiener-Hopf design of optimal controllers, part 2. IEEE Trans Automat Contr AC 21:319–228
123. Zhang Z, Freudenberg JS (1990) Loop transfer recovery for nonminimum phase plants. IEEE Trans Automat Contr AC 35:547–553
124. Zhang Z, Freudenberg JS (1991) On discrete-time loop transfer recovery. In: Proceedings American Control Conference, June 1991
125. Zhou K (1995) Frequency-weighted $\mathscr{L}_\infty$ norm and optimal Hankel norm model reduction. IEEE Trans Automat Contr AC 40:1687–1699
126. Zhou K, Doyle J, Glover K (1996) Robust and optimal control. Printice-Hall, Englewood Cliffs, NJ

Index

G. Gu, *Discrete-Time Linear Systems: Theory and Design with Applications*,
DOI 10.1007/978-1-4614-2281-5, © Springer Science+Business Media, LLC 2012